Perspectives on Statistical Thermodynamics

This original text develops a deep, conceptual understanding of thermal physics, highlighting the important links between thermodynamics and statistical physics, and examining how thermal physics fits within physics as a whole, from an empirical perspective. The first part of the book is devoted to elementary, mesoscopic topics such as Brownian motion, which leads to intuitive uses of large deviation theory, one of the pillars of modern probability theory. The book then introduces the key concepts behind statistical thermodynamics and the final part describes more advanced and applied topics from thermal physics such as phase transitions and critical phenomena. This important subject is presented from a fresh perspective and in a pedagogical manner, with numerous worked examples and relevant cultural side notes throughout, making it suitable as either a textbook for advanced thermal physics courses or for self-study by undergraduate and graduate students in physics and engineering.

Yoshitsugu Oono is a Professor of Physics at the University of Illinois at Urbana-Champaign, where he currently teaches the statistical physics courses. His research interests focus on nonequilibrium statistical thermodynamics and theory of biological systems.

Perspectives on Statistical Thermodynamics

YOSHITSUGU OONO

University of Illinois at Urbana-Champaign

CAMBRIDGE
UNIVERSITY PRESS

CAMBRIDGE
UNIVERSITY PRESS

University Printing House, Cambridge CB2 8BS, United Kingdom

One Liberty Plaza, 20th Floor, New York, NY 10006, USA

477 Williamstown Road, Port Melbourne, VIC 3207, Australia

314–321, 3rd Floor, Plot 3, Splendor Forum, Jasola District Centre, New Delhi – 110025, India

79 Anson Road, #06–04/06, Singapore 079906

Cambridge University Press is part of the University of Cambridge.

It furthers the University's mission by disseminating knowledge in the pursuit of
education, learning, and research at the highest international levels of excellence.

www.cambridge.org
Information on this title: www.cambridge.org/9781107154018
DOI: 10.1017/9781316650394

First published 2017

A catalogue record for this publication is available from the British Library.

Library of Congress Cataloging-in-Publication Data
Names: Oono, Yoshitsugu, 1948– author.
Title: Perspectives on statistical thermodynamics / Yoshitsugu Oono,
University of Illinois at Urbana-Champaign.
Description: Cambridge, United Kingdom ; New York, NY : Cambridge University Press, 2017. |
Includes bibliographical references and index.
Identifiers: LCCN 2017014907| ISBN 9781107154018 (hardback ; alk. paper) |
ISBN 1107154014 (hardback ; alk. paper)
Subjects: LCSH: Statistical thermodynamics.
Classification: LCC QC311.5 .O56 2107 |
DDC 536/.7015195–dc23 LC record available at https://lccn.loc.gov/2017014907

ISBN 978-1-107-15401-8 Hardback

To the memory of Robert M. Clegg (1945–2012)
Sharing critical points of view with Bob was extremely encouraging. . .

Contents

Part II Statistical Thermodynamics: Basics

Preface

This book is for a motivated undergraduate student to learn elements of thermal physics on her own after a rudimentary introduction to the subject (at the highest, 200 level in the US). This is a book I wished I had when I was an undergraduate student, struggling to learn physics by myself (I was never taught what I am teaching now; I was a wet chemist). Thus, most intermediate formulas are explicitly given and all the problems are with detailed solutions except for extremely elementary ones (even they come with answers).

Every student graduating from physics should know at least the existence of the topics covered in this book. This book can be used as a course textbook for a 'second introduction' to thermal physics or as a bridge between traditional thermal physics undergraduate and graduate courses. All the key elements of thermal physics are explained without demanding any prior experience of advanced physics as a second introduction, but we go far beyond the elementary introduction.

I intended to write a thermal physics book that can connect rudimentary gas kinetics and Brownian motion to statistical thermodynamics. Traditionally, equilibrium statistical mechanics books do not tell the reader how to count atoms, even though atomism should be the key idea of the traditional exposition. I explain Einstein's Brownian motion theory before thermodynamics and statistical mechanics. The explanation of Brownian motion and Langevin equations naturally leads us to large deviation theory, one of the pillars of the modern probability theory. Thus, this book emphasizes three levels of descriptions of our world: microscopic, mesoscopic, and macroscopic. Accordingly, the law of large numbers and the large deviation principle are used as the mathematical backbone of thermal physics.

Although this book does not hesitate to point out the key mathematical ideas that physicists should know, "back of the envelope" calculation and intuitive understanding are emphasized whenever possible (e.g., dimensional analysis is stressed). The required mathematics is restricted to a minimum throughout this book. Still, some minimal prerequisites are desirable to read this book. They are linear algebra, calculus, and (classical and quantum) mechanics (and perhaps a bit of electromagnetism). There is an appendix explaining (quantum) mechanics for the reader who may be learning quantum mechanics concurrently, but it is only a brief summary and should not be regarded as a substitute of an elementary mechanics course.

Theoretically oriented people, including myself, tend to forget that science is an empirical endeavor. For example, thermodynamics textbooks these days often start with a demand to accept the existence of entropy S. This book is critical about such trends. Thus, when some theory is explained, its empirical bedrock is discussed to reduce metaphysical beliefs as much as possible. If the reader wishes to be a builder of a theoretical system rather than a mere consumer of theories, they must judge what empirical facts they should respect. The reader should not forget that the belief that microscopic laws are more fundamental than macroscopic empirical laws has no empirical ground. For example, although many readers may well stop reading this book at this point, we should candidly admit that there is no empirical demonstration that many-body systems obey many-body equations of motion. Therefore, in this book I try to distinguish what is empirically (relatively) certain and what is not. This is why learning how to count atoms is discussed before statistical thermodynamics; within the usual equilibrium statistical thermodynamics there is no way to count atoms. I will discuss the consistency of thermodynamics and statistical mechanics, but I do not assert that the latter furnishes the basis of the former. The idea that smaller scale physics is more fundamental may well be a mere Zeitgeist of the so-called modern era. In any case, to minimize metaphysics is a policy of this book.

This book intends to give a continuous narrative about thermal physics, so it consists of a series of (rather short) chapters, in three parts: Chapters 1–10, 11–30, and the rest. The first part is devoted to the introduction and elementary atomic and mesoscopic discussions, including Brownian motion. The second part is the usual statistical thermodynamics, and the third part is devoted to the rudiments of phase transition. If the reader wishes to follow the traditional storyline, they could start from Part II.

Perhaps some features hard to find in other books of this level may be mentioned. The large deviation theoretical framework is used to study fluctuations and Langevin equations (Chapters 1, 10, 26). The thermodynamic background of Boltzmann's principle and the principle of equal probability is discussed (Chapters 17, 19). The Legendre transformation is explained properly as a tool of convex analysis (Chapters 16, 36). Accordingly, ensemble equivalence is stressed. Maxwell's relations are unified in terms of Jacobians (Chapter 24; the most practically convenient formulation). The concept of probability is introduced in a basic measure-theoretical way (Chapter 3, but intuitively without formal measure theory). Conspicuously missing topics are imperfect gas, electrochemical potential, and numerical approaches.

Acknowledgements

Joel Cannon, Chuck Yeung, Nobuto Takeuchi, and Barry Friedman gave detailed advice on the introductory part to make the manuscript accessible to undergraduate students. I benefitted from discussing some parts of the manuscript with Hal Tasaki, Akira Shimizu, Hisao Hayakawa, Ken Sekimoto, Kaz Takahashi, and Glenn Paquette. These people have been critical to check my rather heretic point of view about physics. Tom Issaevitch critically read the entire manuscript and gave me numerous corrections and suggestions. Last but not least Bob Clegg urged me to pay very serious attention to Einstein about the foundation of statistical physics.

When I got an offer from the University of Illinois, I went to see Leo Kadanoff for his advice, because I had never been in any physics department. Leo said, "Since the Illinois physics is big, it must be able to accomodate one or two oddballs like you." I have been grateful to the Department of Physics of Illinois for tolerating my presence ever since.

Self-Study Guide

The best way to study is not to study when you do not wish to.

This book grew from course lecture notes for an advanced undergraduate level thermal physics course called "Statistical Physics, a jogging course," aimed at the top third of students. However, I believe every student graduating from physics should have some understanding of the topics in this book. Therefore, I tried to give detailed derivations of all the formulas. Since I learned mathematics and physics without attending any course (beyond 300 in the US level), I certainly wished to have books filled with details and with all the problems solved.

However, I learned some willpower was needed to use such books effectively, because learning is always "active learning";[1] If one wishes to build one's muscle, some load is required. Therefore, always try to guess the next line or step in the derivation/transformation of formulas before reading the lines. Think what you would do if you were to encounter the problem as the first person in the world.

Each chapter has a **Summary**, **Key words**, and **The reader should be able to** at the beginning. "Summary" tells the reader what the section will discuss. Since it is placed at the beginning of each chapter, the reader will not be able to understand fully what it summarizes, but still some flavor of the section can be sensed. The summary should be checked again after finishing the section. "Key words" is a list of concepts/technical terms the reader must be able to explain to her intelligent lay friends. "The reader should be able to" is a list of minimal (mostly practical) items the reader should be able to do with comfort.

Each chapter has a modular structure, consisting of about ten numbered short units with titles such as "**21.14** Sanov's Theorem and Entropy Maximization." This should make the main line of each unit explicit and easy to grasp. A few units with dagger marks (†) are slightly advanced. There are other units which are details or side issues that the reader can skip or browse through, these are indicated by an asterisk (∗). Even in a single unit there may be a portion which gives technical details or extra discussions – these will be indicated by an open square at the beginning (□). Units are cross-referenced in the text in bold as **21.14**. Perhaps, it may be a good practice to try to recall the contents of the referenced units before actually looking at them.

Most formulas are with detailed derivations, so the reader can often follow them even without a pencil and paper. However, before reading the details I urge the reader to try to demonstrate them by herself (to be an active learner). Footnotes with ∗ are devoted to

---

[1] Read: Brown, P. C., Roediger III, H. L. and McDaniel, M. A. (2014). *Make it Stick: The Science of Successful Learning*, Cambridge (MA): The Belknap Press.

the derivation of marked formulas. The reader can regard them as solutions to technical quizzes.

Most chapters end with fully solved problems with titles as **Q26.4 Fluctuation and Le Chatelier–Braun Principle**, some of which may introduce advanced topics. They are quoted as **Q26.4** in the text. The solutions are clearly separated, so they are easy to hide, if you wish to be an active learner. Needless to say, you can read a line or two for a hint if you have no idea how to answer the question. Use the solutions sparingly (and wisely).

However, there is not enough space to give all the representative elementary problems. Therefore, to augment the book, I urge the reader to consult the following two problem books:

R. Kubo, H. Ichimura, T. Usui and N. Hashitsume, *Thermodynamics* (North Holland, 1968),

R. Kubo, H. Ichimura, T. Usui and N. Hashitsume, *Statistical Mechanics* (North Holland, 1990 paperback).

I learned thermal physics from these books. All the problems are fully solved, but many of them are not very easy. Try to solve at least the problems in Part [A] of these books. These books will be (collectively) quoted as Kubo's problem book (because the original Japanese version is a single book).

There are numerous footnotes, because I am too impatient a reader to search for endnotes. As noted earlier the footnotes with ∗ explain detailed derivations. Most of the relatively long footnotes (without ∗) are with titles such as **Carnot was totally forgotten** or **Can we tell if two phases coexist?** and may be understood as boxed short articles. One or two line footnotes are usually for references or for small remarks (e.g., "Do not forget the negative sign.").

Some footnotes are cultural to exhibit the events of (cultural) importance contemporary to the important thermophysics developments. For example, it is noted that Beethoven died in the year Brown clearly recognized the universality of Brownian motion. Such footnotes were quite unpopular to some referees of this book; since students tend to remember only unimportant gossips, I was urged to minimize such "distracting footnotes." Still, I continue to keep these footnotes. Isn't it wonderful if we think about mesoscopic dynamics sustaining our lives, when we listen to the "Choral"? Physics must be an important part of the human culture. Besides, not all the readers will be specialists of statistical physics.

Symbols

A Helmholtz's free energy

B magnetic field
β $1/k_B T$

C (various) specific heat
$\mathbb{C}$ complex numbers
χ indicator, susceptibility

D diffusion constant
$\mathcal{D}$ one-particle density of states
$\eth$ spatial dimensionality

E internal energy; expectation value; electric field

G Gibbs' free energy; group (in Chapter 35)

H Hamiltonian; information; enthalpy
$\mathcal{H}$ Hilbert space
h magnetic field

I large deviation function (rate function); mutual information

K kinetic energy

L the dimension of length in dimensional analysis

M magnetization; the dimension of mass in dimensional analysis
μ chemical potential; magnetic dipole moment

N number of particles
n number density
$\mathbb{N}$ non-negative integers

O Landau's O
o Landau's o

P pressure; probability; projection

Q heat
$\mathbb{Q}$ rational numbers

R the gas constant
$\mathbb{R}$ real numbers
ρ density operator

S entropy
s (Ising) spin

T temperature; the dimension of time in dimensional analysis
$\mathcal{T}$ time-evolution operator

U potential energy; total energy

V volume, variance

W work; phase volume
w microcanonical partition function
$\tilde{w}$ microcanonical ensemble
Ω sample space

X generic extensive quantity (generic work coordinate)
x generic intensive quantity conjugate to X (with respect to energy)
Ξ grand canonical partition function

Z canonical partition function
$\mathbb{Z}$ integers

Mathematically Standard Notations

The standard symbols $\mathbb{N}, \mathbb{Z}, \mathbb{Q}, \mathbb{R}, \mathbb{C}$ are listed above.

$a \in A$ implies that a is an element of set A.

$A = \{\omega \,|\, R(\omega), \omega \in B\}$ implies the set of all the elements of set B satisfying the condition R.

(a, b) denotes an open interval: $(a, b) = \{x \,|\, a < x < b, x \in \mathbb{R}\}$.

Similarly, $[a, b] = \{x \,|\, a \leq x \leq b, x \in \mathbb{R}\}$, $(a, b] = \{x \,|\, a < x \leq b, x \in \mathbb{R}\}$, and $[a, b) = \{x \,|\, a \leq x < b, x \in \mathbb{R}\}$.

PART I

ATOMIC AND MESOSCOPIC VIEWPOINT

biology (the other is Darwinism). This indicates the limitation of philosophers who are not empirical enough. The lack of the idea of "molecule" in the ancient atomism is also an example of this limitation. Perhaps, it is a sign of progress to recognize that the world does not have the structures we "naturally" expect. Kepler's discovery that the circular orbit is not natural may be an example; this was never accepted by Galileo.

Mechanics was also beyond philosophers' grasp. Therefore, modern atomism was beyond the reach of any philosopher. We must respect empirical facts. Science is an empirical endeavor. At the same time, however, as the reader recognizes from the works of Newton, Maxwell, Darwin, and others, 'pure empiricism' is not enough at all to do good science.

1.3 How Numerous are Atoms and Molecules?

How many water molecules are there in a tablespoonful ($15\,\text{cm}^3$) of water? Although we should discuss how to determine the size or mass of an atom empirically before discussing the question (see Chapter 9), let us preempt the result.

Suppose one person removes one molecule of water at a time from the tablespoonful of water, and another person uses the tablespoon to scoop out the ocean's water to the outer space (although no one would dare to start scooping out water of even a $50\,\text{m}$ swimming pool). If they perform their operations synchronously, starting simultaneously, which person will finish first? The second person. However, the number of molecules in a spoonful of water is comparable to the amount of ocean water measured in tablespoons (the ratio is about 5; the number of molecules wins).

1.4 Why are Molecules so Small?

Molecules are numerous. They are numerous because they are tiny. Why is an atom so tiny? Because we are so big. Thus, the title question properly understood is: why is the size ratio between atoms and us so big? Do not forget that we human beings are products of Nature. To compare us with atoms does not necessarily imply that we subscribe to anthropocentric prejudices.

Large animals are often constructed with repetitive units such as segments. The size of the repeating unit is at least about one order (base 10) larger than the cell size. Consequently, the size of "advanced" organisms must be at least 2–3 orders as large as the cell size.

Thus, the problem is the cell size. Since we are complex systems,[9] the crucial information and materials required to build us come from the preceding generation. Since there is no ghost in the world, information must be carried by a certain thing (we should call

---

[9] See, e.g., chapter 5 of Oono, Y. (2013). *The Nonlinear World*, Tokyo: Springer.

this *no ghost principle*). Stability of the thing requires that information must be carried by polymers. What polymer should be used? Such a question is a hard question, so we simply imagine something like DNA. The "no ghost principle" tells us that organisms require a certain minimal DNA length. This seems to be about 1 mm $\sim$ 1 m (notice that the question is almost a pure physics question; to answer this we must understand the amount of the needed information as physicists, since physics can explain the atom size: 0.1–0.2 nm). As a ball its radius is about $0.1 \sim 1$ μm. This implies that our cell size is ~ 10 μm ($= 10^{-5}$ m).[10]

Thus, the segment size is about 1 mm, and the whole body size is at least about 1 cm (this is actually about the size of the smallest vertebrates). If we require good eyesight, the size becomes easily one to two orders more, so intelligent creatures cannot be smaller than ~ 1 m. That is, the atom size must be 10^{-10} as large as our size.

We have, at least roughly, understood why atoms are small.

1.5 Our World is Lawful to the Extent of Allowing the Evolution of Intelligence

We have discussed, with the aid of atomism and the cell theory, that science is an empirical endeavor. We observe the world and are making science, so we must be at least slightly intelligent. To be intelligent we need to be at least $10^9 \sim 10^{10}$ times as large as the atom (in linear length). A large size is not enough, however. The world must have allowed intelligence to evolve.

If there is no lawfulness at all, or in other words, there is no order in the world,[11] then intelligence is useless. We use our intelligence to guess what happens next from the current knowledge we have. If organisms' guesses with the aid of their intelligence were never better than simple random choices (say, following a dice), then intelligence would not evolve. Random grading would not give an incentive for students to study. Recall that the human brain is the most energy consuming very costly organ.[12] This means that the macroscopic world (the world we observe directly at our space-time scale) must be at least to some extent lawful with some regularity. This is consistent with our superstitiousness, because mistaking correlation as causality is an important ingredient of superstition.

However, if the law or regularity is too simple, then again no intelligence is useful. If the world is dead calm, no intelligence is needed. The world must be "just right." The macroscopic world we experience is neither violent nor dull for most of the time (though it is punctuated by catastrophes that do not contribute to our evolution except for resetting the stage).

---

[10] We may safely claim that the lower bound of the cell size is determined by the amount of DNA.
[11] "Order" may be understood as redundancy in the world; knowing one thing can tell us something about other things, because everything is not totally unrelated.
[12] Its weight is 2% of the body weight, but it consumes about 20% of the whole body energy budget.

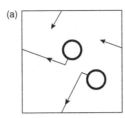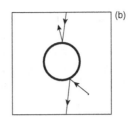

Fig. 1.1 Sinai billiard: (a) a motivation. Two hard elastic disks (pucks) are running around on the table with a periodic boundary condition (if a disk disappears from one edge, it reappears from the opposite edge with the same velocity), colliding from time to time with each other. This is a toy model of a confined gas. (b) If the dynamics of the center of mass (CM) of one disk is observed from the CM of the other disk, the former may be understood as a ballistic motion of a point mass with occasional collisions with the central circular obstacle. This is called the *Sinai billiard*, and is known to be maximally chaotic.

1.6 Microscopic World is Unpredictable

In contrast, we know the world of atoms and molecules (the microscopic world) is a busy and bustling world. They behave quite erratically and unpredictably. Atoms and molecules are believed to be rigorously governed by (quantum) mechanics. If we consider classical approximation for simplicity, there are at least two reasons for unpredictability: chaos and external disturbances. Quantum mechanically, the situation might be better intrinsically, since there is no real chaos in quantum mechanics. Still, extreme difficulties in the fundamental experiments to verify quantum effects attest eloquently to the fragile nature of intrinsic quantum mechanical evolution of the system.

Maxwell clearly recognized that molecules behave erratically due to collisions. Perhaps the simplest model to illustrate the point is the *Sinai billiard*. A hard ball (or rather, imagine an ice hockey puck) is moving without friction on a flat table, which has a circular obstacle on it. The ball hits the obstacle and is bounced back specularly (see Fig. 1.1(b)). Roughly speaking, a small deviation of the direction of the particle is doubled upon specular reflection by the central circle, so, for example, to predict the direction of the particle after 100 collisions is prohibitively hard.[13] Imagine what happens if there are numerous such particles colliding with each other. Predictions would be absolutely impossible. Worse still, it is very hard to exclude the effects of the external world, where we do not know what is going on at all. Notice that we cannot even breathe if we wish to study the "intrinsic behavior" of a collection of atoms in front of us.

1.7 Why our Macroscopic World is Lawful: the Law of Large Numbers

The world at the scale of atoms is full of (apparently) unpredictable behaviors. We know our scale is quite remote from the atomic scale. The time scales are also disparate; the

---

[13] It is convenient to remember that $2^{10} \simeq 10^3$, so $2^{100} \simeq 10^{30}$.

time scale required to describe molecular dynamics is $0.1 \sim 1$ fs (1 fs $= 10^{-15}$ s), but the shortest time span we can recognize must be much longer than 1 ms $= 10^{-3}$ s, because our lowest audible frequency is about 20 Hz. Lawfulness must come from suppression of unpredictability. Our size is crucial to this; even if molecules in a small droplet undergo quite erratic movements, their effect would not be detected easily in the droplet motion as a whole. This statement may be formally expressed as follows.

Let us consider a collection of numerous ($N \gg 1$) random variables.[14] X_n is the nth among them. Then,[15]

$$\sum_{n=1}^{N} X_n = Nm + o[N], \tag{1.1}$$

where m is the average value of X_n. This is the *law of large numbers*, the most important pillar of probability theory and the key to understanding the macroscopic world (Chapter 4). Imagine outcomes of coin tossing as an example: $X_n = 1$ if the nth outcome is a head; otherwise, $X_n = 0$. By throwing a coin N times, we get a 01 sequence of length N, say, $0100101101110101\cdots001$. We guess the sum is roughly $N/2$, if N is large enough. This is the law of large numbers. We clearly see the importance of being big (relative to atoms).

1.8   We Live in a Rather Gentle World

The reader might object, however, that being big may not be enough; we know about violent phenomena in the macroscopic world like turbulence or perhaps the cores of galaxies. If the variances are too big, perhaps we might not be able to expect the expectation value to settle down within a reasonable narrow range. Also even if the expectation value eventually converges, the needed N in (1.1) should not be too big; if we can recognize the regularity of the world only after averaging the observations during 1000 generations, probably the law of large numbers cannot favor intelligence very much. Thus, as already discussed, the world in which intelligence can emerge cannot be too violent. We have emerged in the world in which the law of large numbers holds rather easily at large scales to allow macroscopic laws. We live in the world whose space-time scale is not only quite remote from the microscopic world of atoms and molecules, but also whose "extent of nonequilibrium" is not too large.[16]

---

[14] **Random variables** We will discuss what we wish to mean by "random variables" more carefully later (Chapter 3), but here, the reader has only to understand them as variables that take various values within a certain range in an unpredictable fashion.

[15] **Landau symbol o** This standard symbol o means higher order small quantities. In the limit being discussed, if $X/Y \to 0$, then we write $X = o[Y]$, which is read as: compared with Y, X is a higher order small quantity in the limit being discussed. This does not mean X and Y themselves are infinitesimal. For example, $N^{0.99}$ is $o[N]$, if N is large (in the $N \to \infty$ limit), because $N^{0.99}/N = N^{-0.01} \to 0$ in this limit.

[16] We need stable and simple laws for feeble minds to work (recall the intelligence must evolve).

1.9 Thermodynamics, Statistical Mechanics, and Phase Transition

The macroscopic world close to equilibrium[17] is a world governed by the law of large numbers. It can be described *phenomenologically* by *thermodynamics* (Chapters 11–16). Here, "phenomenologically" implies that what we observe directly can be organized into a single theoretical system without assuming any entities beyond direct observations. Thermodynamics is distilled from empirical facts observable at our scale, so it is one of the most reliable theoretical systems we have in physics. Thus, thermodynamics will be explained through the basic facts; they are used to construct the key concept: entropy (roughly, a measure of diversity of microstates compatible with a macroscopic situation; Chapter 13). Its microscopic interpretation by Boltzmann, which is the key ingredient of statistical mechanics, will also be explained based on thermodynamics (Chapter 17) following Einstein's logic. It is quite advantageous to understand entropy intuitively in terms of information, so information theory and its relevance to thermal physics will be outlined (Chapters 21, 22).

We will learn that if we know a "thermodynamic potential," we can compute any equilibrium properties of a macroscopic system. In particular, the Helmholtz free energy A is a convenient thermodynamic potential (Chapter 16). Once A is known, for example, we can compute the amount of work we can obtain from a system. However, thermodynamics cannot give A; we need a more microscopic approach to obtain it. Thus, we need a bridge between the microscopic and the macroscopic worlds called statistical mechanics. Statistical mechanics gives the following expression of the Helmholtz free energy A (Chapter 18):

$$A = -k_B T \log Z, \tag{1.2}$$

where k_B is the Boltzmann constant, T is the absolute temperature, and Z is the (canonical) *partition function*

$$Z = \sum e^{-H/k_B T}. \tag{1.3}$$

Here, H is the system Hamiltonian and the summation is over all the *microstates* (microscopically described states of the system distinguishable by mechanics). Each term $e^{-H/k_B T}$ is a smooth function of T (> 0), so if the number of terms summed in (1.3) is finite, nothing very singular can happen for A as a function of T. However, if the system is very big (ideally, infinitely big, in the so-called *thermodynamic limit*), A can lose its smoothness as a function of T. Thus, *phase transitions* can occur, if the system is large enough (Chapters 31–36).

---

[17] Intuitively, the reader may regard a system to be close to equilibrium, if all the rapid changes in it have subsided.

1.10 The Mesoscopic World

What does the world look like if we observe it at the scale intermediate between the microscopic and the macroscopic scales? In (1.1) the $o[N]$ term becomes unignorable. That is, *fluctuation* cannot be ignored. This is the world where *Brownian motion* dominates, where unicellular organisms live, and where the cells making our bodies function. Intelligence is useless, because fluctuation is still too large and prevents agents from predicting what would happen. Often the best strategy is to wait patiently for a "miracle" to happen, and if it happens, to cling to it. Many molecular machines such as ribosomes follow just this strategy. We will discuss Brownian motion and will give an informal discussion of transport phenomena in Chapters 8–10. The study of Brownian motion substantiates the reality of atoms.

1.11 Large Deviation and Fluctuation

In the mesoscopic world, the average of what we observe is consistent with our macroscopic observation results. However, if we observe individual systems, observables fluctuate a lot around the expected macroscopic behaviors. What is the natural framework to understand the mesoscopic world, or $o[N]$ in (1.1)? It is the *large deviation principle* (Chapter 10) that refines the law of large numbers in the following form:

$$P\left(\frac{1}{N}\sum_{i=1}^{N}X_i \in v(x)\right) \approx e^{-NI(x)}, \tag{1.4}$$

where $\approx$ means that the ratio of the logarithms of the both sides is unity asymptotically for large N, $v(x)$ denotes the volume element around x, P is the probability (Chapter 3) of the event in the parentheses, and I is called the *large deviation function* (or *rate function*). I may be approximated by a quadratic function when x is close to the true expectation value m:

$$I(x) \simeq \frac{1}{2V}(x - m)^2. \tag{1.5}$$

Here, V is a positive constant (corresponding to the variance) and m is the expectation value. If N is large, the probability is positive only if x is very close to m: $I(m) = 0$ implies the law of large numbers. Equation (1.5) means that mesoscopic noise is usually Gaussian. That is, with the aid of a Gaussian noise v whose average is zero and whose variance is V/N, we can write

$$\frac{1}{N}\sum_{i=1}^{N}X_i = m + v + o[1/\sqrt{N}]. \tag{1.6}$$

As we will see later (Chapter 26), I is related to the decrease of entropy from equilibrium due to fluctuations. Applying the idea of large deviation to Brownian motion allows us to study time dependence in the mesoscopic world (Chapters 10, 26).

To understand thermal physics is to understand the interrelationships among three levels of description of our world: the macroscopic, the mesoscopic, and the microscopic descriptions.

Problems

Q1.1 Big Numbers

(1) How many cells in a human body?
(2) How many atoms in a cell?
(3) How many seconds in 1 billion years?
(4) How many stars in the Universe?

Find the reader's own natural and interesting examples of big numbers relevant to the topics of this book.

Solution

(1) There are about 3.7×10^{13} of our own cells in our body.[18] We have about 1.5 kg of symbiotic prokaryotes (bacteria and archaea), whose number of cells is about the same as the total number of our own cells.[19] Therefore, we have at most 10^{14} cells. Now the world population is 7.4×10^9. This means that there are less than 1 mole $(= 6.022 \times 10^{23})$ of human cells on the Earth.

(2) There are about 2×10^{14} atoms in a cell; about a few 10^9 atoms in *Escherichia coli*. Thus cells are definitely mesoscopic.

(3) There are 3.1536×10^{16} seconds in 1 billion years = 1 Ga. By the way, 1 billion seconds is about 32 years. Such time-scale discrepancies are crucial to understanding nonequilibrium phenomena as we will see in Chapter 10.

(4) Our Milky Way is estimated to have about 10^{11} stars; the total Universe is said to have 10^{29} stars. This is about the same as the number of prokaryotes produced in the oceans in a year. The total number of electrons in our body is also of this order.

Q1.2 Atoms

At the beginning of this section, we read, "We could even see them, for example, with the aid of atomic force microscopes. However, only 50 years ago no one could see atoms." Are we really sure we can see them today? The reader should think over what we mean when we say we see something.

---

[18] There is a website called Bionumbers (the database of useful biological numbers) http://bionumbers.hms.harvard.edu/default.aspx.

[19] The latest may be: Sender, R., Fuchs, S., and Milo, R. (2016). Are we really vastly outnumbered? Revisiting the ratio of bacterial to host cells in humans. *Cell*, **164**, 337–340.

Summary

* Gay-Lussac gave the key empirical facts: $PV \propto T$, the law of constant temperature, and the law of combining volumes.
* Bernoulli's dynamical theory of gases demonstrated $\langle mv^2/2 \rangle \propto T$.
* In equilibrium all gas particles have the same average kinetic energy (equipartition of energy).

Key words

law of partial pressure, the law of constant temperature, the law of combining volumes, equipartition of energy

The reader should be able to:[1]

* Explain the law of constant temperature.
* Understand Bernoulli's key ideas (i.e., be able to reconstruct his logic).
* Derive the equipartition of energy (for the translational motion).

2.1 Aristotelian Physics and Galileo's Struggle[2]

According to Aristotle's (384–322 BCE) physics, the four properties, hot, cold, dry, and wet were irreducible properties, which were related to the four elements of Empedocles (c. 490–430 BCE), fire, water, earth, and air (as fire is both hot and dry, air is hot and wet, etc.). The crucial point is that what we observe directly by our senses has direct material basis in the sense that there are materials or elements corresponding to wetness, hotness, etc.

---

[1] "**The reader should be able to:**" generally summarizes what the reader should be able to do with comfort. Most things required are basic and practical.

[2] Historical comments in this chapter and Chapters 11 and 13 are heavily dependent on Yamamoto, Y. (2007–2008). *Historical Development of Thoughts of Heat Theory*. Tokyo: Chikuma Shobo.

This type of idea is called "thingification" or "reification." Even *Galileo* (1564–1642) was initially under this influence, but mechanics was a key factor for him to overcome reification.[3] Later, he clearly established the mechanical view of Nature, asserting that what we could feel (e.g., color, odor, etc.) existed only in the relation between the sensing subjects and the sensed objects and was thus subjective and secondary; only the (geometrical) shapes, numbers, configurations (positions) and movements (position changes) of substances were objective and were primary properties.

2.2 Boyle: The True Pioneer of Kinetic Theory of Heat[4]

Boyle (1627–1691) was the first to accept the principle that matter and motion were the primary things, and was truly free from the Aristotelian "reificationism." He correctly asserted that there were microscopic and macroscopic motions. The former was sensed as heat but could not be sensed by us as motion; the only motion we could sense as such was the "progressive motion of the whole" (i.e., the systematic motion), which could not be felt by us as warmth even if it was vigorous. Boyle paved the path to the discussion of mutual convertibility of heat and (macroscopic) motion.

2.3 Discovery of Atmospheric Pressure and Vacuum

The biggest discovery of modern physics about gases was the discovery of atmospheric pressure and vacuum by Torricelli (1608–1647), Pascal (1623–1662) and von Guericke (1602–1686). This was a discovery demarcating the medieval and the modern ages, its importance only second to heliocentrism. Within the Aristotelean system, air and fire were regarded as essentially light elements, having the tendency to go away from the Earth. Therefore, the idea of the mass (or weight) of air could not possibly be born. The discovery of vacua decisively discredited Aristotle.

---

[3] **Archimedean mechanics was crucial** Mechanics, or more precisely, studying Archimedes (*c.* 287–212 BCE) was the key for Galileo and Descartes (1596–1650) to overcome the Aristotelian "physics." Archimedes gave them the conviction that the natural laws could be formulated mathematically; indeed the world is mathematically constructed. Descartes' famous "*Ju pense, donc ju suis,*" was the foundation of his attempt to establish mathematical physics free from empiricism.

[4] **Thermal physics history reading** The following books are recommended for the historical background: Brush, S. G. (1983). *Statistical Physics and the Atomic Theory of Matter: From Boyle and Newton to Landau and Onsager.* Princeton: Princeton University Press (esp., Chapter 1).

  Lindley, D. (2001). *Boltzmann's Atom: The Great Debate that Launched a Revolution in Physics.* New York: The Free Press.

2.4 Daniel Bernoulli's Modern Dynamic Atomism and a 100 Year Hiatus

Thus, a modern dynamic atomic theory should be possible at any time, and indeed, Daniel Bernoulli's (1700–1782) gas model (1738) was the first fully kinetic model. We will discuss it in a modern guise shortly in **2.6**.

The success of Newtonian universal gravity, however, almost derailed the modern atomism based on mechanics. Newton (1643–1727) tried to explain Boyle's law (i.e., $PV = $ constant, where P is the pressure, and V the volume) in terms of (repulsive) forces acting between particles. For Newton's contemporary scientists (and also for himself) introduction of the gravitational force explaining the solar system was so impressive that it led to a program to find forces to explain other phenomena as well.[5] For about 100 years, Newton's program stifled the kinetic attempts to explain the pressure of gases.[6]

2.5 Between Bernoulli and Maxwell

Between Daniel Bernoulli (*c.* 1740) and the birth of the modern kinetic theory (due to Maxwell (1831–1879) *c.* 1860) were the general acceptance of chemical atomic theory (*c.* 1810) and the birth of physics in the *modern sense*.[7] Also during this period crucial empirical facts were accumulated, making kinetic theory almost the sole consistent explanation of gasses.

Dalton (1766–1844) asserted the *law of partial pressure*: the total pressure of a gas mixture is simply the sum of the pressures each kind of gas would exert if it was occupying the space by itself (Fig. 2.1). It is very naturally explained from the atomic point of view (see **Q2.3**), because the pressure is due to individual bombarding of the wall by molecules.

Gay-Lussac (1778–1850) then established three important laws (*c.* 1810[8]):

(i) The law of thermal expansion of gases (also called *Charles' law*; $P \propto T$ if V is constant).

---

[5] **From the preface to *Principia*** Newton wrote in his preface to *Principia*, "I wish we could derive the rest of the phenomena of nature by the same kind of reasoning from mechanical principles; for I am induced by many reasons to suspect that they may all depend upon certain forces by which the particles of bodies, by some causes hitherto unknown, are either mutually impelled towards each other, and cohere in regular figures, or are repelled and recede from each other; which forces being unknown, philosophers have hitherto attempted the search of nature in vain; but I hope the principles here laid down will afford some light either to this or some truer method of philosophy"(*Principia*, author's preface, May 8, 1686).

[6] The repulsive force theory was decisively discredited by Joule and Thomson's throttling process experiments in 1852–1854. See **Q16.2**(3).

[7] For instance, experimental and mathematical rather than speculative, with dedicated laboratories and professional scientists.

[8] [In 1810: Napoleon married Marie Louise of Austria] Notice that Gay-Lussac was an example of the first generation of professional scientists trained professionally to be scientists. He was a product of the French Revolution.

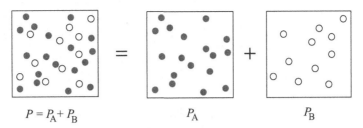

$$P = P_A + P_B \qquad\qquad P_A \qquad\qquad P_B$$

Fig. 2.1   Dalton's law of partial pressure.

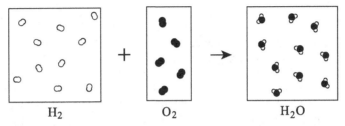

$$H_2 \qquad\qquad O_2 \qquad\qquad H_2O$$

Fig. 2.2   The law of combining volumes indicating that generally gases are made of molecules instead of atoms. The figure illustrates $2H_2 + O_2 \rightarrow 2H_2O$.

(ii) The *law of constant temperature* under "adiabatic expansion": if a gas is suddenly allowed to occupy a much larger space, there is practically no temperature change.

(iii) The *law of combining volumes*: in gas phase reactions the volumes of reactants and products are related to each other by simple rational ratios (Fig. 2.2). This suggests that the reactions occur among discrete units, but that these "units" cannot generally be atoms.

In 1811,[9] Avogadro (1776–1856) proposed *Avogadro's hypothesis*: every gas contains the same number of *molecules* at the same pressure, volume, and temperature.[10]

2.6 Daniel Bernoulli's Kinetic Theory

Let us look at Daniel Bernoulli's work. The (kinetic interpretation of) pressure P on the wall is the average momentum given to the wall per unit time and area by the gas. Consider the wall perpendicular to the x-axis (see Fig. 2.3). Assume that the mass of each particle

---

[9] In 1811 Stevens began operating the first steam-powered ferry service between New York City and Hoboken; The Luddite uprising began.

[10] **Acceptance of molecular theory** According to Cercignani, the molecular theory was not generally accepted until 1860, when Cannizzaro (1826–1910) advocated Avogadro's proposal in the Karlsruhe Congress; Clausius accepted this before; actually, Cannizzaro was triggered by Clausius' kinetic theory paper a year before. See Cercignani, C. (2001). The rise of statistical mechanics. In *Chance in Physics*, Lecture Notes Physics **574**. eds J. Bricmont, et al. Berlin: Springer, pp. 25–38. This chapter also gives a good summary of Boltzmann's progress.

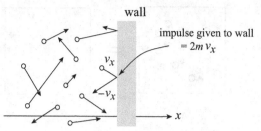

Fig. 2.3 Bernoulli's theory (or the mechanical model of gas). Particles are so small that they are assumed not to collide with each other.

is m, and that the number density of the particles is $n = N/V$, where V is the volume of the (uniform) gas and N the total number of particles:

(i) For a single particle with velocity $v = (v_x, v_y, v_x)$ hitting the wall ($v_x > 0$) in the figure, the momentum given to the wall upon collision must be $2mv_x$.

(ii) The total momentum (or the impulse) given to the wall in one second is the force on the wall in the x-direction, whose magnitude is equal to PA, where A is the area of the wall. For a particle moving toward the wall to hit it within the next 1 second, it must be within distance v_x ($\times 1$ second) from the wall. Therefore, to contribute to the pressure the particles with the x-component velocity close to v_x (> 0) must be in the volume of $A \times v_x(\times 1 \text{ s})$.[11]

(iii) Let $n(v_x)$ be the number density of the particles with its x-component velocity close to v_x. Then, the contribution of such particles to the pressure (times the wall area) must be $n(v_x) \times Av_x \times 2mv_x$ according to (i) and (ii).

(iv) Therefore, summing over all the incoming (x-direction) speeds, we get

$$PA = \sum_{v_x > 0} 2n(v_x)Amv_x^2. \tag{2.1}$$

That is,

$$P = \sum_{v_x > 0} 2n(v_x)mv_x^2 = \frac{\sum_{v_x>0} 2n(v_x)mv_x^2}{\sum_{v_x>0} n(v_x)} \times \sum_{v_x > 0} n(v_x) = \langle 2mv_x^2 \rangle_+ \times n_+, \tag{2.2}$$

where $n_+ = \sum_{v_x>0} n(v_x)$ is the number of particles with positive v_x, and $\langle \ \rangle_+$ means the average over molecules with positive v_x (to hit the wall).

(v) We do not expect the mean square velocity of the left-moving and right-moving particles are different, so $\langle v_x^2 \rangle_+ = \langle v_x^2 \rangle$ (henceforth $\langle \ \rangle$ generally implies averaging, or calculation of expectation values) and $n_+ = n/2$ (just half of the particles move to the right; notice that we have used the law of large numbers intuitively). Therefore,

$$P = nm\langle v_x^2 \rangle. \tag{2.3}$$

---

[11] To be precise we must write v_x to be in $(v_x, v_x + dv_x]$ and the sum $\sum_{v_x>0}$ in the following formulas is the integration $\int_0^{+\infty} dv_x$ (see Chapter 5), but let us proceed as informally as possible.

(vi) Using the isotropy of the gas, we expect $\langle v_x^2 \rangle = \langle v_y^2 \rangle = \langle v_z^2 \rangle$, so $\langle v^2 \rangle = \langle v_x^2 \rangle + \langle v_y^2 \rangle + \langle v_z^2 \rangle = 3\langle v_x^2 \rangle$. Therefore,

$$P = \frac{1}{3}mn\langle v^2 \rangle. \tag{2.4}$$

Or, recalling $n = N/V$, we have

$$PV = \frac{2}{3}N\langle K \rangle, \tag{2.5}$$

where $K = mv^2/2$ is the kinetic energy of a single gas particle.

Comparing this with the (modern) equation of state of an ideal gas $PV = Nk_BT$,[12] where $k_B = 1.3806 \times 10^{-23}$ J/K is the Boltzmann constant, we obtain

$$\langle K \rangle = \frac{3}{2}k_BT. \tag{2.6}$$

They recognized that temperature T is related to the kinetic energy of the molecule.

2.7 Equipartition of Kinetic Energy

Let us see that all the particles in a gas consisting of particles with different masses have, on average, identical translational kinetic energies. That is, (2.6) holds for any particle in a gas mixture (if it is "in equilibrium"). As we will learn later, this is almost self-evident from statistical mechanics (Chapter 20), but we should also be able to have an elementary understanding.

Consider a two-particle collision process. In equilibrium (i.e., if, on average, we cannot discern any change),

$$\langle \mathbf{w} \cdot \mathbf{V} \rangle = 0, \tag{2.7}$$

where $\mathbf{w}$ is the relative velocity and $\mathbf{V}$ the center of mass velocity. If we write these in terms of the velocities of two particles $\mathbf{v}_1$ and $\mathbf{v}_2$ and their respective masses m_1 and m_2, we have

$$\mathbf{w} \cdot \mathbf{V} = (\mathbf{v}_1 - \mathbf{v}_2) \cdot \frac{(m_1\mathbf{v}_1 + m_2\mathbf{v}_2)}{m_1 + m_2} = \frac{(m_1 v_1^2 - m_2 v_2^2) + (m_2 - m_1)\mathbf{v}_1 \cdot \mathbf{v}_2}{m_1 + m_2}. \tag{2.8}$$

We know $\langle \mathbf{v}_1 \cdot \mathbf{v}_2 \rangle = 0$ (the motions of two particles are unrelated, and their averages must be zero),[13] so we get the equality of the average kinetic energies.

Notice that Bernoulli's formula and equipartition of translational kinetic energy imply that even if all the particles in an ideal gas (noninteracting particle system) are with different masses, still the ideal gas law holds. "Avogadro's hypothesis" is true even for any gas mixture.

---

[12] Here, the modern notations are used; what they knew those days was that $PV \propto NT$ (in terms of a certain empirical temperature instead of absolute temperature T, precisely speaking), but they could not find N. Needless to say, k_B, the Boltzmann constant, was not known.

[13] More precisely, we say $\mathbf{v}_1$ and $\mathbf{v}_2$ are statistically independent as we will see in Chapter 3.

We have demonstrated the equipartition law, but we can give any initial condition to the gas. Does the equipartition law eventually hold, even if the initial condition does not satisfy the law?

Problems

Q2.1 Elementary Questions

(Pretend that we already know $k_B = 1.38 \times 10^{-23}$ J/K, and also Avogadro's constant $N_A = 6.02 \times 10^{23}$.)

(1) In a $1\,m^3$ box is neon gas at pressure 5.0×10^5 Pa and temperature 220 K. Find the number of neon atoms and their root-mean-square velocity. ($N = 1.65 \times 10^{26}$ and $\sqrt{3P/M} = 521$ m/s.)

(2) Find the density of dry air at 300 K (78% volume N_2, 21% O_2, and 1% Ar). If O_2 is at 35% volume, what is the density of the air?[14]

(3) On a planet it requires the surface temperature 320 K for hydrogen molecules to escape appreciably from its surface (to infinity). What is the temperature required for methane to escape from the planet surface at the same appreciable rate? ($T = T_{H_2} m'/m = 8T_{H_2} = 2560$ K.)

Q2.2 Effusion and Graham's Law

Effusion is the process in which gas particles escape from a small hole.[15] An empirical law called *Graham's law* tells us that the leak rate is inversely proportional to the square root of the mass of the leaking molecule at a given temperature T.

(1) Explain Graham's law.
(2) What is the average kinetic energy of the effusing atoms? (This will be easy after reading Chapters 5 and 6, since we need Maxwell's distribution, so come back later.)

Solution

(1) Since there is no macroscopic flow, basically the hitting rate determines the flux, that must be proportional to $n\langle v \rangle$, where $\langle v \rangle$ is the average speed.[16] This is proportional to $\sqrt{T/m}$.

(2) Suppose the hole cross section is perpendicular to the x-axis. From Bernoulli's argument it follows that the number of particles going out from the hole is proportional

---

[14] **Carboniferous atmospheric oxygen** According to Graham, J. B., et al., (1995). Implication of the late Paleozoic oxygen pulse for physiology and evolution, *Nature*, **375**, 117–120, during the Carboniferous, the atmospheric O_2 level may have reached a max of 35%. The authors suggest that this would favorably influence the evolution of insect flight. What does the reader think?

[15] This is (much) smaller than the mean free path (Chapter 7)

[16] More precisely, we can write $n\langle v \rangle/4$, 1/4 is the product of 1/2 (because one half of the particles are running in the wrong direction) and the other 1/2, which is the average of the normal projection of the velocity $\langle \cos\theta \rangle = \int_0^{\pi/2} \sin\theta\, d\theta \cos\theta / \int_0^{\pi/2} \sin\theta\, d\theta = 1/2$.

to the x-component of its velocity. Therefore, the contribution of the x-component to the energy is enhanced:

$$\int_0^\infty dv_x\, v_x \left(\frac{mv_x^2}{2}\right) e^{-mv_x^2/2k_BT} \Bigg/ \int_0^\infty dv_x\, v_x e^{-mv_x^2/2k_BT} = k_BT. \qquad (2.9)$$

Since the components perpendicular to the leaking direction does not change, $k_BT + (1/2)k_BT + (1/2)k_BT = 2k_BT$ is the average kinetic energy of the outgoing atoms.

Q2.3 Dalton's Law by Bernoulli's Theory

Show Dalton's law of partial pressure (i.e., $PV = (N_1 + N_2)k_BT$ for a two-component gas consisting of N_1 and N_2 particles) with the aid of Bernoulli's kinetic theory. (Do not repeat everything. If the reader thinks this is obvious, skip the question.)

Solution

What we have to change is that in (2.1) the mass m can be m_1 or m_2 for different species. Then, obviously, (2.2) must be modified as

$$P = 2n_{+1}m_1\langle v_x^2\rangle_{+1} + 2n_{+2}m_2\langle v_x^2\rangle_{+2}. \qquad (2.10)$$

Here, subscripts 1 and 2 denote different species. The average velocities are of course different. Thus, we get

$$PV = \frac{2}{3}N_1\langle K\rangle_1 + \frac{2}{3}N_2\langle K\rangle_2 \qquad (2.11)$$

Finally, we use the equipartition of energy to get Dalton's law of partial pressure:

$$PV = N_1 k_BT + N_2 k_BT. \qquad (2.12)$$

Q2.4 Relative Velocities

Let $\boldsymbol{w}$ be the relative velocity of the molecule of mass m and the molecule of mass $3m$ in a gas mixture (in equilibrium), and $\boldsymbol{v}$ the velocity of the molecule with mass m. Give the ratio between the root mean squares of $\boldsymbol{v}$ and $\boldsymbol{w}$.

Solution

Two different particles are statistically unrelated (we say, as we will learn in Chapter 3, statistically independent), so $\langle \boldsymbol{v} \cdot \boldsymbol{v}'\rangle = \langle \boldsymbol{v}\rangle \cdot \langle \boldsymbol{v}'\rangle = 0$, because the gas does not move as a whole, where $\boldsymbol{v}'$ is the velocity of the particle with mass $3m$. Thus,

$$\langle \boldsymbol{w}^2\rangle = \langle \boldsymbol{v}^2\rangle + \langle \boldsymbol{v}'^2\rangle. \qquad (2.13)$$

Due to the equipartition of kinetic energy

$$\frac{1}{2}m\langle \boldsymbol{v}^2\rangle = \frac{1}{2}3m\langle \boldsymbol{v}'^2\rangle \implies \langle \boldsymbol{v}^2\rangle = 3\langle \boldsymbol{v}'^2\rangle. \qquad (2.14)$$

Therefore,

$$\langle \boldsymbol{w}^2\rangle = \langle \boldsymbol{v}^2\rangle + \frac{1}{3}\langle \boldsymbol{v}^2\rangle = \frac{4}{3}\langle \boldsymbol{v}^2\rangle \Rightarrow \sqrt{\langle \boldsymbol{w}^2\rangle} = \frac{2}{\sqrt{3}}\sqrt{\langle \boldsymbol{v}^2\rangle}. \qquad (2.15)$$

Summary

* The probability $P(A)$ of event A is, in essence, the "volume" of our confidence in the occurrence of event A measured on the 0–1 scale.
* Events are expressed as sets. Then, probability is a sort of normalized volume.
* Probability must satisfy additivity: $P(A \cup B) = P(A) + P(B)$, if $A \cap B = \emptyset$.
* The gamble called survival race forces our subjective probability to agree fairly well with empirical probability.

Key words

probability, elementary event, sample space, event, mutually exclusive events, conditional probability, independence, stochastic variable, expectation value, variance, standard deviation, indicator

The reader should be able to:

* Understand how to express events in terms of sets.
* Understand that probability is a sort of volume to measure one's confidence level, so additivity must be satisfied.
* Obtain expectation values and variances for simple examples.
* Understand the relation between the indicator and the probability: $P(A) = \langle \chi_A \rangle$.
* Review elementary combinatorics (e.g., binary expansion).

To go beyond Bernoulli, we need the idea of probability. What follows is an introduction to (the spirit of) *measure-theoretical probability theory* (i.e., the mathematically standard probability theory[1]) although no formal introduction of measures will be given. The reader can simply understand that a "measure" is a precise version of volume.[2]

---

[1] **Introduction by Kolmogorov** There is a wonderful exposition by the founder of the theory (1933), A. N. Kolmogorov, himself: Kolmogorov, A. N. (2004; reprint of a book chapter of 1956) The theory of probability, *Theor. Prob. Applic.*, **48**, 191–200. If the reader has time, read this article which is much kinder than this chapter. It covers the law of large numbers, the central limit theorem and large deviation theory (though not explicitly named).

[2] **Measure** An introductory exposition of *measure* may be found in appendix 2.4A of Oono, Y. (2013). *The Nonlinear World*, Tokyo: Springer, p. 66.

3.1 Probability is a Measure of Confidence Level

Suppose we have a jar containing five white balls and two black balls. What is the degree C_w on the 0–1 scale of our confidence in our picking a white ball out of the jar? Let us interpret that to say we have confidence C_w is equivalent to believing NC_w white balls can be picked up in the long run, if we repeat the sampling N times (with replacement). Suppose we can obtain a dollar if we pick a white ball out, but otherwise must pay X dollars. Whether we wish to participate in this gamble or not depends on X. What is the wise choice? With our confidence level C_w we expect our gain G_N after N repetitions would be

$$G_N = (NC_w) \times 1 + [N(1 - C_w)] \times (-X). \tag{3.1}$$

If this is non-negative, we may participate in this gamble. Since we are supposedly free to have any thoughts, we may freely assume C_w to be any number between 0 and 1, say 4/5. However, there is no freedom of action, if we wish to stay in this world (the essence of Darwinism). So C_w must be realistic, if we wish to use it to choose our action. For our jar game $C_w = 5/7$ would actually be forced upon us. We can check whether our confidence level is rational or not empirically by repeating the gamble.

Probability seems to show up even in cases where we cannot repeat events. For example, the statement that the probability of rain tomorrow is 70% implies that we should have a confidence level of 0.7 in it raining tomorrow (if we bet money on the weather, we'd better follow this confidence estimate). However, we cannot repeat "tomorrow," so how can we check that the choice is good? In practice, the extent of confidence is estimated by relying on the past experiences of similar events.[3]

3.2 Events and Sets

To make a mathematical theory of probability, we must express events (or what can happen) as sets. An event which cannot be analyzed further, or need not be analyzed further, is called an *elementary event*. Elementary events need not be atomic events that cannot be dissected further into more basic events. For example, when we cast a dice, usually we regard a particular face, 1, 2, 3, 4, 5, or 6 to be up as an elementary event. However, if we pay attention only to the even–odd properties of the numbers, the elementary events could be "even" and "odd." On the other hand, if we wish to use the direction of the edges as well as the faces, then $1, \ldots, 6$ are no longer elementary events.

---

[3] **Evolutionary or phylogenetic learning** Why does the estimated confidence level often match reality? It is thanks to the totality of our 4-billion-year experiences (this is called *phylogenetic learning*). Even if an event does not seem to be repeatable, sufficiently many very similar events happened in the past. "What has been is what will be, and what has been done is what will be done; there is nothing new under the sun." (Ecclesiastes 1:9)

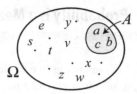

Fig. 3.1 In this illustration the sample space is $\Omega = \{a, b, c, \ldots, x, y, z\}$, where lower case letters denote elementary events; the elementary events are regarded as what actually happens. Event $A = \{a, b, c\}$ is said to occur, if a, b or c actually happens.

In statistical mechanics, elementary events are events that are elementary events of mechanics; the events specified maximally in detail in mechanics.[4]

The totality of elementary events allowed in the situation or to the system under study is called the *sample space* and is denoted by Ω. Any (compound) *event* is identified with a subset of Ω (Fig. 3.1). When we say an event corresponding to a subset A of Ω *occurs*, we mean that exactly one of the elements (or elementary events) in A actually occurs.

3.3 Probability

Let us denote the *probability* of event $A \subset \Omega$ by $P(A)$. Since probability should measure the degree of our confidence on a 0–1 scale, we demand that

$$P(\Omega) = 1: \text{we are sure that something happens.} \tag{3.2}$$

Then, since some event surely happens, it is also sensible to demand

$$P(\emptyset) = 0: \text{we are sure that "nothing happens" never happens.} \tag{3.3}$$

3.4 Additivity

Consider two *mutually exclusive events*, A and B: when A occurs, B never occurs and vice versa. The occurrence of event A implies that one of the elementary events in A actually occurs. Since A and B never occur simultaneously, no elementary event in A should be in B (and vice versa). That is, $A \cap B = \emptyset$. It is sensible to demand

$$P(A \cup B) = P(A) + P(B), \quad \text{if } A \cap B = \emptyset. \tag{3.4}$$

This is the *additivity* of probability.

For example, for a dice with the probability (or our confidence) for face 1 to be 0.15 and that for face 2 or 3 to be 0.4 (needless to say, this dice is not fair or we believe it is not fair), the probability to observe faces with values not more than 3 should be $0.15 + 0.4 = 0.55$.

---

[4] Quantum mechanically, it is a single vector (ket); classical mechanically it is a state specified by a complete set of the canonical coordinates of the system, i.e., the phase point (see **17.5**).

3.5 Probability Measure

We know other quantities for which additivity (3.4) holds: area, volume, mass (if discrete, number), etc. If the probability measures the amount of our confidence, it should be something like volume.

A function that assigns numbers to sets (or a map from sets to numbers) is called a *set function*. Roughly speaking, an additive non-negative set function is a *measure*. These examples such as area, volume, etc., are mathematically refined as measures. A measure whose value on the total set is normalized to unity is a probability (a *probability measure*).

Suppose a shape is drawn in a square of area 1 (a unit square; Fig. 3.2). If we pepper it with points uniformly, then the probability (equal to our confidence level) of a point to land on the shape should be proportional to its area. Thus, again it is intuitively plausible that probability and area or volume are closely related.

If an event is given, a certain confidence level in the occurrence of this event may be expressed as a certain probability measure. Whether the confidence level is useful/rational or not is not a concern of probability theory. In probability theory we are interested in the conclusions we can deduce from the conditions any confidence belief must satisfy.

As an example, let us consider a series of experiments consisting of tossing a coin three times. The sample space is

$$\Omega = \{HHH, HHT, HTH, THH, HTT, THT, TTH, TTT\}. \tag{3.5}$$

The word "fair" means that all elementary events are equally likely. Or, it means that we may live without any particular penalty even if we have the same confidence level 1/8 for the occurrence of any elementary event in Ω. However, if one firmly believes that the world is created to favor consecutive Hs, the confidence level in the occurrence of HHH may be 0.5, and HHT or THH may be 0.2, respectively. Even for such a person the totality of probability must be 1, and the probability of the remaining five possibilities must be $1 - 0.5 - 2 \times 0.2 = 0.1$.

The event A that at least two H appear is $A = \{HHH, HHT, HTH, THH\}$. Since all the elementary events are mutually exclusive, $P(A) = 1/2$ for a person who believes that the coin is fair, but it is obviously not smaller than 0.9 for the person with the peculiar H belief. The difference between these two confidence levels is so large that very quickly we can check (experimentally) which is more realistic.

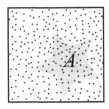

Fig. 3.2   Peppering the unit square evenly with points, we can estimate the area of A.

3.6 Relation to Combinatorics

In many cases especially when we believe in the fair event probabilities (as just seen), to count the number of cases satisfying a certain condition is the technical core of probability calculation. It is, however, just a technical aspect, and is not a crucial part of probability theory. Still, we should be able to do practical calculations, so elementary combinatorics is outlined in Appendix 3A. If the reader is not familiar with the binomial theorem, study the Appendix.

3.7 Objectivity of Subjective Probability

Since probability is introduced as the confidence level, the reader might have thought that probability is only subjective. Indeed, in the sense that probability theory is indifferent to whether a particular probability (or confidence level) assignment is useful or not to live in this world, probability may be subjective and not objective. Then, such a subjective concept should not be relevant to objective science such as physics. However, our subjective feeling (emotion underlying decisions) has been molded by natural selection during the past 4 billion years, so our subjective probability estimates (confidence levels) are often consistent with the objective probability.

Probabilities appearing in physics should be objective. If we say they are objective, there must be a means to measure them. To this end, we must learn elementary probability theory a bit further.

Let us begin with elementary facts. Since probability is something like volume, for (any) events A and B

$$P(A \cup B) \leq P(A) + P(B), \tag{3.6}$$

and

$$A \subset B \Rightarrow P(A) \leq P(B). \tag{3.7}$$

Denoting $\Omega \setminus A$ by A^c (complement), we get

$$P(A^c) = 1 - P(A), \tag{3.8}$$

since A and A^c are mutually exclusive.

Show, if $P(A) = 1$, then $P(A \cap B) = P(B)$.

☐ It is important to feel that this is obviously true, but the reader should also be able to give a logical proof. Notice that $P(B) = P(A \cap B) + P(A^c \cap B)$ and also that $P(A^c \cap B) \leq P(A^c) = 0$.

3.8 Conditional Probability

Suppose we know for sure that event B has occurred. Under this condition what is the probability of the occurrence of event A? To answer this, we need the concept of *conditional probability*. We write this conditional probability as $P(A \mid B)$, and define it as

$$P(A \mid B) = \frac{P(A \cap B)}{P(B)}, \tag{3.9}$$

so $P(B \mid B) = 1$ should hold. This is essentially the "volume fraction" of the overlapping portion of A with B in B.

3.9 Statistical Independence

When the occurrence of event A does not tell us anything new about event B and vice versa, we say two events A and B are (statistically) independent. Do not confuse "independent events" and "mutually exclusive events." Since knowing about event B does not help us to obtain more information about event A if A and B are independent, we should get

$$P(A \mid B) = P(A), \tag{3.10}$$

where $P(A \mid B)$ is the conditional probability just introduced. Therefore, the following formula must be an appropriate definition of *statistical independence* of events A and B:

$$P(A \cap B) = P(A) \cdot P(B). \tag{3.11}$$

See **Q3.3** for an example and an important warning.

3.10 Stochastic Variables

Let Ω be a sample space and a probability P is given on it.[5] Then, a function (map) from Ω to some mathematical entity (real numbers, vectors, etc.) is called a *stochastic variable* or *random variable*.

---

[5] **Probability space** Here, that P is given on Ω implies that the value of P is given for all the elementary events in Ω (in case Ω is discrete; if it is continuous, then P must be defined on an appropriate family of subsets of Ω), and (Ω, P) is called a *probability space*. If the reader reads a respectable probability book, they will encounter something like $(\Omega, \mathcal{B}, P)$, where $\mathcal{B}$ is a family of "measurable sets." We will not discuss this in this book. (Not all the events should have probabilities to avoid something like $1 + 1 = 3$, so we must specify what events can have probabilities. This is the role of $\mathcal{B}$.)

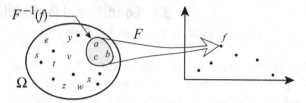

$$\text{Prob}(F = f) = P(F^{-1}(f)) = P(\{a, b, c,\}) = P(a) + P(b) + P(c)$$

Fig. 3.3 The probability for a stochastic variable F to assume a particular value f. For example, suppose we cast a dice ($\Omega = \{1, 2, 3, 4, 5, 6\}$), and we obtain \$1 if the face is odd; otherwise, we must pay \$1. Then, our gain F is a random variable $F : \Omega \to \{-1, +1\}$ such that $F^{-1}(-1) = \{2, 4, 6\}$ and $F^{-1}(+1) = \{1, 3, 5\}$. Therefore, $\text{Prob}(F = +1) = P(\{1, 3, 5\})$ and $\text{Prob}(F = -1) = P(\{2, 4, 6\})$.

Let $\Omega = \{\omega_i\}$. A real-valued stochastic variable F is a map $F : \Omega \to \mathbb{R}$ (the set of real numbers). It is rational to write the probability for this stochastic variable to take a particular value f as

$$\text{Prob}(F = f) = P(\{\omega \,|\, F(\omega) = f\}) = P(F^{-1}(f)). \tag{3.12}$$

Since $F^{-1}(f)$ is the totality of the elementary events ω such that $F(\omega) = f$, summing all the probabilities for these elementary events should give the probability of the set $F^{-1}(f) =$ "event such that $F = f$" (see Fig. 3.3). Therefore, the above definition is very reasonable.

In short, if we write the probability for a stochastic variable F to assume a value f as $P_F(f)$, then

$$P_F(f) = P(F^{-1}(f)). \tag{3.13}$$

3.11 Expectation Value

The *expectation value* (equal to average) of F is written as (if one wishes to express the underlying probability P explicitly) $E_P(F)$ or $\langle F \rangle_P$ and is defined by

$$E_P(F) \equiv \langle F \rangle_P \equiv \sum_{\omega \in \Omega} P(\omega) F(\omega) = \sum_f P_F(f) f. \tag{3.14}$$

Often the subscript P is omitted. The last equality can be checked by a straightforward calculation (also see Fig. 3.3). Let us denote the event $F = f$ as $\text{ev}(F = f) = \{\omega \,|\, F(\omega) = f\}$:

$$\sum_{\omega \in \Omega} P(\omega) F(\omega) = \sum_f \left(\sum_{\omega \in \text{ev}(F=f)} P(\omega) f \right) = \sum_f P(\text{ev}(F = f)) f = \sum_f P_F(f) f. \tag{3.15}$$

At the last step the definition of P_F has been used.

The sum becomes integration when we study events which are specified by a continuous parameter. In this case,

$$E_P(F) \equiv \langle F \rangle_P \equiv \int_{\omega \in \Omega} F(\omega) P(d\omega) = \int_{\omega \in \Omega} F(\omega) dP(\omega), \tag{3.16}$$

where $P(d\omega)$ is the probability of the volume element $d\omega$; often $P(d\omega)$ is written as $dP(\omega)$. We may simply interpret this integral just as the Riemann integral.

The symbol E_P, or E with P omitted, may be understood as an operator.[6] Let f and g be stochastic variables, and a and b real numbers. Then, we have the following equality

$$E(af + bg) = aE(f) + bE(g). \tag{3.17}$$

An operator with this property is called a *linear operator*. The expectation value operator E is a linear operator (irrespective of the probability law P used to compute E).

3.12 Variance

We are also interested in the "spread" of the variables. By far the most common measure of this spread is the *variance* defined as

$$V(X) = E([X - E(X)]^2) = E(X^2) - E(X)^2. \tag{3.18}$$

Its square root $\sigma(X) = \sqrt{V(X)}$ is called the *standard deviation* of X.

3.13 Indicator

The *indicator* χ_A of a set (equal to event in our context) A is defined by

$$\chi_A(\omega) \equiv \begin{cases} 1 & \text{if } \omega \in A, \\ 0 & \text{if } \omega \notin A. \end{cases} \tag{3.19}$$

This indicates the answer "yes" or "no" to the question: is an elementary event ω in A? If $\chi_A = 1$, we may regard that event A happens, because the occurrence of event A is defined by the occurrence of any elementary event in A.

Notice that (apply (3.14) straightforwardly)

$$\langle \chi_A \rangle_P = \sum_\omega \chi_A(\omega) P(\omega) = \sum_{\omega \in A} P(\omega) = P(A). \tag{3.20}$$

This is a very important relation for the computation of probabilities.

Let $X(\omega)$ ($\omega \in \Omega$) be a stochastic variable defined on Ω. If we denote the event $X = x$ as a set $\text{ev}(X = x) = \{\omega \mid X(\omega) = x\}$, then X defined on Ω may be written as

$$X(\omega) = \sum_x x \chi_{\text{ev}(X=x)}(\omega). \tag{3.21}$$

Equation (3.14) follows from this and (3.20):

$$\langle X(\omega) \rangle = \sum_x x \langle \chi_{\text{ev}(X=x)}(\omega) \rangle = \sum_x x P_X(x), \tag{3.22}$$

---

[6] **Operator** "Operator" is a map that maps a function to another function or number. For example, the differential operator d/dx maps a differentiable function f to its derivative f' and is a linear operator.

where $P_X(x)$ is the probability for X to be x. Here, we have used the fact that the expectation value operator $\langle \ \rangle$ (written as E previously) is a linear operator and a similar calculation as (3.15):

$$\langle \chi_{\mathrm{ev}(X=x)}(\omega) \rangle = \sum_{\omega \in \Omega} \chi_{\mathrm{ev}(X=x)}(\omega)P(\omega) = P(\mathrm{ev}(X=x)) = P_X(x). \tag{3.23}$$

3.14 Independence of Stochastic Variables

How should we define "independence" (*statistical independence*) of two stochastic variables X_1 and X_2? An answer is that

$$E(F(X_1)G(X_2)) = E(F(X_1))E(G(X_2)) \tag{3.24}$$

holds for any "reasonable" functions[7] F and G of the stochastic variables. In particular, if stochastic variables X_1 and X_2 are statistically independent,

$$E(X_1 X_2) = E(X_1)E(X_2). \tag{3.25}$$

□ Why is (3.24) reasonable? The event $F(X_1)=a$ and event $G(X_2)=b$ should be statistically independent for any (admissible) values a and b. This requires (we denote event $\{\omega \mid F = a\}$ as $ev(F=a)$, etc.) (notice that $\chi_A \chi_B = \chi_{A \cap B}$)

$$E(\chi_{\mathrm{ev}(F=a) \cap \mathrm{ev}(G=b)}) = E(\chi_{\mathrm{ev}(F=a)}\chi_{\mathrm{ev}(G=b)}) = E(\chi_{\mathrm{ev}(F=a)})E(\chi_{\mathrm{ev}(G=b)}), \tag{3.26}$$

but this follows from (3.24). Conversely, (3.26) implies (3.24) as follows, as (3.21) we can write $F(X_1) = \sum_a F(a)\chi_{\mathrm{ev}(X_1=a)}(\omega)$ and $G(X_2) = \sum_b G(b)\chi_{\mathrm{ev}(X_2=b)}(\omega)$. Therefore,

$$E(F(X_1)G(X_2)) = \sum_{a,b} F(a)G(b)E(\chi_{\mathrm{ev}(X_1=a)}\chi_{\mathrm{ev}(X_2=b)}) \tag{3.27}$$

$$= \sum_{a,b} F(a)G(b)E(\chi_{\mathrm{ev}(X_1=a)})E(\chi_{\mathrm{ev}(X_2=b)}) \tag{3.28}$$

$$= \left(\sum_a F(a)E(\chi_{\mathrm{ev}(X_1=a)}) \right) \left(\sum_b G(b)E(\chi_{\mathrm{ev}(X_2=b)}) \right) \tag{3.29}$$

$$= E(F(X_1))E(G(X_2)). \tag{3.30}$$

3.15 Independence and Correlation

If random variables X and Y are independent, then variance V obeys

$$V(X + Y) = V(X) + V(Y). \tag{3.31}$$

---

[7] "Any 'reasonable' functions" here means "any (Lebesgue) integrable functions." The author wishes to be mathematically "reasonably" unsophisticated throughout this book, so a footnote like this will not be added. Needless to say, no experience with the Lebesgue integral is required.

If we have two stochastic variables X and Y, we might wish to know how they are correlated. The most common measure of their relation is

$$C(X, Y) = E([X - E(X)][Y - E(Y)]) = E(XY) - E(X)E(Y), \qquad (3.32)$$

which is called the *covariance* between X and Y. This shows up when we wish to study fluctuations.

If X and Y are statistically independent stochastic variables, then $C(X, Y) = 0$, but the converse is not true. Let $Y = \pm X$, where $\pm$ is randomly chosen by coin-tossing. Then, $C(X, Y) = 0$, but we always have $X^2 = Y^2$, so they cannot be statistically independent; they violate the definition of statistical independence (3.24), and also intuitively we cannot say X and Y are unrelated.

3.16 Stochastic Process

Now that we have defined stochastic variables, we can define a stochastic process fairly precisely. Suppose a probability space (Ω, P) is set up. A real-valued stochastic process defined on (Ω, P) is a real function $f(t, \omega) : \mathcal{T} \times \Omega \to \mathbb{R}$ parameterized with time $t \in \mathcal{T}$. A sample process (path) is specified by ω and its value at time t is given by $f(t, \omega)$.

□ A trajectory of a 1000 step walk starting from the origin of the 2D square lattice $\mathbb{Z}^2$ may be expressed as $\{r_0, r_1, r_2, \ldots, r_{1000}\}$, where r_n is the location of the walker after n steps ($r_0 = 0$); each step is a bond vector between nearest-neighbor lattice points. Many different walk tracks are possible, so we introduce a parameter ω distinguishing different tracks. The totality $\Omega = \{\omega\}$ of the 1000 step walks starting from the origin of $\mathbb{Z}^2$ may be regarded as a sample space. If we choose ω "arbitrarily," the location of the walker after n steps may be understood as a stochastic process $r(n, \omega)$. That is, if a person begins a walk from the origin, they "unconsciously" choose (free will is not engaged) $\omega \in \Omega$. To execute the walk is to follow $r_n = r(n, \omega)$ for a fixed ω. Usually, we assume that all the tracks are equally likely, so $P(\omega)$ does not depend on ω. This stochastic process $r : \{0, 1, \ldots, 1000\} \times \Omega \to \mathbb{Z}^2$ is called the *random walk* (with 1000 steps) on a 2D square lattice (see Fig. 3.4).

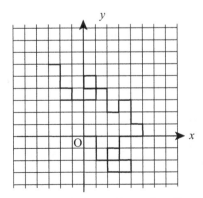

Fig. 3.4   A lattice random walk starting from the origin on the 2-square lattice; the walker can retrace their steps.

Problems

Q3.1 Elementary Problems

(1) Throw five fair dice at once. What is the probability that at least two dice show the same face?

(2) Two fair coins are thrown but we cannot see them. We are told that at least one coin exhibits a head (H) and that if there is a coin exhibiting a tail (T), we will be awarded $900. However, to participate in this game, we must pay a participation fee of $500. Should we still play the game, expecting some monetary gain?

(3) There are two shooters A and B aiming at a target. The success rate of A is 30% and B is 60%. They shoot one bullet each simultaneously, and only one bullet hits the target. What is the probability P_A that the bullet is shot by A? More generally, suppose the hitting probability of B is p. How does P_A depend on p?

Solution

(1) The probability that all the faces are distinct is $(6)_5/6^5 = 1 \times (5/6) \times (4/6) \times (3/6) \times (2/6) = 0.015$, so $1 - 0.015 = 0.985$ is the probability for at least two faces agree.

(2) Since we know at least one coin is in H state, the possible outcome must be one of HH, HT, and TH. Do not forget that coins can be distinguished, so HT and TH are distinct. Since the coins are fair, $P(H) = P(T) = 1/2$. Two coins are unrelated, so generally $P(XY) = P(X)P(Y)$. Therefore, all the possibilities are equally probable and (symbolically) $P(T \,|\, \text{one } H) = 2/3$. That is, the probability we get $900 must be 2 out of 3 trials: $600 is our expected gain. This is definitely larger than $500, so the game is worth participating in.

(3) This is a problem of conditional probability. The original sample space is {AB, aB, Ab, ab}, where the uppercase X implies the event that shooter X hits the target, and the event x that shooter X fails. Under the condition in the problem the sample space shrinks to {Ab, aB}. Therefore,

$$P_A = 0.3 \times 0.4/(0.3 \times 0.4 + 0.7 \times 0.6) = 0.22 < 0.3. \tag{3.33}$$

If the success rate of B is p, we get $P_A = 0.3/(0.3 + 0.7[p/1 - p])$. Therefore, it is a monotone decreasing function of p.

Q3.2 Fun Problems

(1) Which probability is larger, (a) or (b), assuming the 6-sided dice are fair?

(a) At least one "1" face appears in one throw of 4 dice.

(b) Two "1" faces appear simultaneously at least once in 25 throws of 2 dice.

(2) There are two kittens. We are told that at least one of them is a male. What is the probability that the two kittens are both males? (Assume that the sex ratio of kittens is 1 to 1.)

(3) There are five boxes A–E of which one contains a prize of $1000. A game participant is asked to choose one box. Before they open the chosen box, the "coordinator" of the gamble opens three of the remaining four boxes to show all three are empty. Then, they are told that if they pay $250 they may switch their choice. What is a good choice (assuming that they wish to get more money), and what is their expected gain?

Solution

(1) (a) The complement of the event "at least one" is "none." That is, if we compute the probability p for the event that no "1" face appears in one throw of four dice, $1 - p$ must be the answer ($P(A) = 1 - P(A^c)$). This is $1 - (5/6)^4 = 0.5177$.

   (b) This is very similar to (a). The probability p of the complement (no simultaneous "1" is $(35/36)^{25}$). Therefore, $1 - (35/36)^{25} = 0.5055$. Thus, (a) is slightly more likely.

   The French nobleman and gambler Chevalier de Méré suspected (purely empirically, of course) that (a) was higher than (b) with 24 throws (in this case the probability is 0.4914; about 5% difference) instead of 25, but his mathematical skill was not great enough to demonstrate why this should be so. He posed the question to Pascal, who solved the problem and proved de Méré correct.[8] We did better: even if we throw 25 times, still (a) is more likely.

(2) The reader should have realized that this is the same situation as **Q3.1**(2). For two kittens (we must recognize kittens can be distinguished), there are four different sex combinations: mm, mf, fm, ff. We know one of the three occurred: mm, mf or fm, since one is male. These three cases occur with equal probability. Therefore, with 1/3 of the probability the other is male.

(3) If they do not switch their choice, obviously the expected gain will be $200 = 1000 \times (1/5)$. The remaining four boxes contain the prize with probability 4/5. After the coordinator opens three empty boxes, this probability is "concentrated in" the remaining box. Therefore, if they switch choices, the expected gain would be $800. Thus, definitely they should pay $250 and switch, since $800 - 250 > 200$!

In this case, the new information changes the condition under which they should reconsider their "confidence level." It may be fun to read: Rosenhouse, J. (2009). *The Monty Hall Problem*, Oxford: Oxford.

Q3.3 Independence of Multiple Events: Pairwise Independence is not Enough

Suppose we throw a fair coin twice. Let A be the event that the first result is a head, B the event that the second result is a head, and C the event that the first and the second results agree (both heads or both tails).

(1) Show that any pair of the events A, B and C is statistically independent.
(2) Show that $P(A \cap B \cap C) \neq P(A)P(B)P(C)$.

---

[8] For example, look up Wikipedia for Antoine Gombaud and the "Problem of Points."

Solution

(1) Event $A = \{(H, H), (H, T)\}$, event $B = \{(H, H), (T, H)\}$ and event $C = \{(H, H), (T, T)\}$. Therefore, all the events have the probability 1/2. The common event of any two events is a single elementary event, so these probabilities are all 1/4. Therefore, $P(A \cap B) = P(A)P(B)$, etc. holds for any pair of events.

(2) $A \cap B \cap C = \{(H, H)\}$, so $P(A \cap B \cap C) = 1/4 \neq 1/8$. Thus, we have shown that $P(A \cap B \cap C) \neq P(A)P(B)P(C)$.

This example clearly tells us that pairwise independence is not enough to define the statistical independence of all the events. The statistical independence of all the events in the collection of events $\{A_i\}_{i \in \Lambda}$ must be defined by $P(\cap_\alpha A_i) = \prod_\alpha P(A_i)$ for all the subsets $\alpha \subset \Lambda$.

Q3.4 Bayes' Theorem and P-value Paradox[9]

(1) There are two events A and B. Show the following relation (called *Bayes' theorem*):

$$P(A \mid B) = P(A)P(B \mid A)/P(B). \qquad (3.34)$$

(2) Under a certain hypothesis (N) the probability to obtain the data D is found to be $P = 10^{-2}$ (a typical example is that under the no regularity condition in the observables (N) P is very small). Can we conclude that the hypothesis N is unlikely to be true (that is, there is some regularity in the data)?

Solution

(1) Go back to the definition of the conditional probability (3.9), $P(A, B) = P(A \mid B)P(B) = P(B \mid A)P(A)$ tells us the theorem.

(2) What we have observed is that $P = P(D \mid N) = 10^{-2}$. This and $P(N \mid D)$ has no direct relation as can be seen from Bayes' theorem, so without more information we cannot reject N, merely because D is improbable. In Massachusetts the frequency of women (event W) that are pregnant (event π) is approximately 2%: $P(\pi \mid W) \simeq 0.02$, so a pregnant person is unlikely to be a woman.

Q3.5 Interatomic Space Distribution

On the 2D plane atoms (ignore their exclusion volumes) are statistically independently and uniformly distributed with the number density n. What is the probability density $P(r)$ of the nearest-neighbor distance r?[10]

---

[9] Cf. Theobald, D. L. (2011). On universal common ancestry, sequence similarity, and phylogenetic structure: the sins of *P*-values and the virtues of Bayesian evidence, *Biology Direct*, **6**: 60.

[10] **Probability density** For the probability density see Chapter 6. Here, we only need: $P(r)dr$ is the probability that the nearest-neighbor atomic distance is between r and $r + dr$.

Solution

We may choose one particle and declare its position the origin. Let $p(r)$ be the probability that there is no particle on the disc D of radius r around the origin. Then, $p(r + dr)$ must be the probability of the joint event of the event that there is no particle on D and the event that no particle lands on the annulus with width dr between the rings of radius r and of $r + dr$ both centered at the origin. These two events are statistically independent, because the atoms distribute statistically independently. The latter probability is $(1 - n2\pi r dr)$,[11]* so

$$p(r + dr) = p(r)(1 - n2\pi r dr) \tag{3.35}$$

or

$$\frac{d}{dr} \log p(r) = -2\pi nr. \tag{3.36}$$

Thus, $p(r) = e^{-\pi nr^2}$, because $\lim_{r\to 0} p(r) = 1$. $1 - p(r)$ is the cumulative distribution function of $P(r)$:

$$\int_0^r dr\, P(r) = 1 - p(r), \tag{3.37}$$

so $P(r) = 2\pi nr e^{-\pi nr^2}$. (See the Poisson distribution **Q28.2**.)

Q3.6 Bertrand's Paradox

Explain Bertrand's paradox in about 10 lines (without using outrageous fonts). What lesson can we learn? (The reader can easily find a detailed account on the Internet.)

Solution

Wikipedia gives a good account of this topic.[12] In short, being random or uniform sampling is a rather tricky idea (probabilities may not be well defined if the method that produces the random variable is not clearly specified). We need a detailed empirical analysis of what we mean by "uniform" or "random." This is the lesson.

However, this article may have given the reader an idea that there is a general principle to "rescue" the ambiguity inherent in the concept of "lack of knowledge" following Jaynes ("maximum ignorance principle" or, in essence, to use fully the symmetry in the problem). This implies that we must perform a detailed analysis of what is not known. If symmetry principles are used inadvertently, we can easily get a nonsensical result. A classic example is *von Mises' wine/water paradox*. The reader can look this up on the Internet, and perhaps will see proposals to resolve the paradox. The resolutions require more detailed analysis of what is not known.

---

[11] *The area of the annulus $2\pi r dr$ is sufficiently small, so at most one particle can land on it. Therefore, $n\times$ this small area is the probability for the small area to be hit. Hence, $1 - n2\pi r dr$ is the probability that this annulus is not hit.

[12] See http://en.wikipedia.org/wiki/Bertrand's_paradox_(probability)

Appendix 3A Rudiments of Combinatorics[13]

As noted earlier, often evaluation of elementary probabilities boils down to counting the number of ways to arrange objects. In statistical mechanics we must be able to count the number of elementary events (i.e., microscopic events called microstates) under various constraints.

3A.1 Sequential Arrangement (without Repetition) of r Objects from n Distinguishable Objects: $_nP_r$

Suppose there is a set of n distinguishable objects. How many ways are there to make sequential arrangements of r objects taken from this set (without repetition) $(r \leq n)$? This number is denoted by $_nP_r \equiv P(n, r)$.

There are n ways in selecting the first object. To choose the second object, there are $(n - 1)$ ways, because we have already taken out the first one. Here, the distinguishability of each object is crucial. In this way we arrive at

$$P(n, r) = n \cdot (n - 1) \cdots (n - r + 1) = \frac{n!}{(n - r)!}, \tag{3.38}$$

where $n! = 1 \cdot 2 \cdot 3 \cdots (n - 1) \cdot n$, which is called the n *factorial*, is the number of ways n distinguishable objects can be arranged in a sequence. The following symbol is also often used:

$$(n)_r \equiv n \cdot (n - 1) \cdots (n - r + 1). \tag{3.39}$$

Also from the logic we see that the number of ways to arrange r objects taken from n distinguishable objects with repetition allowed is n^r. We can show $(n)_r/n^r \to 1$, if n becomes large with fixed r. That is, asymptotically for large n the samplings with and without replacement are the same (as intuitively expected).

3A.2 Selection (without Repetition) of r Objects from n Distinguishable Objects: Binomial Coefficient, $_nC_r$

Under the same distinguishability condition, we now disregard the order in the arrangement of r objects. That is, we wish to answer the question: how many different subsets can we make, if we choose r elements without repetition from a set consisting of n distinguishable elements?

Since we disregard the ordering in each arrangement of r distinguishable objects, the answer should be

$$_nC_r \equiv \binom{n}{r} \equiv \frac{_nP_r}{r!} = \frac{n!}{(n - r)!\, r!}. \tag{3.40}$$

The number $\binom{n}{r}$ is called the *binomial coefficient* due to the reason made clear in (3.42).

---

[13] Chapter II of Feller, W. (1957). *An Introduction to Probability Theory and its Applications*, New York: Wiley, volume 1 is a useful reference.

Show the following equality and give its combinatorial explanation:

$$\binom{n}{r} = \binom{n-1}{r-1} + \binom{n-1}{r}. \tag{3.41}$$

3A.3 Binomial Theorem

Consider the nth power of $x + y$. There exists an expansion formula called the *binomial expansion*:

$$(x+y)^n = \sum_{r=0}^{n} \binom{n}{r} x^{n-r} y^r. \tag{3.42}$$

This can be seen easily as follows: We wish to expand the product of n $(x+y)$:

$$\overbrace{(x+y)(x+y)(x+y)\cdots(x+y)\cdots(x+y)}^{n}. \tag{3.43}$$

Take the term $x^r y^{n-r}$. To produce this term by expanding the above product, we must choose r $x + y$s, in order to get xs, disregarding their order from n ordered (or numbered $1, \ldots, n$) $(x+y)$s. There are $\binom{n}{r}$ ways to do this, so the coefficient must be $\binom{n}{r}$.

3A.4 Multinomial Coefficient

Suppose there are k species of particles. There are q_i particles for the ith species. We assume that the particles of the same species are not distinguishable. The total number of particles is $n \equiv \sum_{i=1}^{k} q_i$. How many ways are there to arrange these particles in one-dimensional array?

If we assume that all the particles are distinguishable, the answer is $n!$. However, the particles of the same species cannot be distinguished, so we need not worry which ith particle is chosen first. Hence, we have over-counted the number of ways by the factor $q_i!$ for the ith species. The same should hold for all species. Thus we arrive at

$$\frac{n!}{q_1! \, q_2! \cdots q_{k-1}! \, q_k!}. \tag{3.44}$$

This is called the *multinomial coefficient*.

3A.5 Multinomial Theorem

There is a generalization of (3.42) to the case of more than two variables and is called the *multinomial expansion*:

$$(x_1 + x_2 + x_3 + \cdots + x_m)^n = \sum_{q_1 + q_2 + \cdots + q_m = n, \, q_i \geq 0} \frac{n!}{q_1! \, q_2! \cdots q_m!} x_1^{q_1} x_2^{q_2} \cdots x_m^{q_m}, \tag{3.45}$$

whose demonstration is very similar to that explained around (3.43). We could use the binomial theorem (3.42) repeatedly to show this as well.

Arranging n indistinguishable objects into r distinguishable boxes

3A.6 Arrangement of Indistinguishable Objects into Distinguishable Boxes

Consider n indistinguishable objects. We wish to distribute them into r distinguishable boxes. How many distinguishable arrangements can we make?

Since the boxes are distinguishable, we arrange them in a fixed sequence, and then distribute the indistinguishable objects. Figure 3.5 tells us that the problem is equivalent to counting the number of arrangements of n indistinguishable balls and $r - 1$ indistinguishable bars on a line. Apply (3.40) to obtain the answer:

$$\frac{(n+r-1)!}{n!\,(r-1)!} = \binom{n+r-1}{n}. \tag{3.46}$$

The problem is equivalent to choosing n objects with repetition from a set of r distinguishable objects. For example, if $r = 3$ and $n = 5$, we choose five letters with repetition to make a set of five letters from $\{a, b, c\}$ (disregarding the order of the choices). In this context the number of arrangements is called the *binomial coefficient of the second kind* and denoted as

$$\left(\!\!\binom{r}{n}\!\!\right) = \frac{(n+r-1)!}{n!\,(r-1)!}. \tag{3.47}$$

If $r \gg n$, the allowing repetition or not should not matter, so (3.47) is almost $\binom{r}{n}$.

For the binomial coefficient of the second kind, we have[14]

$$(1-x)^{-r} = \sum_{n=0}^{\infty} \left(\!\!\binom{r}{n}\!\!\right) x^n. \tag{3.48}$$

To show this we use[15]*

$$(1-x)^{-r} = \sum_{n=0}^{\infty} \binom{-r}{n}(-x)^n \tag{3.49}$$

---

[14] Related topics are in Konvalina, J. (2000). A unified interpretation of the binomial coefficients, the Stirling numbers, and the Gaussian coefficients, *Am. Math. Month.*, **107**, 901–910.

[15] ***Binomial series** More generally,

$$(1+x)^{\gamma} = \sum_{k=0}^{\infty} \binom{\gamma}{k} x^k = 1 + \gamma x + \frac{\gamma(\gamma-1)}{2!} x^2 + \cdots$$

with the (generalized) binomial coefficient

$$\binom{\gamma}{k} = \frac{\gamma(\gamma-1)\cdots(\gamma+k-1)}{k!}$$

is called the binomial series, where γ can be any real number. This generalization of the binomial theorem was initially found by Newton for positive rational γ. The formula may be demonstrated by the Taylor expansion.

and an explicit calculation of the binomial coefficients:

$$\binom{-r}{n} = \frac{(-r)(-r-1)(-r-2)\cdots}{(-r-n)(-r-n-1)(-r-n-2)\cdots n!} = \frac{(-r)(-r-1)\cdots(-r-n+1)}{n!}$$

(3.50)

$$= (-1)^n \frac{(r+n-1)!}{n!\,(r-1)!} = (-1)^n \left(\left(\begin{array}{c} r \\ n \end{array} \right) \right)$$

(3.51)

How about the arrangement of the distinguishable **n** into **r** distinguishable boxes? The first particle can be put into one of r boxes. Then, the second, etc. Thus, there are r^n ways. There are two more conceivable cases:

(i) How about the arrangement of n distinguishable particles into r indistinguishable boxes? This is not easy.[16]

(ii) How about the arrangement of n indistinguishable particles into r indistinguishable boxes? (This is not easy, either. This is related to the decomposition of n into r positive integers equal to the *integer partition problem*.)

[16] See J. Konvalina (2000) for some relevant information.

4   The Law of Large Numbers

Summary

* Using many measurement results of the same observable, its very accurate expectation value may be empirically obtained (the law of large numbers).
* The law of large numbers allows us to measure probability; it is a very crucial theorem for empirical sciences.
* We can use the law of large numbers to estimate high-dimensional integrals.

Key words

Chebyshev's inequality, law of large numbers, Monte Carlo integration

The reader should be able to:

* Understand why the law of large numbers is plausible.
* Recognize that the law of large numbers makes various probabilities observable.
* However, remember the fundamental conditions for the law of large numbers to hold.
* Use the law of large numbers to estimate the needed number of samples for "safe" inference.
* Appreciate the power of randomness.

4.1   How can we Measure Probability?

Let us return to the problem in Fig. 3.2. Suppose χ_A is the indicator of the area A in the unit square. We pepper dots on it evenly. If the ith dot (location x_i) is on A, $\chi_A(x_i) = 1$, otherwise, 0. If we count the number N_1 of points for which $\chi_A = 1$ in the total trial with N dots, N_1/N should be close to the area, which is the probability P for the dot to land on A. We expect

$$\frac{1}{N} \sum_{i=1}^{N} \chi_A(x_i) \to E(\chi_A) = P(A) \tag{4.1}$$

in the large N limit. This can be verified as the most important theorem of probability theory: *the law of large numbers*.

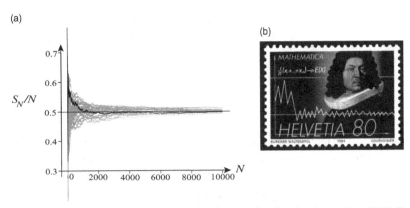

(a) The law of large numbers illustrated for coin-tossing: the fraction of heads in N trials up to $N = 10000$. There are 30 runs superposed; one run is highlighted. Note the slow convergence rate. (b) Jacob Bernoulli (1654–1705) proved the law of large numbers (published posthumously in 1713) (Swiss stamp in 1994. Photo (original in color) courtesy of Professor M. Börgens of Technische Hochschule Mittelhessen).

For a fair coin, let X_n be the indicator of head for the nth tossing of the coin (i.e., $X_n = 1$, if the outcome of the nth tossing is a head, otherwise, 0). Then, we expect, in the large N limit,

$$\frac{1}{N}\sum_{n=1}^{N} X_n \to \frac{1}{2}.$$ (4.2)

Jacob Bernoulli (1654–1705) proved this (Fig. 4.1).[1]

The most important message is that probabilities of events may be observed experimentally. Although we introduced probability $P(A)$ as a measure of our (subjective) confidence in the occurrence of event A, whether the probability is realistic or not can be determined empirically in many cases. Do not forget that our intuition/emotional judgement is based on our nervous systems, which have been subjected to rigorous selection processes in the past 1 billion years. Thus, inevitably, our subjective judgements tend to be consistent with the objective world.

4.2 Precise Statement of the Law of Large Numbers

A precise statement of the law of large numbers (LLN) is as follows:

Let $\{X_i\}_{i=1}^{n}$ be a collection of independently and identically distributed (often abbreviated as *iid*) stochastic variables. For any $\varepsilon > 0$,

$$\lim_{N\to\infty} P\left(\left|\frac{1}{N}\sum_{n=1}^{N} X_n - E(X_1)\right| > \varepsilon\right) = 0$$ (4.3)

---

[1] The proof was published posthumously in *Art Conjectandi* (1713). The base of the natural logarithm, e, was introduced by him as well.

holds under the condition that the distribution of X_i is not too broad: $E(|X_1|) < \infty$. If $V(X_1) < +\infty$, the condition is satisfied.[2] In the following, the law of large numbers is demonstrated under this assumption. The following is also a precise expression:[3] for large N

$$\sum_{n=1}^{N} X_n = NE(X_1) + o[N].$$ (4.4)

The interpretation of (4.3) is as follows: we perform a series of N experiments to produce the *empirical expectation value* $(1/N)\sum_{n=1}^{N} x_n$ (lower case x_i here means the actually obtained value). This set of N experiments (trials/samplings) is understood as a single "run," and we imagine many such runs. Then, (4.3) tells us that the probability that these runs produce empirical averages S_N/N deviating from the true mean $E(X_1)$ by more than (any positive number) ε goes to zero in the limit of the infinite sample size $N \to \infty$.

Remark: Suppose we find an empirical average S_N/N larger than $E(X_1)$. Then, we might (erroneously) expect more outcomes smaller than $E(X_1)$ in the near future. This is the famous *gambler's fallacy* (or fallacy of the maturity of chances).

4.3 Why is the Law of Large Numbers Plausible?

Before going to a demonstration of the law of large numbers, let us understand why it is plausible. We could expect that the average of S_N/N (the empirical average) should fluctuate around $E(X_1)$. Its width of fluctuation may be evaluated by the variance (notice that $V(cX) = c^2 V(X)$ and "additivity" (3.31) in the following calculation):

$$V\left(\frac{1}{N}\sum_{n=1}^{N} X_n\right) = \frac{1}{N^2}V\left(\sum_{n=1}^{N} X_n\right) = \frac{1}{N^2}\sum_{n=1}^{N} V(X_n) = \frac{1}{N}V(X_1).$$ (4.5)

Thus, the width of the distribution shrinks as N is increased. That is why S_N/N clusters tightly around $E(X_1)$ as $N \to \infty$. This is the essence of the law of large numbers.

4.4 Chebyshev's Inequality and a Proof of LLN

The key to a proof of the law of large numbers is *Chebyshev's inequality*

$$a^2 P(|X - E(X)| \geq a) \leq V(X).$$ (4.6)

---

[2] Since all X_n are distributed identically, we use X_1 as a representative, so $E(|X_1|)$, etc., show up in the statement.
[3] **Two kinds of law of large numbers** Equations (4.3) and (4.4) are mathematically different but in this book we do not distinguish them; the former is called the *weak law* of large numbers and the latter the *strong law* of large numbers. In statistical mechanics, the author feels that in most cases when the weak law holds, the strong law holds as well.

This can be shown as follows (let us redefine X by shifting as $X - E(X)$ to get rid of $E(X)$ from the calculation). We start from the definition of the variance:

$$V(X) = \int_\Omega X^2 dP(\omega). \qquad (4.7)$$

Here, the integration range is over all values of X. Now, let us remove the range $|X| < a$ from this integration range. The contribution of the removed portion to the original integrand is positive, so obviously

$$V(X) = \int_\Omega X^2 dP(\omega) \geq \int_{\{\omega : |X| \geq a\}} X^2 dP(\omega). \qquad (4.8)$$

Here, the integral on the right-hand side is over the elementary events satisfying the condition $|X| \geq a$. Under this condition $X^2 \geq a^2$, so

$$\int_{\{\omega : |X| \geq a\}} X^2 dP(\omega) \geq \int_{\{\omega : |X| \geq a\}} a^2 dP(\omega) = a^2 \int_{\{\omega : |X| \geq a\}} dP(\omega). \qquad (4.9)$$

This implies

$$V(X) \geq a^2 P(|X| \geq a). \qquad (4.10)$$

Since we have shifted X by $E(X)$, this implies Chebyshev's inequality (4.6).

We wish to apply Chebyshev's inequality (4.6) to the sample average $(1/N)\sum X_n$. Replacing corresponding quantities in (4.6) ($X \to (1/N)\sum X_n$, $a \to \varepsilon$), and using (4.5), we get

$$P\left(\left| \frac{1}{N} \sum_{n=1}^N X_n - E(X_1) \right| \geq \varepsilon \right) \leq \frac{V(X_1)}{\varepsilon^2 N}. \qquad (4.11)$$

Taking $N \to \infty$, we arrive at the law of large numbers.

☐ **Warning on the reliability of statistical tests**: Statistical tests are ultimately secured by the law of large numbers. We must never forget the conditions for the theorem to hold, i.e., the independence and the identity of the distributions. The independence might be realized relatively easily by making uncorrelated sampling, but the identity of the distribution is not very simple. For example, it is very hard to obtain a large number of samples obeying the same statistical law in biology; even clones depend on their living environments, and on the timing of measurement or sampling, etc. Thus, large sample size does not necessarily guarantee the quality of the estimated results. It is not surprising that statistically apparently rigorous genomics results often lack reproducibility.

4.5 Almost no Fluctuation of Internal Energy, an Example

The totality of mechanical energy of a macroscopic system is called the internal energy (Chapter 11). The law of large numbers and the equipartition of energy (Chapter 2) imply that for an ideal gas the internal energy does not fluctuate macroscopically.

Let us take an ideal gas consisting of N particles in a volume V kept at temperature T (with the aid of a thermostat). We have demonstrated that all the particles have the same average kinetic energy $3k_BT/2$. Hence, the law of large numbers tells us (see **2.6** for the notation)

$$P\left(\left|\frac{1}{N}\sum_{n=1}^{N}\frac{1}{2}mv_n^2 - \frac{3}{2}k_BT\right| > \varepsilon\right) < \frac{V(K)}{\varepsilon^2 N}, \qquad (4.12)$$

where $V(K)$ is the variance of the kinetic energy of each particle. Or, since $E = \sum(1/2)mv_n^2$ is the internal energy,

$$P\left(\left|\frac{E}{N} - \frac{3}{2}k_BT\right| > \varepsilon\right) < \frac{V(K)}{\varepsilon^2 N}. \qquad (4.13)$$

Since we measure a macroscopic quantity, the error magnitudes of typical interest have only to be small macroscopically, say, 1% of the actual value. This implies that $\varepsilon = 0.01(3k_BT/2)$, so the upper bound of the probability in the above formula reads $[V(K)/(3k_BT/2)^2]/(0.01^2 N)$. Notice that $V(K)$ is of the same order of $(k_BT)^2$, so the probability to obtain the internal energy with a relative error of 1% is larger than $1 - c/(0.01^2 N)$, where $c = V(K)/(3k_BT/2)^2$ is a constant of order unity. Thus, we have realized that "surely" E is constant within 1% for any macroscopic N. Actually, even if we increase the observation accuracy to 10^{-5}% still the situation does not change practically very much for a 1 liter of gas (at 1 atm at 300 K).

4.6 Monte Carlo Integration

Let us consider the problem of numerically evaluating a high-dimensional integral (the *Monte-Carlo integration* method):

$$I = \int_0^1 dx_1 \cdots \int_0^1 dx_{1000} f(x_1, \ldots, x_{1000}). \qquad (4.14)$$

Assume $|f| \leq 1$. If we wish to sample (only) two values for each variable, we need to evaluate the function at $2^{1000} \sim 10^{300}$ points (recall $2^{10} \simeq 10^3$). Such sampling is of course impossible.

This integral can be interpreted as the average of f over a 1000-dimensional unit hypercube:

$$I = \frac{\int_0^1 dx_1 \cdots \int_0^1 dx_{1000} f(x_1, \ldots, x_{1000})}{\int_0^1 dx_1 \cdots \int_0^1 dx_{1000}}. \qquad (4.15)$$

Therefore, randomly sampling the points r_n in the hypercube, we can obtain

$$I = \lim_{N \to \infty} \frac{1}{N}\sum_{n=1}^{N} f(r_n). \qquad (4.16)$$

How many points should we sample to estimate the integral within a 10^{-2} error, if we tolerate larger errors than this at most once out of 1000 such calculations? We can readily know the answer from (4.11): $V(f(X_1))/\varepsilon^2 N < 1/1000$ or $1000 \times 10^4 V(f(X_1)) < N$.[4] The variance of the value of f is of order $\max |f|^2$, a constant. Compare this number with 10^{300} earlier and appreciate the power of randomness. This is the principle of the Monte Carlo integration. Notice that the computational cost does *not* depend on the dimension of the integration range.

4.7[†] Central Limit Theorem

When we obtain the law of large numbers, we divide the empirical sum $\sum_{i=1}^{N} X_i$ with N (we assume X_i to be iid stochastic variables as previously). Consequently, we cannot detect deviations of this sum from its representative behavior. However, without this division the sum diverges and cannot tell us anything useful. Can we make a quantity that obeys a distribution that does not spread all over nor shrink to a point by dividing the empirical sum with some factor that is smaller than N but still increasing with N? This is the basic motivation for the central limit theorem. As can be seen from (4.5), the spread of the empirical sum is $O[\sqrt{N}]$.[5] Therefore, let us divide the empirical sum with $\sqrt{N}$ instead of N. We compute the characteristic function,[6] $g_N(\xi) \equiv \langle e^{i\xi Y_N} \rangle$, of

$$Y_N = \frac{1}{\sqrt{N}} \sum_{i=1}^{N} X_i. \tag{4.17}$$

For simplicity, we assume that the expectation value of X_i is zero, and its variance unity. Furthermore, we assume that $E(|X_1|^3) < \infty$. The characteristic function g_N reads:[7*]

$$g_N(\xi) = \omega(\xi/\sqrt{N})^N, \tag{4.18}$$

where ω is the characteristic function of X_1, that is, $\omega(\xi) = \langle e^{i\xi X_1} \rangle$. Since we have assumed that the third moment of X_i exists, ω is three times differentiable, so we may write

$$\omega(\xi/\sqrt{N}) = 1 - \xi^2/2N + O[1/N^{3/2}]. \tag{4.19}$$

Therefore, we obtain

$$\lim_{N\to\infty} g_N(\xi) = e^{-\xi^2/2}, \tag{4.20}$$

---

[4] Here, the inequality gives a sufficiently safe estimate. In practice, a smaller number of samples may well be admissible.

[5] **Symbol O** $Y = O[X]$ in a certain limit of X implies that there is a positive number a such that $|Y| < a|X|$ in this limit. We read that Y is the same order as X.

[6] It is the Fourier transform of the density distribution function (see Chapter 5).

[7*] $g_N(\xi) = \langle \exp[(\xi/\sqrt{N})(X_1 + \cdots + X_N)] \rangle = \langle \exp[(\xi/\sqrt{N})X_1] \rangle^N = \omega(\xi/\sqrt{N})^N$, because $X_1, \ldots, X_N$ are iid. (4.19) is roughly due to $\omega(\xi) = \langle (1 - i\xi X_1 - (\xi^2/2)X_1^2 + \cdots) \rangle = 1 - (\xi^2/2) + \cdots$ (recall $\langle X_1 \rangle = 0$, $\langle X_1^2 \rangle = 1$).

which is the characteristic function of a Gaussian distribution (see Chapter 5) with average 0 and variance 1. Thus, in the general case of X_i with expectation m and variance V, asymptotically in the $N \to \infty$ limit

$$P\left(\frac{1}{\sqrt{NV}}\sum_{i=1}^{N}(X_i - m) \leq y\right) \to \int_{-\infty}^{y}\frac{1}{\sqrt{2\pi}}e^{-x^2/2}dx \qquad (4.21)$$

holds, which is called the *central limit theorem* (CLT).[8] This central limit theorem shows us small fluctuations that cannot be detected if we normalize the observables with the system size.

4.8[†] There is no Further Refinement of LLN

The reader might hope that $Y_N = Nm + \sqrt{NV}\xi +$ further corrections of order $N^0, N^{-1/2}$, etc., where ξ is a Gaussian noise with mean 0 and variance 1, describing the CLT. Beyond LLN and CLT or the large deviation principle no correct general refinement can exist.[9]

Problems

Q4.1 Very Elementary Questions

In a cubic box with edge size 20 cm are ideal gas particles. Let us place the geometrical center of the cube at the origin and choose a Cartesian coordinate system whose axes are parallel to the edges. Due to the roughness of the wall, the directions of movement of the particles are, even without any mutual collision, quite random.

(1) What is the average position of a particle?
(2) What is the variance of the position of a particle?
(3) What is the variance of the center of mass of the particles, if there are 1000 of them?
(4) How many particles do you wish to place in the box if you wish to confine the center of mass of the whole gas within 1 μm of the origin for 99% of time? Give your rational estimate.

---

[8] Here, the convergence is for each y.
   Central limit theorem references The introductory article by Kolmogorov quoted in Chapter 3 is a gentle introduction. The bible of the central limit theorem is chapters III and XVII of Feller, W. (1971). *An Introduction to Probability Theory and Its Applications*, vol. II, New York: Wiley. Chapter 2 of R. Durrett, R. (1991). *Probability: Theory and Examples*, Pacific Grove: Wadsworth & Brooks/Cole may be more accessible.
[9] McKean, H. (2014). *Probability: The Classical Limit Theorems*. Cambridge: Cambridge University Press. Chapters 2 and 11.

Solution

(1) Obviously, at the origin by symmetry.

(2) In the x direction the distribution is uniform in $[-0.1, 0.1]$, so

$$\langle x^2 \rangle = \frac{1}{0.2} \int_{-0.1}^{0.1} x^2 dx = \frac{1}{0.2} 2 \times \frac{0.1^3}{3} = 0.01/3. \tag{4.22}$$

All three directions are independent, so $\langle r^2 \rangle = 0.01\,\text{m}^2$.

(3) Using the statistical independence of the particles, we have generally

$$V\left(\frac{1}{N}\sum_i \boldsymbol{r}_i\right) = \frac{1}{N^2}V\left(\sum_i \boldsymbol{r}_i\right) = \frac{1}{N^2}\sum_i V\left(\boldsymbol{r}_i\right) = \frac{1}{N}V(\boldsymbol{r}_1). \tag{4.23}$$

Therefore, for $N = 1000$, $1 \times 10^{-5}\,\text{m}^2$.

(4) This is a typical law of large numbers question: if

$$P\left(\left|\frac{1}{N}\sum_i \boldsymbol{r}_i\right| > 10^{-6}\right) < 0.01, \tag{4.24}$$

then we are safe. That is, $\varepsilon = 10^{-6}$. Therefore, $0.01/N(10^{-6})^2 < 0.01$ or $N > 0.01 \times 10^{12}/0.01 = 1 \times 10^{12}$ is sufficient.

The distribution of the center of mass position actually obeys a Gaussian distribution of variance V_1/N, where V_1 is the one-particle position variance (see **4.7** Central Limit Theorem (CLT)). If the reader uses this accurate knowledge, a better estimate is obtained. However, here, the problem is interpreted as an elementary question as to LLN. Its estimate gives us a sufficient number.

Q4.2 LLN does not Hold if the Distribution is too Broad

The *Cauchy distribution* is defined by the following density distribution function (i.e., $P((x, x + dx]) = p(x)dx$. See Chapter 5 for density distribution functions):

$$p(x) = \frac{1}{\pi}\frac{a}{x^2 + a^2}. \tag{4.25}$$

This distribution does not satisfy $E(|X|) < +\infty$ (needless to say, the variance is infinite). Actually, the density distribution of

$$Y_n = \frac{X_1 + \cdots + X_n}{n} \tag{4.26}$$

has exactly the same distribution function as X_1, if $\{X_j\}$ all obey the same Cauchy distribution and are statistically independent. Let us demonstrate this.

(1) What is the characteristic function $\omega(k) = \langle e^{ikX} \rangle$ of the Cauchy distribution? The reader can look up the result, but even in that case you must explain why the result is correct.

(2) Show what we wish to demonstrate.

Solution

(1) We have only to compute

$$\omega(k) = \int_{-\infty}^{\infty} dx\, e^{ikx} p(x) = \frac{1}{2\pi i} \int_{-\infty}^{\infty} dx \left(\frac{1}{x - ia} - \frac{1}{x + ia} \right) e^{ikx} = e^{-a|k|}. \qquad (4.27)$$

It may be a good occasion to review contour integration, Cauchy's theorem, etc.

(2) The characteristic function for Y_n is given by $\omega(k/n)^n$ for any $n > 0$. This is in our case exactly $\omega(k)$ itself.

Q4.3 St. Petersburg Paradox by Daniel Bernoulli

Let $\{X_i\}$ be independently and identically distributed (iid) with

$$P(X_1 = 2^n) = 2^{-n} \qquad (4.28)$$

for all positive integers n. This models a gamble in which with probability 2^{-n} the gambler receives a payoff 2^n.

(1) Show that $E(X_1) = \infty$. Thus, it seems that if X_1 is the gambler's gain, the reader can participate in this gambling game with any entry price and still can expect a positive gain. However, any sensible person would never pay \$1000 as a fair price for playing. Why? This is the "paradox."

(2) Needless to say, the law of large numbers does not hold for $Y_n = (X_1 + X_2 + \cdots + X_n)/n$. This implies that empirically obtainable expectation and theoretical one should have some discrepancy. Indeed, it can be proved (the reader need not show this; not very easy) that for any positive ε

$$P\left(|Y_n / \log_2 n - 1| > \varepsilon \right) \to 0 \qquad (4.29)$$

in the $n \to \infty$ limit. Recall that Y_n is the expected payoff for the first n trials. Explain why the reader does not wish to pay \$1000. (Or for this to be a fair price how many times does the reader have to play?)

Solution

(1) This is obvious: $E(X_1) = \sum_m 2^m 2^{-m}$.

(2) The above estimate implies with high probability (asymptotically) $Y_n \sim \log_2 n$. That is, we must wait until $\log_2 n = 1000$ to break-even. That is $n = 2^{1000} \simeq 10^{300}$.

5 Maxwell's Distribution

Summary

∗ The probability density of the particle velocity in equilibrium states (Maxwell's distribution) is $\propto e^{-\beta m v^2/2}$.
∗ How to calculate Gaussian integral and averages is explained.
∗ Boltzmann factor $e^{-\beta U}$ is derived and used to "re-derive" Maxwell's distribution.

Key words

density distribution function, Maxwell's distribution, Boltzmann factor, Gaussian integral

The reader should be able to:

∗ Use distribution functions to estimate expectation values.
∗ Use the Boltzmann factor intuitively.
∗ Recognize that the molecular speed in a gas is of the same order of the sound speed in it.

5.1 Density Distribution Function

To make the kinetic theory quantitative, we must know the probability of a particle to assume various velocities. For the velocity of a particle to be $\boldsymbol{v}$ exactly is obviously with probability zero. In the present case, the sample space (the totality of all the possible cases) is $\Omega = \{\boldsymbol{v} \,|\, v_x, v_y, v_z \in \mathbb{R}\} = \mathbb{R}^3$.[1] Thus, we need a probability (i.e., a probability measure) P defined on Ω. In the present case, $P(A) \to 0$ as the volume[2] of $A \to 0$, so we may define the probability density symbolically as (see Fig. 5.1),

$$f(\boldsymbol{v}) = \frac{P(d\tau(\boldsymbol{v}))}{d\tau(\boldsymbol{v})}, \tag{5.1}$$

where $d\tau(\boldsymbol{v})$ is the volume element (of the 3-space) around $\boldsymbol{v}$. Here, its volume is also denoted by the same symbol $d\tau(\boldsymbol{v})$, which may also be written as $d^3\boldsymbol{v} = dv_x dv_y dv_z$. The probability $P(A)$ of event $A \subset \Omega$ may be expressed as

---

[1] **n-object** $\mathbb{R}^3$ denotes the 3-real vector space; generally, "n-object" implies n-dimensional object.
[2] This is the actual volume of A as a subset of 3-space, which mathematicians call the Lebesgue measure of A.

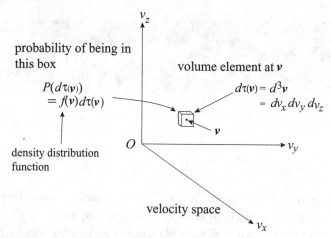

The volume element of the velocity space and the density distribution function f.

$$P(A) = \int_A d^3v\, f(v).\tag{5.2}$$

f is called the *density distribution function*.

5.2 Maxwell's Derivation of the Distribution Function

In his "Illustrations of the dynamical theory of gases" Maxwell (1860) introduced the density distribution function $f(v)$ of the velocity of a gas particle. He assumed that in equilibrium[3] orthogonal components of the velocity of particles are statistically independent (cf. (3.11)). This implies that we may write

$$f(v) = \phi_x(v_x)\phi_y(v_y)\phi_z(v_z),\tag{5.3}$$

where ϕ_x, etc., are density distribution functions for individual components. Maxwell also assumed isotropy (all space directions are indistinguishable), so f is a function of $v^2 \equiv |v|^2$, $f(v) \equiv F(v^2)$ (i.e., some function of v^2) and ϕ_x, etc., do not depend on the subscripts specifying the coordinates: $\psi(v_x^2) \equiv \phi_x(v_x)$, $\psi(v_y^2) \equiv \phi_y(v_y)$ and $\psi(v_z^2) \equiv \phi_z(v_z)$. Therefore, (5.3) reads $F(v_x^2 + v_y^2 + v_z^2) = \psi(v_x^2)\psi(v_y^2)\psi(v_z^2)$ or

$$F(x + y + z) = \psi(x)\psi(y)\psi(z).\tag{5.4}$$

Let us write $\psi(0) = a$. Then, $F(x) = a^2\psi(x)$, etc., so

$$F(x + y) = a\psi(x)\psi(y) = F(x)F(y)/a^3.\tag{5.5}$$

---

[3] **Equilibrium** What is *equilibrium*? It is a state reached by a gas (in the present case) isolated in a box sufficiently long after its preparation. There is no macroscopic flow in it and the gas is spatially uniform and time-independent (if observed at the macroscale).

Let $G(x) = \log[F(x)/a^3]$. Then, (5.5) reads

$$G(x + y) = G(x) + G(y), \tag{5.6}$$

which is called the *Cauchy functional equation*, whose general solution is $G(x) = -cx$, where c is a constant, if we assume G is continuous.[4*] Therefore, $F(x) \propto e^{-cx}$. That is,

$$f(v) \propto e^{-cv^2}. \tag{5.7}$$

We should not forget, however, that Maxwell actually did not like this derivation that assumed statistical independence of three orthogonal directions. He re-derived it later with a different logic. We will do so as well.

5.3 Gaussian Integral

We must compute the normalization constant and c (written as $1/2\sigma^2$ below with $\sigma > 0$) in (5.7). An easy way to compute the normalization constant is the following elegant method. Since the integral is positive, let us compute the square of what we want:

$$\left[\int_{-\infty}^{\infty} dx \, e^{-x^2/2\sigma^2} \right]^2 = \int_{-\infty}^{\infty} dx \, e^{-x^2/2\sigma^2} \int_{-\infty}^{\infty} dy \, e^{-y^2/2\sigma^2} = \int_{\mathbb{R}^2} dx dy \, e^{-(x^2+y^2)/2\sigma^2}, \tag{5.8}$$

because integration variables are dummy variables. Now, we go to the polar coordinates $(x, y) \rightarrow (r, \theta)$:

$$\int_{\mathbb{R}^2} dx dy \, e^{-(x^2+y^2)/2\sigma^2} = 2\pi \int_0^{\infty} e^{-r^2/2\sigma^2} r dr = 2\pi \int_0^{\infty} dz \, e^{-z/\sigma^2} = 2\pi\sigma^2. \tag{5.9}$$

Hence,

$$\int_{-\infty}^{\infty} dx \, e^{-x^2/2\sigma^2} = \sqrt{2\pi}\sigma. \tag{5.10}$$

5.4 Gaussian Distribution Function and Maxwell Distribution

The *Gaussian density distribution function* $g(x)$ generally has the following form:

$$g(x) = \frac{1}{\sqrt{2\pi}\sigma} e^{-(x-m)^2/2\sigma^2}, \tag{5.11}$$

[4*]**How to solve Cauchy's functional equation** $G(2x) = 2G(x)$ is immediately obtained from (5.6). Repeating this, we get $G(nx) = nG(x)$ for positive integer n. This implies $nG(1/n) = G(1)$ or $G(1/n) = G(1)/n$. Therefore, $G(m/n) = mG(1/n) = (m/n)G(1)$ for any positive integers m and n. Also, $G(0) = 2G(0)$, so $G(0) = 0$. This with (5.6) implies $G(-x) = -G(x)$. Thus, we have demonstrated that for $q \in \mathbb{Q}$ (rational numbers) $G(q) = -cq$, where $c = -G(1)$ is a constant. If we assume G is continuous, $G(x) = -cx$ must hold for any real x.

where $E(x) = m$ and $V(x) = \sigma^2$; we can write down a Gaussian (density) distribution function, if we know its expectation value and variance.

Since $\langle v_x \rangle = 0$ and $V(v_x) = (2/m)(k_B T/2) = k_B T/m$ (thanks to the equipartition of kinetic energy; see **2.7**), we have

$$\phi(v_x) = \sqrt{\frac{m}{2\pi k_B T}} e^{-mv_x^2/2k_B T}. \tag{5.12}$$

That is,

$$f(v) = \left(\frac{m}{2\pi k_B T}\right)^{3/2} e^{-mv^2/2k_B T}. \tag{5.13}$$

This is *Maxwell's distribution* function (actually, a density distribution function). The reader must be able to compute various probabilities and expectation values with the aid of Maxwell's distribution.

5.5* Moment Generating Function

At this juncture, let us practice a basic trick. The expectation value of $e^{\alpha x}$, where α is generally a complex number, is called a *moment generating function*:

$$\langle e^{\alpha x} \rangle = \frac{1}{\sqrt{2\pi}\sigma} \int_{-\infty}^{\infty} dx \, e^{\alpha x - (x-m)^2/2\sigma^2}. \tag{5.14}$$

If $\alpha = -s$, it is the Laplace transform of the distribution function; if $\alpha = ik$, where k is real, $\langle e^{ikx} \rangle$ is the Fourier transform of the density distribution, and is called the *characteristic function*.

The standard trick to compute this integral is to complete the square as follows (to rewrite the quadratic form $ax^2 + bx + c$ as $A(x-B)^2 + C$):

$$\alpha x - \frac{1}{2\sigma^2}(x-m)^2 = \alpha(x-m) + \alpha m - \frac{1}{2\sigma^2}(x-m)^2 = -\frac{1}{2\sigma^2}(x - m - \sigma^2\alpha)^2 + \frac{\sigma^2\alpha^2}{2} + \alpha m. \tag{5.15}$$

Therefore, (5.14) can be rewritten as

$$\langle e^{\alpha x} \rangle = \frac{1}{\sqrt{2\pi}\sigma} \int_{-\infty}^{\infty} dx \, e^{-(x - m - \sigma^2\alpha)^2/2\sigma^2 + \sigma^2\alpha^2/2 + \alpha m}. \tag{5.16}$$

We shift the integration range by $m + \sigma^2\alpha$ (or we introduce a new integration variable $x' = x - m - \sigma^2\alpha$ and rewrite the integral). Then, the integration just gives the normalization factor, so

$$\langle e^{\alpha x} \rangle = \frac{1}{\sqrt{2\pi}\sigma} \int_{-\infty}^{\infty} dx \, e^{-x^2/2\sigma^2} e^{\sigma^2\alpha^2/2 + \alpha m} = e^{\sigma^2\alpha^2/2 + \alpha m}. \tag{5.17}$$

Notice that

$$\frac{d}{d\alpha} \langle e^{\alpha x} \rangle \bigg|_{\alpha=0} = E(x) = m, \tag{5.18}$$

and

$$\frac{d^2}{d\alpha^2}\langle e^{\alpha(x-m)}\rangle\bigg|_{\alpha=0} = V(x) = \sigma^2. \qquad (5.19)$$

Up to this point our argument is in 3-space, but the general $\mathfrak{d}$-Maxwell distribution[5] should be obtained easily by multiplying $\mathfrak{d}$ 1D Maxwell results (5.12).

5.6 Bernoulli Revisited

Using Maxwell's distribution function $f(\mathbf{v})$, let us review Bernoulli's kinetic interpretation of pressure. The (kinetic interpretation of) pressure on the wall is the average momentum (impulse) given to the wall per unit time and area. Consider the wall perpendicular to the x-axis (just as was in the elementary discussion in Chapter 2; see especially (i)–(iii) in **2.6**: Daniel Bernoulli's Kinetic Theory). Then, the number of particles with its x-component of the velocity being between v_x and $v_x + dv_x$ that can impinge on the unit area on the wall per unit time is given by

$$nv_x dv_x \int_{-\infty}^{\infty} dv_y \int_{-\infty}^{\infty} dv_z f(\mathbf{v}), \qquad (5.20)$$

where n is the number density of the gas molecules. Each particle gives the momentum $2mv_x$ upon collision to the wall, so the pressure, P, is given by

$$P = \int_{v_x \geq 0} d^3v\, 2mnv_x^2 f(\mathbf{v}) = \int_{\text{all } \mathbf{v}} d^3v\, mnv_x^2 f(\mathbf{v}) = mn\langle v_x^2 \rangle = \frac{1}{3}\frac{N}{V}m\langle v^2 \rangle, \qquad (5.21)$$

where we have used the symmetry $f(\mathbf{v}) = f(-\mathbf{v})$, and the isotropy, $\langle v_x^2 \rangle = \langle v_y^2 \rangle = \langle v_x^2 \rangle = (1/3)\langle v_x^2 + v_y^2 + v_x^2 \rangle$. We have arrived at

$$PV = \frac{2}{3}N\langle K \rangle = Nk_B T, \qquad (5.22)$$

where K is the kinetic energy of the single gas particle, and N the number of particles in the volume V. Thus, Bernoulli's result has been derived once more (but rather mechanically).

Important Remark. Although the Boyle–Charles law is obtained, this does not tell us anything about molecules, because we do not know N. We cannot tell the mass of the particle, either. Remember that k_B was not known when the kinetic theory of gases was being developed.

A notable point is that with empirical results that can be obtained from strictly equilibrium and macroscopic studies, we cannot tell anything about molecules, even about their existence. Remember that the law of combined volumes for chemical reactions (**2.5**(iii)) crucial to demonstrating the particular nature of chemical substances is about (often violent) irreversible processes converting the reactants into the products. If we

---

[5] $\mathfrak{d}$: **spatial dimensionality notation** Throughout this book $\mathfrak{d}$ (lower case d in Fraktur) is used to indicate the spatial dimensionality to distinguish it from differential d, D (diffusion constant), $\mathcal{D}$ (density of state), etc. The reader simply may use d instead if there is no confusion.

make a tiny hole through the wall of the container, we could make a molecular beam, so in principle, we can measure the particle speed, but only through creating an extreme nonequilibrium (and not macroscopic) situation.

The (root-mean-square) speed of the molecules can be computed correctly, however, because we only need PV/Nm, where Nm is the total mass of the gas. Thus, in 1857, Clausius calculated the speed of molecules at $0\,°C$: oxygen 461 m/s, nitrogen 492 m/s, and hydrogen 1844 m/s. Notice, that these speeds are not very different from the sound speeds in respective gases.[6]

5.7 How to Derive the Boltzmann Factor

Let us derive Maxwell's distribution in a more elementary fashion, following Feynman. We start with the derivation of the *Boltzmann factor*. This first part should be clearly understood. The rest may require some maturity of the reader; elementary approaches need not be very simple, so do not worry if you do not understand the derivation too well.

Take a vertical column of gas with cross section A in the gravitational field just around us (Fig. 5.2). Consider the force balance on the slice between heights h and $h + dh$ of a cylinder. If n is the number density and m the mass of the molecule, we have as a force balance equation with the aid of $P = nk_BT$

$$n \times Adh \times mg = -AdP = -Ak_BTdn \tag{5.23}$$

or

$$\frac{dn}{dh} = -\beta nmg, \tag{5.24}$$

where $\beta = 1/k_BT$ (a standard abbreviation we will use often). That is,

$$n = n_0e^{-\beta mgh}, \tag{5.25}$$

where n_0 is a positive constant corresponding to n at $h = 0$. This can describe the *sedimentation equilibrium* of colloidal particles.

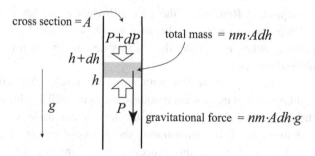

Fig. 5.2   Force balance along a gas column in gravity (see Equation 5.23).

[6] Sound speeds (in the standard state): oxygen 317 m/s, nitrogen 337 m/s, hydrogen 1270 m/s. These are about 2/3 of the molecular speeds. For ideal gases, this ratio is exact.

This equation suggests how the (relative) number of molecules depends on the potential energy difference. More generally, the same logic derives for a conserved force with potential U

$$n(r) = n(0)e^{-\beta[U(r)-U(0)]}. \tag{5.26}$$

That is, the factor $e^{-\beta U}$ called the *Boltzmann factor* tells us the ratio of particle densities (or the probabilities to find particles) at different locations, when there is a position-dependent potential energy U.[7]

☐ **Remark**: We have assumed that in equilibrium the temperature is everywhere the same independent of the height along the column. Is this correct? A more detailed argument can establish that the temperature must be uniform throughout the column in equilibrium.

5.8 Elementary Derivation of Maxwell's Distribution

Consider a column of an ideal gas which is in equilibrium with gravity. Let $n_{>u}(h)$ be the number of particles with $v_z > u > 0$ passing through the unit area at height h upward per second (Fig. 5.3). Since stationarity of the distribution implies $n_{>u}(0) = n_{>0}(h)$, if $mgh = (1/2)mu^2$,

$$\frac{n_{>0}(h)}{n_{>0}(0)} = \frac{n_{>u}(0)}{n_{>0}(0)}. \tag{5.27}$$

Since $n_{>0}(h)/n_{>0}(0) = n(h)/n(0)$, we can use the Boltzmann factor just obtained:[8]

$$\frac{n_{>u}(0)}{n_{>0}(0)} = e^{-\beta mgh} = e^{-\beta mu^2/2}. \tag{5.28}$$

Let $n(0, u)du$ be the number density of particles at height 0 with the z-component of the velocity in $(u, u+du]$. Notice that more numerous faster particles pass height 0 than slower ones, so we must take care of the speed in the z-direction:

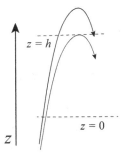

Fig. 5.3 If $mu^2/2 = mgh$, then $n_{>u}(0) = n_{>0}(h)$.

---

[7] This is true even if U is extremely complicated. Thus, even if U is not due to an external effect but due to other molecules in the system, this relation holds.

[8] The derivation is for the case without collisions, but since we have only to track energies that are conserved, collisions do not change the situation at all.

$$n_{>u}(0) = \int_u^\infty u'\, n(0, u')\, du' \propto e^{-\beta m u^2/2}. \tag{5.29}$$

Differentiating this equation, we obtain $n(0, u) \propto e^{-\beta m u^2/2}$. Thus, we have derived Maxwell's distribution again.

Problems

Q5.1 Doppler Observation of Maxwell's Distribution

A gas at temperature T consists of atoms that can emit light of wavelength $\lambda_0 = 2\pi c/\omega_0$. This line spectrum is observed through a window perpendicular to the x-axis. Since there is a distribution of the x-component of the velocity, the line broadens due to the Doppler shift (Doppler broadening): $\omega = \omega_0(1 + v_x/c)$. Thus, observing the line shape, we can get the distribution of v_x. Assuming that the gas is in equilibrium, find the density distribution function of the frequency $I(\omega)$.

Solution

The density distribution function of v_x is the Maxwell distribution:

$$f(v_x) = \left(\frac{m}{2\pi k_B T}\right)^{1/2} e^{-m v_x^2/2k_B T}. \tag{5.30}$$

From the Doppler formula $v_x = (\omega - \omega_0)c/\omega_0$, so

$$I(\omega) = \left(\frac{m}{2\pi k_B T}\right)^{1/2} e^{-mc^2(\omega-\omega_0)^2/2k_B T\omega_0^2}(dv_x/d\omega) \tag{5.31}$$

$$= \left(\frac{mc^2}{2\pi k_B T\omega_0^2}\right)^{1/2} e^{-mc^2(\omega-\omega_0)^2/2k_B T\omega_0^2}. \tag{5.32}$$

Perhaps, ignoring the normalization and then calculating the normalization factor later may be simpler. In the actual experiment, collisions among molecules can blur the line shape, so dilute gas is desirable.[9]

Q5.2 The Mode of the Molecular Speed

Consider a $\eth$-dimensional ideal gas ($\eth > 1$). That is, the velocity is a $\eth$-vector $v = (v_1, v_2, \ldots, v_\eth)$, and the kinetic energy of a single particle is given by $(m/2)v^2 = (m/2)(v_1^2 + v_2^2 + \cdots + v_\eth^2)$, where m is the mass of the particle.

(1) Write down the $\eth$-dimensional Maxwell distribution function (i.e., find the density distribution function of the velocity in $\eth$-space).

---

[9] The method can be used to measure k_B. See Daussy, C., et al. (2007). Direct determination of the Boltzmann constant by an optical method, *Phys. Rev. Lett.*, **98**, 250801.

(2) What is the most probable speed v_∂ in ∂-space? That is, what is the mode speed (the speed for which the density distribution function for the speed is maximal)?

(3) What is the ratio of v_∂ obtained in (2) and the root-mean-square velocity in ∂-space in the $\partial \to \infty$ limit?

Solution

(1) As the reader can guess from the 3D result, we have only to multiply ∂ 1D results as

$$f(v) = \left(\frac{m}{2\pi k_B T}\right)^{\partial/2} e^{-m|v|^2/2k_B T}. \tag{5.33}$$

(2) We need the density distribution function $F(v)$ of the speed $v = |v|$. We should go to the polar coordinate system in ∂-space:

$$d^\partial v = S_{\partial-1} v^{\partial-1} dv, \tag{5.34}$$

where $S_{\partial-1}$ is the volume of the $\partial - 1$-unit sphere (corresponding to 4π in 3-space), but we do not need its explicit form (see **Q17.5**, however). We get

$$F(v) = S_{\partial-1} \left(\frac{m}{2\pi k_B T}\right)^{\partial/2} v^{\partial-1} e^{-mv^2/2k_B T}. \tag{5.35}$$

This is the density distribution function for the speed. To find its peak, we have only to maximize $mv^2/2k_B T - (\partial - 1)\log v$:

$$\frac{mv}{k_B T} - (\partial - 1)\frac{1}{v} = 0, \tag{5.36}$$

or $v^2 = (\partial - 1)k_B T/m$. That is,

$$v_\partial = \sqrt{\frac{(\partial - 1)k_B T}{m}}. \tag{5.37}$$

(3) The equipartition of kinetic energy tells us that

$$\left\langle \frac{1}{2}mv^2 \right\rangle = \frac{m\partial}{2}\langle v_x^2 \rangle = \frac{\partial}{2}k_B T. \tag{5.38}$$

That is, the root-mean-square velocity is $\sqrt{\partial k_B T/m}$. The ratio obviously converges to unity.

Q5.3 Density Distribution of Relative Velocity, an Elementary Approach[10]

(1) There are two particles 1 and 2 in an equilibrium pure ideal gas. Write down the simultaneous density distribution function $f(v_1, v_2)$ of their velocities v_1 and v_2. The reader may assume the temperature of the gas is T, and the mass of the individual particles is m.

(2) Now, introduce the velocity V of the center of mass of these two particles and the relative velocity $w = v_1 - v_2$. Write down the density distribution function of w.

---

[10] Here, our approach is elementary, but the reader can use the δ-function technique that will be explained in Chapter 6.

Solution

(1) Maxwell's (density) distribution function $f(v)$ implies

$$P(d^3 v) = f(v)d^3 v, \tag{5.39}$$

where $P(d^3 v)$ is the probability to find a particle with the velocity in a volume element $d^3 v$ of the velocity space. Since we know two particles are statistically independent,

$$P(d^3 v_1, d^3 v_2) = P(d^3 v_1)P(d^3 v_2), \tag{5.40}$$

the density $f(v_1, v_2)$ must be a product of two Maxwellian distributions. Therefore,

$$f(v_1, v_2) = \left(\frac{m}{2\pi k_B T}\right)^3 e^{-m(v_1^2 + v_2^2)/2k_B T}. \tag{5.41}$$

(2) $v_1 = V + w/2$ and $v_2 = V - w/2$, so

$$v_1^2 + v_2^2 = 2V^2 + \frac{1}{2}w^2. \tag{5.42}$$

Since we are computing the density distribution $g(V, w)$, we must demand

$$f(v_1, v_2)d^3 v_1 d^3 v_2 = g(V, w)d^3 V d^3 w, \tag{5.43}$$

or

$$f(v_1, v_2)\left|\frac{\partial(v_1, v_2)}{\partial(V, w)}\right| = g(V, w). \tag{5.44}$$

The Jacobian appearing in (5.44) is unity, so

$$g(V, w) = \left(\frac{m}{2\pi k_B T}\right)^3 e^{-(2mV^2 + (m/2)w^2)/2k_B T}. \tag{5.45}$$

Notice that the center of mass kinetic energy is $(1/2)(2m)V^2$, and the kinetic energy of the relative motion is $(1/2)(m/2)w^2$, where $m/2$ is the reduced mass. We can read them off from (5.45).

The marginal distribution $g(w)$ is obtained by integrating V out, or simply splitting $g(V, w)$ using statistical independence of V and w:

$$g(w) = \left(\frac{m}{4\pi k_B T}\right)^{3/2} e^{-mw^2/4k_B T}. \tag{5.46}$$

Q5.4 Probability of the Kinetic Energy Larger than the Average

We know, in equilibrium, the mean kinetic energy is given by the equipartition of energy. What is the probability of a particle to have the kinetic energy K more than the average kinetic energy $3k_B T/2$ (in equilibrium)? Before any calculation discuss whether the probability is less than 1/2 or not.

Solution

K can be indefinitely large and its density distribution has only one peak, so the probability should be less than $1/2$.

$$P \equiv P(K \geq 3k_B T/2) = \langle \chi_{\{v:v^2 \geq 3k_B T/m\}} \rangle = \int_{\{v:v^2 \geq 3k_B T/m\}} f(v) d^3 v, \qquad (5.47)$$

where f is Maxwell's distribution. Therefore,

$$P = 4\pi \int_{\sqrt{3k_B T/m}}^{\infty} \left(\frac{m}{2\pi k_B T} \right)^{3/2} e^{-mv^2/2k_B T} v^2 dv = \frac{4}{\sqrt{\pi}} \int_{\sqrt{3/2}}^{\infty} e^{-x^2} x^2 dx \simeq 0.39. \quad (5.48)$$

6   Delta Function and Density Distribution Functions

Summary

∗ Although slightly advanced, it is a good occasion to learn how to use the δ-function. For one-dimension $\int dx\, \delta(x)\varphi(x) = \varphi(0)$ for a continuous function φ.
∗ Practically important formulas are

$$\delta(\alpha(x - b)) = \delta(x - b)/|\alpha|,$$
$$\delta(x^2 - a^2) = [\delta(x - a) + \delta(x + a)]/2|a|.$$

∗ The density distribution function $f(y)$ of a stochastic variable $y = \varphi(X)$ reads $f(y) = \langle \delta(y - \varphi(X)) \rangle_X$.

Key word

δ-function

What the reader should be able to do:

∗ Intuitively explain what the δ-function is.
∗ Use $f(y) = \langle \delta(y - \varphi(X)) \rangle_X$ (to compute, say, the density distribution of the kinetic energy).

How to Read this Chapter

Now, we wish to go to a technical topic that will be crucial in more advanced physics, and also that makes many calculations of (density) distributions quite mechanical.

With this in mind the reader can choose to go on to the next chapter, after understanding what the δ-function is like, without following actual calculations of density distribution functions.

6.1   What Average Gives the Density Distribution?

We have learned (in Chapter 3) that the probability $P(A)$ is given by the expectation value of the indicator χ_A (recall **3.13**) of event A:

$$P(A) = \langle \chi_A \rangle_P. \tag{6.1}$$

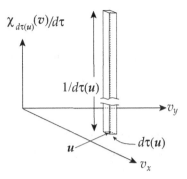

Fig. 6.1   Here $\chi_{d\tau(u)}(v)/d\tau(u)$, where the indicator of the volume element is concentrated around u. Its value is $1/d\tau(u)$ on the volume element around u, but otherwise zero.

We also know (see (5.1)) that the density distribution function may be written as

$$f(u) = \frac{P(d\tau(u))}{d\tau(u)}. \tag{6.2}$$

Here, $d\tau(u)$ is the volume element located at u (and also denotes its volume in the denominator). Combining these two relations, we can formally[1] write

$$f(u) = \left\langle \frac{\chi_{d\tau(u)}(v)}{d\tau(u)} \right\rangle_v. \tag{6.3}$$

Here, $\chi_{d\tau(u)}$ is the indicator of the volume element $d\tau(u)$ around u that is a particular location in the velocity space (see Fig. 6.1). In other words, scanning the velocity space with a scanner, if its location v is in $d\tau(u)$, $\chi_{d\tau(u)} = 1$, otherwise, 0.

6.2 Introducing the δ-function

How does the quantity $\chi_{d\tau(u)}(v)/d\tau(u)$ we have formally obtained look like as a function of v? See Fig. 6.1 for the two-dimensional case. In the figure the infinitesimal volume element $d\tau(u)$ is depicted as a tiny finite square for illustration's sake.

Informally,

$$\frac{\chi_{d\tau(u)}(v)}{d\tau(u)} = \begin{cases} 1/d\tau(u) = \infty, & \text{if } v \in d\tau(u). \\ 0, & \text{otherwise.} \end{cases} \tag{6.4}$$

Here, ∞ appears because $d\tau(u)$ is infinitesimally small. Then, following Dirac, let us introduce the δ-function (*delta function*) (in $\eth$-space[2]) concentrated at u as (here v is the running variable, and u is a fixed constant vector; see Fig. 6.1):

---

[1] **Formally** This word is used to indicate that the formula holds or is derivable, if we follow the mechanical rules of computation without mathematically justifying every step. It does not mean the result is "officially sanctioned." It is a standard usage in mathematics. It is often the case that formal results have already been or will be mathematically justified, if the results look natural and useful.

[2] **$\eth$-object** As noted already, throughout this book "$\eth$-object" means $\eth$-dimensional object, following the usual practice in mathematics.

$$\chi_{d\tau(u)}(v) = \delta(v-u)\, d^{\partial}v = \begin{cases} 0, & \text{if } v \neq u, \\ 1, & \text{if } v = u. \end{cases} \tag{6.5}$$

This implies that for any continuous function φ of v

$$\int \varphi(v)\delta(v-u)\, d^{\partial}v = \varphi(u). \tag{6.6}$$

As can be seen in Appendix 6A mathematically we may use this as the definition of the ∂-dimensional δ-function. Intuitively, as a function of v the reader may imagine $\delta(v-u)$ as an infinitely thin but infinitely long needle vertically standing at $v = u$ whose total volume is unity.

6.3 A Formal Expression of Density Distribution

Suppose we know the probability measure P for a vector v. Then the density distribution function $f(v)$ for v at $v = u$ may be written as (or (6.3) can be written as):

$$f(u) = \langle \delta(v-u) \rangle_P = \langle \delta(u-v) \rangle_P, \tag{6.7}$$

where subscript P denotes the distribution (the probability measure) for v. Notice that "δ-function" may be regarded as an even function (see Fig. 6.3 later). We must keep in mind with respect to what variable we are taking the expectation value. In this case the averaging is with respect to v. Here u is a mere parameter while we compute the expectation value with respect to v.

We wish to obtain the density distribution functions of various functions of v. To this end we must understand the δ-function a bit more.

6.4 Distribution Function: The 1D Case

Let us look at the 1D case in more detail.

Let $X(\omega)$ be a real stochastic variable defined on a probability space (Ω, P) (see **3.10**). The probability, denoted here by $F(x)$, for the event $X \leq x$ is called the *distribution function* of X:[3*]

$$F(x) = P(X \leq x). \tag{6.8}$$

The indicator of the event $X \leq x$ may be written in terms of the unit step function Θ:

$$\Theta(y) = \begin{cases} 1, & \text{for } y \geq 0, \\ 0, & \text{for } y < 0 \end{cases} \tag{6.9}$$

[3*]In the following expression, if the reader wishes to be very explicit, write $P(X \leq x) = P(\{\omega \,|\, X(\omega) \leq x, \omega \in \Omega\})$. That is, the right-hand side reads the probability measure P ($\simeq$ the normalized volume) of the subset of the sampling space Ω satisfying the condition $X(\omega) \leq x$ (i.e., the totality of the elementary events ω satisfying this condition).

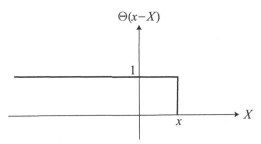

$\Theta(x-X)$

Illustration of $\Theta(x - X)$. Notice that the plot is with respect to X; x is regarded as a parameter in (6.10), because averaging is over X.

as $\Theta(x - X)$, so we have:[4*]

$$F(x) = \langle \Theta(x - X) \rangle. \tag{6.10}$$

Here, the average is over X.

See Fig. 6.2 for an illustration of $\Theta(x - X)$.

6.5 Density Distribution Function: The 1D Case

The derivative $f(x)$ of $F(x)$ is called the *density distribution function* of X. Using the fundamental theorem of calculus, we can write as

$$F(x) = \int_{-\infty}^{x} f(x)dx. \tag{6.11}$$

Physicists often call f the distribution function.[5]

If we can "differentiate" the step function $\Theta(x)$, we should be able to write the density distribution function in a neat form. An analogy to (6.7) suggests

$$f(x) = \langle \delta(x - X) \rangle_P, \tag{6.12}$$

using the delta function adapted to 1D. Needless to say, here the average is with respect to X. Therefore, $\delta(x) = \Theta'(x)$ must be true.

6.6 The 1D δ-function

Indeed, $\delta(x) = \Theta'(x)$ is true. The step function $\Theta(x)$ is constant except at $x = 0$, so its "derivative" $\delta(x) = 0$ if $x \neq 0$. However, it cannot be 0 everywhere. It is impossible to "differentiate a vertical wall," but we intuitively see it to be $+\infty$; $\delta(0) = \infty$. The area

---

[4*] If the reader wishes to be more explicit, write $\langle \Theta(x-X) \rangle = \langle \Theta(x - X(\omega)) \rangle$. The average is over random events $\omega \in \Omega$.

[5] Therefore, to distinguish it from the more standard usage of the terminology F is sometimes called the *cumulative distribution function*.

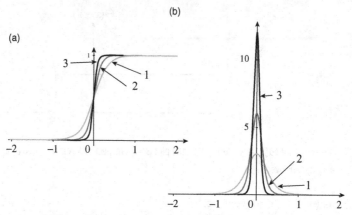

Fig. 6.3 The derivatives (b) of increasingly steep cliffs 1 → 3 (a). Gray levels are correspondent. Imagining such a limiting process, we can understand the δ-function as the derivative of the step function. Notice that the areas below the graphs in b are always unity.

between $\delta(x)$ and the x-axis must be unity, because Θ jumps exactly by 1 at $x = 0$. The reader may intuitively imagine an infinitely long needle at $x = 0$ whose "area" is $0 \times \infty =$ unity as noted in the $\eth$-dimensional general case earlier (recall Fig. 6.1).[6] Our intuition, just explained, may be illustrated as in Fig. 6.3, which may be mathematically justified as explained in Appendix 6A.

Therefore, for any continuous function $\varphi(x)$ on the real line (see (6.6))

$$\int_{-\infty}^{+\infty} \varphi(x)\delta(x - a)dx = \varphi(a). \tag{6.13}$$

Thus, the Fourier transformation of $\delta(x)$ is

$$\int_{-\infty}^{+\infty} e^{ikx}\delta(x)dx = 1. \tag{6.14}$$

Inverse Fourier transforming this, we get[7]

$$\delta(x) = \frac{1}{2\pi} \int_{-\infty}^{+\infty} e^{-ikx}dk. \tag{6.15}$$

This is perhaps the most useful identity in theoretical physics (cf. **A.4**).

6.7 Key Formula 1

Let $\alpha \, (\neq 0)$ be a real number. We have the following important identity:

$$\delta(\alpha(x - a)) = \frac{1}{|\alpha|}\delta(x - a). \tag{6.16}$$

---

[6] **Theory of distributions** The reader would say this cannot be mathematics. However, we can almost completely rationalize our "intuitive picture" with the aid of the theory of distributions due to Schwartz (see Appendix 6A for the rudiments).

[7] All these formal calculations are justified by the theory of distribution. See Appendix 6A for its rudimentary portion.

This may be shown as follows:

$$\int_{-\infty}^{\infty} dx\, \delta(\alpha x - \alpha a)\varphi(x) = \int_{-\infty}^{\infty} d(y/|\alpha|)\, \delta(y - \alpha a)\varphi(y/\alpha) = \frac{1}{|\alpha|}\varphi(a). \tag{6.17}$$

6.8 Key Formula 2

How about $\delta(g(x))$ for a differentiable function g? Suppose x_0 is a simple real zero of $g(x)$. That is, assume $g(x) \simeq g'(x_0)(x - x_0)$ near x_0. Then, (6.16) suggests that near $x = x_0$, $\delta(g(x))$ must be

$$\delta(g(x)) = \frac{1}{|g'(x_0)|}\delta(x - x_0). \tag{6.18}$$

There might be more than one simple real zero of $g(x)$ (i.e., $g(r_1) = g(r_2) = \cdots = 0$). Thus, we obtain

$$\delta(g(x)) = \sum_i \frac{1}{|g'(r_i)|}\delta(x - r_i) = \frac{1}{|g'(x)|}\sum_i \delta(x - r_i), \tag{6.19}$$

where the summation is over all the simple real zeros of g. For example,

$$\delta(x^2 - a^2) = \frac{1}{2|a|}\big[\delta(x - a) + \delta(x + a)\big]. \tag{6.20}$$

6.9* Some Practice

Let us have some practice:

$$\int_{-1}^{1} dx\, \delta(\pi - 6x)\cos x = \frac{1}{6}\cos\frac{\pi}{6} = \frac{\sqrt{3}}{12}, \tag{6.21}$$

$$\int_{3}^{4} dx\, \delta(\pi - 6x)\cos x = 0, \tag{6.22}$$

$$\int_{-\infty}^{\infty} dx\, \delta\big(x^2 - 3x - 10\big)x^3 = \frac{1}{7}(125 - 8) = \frac{117}{7}. \tag{6.23}$$

6.10 How to Compute Density Distribution Functions

Suppose we know the density distribution function $g(X)$ of a stochastic variable X. What is the density distribution function $f(x)$ of $x = \varphi(X)$? Let us go back to our starting point. According to (6.3), since the fixed location for x is $\varphi(X)$

$$f(x) = \left\langle \frac{\chi_{dx}(\varphi(X))}{dx} \right\rangle_X = \langle \delta(x - \varphi(X)) \rangle_X = \int dX \, g(X) \delta(x - \varphi(X)), \tag{6.24}$$

where the integration range is the domain on which g is defined (the support of g). Equation (6.24) is a very useful formula as we will see next.

6.11* Kinetic Energy Density Distribution of 2D Ideal Gas

Let us obtain the density distribution function of the kinetic energy of a 2D gas. Our starting point is the formal expression (6.24) of the density distribution:

$$f(K) = \langle \delta(K - mv^2/2) \rangle_v, \tag{6.25}$$

where the average is over the 2D Maxwell distribution. Therefore, we must calculate

$$f(K) = \int_{\mathbb{R}^2} dv_x dv_y \left(\frac{m}{2\pi k_B T} \right) e^{-m\left(v_x^2 + v_y^2\right)/2k_B T} \delta\left((m/2)\left(v_x^2 + v_y^2\right) - K \right). \tag{6.26}$$

Rewrite this into a 1D integral with the aid of the polar coordinates (v, θ). We know the distribution must be isotropic, so we can integrate over the direction θ to get 2π. We get

$$f(K) = 2\pi \left(\frac{m}{2\pi k_B T} \right) \int_0^\infty v \, dv \, \delta\left(mv^2/2 - K\right) e^{-mv^2/2k_B T} \tag{6.27}$$

$$= \frac{m}{k_B T} \int_0^\infty v \, dv \, \frac{1}{mv} \delta\left(v - \sqrt{2K/m}\right) e^{-mv^2/2k_B T} = \left(\frac{1}{k_B T} \right) e^{-K/k_B T}. \tag{6.28}$$

We have used the key formula 2 or (6.19). Do not forget that $v > 0$.

Problems

Q6.1 Density Distribution Function of Speed

(1) We already computed this in **Q5.2**, but redo this with the aid of the δ-function.
(2) Compute $\langle v \rangle \langle 1/v \rangle$ in 3-space. Before any explicit computation show that this cannot be smaller than 1 in any $\mathfrak{d}(> 1)$-space.

Solution

(1) The density distribution function $f(u)$ of the speed $u = |v|$ can be written as

$$f(u) = \langle \delta(u - |v|) \rangle = \int_{\mathbb{R}^\mathfrak{d}} d^\mathfrak{d} v \, \delta(u - |v|) \left(\frac{m}{2\pi k_B T} \right)^{\mathfrak{d}/2} e^{-m|v|^2/2k_B T} \tag{6.29}$$

Remark To use the rules of the computation of the integral containing a δ-function, the independent variable in the δ-function (in our case $|v|$) must be the integration variable.

Thus, we must convert the δ-function or convert the integration variable (in our case, we convert the integration variable from $\boldsymbol{v}$ to $v = |\boldsymbol{v}|$).

Now we should go to the polar coordinate system in $\eth$-space:

$$d^{\eth} \boldsymbol{v} = S_{\eth-1} v^{\eth-1} dv, \tag{6.30}$$

where $S_{\eth-1}$ is the volume of the $\eth - 1$-unit sphere (see Q17.5). Integrating over the velocity, we get (u has been replaced by v in the following formula):

$$f(v) = S_{\eth-1} \left(\frac{m}{2\pi k_B T} \right)^{\eth/2} v^{\eth-1} e^{-mv^2/2k_B T}. \tag{6.31}$$

This is the density distribution function for the speed.

(2) Since v is non-negative, the Cauchy–Schwarz inequality applied to $\sqrt{v}$ and $\sqrt{1/v}$ tells us $1 \le \langle v \rangle \langle 1/v \rangle$. The rest is calculation: in 3-space we have

$$\langle v \rangle = \sqrt{\frac{8k_B T}{\pi m}}, \quad \langle 1/v \rangle = \sqrt{\frac{2m}{\pi k_B T}}. \tag{6.32}$$

Thus, $\langle v \rangle \langle 1/v \rangle = 4/\pi > 1$.

Q6.2 Density Distribution Function of Kinetic Energy in 3-space

Find the density distribution function for the kinetic energy E of a particle in a 3D equilibrium ideal gas.

Solution

Similar to the 2D case we already did:

$$f(E) = \left\langle \delta \left(E - mv^2/2 \right) \right\rangle = \int_0^{\infty} 4\pi v^2 dv \, \delta \left(E - mv^2/2 \right) \left(\frac{m}{2\pi k_B T} \right)^{3/2} e^{-m|\boldsymbol{v}|^2/2k_B T}. \tag{6.33}$$

We use (recall **6.8**)

$$\delta \left(E - mv^2/2 \right) = \frac{1}{mv} \delta \left(v - \sqrt{2E/m} \right) \tag{6.34}$$

to get

$$f(E) = \int_0^{\infty} 4\pi v^2 dv \, \frac{1}{\sqrt{2mE}} \delta(v - \sqrt{2E/m}) \left(\frac{m}{2\pi k_B T} \right)^{3/2} e^{-m|\boldsymbol{v}|^2/2k_B T} \tag{6.35}$$

$$= 4\pi \frac{2E}{m} \frac{1}{\sqrt{2mE}} \left(\frac{m}{2\pi k_B T} \right)^{3/2} e^{-E/k_B T} = \frac{2}{k_B T} \sqrt{\frac{E}{\pi k_B T}} e^{-E/k_B T}. \tag{6.36}$$

Q6.3 Relative Velocity Distribution Revisited

Let us obtain the root-mean-square relative velocity of two molecules with different masses m and M in an equilibrium gas at temperature T.

(1) Using the delta function trick, we can write the density distribution function $f(w)$ for the relative velocity w as

$$f(w) = \left(\frac{m}{2\pi k_B T}\right)^{3/2} \left(\frac{M}{2\pi k_B T}\right)^{3/2} \int_{\mathbb{R}^3} d^3 v_1 \int_{\mathbb{R}^3} d^3 v_2 \, \delta(w - (v_1 - v_2)) e^{-mv_1^2/2k_B T - Mv_2^2/2k_B T}.$$
(6.37)

Perform the integration over v_2.

(2) Then, perform the integration over v_1 to obtain $f(w)$.

(3) Find $\langle w^2 \rangle$ and check that the answer agrees with the result obtained by the equipartition of energy.

Solution

(1) This is straightforward. By inspection, we get

$$f(w) = \left(\frac{m}{2\pi k_B T}\right)^{3/2} \left(\frac{M}{2\pi k_B T}\right)^{3/2} \int_{\mathbb{R}^3} d^3 v_1 \, e^{-mv_1^2/2k_B T - M(w - v_1)^2/2k_B T}.$$
(6.38)

(2) To perform the Gaussian integral we use the trick to complete the square (recall **5.5**):

$$mv_1^2 + M(w - v_1)^2 = (m + M)v_1^2 + Mw^2 - 2Mw \cdot v_1$$
(6.39)

$$= (m + M)\left(v_1 - \frac{M}{m + M}w\right)^2 + \frac{mM}{m + M}w^2.$$
(6.40)

Thus, we obtain

$$f(w) = \left(\frac{mM}{2(m + M)\pi k_B T}\right)^{3/2} e^{-\frac{mM}{m+M}w^2/2k_B T}.$$
(6.41)

Notice that the emergence of the reduced mass is quite natural.

(3) The expectation value can be read off from the formula as

$$\langle w^2 \rangle = 3\frac{m + M}{mM}k_B T.$$
(6.42)

The result is of course consistent with the elementary results as follows:

$$\langle w^2 \rangle = \langle v_1^2 \rangle + \langle v_2^2 \rangle = \frac{3k_B T}{m} + \frac{3k_B T}{M} = 3\frac{m + M}{mM}k_B T.$$
(6.43)

Appendix 6A Extreme Rudiments of Distributions

6A.1 Motivation of Theory: Delta Function as Linear Functional

Let $\mathcal{D}$ be a set of real-valued functions (called test functions) on $\mathbb{R}$. Let us define a map $T_\delta : \mathcal{D} \to \mathbb{R}$ as $T_\delta(f) = f(0)$. The most obvious and important property of T_δ is its linearity:

$$T_\delta(af + bg) = aT_\delta(f) + bT_\delta(g),$$
(6.44)

where $a, b \in \mathbb{R}$ and $f, g \in \mathcal{D}$. Therefore, we are tempted to write T_δ in terms of an integral with some integration kernel δ and call it the δ-function:

$$T_\delta(f) = \int dx\, \delta(x) f(x) = f(0),\tag{6.45}$$

where the integration range is the domain of the test functions (henceforth such statements will not be written explicitly). However, for T_δ there is no ordinary function δ satisfying this equality, because its "value" at 0 cannot be finite. Still, T_δ is well-defined. Therefore, we *define* δ through T_δ.

This motivates a whole class of mathematical objects called *generalized functions* (or distributions).

6A.2 Generalized Function

Let T_q be a linear functional defined on a set $\mathcal{D}$ of real-valued functions on $\mathbb{R}$. The formal symbol $q(x)$ such that for $f \in \mathcal{D}$

$$T_q(f) = \int dx\, q(x) f(x)\tag{6.46}$$

is called a *generalized function* (or a distribution). The following notation is also often used for convenience:

$$T_q(f) = \langle q, f \rangle.\tag{6.47}$$

The definition must include the rule for changing the independent variable: $s \to x = \varphi(s)$. Under this transformation $q(x) \to \varphi'(s) q(\varphi(s))$. The rule is exactly the same as in the case of ordinary functions and is motivated by

$$\int dx\, q(x) f(x) = \int ds\, \varphi'(s) q(\varphi(s)) f(\varphi(s)).\tag{6.48}$$

6A.3 Test Functions

The definition depends on the set $\mathcal{D}$, which is called the *test function set* and its elements are called *test functions* as noted already, because we can know the properties of a generalized function only integrating it with them.[8]

---

[8] **On test function sets** From the practitioner's point of view, we need not pay much attention to $\mathcal{D}$, but should remember that very often $\mathcal{D}$ is the set of all the C^∞-functions (infinite-times differentiable functions) with compact domains (i.e., C_0^∞) or the set of all the *functions of rapid decay* (or rapidly decreasing functions, *Schwartz-class functions*):

$$\mathcal{D} = \left\{ f : \mathbf{R} \to \mathbf{C},\ C^\infty \text{ such that } x^n f^{(r)}(x) \to 0 \text{ as } |x| \to \infty \text{ for } \forall n, r \in \mathbb{N} \right\}.$$

(In words, $\mathcal{D}$ consists of infinite times differentiable functions whose any derivative decays faster than any inverse power.) The generalized functions defined on this $\mathcal{D}$ are called *generalized functions of slow growth*.

6A.4 Equality of Generalized Functions

Two generalized functions p and q are said to be equal, if no test function can discriminate them:

$$T_p(f) = T_q(f) \text{ for all the test functions } f \in \mathcal{D} \iff p = q. \tag{6.49}$$

If a generalized function q is equal to some ordinary function, we say q is a *regular distribution*.

6A.5 δ-function: An Official Definition[9]

The symbol $\delta(x)$ such that

$$T_\delta(f) = \int \delta(x)f(x)dx = f(0) \tag{6.50}$$

is called the *δ-function*.

6A.6 Differentiation of Generalized Functions

If q is an ordinary differentiable function, then for $f \in \mathcal{D}$

$$\int q'(x)f(x)dx = -\int q(x)f'(x)dx. \tag{6.51}$$

The right-hand side makes sense, if f is differentiable. If we take C^∞ test functions, we may regard (6.51) as the definition of q'. That is, the derivative q' of a generalized function q is defined as the generalized function satisfying $\langle q',f \rangle = -\langle q,f' \rangle$. Show indeed that $\Theta'(x) = \delta(x)$ or

$$x\delta'(x) = -\delta(x). \tag{6.52}$$

6A.7 All the Ordinary Rules for Differentiation Survive

For example, the chain rule is applicable. When the reader feels uneasy in some use or abuse of generalized functions, always return to the definition (6.46).

6A.8 Value of Generalized Function at each Point is Meaningless

The value of a generalized function at a point is totally meaningless, because changing the value of a function at a point does not affect its integral. Therefore, although the δ-function was originally "defined" such that $\delta(x) = 0$ for $x \neq 0$, according to our official definition shown in **6.6**, we cannot mention anything about the value of $\delta(x)$ for any $x \in \mathbb{R}$. Consequently, the product of delta functions containing common variables

---

[9] It is often called the Dirac δ-function, ignoring the fact that this type of function has been used for well over a hundred years.

(e.g., $\delta(x)\delta(x-1)$) is a very dangerous concept. Needless to say, $f(g)$ does not usually make any sense for generalized functions f and g.

6A.9 Convergence of Generalized Function

A sequence of generalized functions q_n ($n = 1, 2, \ldots$) is said to converge to q, if for any test function f

$$\lim_{n \to \infty} \int dx\, q_n(x)f(x) = \int dx\, q(x)f(x). \tag{6.53}$$

We denote this simply as $\lim_{n \to \infty} q_n = q$. That is, the order of limit and integration can be freely exchanged. We can also show the order of limit and differentiation may be exchanged freely from the definition of this derivative.

6A.10 Definition of Generalized Function using Limits

The last observation allows us to introduce generalized functions using convergent sequences in the previous sense (with respect to a test function set). We already did this for the δ-function in Fig. 6.3; in formulas $(n/\sqrt{\pi})e^{-n^2x^2} \to \delta(x)$. The following sequences all define the delta function:

$$\varphi_n(x) = \frac{\sin nx}{\pi x} = \frac{1}{2\pi} \int_{-n}^{n} e^{ikx} dk = \frac{1}{\pi} \int_{0}^{n} \cos(xk) dk, \tag{6.54}$$

$$\varphi_n(x) = \frac{1}{2\pi} \frac{\sin\left[(n+1/2)x\right]}{\sin(x/2)} \quad \text{(the Dirichlet kernel)}, \tag{6.55}$$

$$\varphi_n(x) = \frac{1}{2n\pi} \left[\frac{\sin(nx/2)}{\sin(x/2)}\right]^2 \quad \text{(the Fejer kernel)}, \tag{6.56}$$

$$\varphi_n(x) = \sum_{j=-n}^{n} e^{2j\pi ix} \quad \text{for } x \in (-2\pi, 2\pi). \tag{6.57}$$

7 Mean Free Path and Diffusion

Summary

* Diffusion in gases is slow because molecular collisions make the trajectories of molecules zigzag.
* The length of the displacement vector of a molecule is reduced by a factor $\propto 1/\sqrt{\#(\text{collisions})}$ compared with the case without any collision.
* To describe the effect of molecular collisions, Clausius introduced the concept of the mean free path.
* Even though there are no forces acting on individual molecules, actual forces are required to suppress their natural motions (entropic force).

Key words

mean free path, random walk, entropic force

The reader should be able to:

* Understand the meaning and the estimation of the mean free path.
* Understand why the molecular displacement distance is drastically reduced by collisions.
* Understand the entropic force and its cause.

7.1 Mean Free Path

Dutch meteorologist Buys-Ballot (1817–1890) noticed that if the molecules of gases really moved as fast as Clausius estimated (see **5.6**), the mixing of gases by diffusion should have been much faster than we actually experienced.

Upon this criticism, Clausius (1858[1]) realized that the gas molecules have large enough diameters so a molecule cannot move very far without colliding with another one. Clausius defined a new parameter called the *mean free path* ℓ of a gas molecule that describes the average distance a molecule can run between two consecutive collisions.

---

[1] **Darwin and Boltzmann** The most important event in 1858 was that the idea of natural selection was officially published by Darwin and Wallace. Physicists should recognize that Boltzmann called the nineteenth century the century of Darwin (not of Maxwell) (see Broda, E. (1955). *Ludwig Boltzmann, Mensch, Physiker, Philosoph.* Wien: F. Deuticke. Part III)).

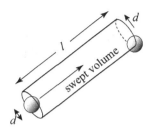

Fig. 7.1   Intuitive explanation of (7.1). The swept volume is illustrated.

We can compute ℓ using the idea of "swept volume": a moving molecule sweeps a volume in space that is cylindrical between collisions (see Fig. 7.1). If the diameter of the molecule is d and the intercollision path length ℓ, then this volume is $\pi d^2 \ell$ (since the collision cross section of the molecule is πd^2). The occurrence of one collision implies that this volume contains one other molecule. Therefore, the swept volume should satisfy $\pi d^2 \ell \times n \sim 1$, where n is the number density. Hence, we guess

$$\ell = 1/n\pi d^2, \tag{7.1}$$

if all other particles are fixed in space.

Actually, all the molecules are moving. When they collide, the average relative speed must be the relevant velocity, which is $\sqrt{2}$ times the mean velocity. That is, the molecule collides $\sqrt{2}$ times more often than the case where all other molecules are fixed in space. Therefore,

$$\ell = \frac{1}{\sqrt{2}\pi n d^2} \tag{7.2}$$

must be the true mean free path (see **Q7.1**, **Q7.2**). Needless to say, Clausius could not estimate it.[2]

The average time τ between successive collisions is called the *mean free time*. This may be estimated by dividing the mean free path with the mean speed:

$$\tau = \ell/v. \tag{7.3}$$

7.2 Effect of Molecular Collisions

Roughly speaking, in air at 1 atm at around room temperature the mean free time is of order ~ 0.1 ns. The moving direction of a molecule changes at every collision. Let us assume crudely that a molecule completely forgets its past direction of movement upon collision. Also, for simplicity, assume that the molecule travels a distance of exactly the mean free path ℓ between collisions and that the time between collisions is exactly the mean

---

[2] **Mean free path in air** We can rewrite (7.2) as $\ell = k_B T/\sqrt{2}P\pi d^2$ using the ideal gas law. $\ell = 6.6/P$ mm is used for air at room temperature by vacuum engineers, where the pressure P is measured in Pa.

free time τ. Let $\boldsymbol{r}_i$ be the displacement vector of the molecule between the $(i-1)$th and the ith collisions. $|\boldsymbol{r}_i| = \ell$ for all i with our simplifying assumption. The total displacement vector $\boldsymbol{R}(t)$ during time t is given by (in the following we ignore the difference between t/τ and $[t/\tau]^3$):

$$\boldsymbol{R}(t) = \sum_{i=1}^{t/\tau} \boldsymbol{r}_i. \qquad (7.4)$$

This is just the random walk we have already encountered in Chapter 3; this time it is not a lattice walk and it is also three-dimensional. This is a stochastic process (recall **3.16**), so $\boldsymbol{R}(t)$ varies every time we repeat the observation (i.e., this must also depend on the stochastic parameter ω denoting the sample explicitly). A 10000 step sample path is shown in Fig. 7.2.

To follow a single realized trajectory of a molecule is to follow $\boldsymbol{R}(t, \omega)$ for a particular ω. We repeat this many times to sample many trajectories (with different ω) and calculate the expectation values (averages). Due to isotropy

$$\langle \boldsymbol{r}_i \rangle = 0. \qquad (7.5)$$

Steps are statistically independent from each other, so

$$\langle \boldsymbol{r}_i \cdot \boldsymbol{r}_j \rangle = \delta_{ij} \ell^2, \qquad (7.6)$$

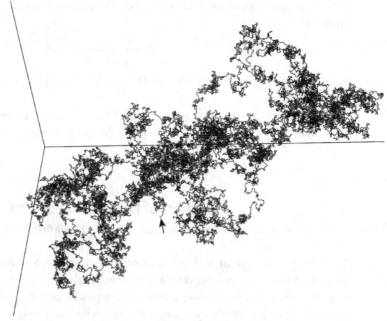

Fig. 7.2   A unit step-length random walk sample path with 10000 steps starting from the arrowhead.

---

[3] Here $[x]$ denotes the largest integer not exceeding x.

where we have used that $\boldsymbol{r}_i \cdot \boldsymbol{r}_i = \ell^2$ for all i. Therefore,

$$\langle \boldsymbol{R}(t) \rangle = \left\langle \sum_{i=1}^{t/\tau} \boldsymbol{r}_i \right\rangle = \sum_{i=1}^{t/\tau} \langle \boldsymbol{r}_i \rangle = 0. \tag{7.7}$$

7.3 Spatial Spread of Random Walks

However, the random walk trajectory spreads in space as Fig. 7.2, so it is better to study the mean-square end-to-end distance $\langle |\boldsymbol{R}(t)|^2 \rangle = \langle R^2(t) \rangle$:[4]

$$\boldsymbol{R}^2(t) = \left(\sum_{i=1}^{t/\tau} \boldsymbol{r}_i \right)^2 = \sum_{i=1}^{t/\tau} r_i^2 + 2 \sum_{i<j} \boldsymbol{r}_i \cdot \boldsymbol{r}_j = (t/\tau)\ell^2 + 2 \sum_{i<j} \boldsymbol{r}_i \cdot \boldsymbol{r}_j. \tag{7.8}$$

Taking the average (the sample average), we see that the second term disappears due to (7.6), and we have

$$\left\langle R^2(t) \right\rangle = (t/\tau)\ell^2. \tag{7.9}$$

If there were no collision at all, during time t a molecule is displaced by a distance $L = (t/\tau)\ell \simeq t\bar{v}$, where $\bar{v}$ is a representative speed of the molecule (say, the root-mean-square velocity). Therefore, L has the same order as the sound speed $\times t$. A molecule can traverse a classroom easily within one second. On the other hand with collisions, the displacement may be given by

$$\sqrt{\left\langle R^2(t) \right\rangle} = L/\sqrt{t/\tau}. \tag{7.10}$$

Here t/τ is the number of collisions a molecule suffers while it runs the total distance L, which is the number of times the trajectory is (randomly) broken. This is larger than 10^{10} in 1 second in air of 1 atm. Thus the actual range over a molecule can move in 1 second is about five orders as small as L. For nitrogen gas it is at most a few mm. Of course, our estimate is very crude, and oversimplified with total loss of memory upon collision. Still, $\langle R^2(t) \rangle \propto t/\tau$ is true even under realistic assumptions, and the size of the displacement range of a molecule would not change very much, either.

☐ Assume that four out of five collisions do not change the moving direction of the molecule at all but once in five collisions the molecule totally loses its memory. How much larger is $\sqrt{\langle R^2(t) \rangle}$ than the simple model discussed here? (It is $\sqrt{5}$ times as large.)

Thus, Clausius correctly answered Buys-Ballot's criticism.

---

[4] The scalar product $\boldsymbol{R} \cdot \boldsymbol{R}$ is often abbreviated as $\boldsymbol{R}^2$, if not confusing.

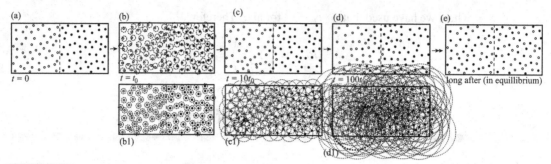

Fig. 7.3 Diffusion in gas occurs due to almost independent random walk of gas molecules. The vertical broken segment indicates the mid line; there is nothing there. The broken circles in the lower figures denote the expected displacement range of each molecule by the time of the picture from its initial position at $t = 0$. Since the molecules are reflected by the walls of the container, we must fold the trajectories back into the container at the wall.

(a) At $t = 0$, the same amount of hard-core molecules are in both halves (each with volume V) of the container. The molecules are only different in "colors."

(b) The broken circles denote the expected displacements of individual molecules from time 0 to t_0 (b1). At t_0 the position of each molecule is chosen randomly from the disk of its own, (b) is the result (this time, the broken circles are kept, since the figure is not overly cluttered).

(c) At $t = 10t_0$ the molecules can move about 3 ($\sqrt{10}$) times as large a distance as the case of (b). Each molecule chooses its position from each disk in (c1) (its center is the initial position in (a); the representative particle near the lower left corner is slightly darker). Needless to say, at the container walls the trajectories must be folded back into the container, (c) is the result.

(d) At $t = 100t_0$ the molecules can move about 10 times as large a distance as the case of (b) (see the dotted circles in (d1)). The result is (d).

(e) After a sufficiently long time molecules can go anywhere they wish, so two gases mix well. This is the equilibrium state and the gases never de-mix again.

7.4 Mixing Process of Two Distinct Gases: Each Molecule Walks Randomly

To prepare for the study of transport phenomena, let us look at the diffusive mixing process of two distinct (almost[5]) ideal gases in detail (Fig. 7.3).

Let us consider a gas consisting of tiny hard spheres. We ignore any interaction other than collisions. Then, we may assume that each molecule executes a statistically independent 3D random walk as in Fig. 7.2. Initially, the right half (volume V) of the container contains all the N white molecules, and the left half (volume V) all the N black molecules. Note that even if a black molecule is at the black–white boundary initially, there is no special preferred direction in which to move. Read the detailed explanation of Fig. 7.3 for how the mixing process proceeds. In short, what happens in the diffusion process is statistically independent random displacement of individual molecules with a constraint

[5] The effects of the small cores on dilute gas equilibrium properties are much smaller than on their dynamics.

that they cannot move a long distance within a short time. There is no correlation among molecules, so no molecules of the same kind try to be away from each other. Although each molecule never experiences any driving force, on the average black molecules move to the left, and the white molecules to the right.

7.5 Characterizing the Equilibrium Macrostate

Figure 7.3 illustrates the states of the system microscopically; individual molecules are depicted in the figure. We call such detailed snapshots "microstates," which are microscopically distinguishable states of the whole (macroscopic) system.[6] We macroscopic observers cannot distinguish individual microstates; we can observe "coarse-grained" observables such as the total number of particles of each species in a macroscopic domain. The states of the system distinguishable through macroscopic observables (observables at the scale we can discern) are generally called *macrostates*. That is, macrostates are the collection of microstates sharing macroobservables.[7]

In Fig. 7.3 (e) is said to be the "equilibrium state": After a sufficiently long time two gases mix well and the gases never de-mix again. As noted previously, we must distinguish this state as a macrostate. Here, let us characterize it macroscopically by the even partition of black and white particles to the both halves.

Let us confirm that this macroscopically characterized equilibrium state is very likely and is stable; the system tends to this state and once it reaches this state, it never leaves it.

Suppose there are two macrostates A and B consisting of $\mathcal{N}_A$ and $\mathcal{N}_B$ microstates (that is, different configurations or patterns as illustrated in Fig. 7.3), respectively. Take a microstate and let us evolve it according to the random walks as in Fig. 7.3. If $\mathcal{N}_A \gg \mathcal{N}_B$, we expect the transition $A \rightarrow B$ is much less likely than $B \rightarrow A$. Thus, we expect the macroscopic equilibrium state consists of the most numerous microstates, and so this state is eventually reached. Also, we expect that this state is surrounded (in state space) by macrostates that have far fewer compatible microstates, because the equilibrium macrostate is stable. That is, any macrostate C that can be reached immediately from the equilibrium state E with $\mathcal{N}_E$ microstates has microstates $\mathcal{N}_C \ll \mathcal{N}_E$; even if rarely some deviation from the equilibrium state into macrostate C occurs, with an overwhelming probability the original state E is restored by fluctuation from C. Thus, the equilibrium state E is stable. Let us confirm these expectations.[8]

---

[6] **Microstate** Microstates are microscopically distinguishable states of the system; we will use this terminology in this sense throughout the book.

[7] **Macrostate** Macrostates are equivalence classes (see footnote 1 of Chapter 17) of microstates according to macroobservable values.

[8] Notice that the temperature is everywhere the same, so we need not consider any change in its velocity distribution for each particle.

7.6 Equilibrium Macrostate Is Most Probable

Since in our case the difference is only "colors" and since these molecules do not interact appreciably except hindering motions, let us consider white molecules.

Notice that every microstate characterized by the particle configuration (pattern) as depicted in Fig. 7.3 seems equally probable. Let us regard each half of the container as having $V^\#$ ($\propto V$) seats, each of which can accommodate at most a particle.

Let N be the total number of white molecules and assume that n are on the left side. If n white molecules are in the left half, since molecules are indistinguishable and have $V^\#$ possible seats upon which to sit, the number of distinguishable configurations is (for reviewing the counting methods, see Appendix 3A):

$$\binom{V^\#}{n} = \frac{V^\#!}{n!\,(V^\# - n)!} \simeq \frac{(V^\#)^n}{n!}. \tag{7.11}$$

Since our gas is close to an ideal gas, we have assumed $V^\# \gg N$, so $V^\#(V^\# - 1)$ $(V^\# - 2)\cdots(V^\# - n + 1) \simeq (V^\#)^n$. Analogously, the right half must have $(V^\#)^{N-n}/(N-n)!$ ways. The total number of configurations $\mathcal{N}_W(n)$ to place n white molecules to the left and $N - n$ to the right is given by

$$\mathcal{N}_W(n) = \frac{(V^\#)^n}{n!}\frac{(V^\#)^{(N-n)}}{(N-n)!} = \frac{1}{N!}\binom{N}{n}(V^\#)^N. \tag{7.12}$$

The total number of configurations $\mathcal{N}_B(m)$ for m black particles to be on the left side is analogous. Black and white molecules may be (almost) independently placed, so

$$\mathcal{N}_W(n)\mathcal{N}_B(m) = \frac{1}{N!^2}\binom{N}{n}\binom{N}{m}(V^\#)^{2N} \tag{7.13}$$

is the total number of microstates for the n white and m black particles to the left. We may consider $\mathcal{N}_W$ and $\mathcal{N}_B$ separately. Thus the problem boils down to the maximization of the binomial coefficient $\binom{N}{n}$. Because $N!$ is extremely large, let us take its logarithm:[9]

$$\log N! = \sum_{k=1}^{N} \log k \simeq \int_0^N dx \log x = N \log N - N. \tag{7.14}$$

Also we may assume $n \gg 1$:

$$\log \mathcal{N}_W(n) = N \log V^\# - n \log n - (N - n)\log(N - n) + N. \tag{7.15}$$

The n that maximizes this is obtained by

$$\frac{d}{dn}\left(n \log n + (N - n)\log(N - n)\right) = \log \frac{n}{N - n} = 0. \tag{7.16}$$

---

[9] **Approximation for $N!$.** The following result is a simplified version of Stirling's formula approximating $N!$, which will be discussed in **18.3** in more detail. This approximation appears again and again, so memorize $\log N \simeq N \log N - N$ or $N! \simeq (N/e)^N$.

Therefore, $n = N - n$ or $n = N/2$ maximizes the number of configurations (microscopic states). That is, $n = m = N/2$ is the most likely state (as expected). Since the peak is unique, the time evolution of the macrostate from $n = N$ or 0 (all on one side) to the $n = N/2$ state is the evolution to the macrostate with a larger number of compatible microstates (to the more likely macrostates).

7.7 The Equilibrium Macrostate Is Stable

Let us confirm our expectation that the macrostates readily reachable from the equilibrium state consist of far fewer microstates. We pretended that the number of particles n can be counted accurately, but this is impossible, so there must be a leeway. Therefore, what we wish to show is that if the number of white (black) particles in the right half is deviated even slightly macroscopically from $N/2$ by ε (i.e., even if $\varepsilon/N \ll 1$ but not excessively small), the number of microstates compatible with this deviation is extremely small, i.e.,

$$\frac{\mathcal{N}_W(N/2 \pm \varepsilon)}{\mathcal{N}_W(N/2)} \ll 1. \tag{7.17}$$

Using (7.15), let us compute

$$\log \frac{\mathcal{N}_W(N/2 \pm \varepsilon)}{\mathcal{N}_W(N/2)} = N\log(N/2) - (N/2 + \varepsilon)\log(N/2 + \varepsilon) - (N/2 - \varepsilon)\log(N/2 - \varepsilon). \tag{7.18}$$

We can use $(A + x)\log(A + x) = A\log A + (\log A + 1)x + x^2/2A + o[x^2]$ to estimate

$$\frac{\mathcal{N}_W(N/2 \pm \varepsilon)}{\mathcal{N}_W(N/2)} \simeq e^{-2\varepsilon^2/N}. \tag{7.19}$$

We see from this that if ε is not less than $\sim \sqrt{5N}$, the ratio is overwhelmingly small. Since $N \gg 1$, say, 10^{20}, $\varepsilon/N \ll 1$. Thus, the equilibrium state is quite stable under natural molecular motions. In other words, the extent of likeliness of macrostates has an extremely sharp peak around the equilibrium state. See **Q17.1**(3) for a very similar situation.

We have learned that, despite individual uncorrelated motions of molecules paying no attention to the whole system, the state of the whole system almost deterministically evolves in time toward the state that is highly robust against fluctuations (the state compatible with the most numerous microstates).

7.8 Entropic Force

We have seen in detail that even though there is no systematic force on individual molecules, by diffusion the center of mass of each gas is displaced (almost) deterministically. For example, black molecules tend to move from their high concentration region to the low concentration region as if they are systematically pushed by a certain driving force. Is this a mere illusion?

How can we obstruct the diffusion of black molecules to the left in the initial state of Fig. 7.3 without placing a barrier at the center? We have already seen in Chapter 5 that the distribution of molecules under an external potential force is governed by the Boltzmann factor (recall **5.7**). Applying a (conservative) force pulling black molecules to the right and white molecules to the left, we could considerably obstruct diffusion; at least the state (e) in Fig. 7.3 would never be reached. Is the potential producing the obstructing force an illusion? Obviously, it is real; real forces are acting on individual molecules. What is the meaning of the necessity of a real force to oppose a "nonexistent driving force"?

If a real force is required to suppress diffusion, can diffusion produce a real force? In Fig. 7.4(a) there is a concentration gradient (of the black particles) and there is a diffusion from the right to the left. If we obstruct this natural tendency as (a) → (b), forces are produced. Thus we must conclude that if we oppose or obstruct the natural tendency of the collective behavior of molecules, we observe a force. We know during diffusion individual molecules never feel any systematic force; however, the collective behavior (or the behavior of each molecule on average[10]) is actually driven toward macrostates compatible with more microstates, and real force is observed when this natural collective tendency is obstructed as illustrated in Fig. 7.4(b).

As we have seen in **7.4–7.7**, the natural tendency is dictated by the direction to increase the diversity of microstates compatible with the macrostates. We will learn in Chapter 17 and subsequent chapters that the "entropy" of a macrostate is a measure of the diversity of microstates compatible with the macrostates (a preview is in **15.7–15.10**), so the force that appears when we oppose or obstruct the natural tendency to increase the diversity of compatible microstates is called *entropic force*.

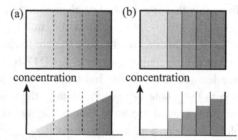

Fig. 7.4 If we "dam" the natural diffusion flow due to a concentration gradient ((a)→(b)), forces appear at the dams. They are entropic forces appearing when the natural evolution of the system to increase diversity of microstates is obstructed.

---

[10] **Entropic force on each molecule** As astute reader should have realized molecules execute almost independent random walks in our system, so the collective behavior of molecules must be the same as the average behavior of each molecule. Look at Fig. 7.3. If we trace a particular molecule, it wanders around the space almost uniformly. Its average position will eventually be the center of the box. To prevent this average position from being realized, we must impose a certain force. We may also say that this is the force opposing the entropic force acting on the particle.

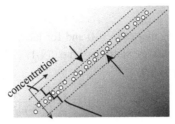

How to measure the entropic force on a particle of a particular chemical species. The dotted lines represent semipermeable membranes blocking only the target chemical species. The pressure gradient times the membrane spacing, i.e., the total force (the sum of big arrows) per area due to the two adjacent membranes acting on the layer sandwiched between them, divided by the number of the particles of the chemical species in the unit area must be the effective entropic force per molecule. However, do not misunderstand that this force is actually acting on the particles; this force appears only when the natural motion of the molecule is hindered (by an obstacle or by other forces).

7.9 How we can Observe Entropic Force

As is clear from the above explanation the entropic force is a collective effect of molecules observed only when their "natural motions/tendencies" are hindered. Thus, to measure entropic force acting on a certain chemical species at least in principle, we prepare two adjacent semipermeable membranes blocking (hindering) only the target particles (thus the membranes must be perpendicular to the concentration gradient[11]), and observe the total force acting on the thin layer sandwiched between the adjacent membranes from these membranes (see Fig. 7.5). If this total force is divided by the number of the particles of the chemical species between the adjacent membranes, we can obtain the entropic force per particle.

We may interpret that this force actually drives diffusion (in a viscous fluid) as we will see in Einstein's theory to compute the diffusion constant (see **9.6**, **9.7**).

Problems

Q7.1 Velocity Dependence of the Mean Free Path

If a particle moves very slowly, while the other molecules move around as usual, its free path length must be short. Therefore, the mean free path is a function of the particle speed v: $\ell = \ell(v)$. What can the reader guess about this function? Is it a bounded function from above? What is it in the $v \to \infty$ limit?

---

[11] Here, the term "gradient" may be understood intuitively, but it can also be interpreted as a technical term we will encounter in **8.5**.

Solution

We guess easily that $\ell(0) = 0$ and for large v $\ell(v)$ should converge to the result when all the other particles are stationary (i.e., (7.1)). The mean free path of a gas molecule running at speed v is given by[12]

$$\ell(v) = \frac{\beta m v^2/2}{\sqrt{\pi} n d^2 \psi(v\sqrt{\beta m/2})}, \tag{7.20}$$

where

$$\psi(x) = xe^{-x^2} + (2x^2 + 1) \int_0^x e^{-y^2} dy. \tag{7.21}$$

Note that $\lim_{x \to 0} x^2/\psi(x) = 0$ as expected. Also $\psi(x)/x^2$ is bounded from above in the large x limit, and we should recover (7.1).

Q7.2 Distribution of Free Path for a Constant Speed Particle

Suppose the mean free path is $\ell(v)$ for a particle with a constant speed v. The density distribution function for the free path length ℓ reads

$$f(\ell) \propto e^{-\ell/\ell(v)}. \tag{7.22}$$

Solution

Let $p(\ell)$ be the probability that there is no collision within the traversed distance between 0 and ℓ. Then

$$p(\ell + d\ell) = p(\ell)(1 - a d\ell) \tag{7.23}$$

where a is a constant (describing the collision frequency per unit length). Therefore, $p(\ell) = e^{-a\ell}$, but the average is $1/a = \ell(v)$, so we get the desired formula.

If we average (7.22) over v, then we get the distribution function for the free path. According to Jeans (p. 145) the result is very close to $f(x) \propto e^{-1.04x/\ell}$, where ℓ is the mean free path.

---

[12] Jeans, J. (1952). *An Introduction to the Kinetic Theory of Gases*. Cambridge: Cambridge University Press, p. 141.

8

Introduction to Transport Phenomena

Summary

* If the flux of a physical quantity is proportional to the (−)gradient of its density, the phenomenon is called a linear transport phenomenon.
* The diffusion equation $\partial_t f = D\triangle f$ is derived with the (intuitive) understanding of divergence, gradient, and the Laplacian operator.
* Various transport coefficients for gases may be approximately estimated with the aid of a simple kinetic consideration.
* Avogadro's constant may be estimated from shear viscosity (first done by Maxwell).

Key words

linear transport phenomenon, density, flux, gradient, divergence, conservation law, Laplacian, transport coefficients, diffusion

The reader should be able to:

* Explain the intuitive meanings of gradient, divergence, and Laplacian.
* Understand how to derive the diffusion equation.
* Understand why the law of large numbers is crucial for transport phenomena.
* Use dimensional analysis.

Clausius did not have any means to estimate the mean free path ℓ. However, as we can expect from the criticism made by Buys-Ballot (see **7.1**) if we could study the so-called transport phenomena, there is a hope to determine ℓ. This is a step toward estimating N, the number of molecules.

8.1 What is a (Linear) Transport Phenomenon?

Suppose a macroscopic system is not far away from equilibrium. The system may be spatially nonuniform, but is macroscopically only gently so. If there is a gentle spatial nonuniformity in some physical quantity X,[1] its density can be described as a space-time dependent field: $\hat{x}(t, r)$. We expect that this physical quantity would change to reduce the spatial nonuniformity; recall the characterization of the equilibrium state in the explanation of diffusion in Chapter 7. This is generally called the *transport phenomenon*. If $\partial \hat{x} / \partial t$ is a linear functional of $\hat{x}$,[2] we say the transport phenomenon is linear.

8.2 Density

Let X be a physical quantity carried by molecules. Its density, $\hat{x}$, around space-time point (t, r) may be expressed as

$$\hat{x}(t, r) = \frac{\sum_{r_i(t) \in d\tau(r)} x_i}{d\tau(r)}, \tag{8.1}$$

where x_i is the quantity X carried by the ith molecule whose spatial location is $r_i(t)$ at time t. Here, $d\tau(r)$ indicates the volume element around r, which is very small ("infinitesimal") from the macroscopic point of view, but is huge from the microscopic molecular point of view. Its volume is also denoted by the same symbol, $d\tau(r)$. The summation in the numerator of (8.1) means that we calculate the summation over particles whose centers of mass are in $d\tau(r)$ at time t. Since $d\tau(r)$ is microscopically huge, the law of large numbers (Chapter 4) tells us that $\hat{x}(t, r)$ thus defined is not fluctuating, so we identify it with the density of X at (around) r at time t.[3]

In short, nonequilibrium systems may be deterministically described in terms of macroscopic transport theories if the law of large numbers holds in the volume over which the macrovariables change infinitesimally.

8.3 Flux

To describe the flow of X, we need the concept of *flux*. A *flux* J_X of X is a vector pointing in the direction of the flow, whose magnitude is given by Q/A, where Q is the amount of X flowing in unit time through a tiny cross section, A, perpendicular to the direction of the flow (see Fig. 8.1).

Often we may interpret J_X as the product of the density of X and the flow velocity carrying it. Indeed, if there is a uniform flow with velocity v, and if the density of the

---

[1] Such as (kinetic) energy, number of particles. In transport phenomena, we are interested in "extensive quantities" (see Chapter 12) that are proportional to the (system) volume if it is uniform.

[2] **Linear functional** F being a "linear functional" of $\hat{x}$ implies the following: $F(a\hat{x}_1 + b\hat{x}_2) = aF(\hat{x}_1) + bF(\hat{x}_2)$.

[3] For the usual LLN (**4.2**) to hold we need iid variables. Since the system is locally uniform, the identity of the distributions of x_i is satisfied. However, the reader may question about the independence; molecules are generally correlated locally, since, especially in condensed phases, molecules are jostling each other. However, molecular correlations are usually very local (across a few molecular sizes), so the sum actually consists of many independent components. Besides, it is known that the independence condition for LLN may be replaced with much weaker conditions.

Fig. 8.1 The flux vector J_X for the quantity X (in the text, its density is denoted by $\hat{x}$): its direction is the transport direction, and its magnitude is given by Q/A, where Q is the amount of X going through an area of cross section, A, perpendicular to the transport direction in unit time. To be precise, we take the $A \rightarrow 0$ limit.

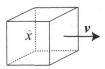

Fig. 8.2 An illustration of the flux formula $J = \hat{x}v$ for a uniform flow: a unit cube is carried by the flow whose velocity is v. Then, the amount of X coming out from the unit square on the right side of the cube is $\hat{x}|v|$ and its direction is parallel to v, so the flux becomes $\hat{x}v$.

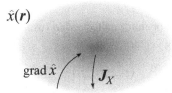

Fig. 8.3 Gentle nonuniformity causes linear transport phenomena. In the figure the density $\hat{x}$ is represented by the gray shade (the reader can imagine a landscape with the gray scale representing the altitudes). The gradient vector grad $\hat{x}$ points the direction of increasing density $\hat{x}$ (darker region), so the flux driven by the gradient points in the $-$grad $\hat{x}$ direction.

carried quantity X by the flow is $\hat{x}$, as is illustrated in Fig. 8.2, its flux is given by $J_X = \hat{x}v$. This expression may be microscopically written as

$$J_X = \frac{\sum_{r_i \in d\tau(r)} x_i v_i}{d\tau(r)}, \tag{8.2}$$

where x_i is the quantity X carried by molecule i whose velocity is v_i. The sum in the numerator is over all the particles whose positions of the centers of mass are in the volume element $d\tau(r)$ (at time t, not explicitly written).

8.4 Linear Transport Law

If the transport phenomenon is linear, the flux J_X of X is proportional to the gradient (see **8.5**) of its density grad $\hat{x}(r)$ (see Fig. 8.3; its direction is opposite, so "$-$" appears in the following formula. Here, time t is suppressed):

$$J_X = -L \operatorname{grad} \hat{x}(\boldsymbol{r}), \tag{8.3}$$

where L is a positive constant called the *transport coefficient*.

8.5 Gradient

Imagine a landscape with altitude $\hat{x}$ given as a function of the position (Fig. 8.3). The gradient, $\operatorname{grad} \hat{x}(\boldsymbol{r})$, is the vector pointing the steepest ascending direction of the landscape at location $\boldsymbol{r}$ with its magnitude being the steepest ascending slope there. The gradient of $\hat{x}$ can be written as

$$\operatorname{grad} \hat{x} \equiv \nabla \hat{x} \equiv \frac{\partial \hat{x}}{\partial x} \boldsymbol{e}_x + \frac{\partial \hat{x}}{\partial y} \boldsymbol{e}_y + \frac{\partial \hat{x}}{\partial z} \boldsymbol{e}_z, \tag{8.4}$$

where $\boldsymbol{e}_k$ is the directional vector (unit vector) in the k-axis direction. In components, we can write

$$\operatorname{grad} \hat{x} = \left(\frac{\partial \hat{x}}{\partial x}, \frac{\partial \hat{x}}{\partial y}, \frac{\partial \hat{x}}{\partial z} \right). \tag{8.5}$$

In (8.4) ∇ is an operator called *nabla* (usually it is read as "del") and may be understood as the following vector:

$$\nabla = \boldsymbol{e}_x \frac{\partial}{\partial x} + \boldsymbol{e}_y \frac{\partial}{\partial y} + \boldsymbol{e}_z \frac{\partial}{\partial z} = \left(\frac{\partial}{\partial x}, \frac{\partial}{\partial y}, \frac{\partial}{\partial z} \right). \tag{8.6}$$

We may understand $\nabla \hat{x}$ as the "product" of a vector ∇ and a scalar $\hat{x}$ (needless to say, their order cannot be switched, since ∇ is a differential operator acting on $\hat{x}$).

8.6 Conservation Law and Divergence

If X is conserved, the amount of change of this quantity in a volume element must be equal to the net influx of X to it. Therefore, if we introduce an operator *div* (read as "divergence") that computes the net efflux of X from around a point per unit volume due to the flux $\boldsymbol{J}_X$, the conservation law for X may be expressed (symbolically for now) as

$$\frac{\partial \hat{x}(\boldsymbol{r})}{\partial t} = -\operatorname{div} \boldsymbol{J}_X(\boldsymbol{r}). \tag{8.7}$$

Here, "div" is an outgoing quantity, so $-$ is used. Let us try to have a mathematical expression of div to substantiate (8.7).

The *divergence* of the flux $\boldsymbol{J}_X$ of X, $\operatorname{div} \boldsymbol{J}_X$, at point P (the total amount of efflux per unit volume per unit time) may be defined as (recall the flux $\boldsymbol{J}_X$ is already a quantity per unit time) :

$$\operatorname{div} \boldsymbol{J}_X = \lim_{V \to P} \frac{\int_{\partial V} \boldsymbol{J}_X \cdot d\boldsymbol{S}}{\int_V d\tau}. \tag{8.8}$$

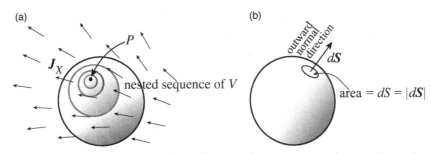

Fig. 8.4 (a) The divergence of the flux J_X at P is defined by the limit along the nested sequence of volumes V converging to a point P: $\mathrm{div}\,J_X = \lim_{V \to P} \int_{\partial V} J_X \cdot dS / \int_V d\tau$. (b) the area element illustrated.

Here, $\lim_{V \to P}$ implies the limit along the sequence of nested (singly connected) volumes V converging to point P (Fig. 8.4(a)) with its surface denoted by ∂V.[4] dS is the surface area element (Fig. 8.4(b)), whose direction is the outward normal direction, and whose magnitude (area) is dS. Thus, the numerator on the right-hand side is the total amount of X going out of the volume V in unit time. The denominator is the volume of V. If we use the Cartesian coordinate system, and consider a nested sequence of cubes V converging to $P = (x, y, z)$ whose edges are parallel to the axes,

$$\lim_{V \to P} \frac{\int_{\partial V} J_X \cdot dS}{\int_V d\tau}$$
$$= \frac{[J_x(x + dx, y, z) - J_x(x, y, z)]dydz + [J_y(x, y + dy, z) - J_y(x, y, z)]dzdx + [J_z(x, y, z + dz) - J_z(x, y, z)]dxdy}{dxdydz}.$$

(8.9)

That is,

$$\mathrm{div}\,J_X = \frac{\partial J_x}{\partial x} + \frac{\partial J_y}{\partial y} + \frac{\partial J_z}{\partial z} = \nabla \cdot J_X. \tag{8.10}$$

The rightmost expression implies that divergence can be formally written as the scalar product of ∇ and the flux vector.

8.7 Local Expression of Conservation Law

Suppose quantity X is conserved. The total amount of X coming into the volume element $d\tau = dxdydz$ per unit time, that is, $-\mathrm{div}\,J_X\,d\tau$ must be the increase rate of X in it (that is, the increase rate of $\hat{x}d\tau$). Therefore, we have

$$\frac{\partial \hat{x}}{\partial t}d\tau = -\mathrm{div}\,J_X\,d\tau. \tag{8.11}$$

---

[4] **Notation for boundary** ∂A is the standard notation for the boundary of the set A. Thus, if V is a bounded volume, ∂V implies its surface.

We have substantiated the conservation equation (8.7). If X can be produced with the rate σ per unit volume (say, due to a chemical reaction), (8.7) is modified to the following general conservation law with production σ:

$$\frac{\partial \hat{x}}{\partial t} = -\mathrm{div}\, \boldsymbol{J}_X + \sigma. \tag{8.12}$$

8.8 The Diffusion Equation

Let us first study the simplest linear transport phenomenon: the diffusion of particles. We know the number of particles is conserved, if there is no chemical reaction. Therefore, if $\boldsymbol{J}$ is the particle number flux, (8.7) is

$$\frac{\partial n}{\partial t} = -\mathrm{div}\, \boldsymbol{J}, \tag{8.13}$$

where n is the number density. We assume a linear transport law for the number of the particles (called *Fick's law*)

$$\boldsymbol{J} = -D\, \mathrm{grad}\, n, \tag{8.14}$$

where D is the *diffusion coefficient*. Combining these two, we get

$$\frac{\partial n(t,\boldsymbol{r})}{\partial t} = -\mathrm{div}(-D\, \mathrm{grad}\, n(t,\boldsymbol{r})) = D\nabla \cdot (\nabla n(t,\boldsymbol{r})). \tag{8.15}$$

This is the conservation law for particles (or for the particle number) called the *diffusion equation*. Introducing the *Laplacian* Δ:

$$\Delta \equiv \nabla \cdot \nabla = \frac{\partial^2}{\partial x^2} + \frac{\partial^2}{\partial y^2} + \frac{\partial^2}{\partial z^2}, \tag{8.16}$$

the diffusion equation reads

$$\frac{\partial n(t,\boldsymbol{r})}{\partial t} = D\Delta n(t,\boldsymbol{r}). \tag{8.17}$$

8.9* The Meaning of the Laplacian

If we understand the meaning of the Laplacian, we can write down the diffusion equation at once, or at least the reader will feel the diffusion equation very natural. Let us consider the 1D Laplacian (1-Laplacian). It is nothing but d^2/dx^2. To estimate the second derivative numerically we use, for example, the following discretization

$$\frac{d^2 f(x)}{dx^2} \quad \leftarrow \quad \frac{f'(x + \Delta x/2) - f'(x - \Delta x/2)}{\Delta x} \tag{8.18}$$

$$= \frac{1}{\Delta x}\left(\frac{f(x + \Delta x) - f(x)}{\Delta x} - \frac{f(x) - f(x - \Delta x)}{\Delta x} \right), \tag{8.19}$$

so

$$\frac{d^2 f(x)}{dx^2} \propto \frac{f(x + \Delta x) + f(x - \Delta x)}{2} - f(x). \tag{8.20}$$

That is, $d^2 f/dx^2 \propto$ "local average of f around x" $- f(x)$. The reader can confirm this conclusion, studying higher dimensional cases. The Laplacian is an operator to compare the central value and the average of the values surrounding it. Thus, in the case of the particle diffusion the Laplacian computes the difference between the average n surrounding r and $n(r, t)$. If this is positive, the diffusion equation increases $n(r, t)$ in order for this quantity to catch up with the neighbors.

8.10 Intuitive Computation of Transport Coefficient

To compute the transport coefficient for a quantity X we need a microscopic description of J_X. Maxwell carried out this step fairly intuitively. Although it is hard to refine his argument quantitatively, as we will see soon, Maxwell's rather crude argument allows a fairly realistic estimation of Avogadro's constant.

Here, we start with a crude microscopic interpretation of a flux as the product of the flow velocity and the density (cf. (8.2)). Basically, we understand that, on average, a molecule brings the physical quantity of our interest adopted at the location of its latest collision to the location r where it is now (see Fig. 8.5). If we write the 'free vector' (a displacement vector of a molecule between successive collisions) as l_i for particle i, the last collision should have occurred at around $r - l_i$. No new collision occurs until the molecule arrives at the volume element around r, so the contribution of this molecule to the flux must be $x_i(r - l_i)v_i$, where $x_i(r - l_i)$ is quantity X the molecule i acquired at its last collision. We use (8.2) to get

$$J_X = \sum_{r_i \in d\tau(r)} x_i(r_i - l_i)v_i \Big/ d\tau(r), \tag{8.21}$$

but usually it is further approximated as

$$J_X = \langle \hat{x}(r - l)v \rangle = \langle \hat{x}(r)v \rangle - \langle (l \cdot \nabla \hat{x}(r))v \rangle + \cdots, \tag{8.22}$$

where the average over l and v is taken for molecules around r, $\langle \hat{x}(r)v \rangle = \hat{x}(r)\langle v \rangle = 0$ and vanishes, because $\hat{x}(r)$ is constant in the volume element.

Fig. 8.5   If a particle moves with velocity v along the free path l ("free vector"), on average, the quantity of interest around $r - l$ transports to r.

To compute the second term in (8.22) let us consider

$$\langle (\boldsymbol{l} \cdot \boldsymbol{A}) \boldsymbol{v} \rangle = \langle \boldsymbol{v} (\boldsymbol{l} \cdot \boldsymbol{A}) \rangle = \left\langle \boldsymbol{v} \sum_{i \in \{x,y,z\}} l_i A_i \right\rangle \tag{8.23}$$

for an arbitrary vector $\boldsymbol{A}$. $\boldsymbol{v}$ and $\boldsymbol{l}$ are parallel and each component of $\boldsymbol{v}$ is statistically independent, so[5*]

$$\langle v_i \ell_j \rangle \simeq \frac{1}{3} \bar{v} \ell \delta_{ij}. \tag{8.24}$$

Here, $\bar{v}$ is the average speed of the particles and ℓ is the mean free path. We have arrived at

$$\langle \boldsymbol{v} (\boldsymbol{l} \cdot \boldsymbol{A}) \rangle = \frac{1}{3} \bar{v} \ell \boldsymbol{A}. \tag{8.25}$$

Thus, we have arrived at

$$\boldsymbol{J}_X(\boldsymbol{r}) = -\frac{1}{3} \bar{v} \ell \, \mathrm{grad}\, \hat{x}(\boldsymbol{r}). \tag{8.26}$$

This is the general formula within Maxwell's approach for the flux.

8.11 The Diffusion Constant

For Fick's law (8.14) $\hat{x} = n$, so the diffusion constant is obtained as

$$D = \frac{1}{3} \bar{v} \ell, \tag{8.27}$$

which may also be written as

$$D = \frac{\ell^2}{3\tau}, \tag{8.28}$$

where τ is the mean free time $\tau = \ell / \bar{v}$.

As can be seen from the derivation earlier, the numerical factor $1/3$ is not quite a definitive number.[6] The main message is that $D/\bar{v}\ell$ is a numerical factor of order unity. Then, this should be derivable dimensional-analytically. Try the suggested dimensional-analytic derivation (see **8.15** which gives an answer).

8.12 Shear Viscosity

Suppose we initially have a shear flow, i.e., a flow with a velocity field $\boldsymbol{V}(\boldsymbol{r})$ solely in the x-direction and its magnitude dependent on the z-coordinate as shown in Fig. 8.6.

[5*]Each component of $\boldsymbol{v}$ is statistically independent, so it assumes $\pm$ independently. Both $\langle \boldsymbol{v} \rangle$ and $\langle \boldsymbol{l} \rangle$ are zero, so if $i \neq j$, $\langle v_i l_j \rangle = 0$. From the isotropy of the space we get $\langle v_1 l_1 \rangle = \langle v_2 l_2 \rangle = \langle v_3 l_3 \rangle = \langle \boldsymbol{v} \cdot \boldsymbol{l} \rangle / 3$, but $\boldsymbol{v}$ and $\boldsymbol{l}$ are parallel vectors, so their scalar product becomes the product of their lengths, i.e., mean speed $\bar{v}$ and mean free path l. Thus, we have arrived at (8.24).

[6] Besides, what average speed $\bar{v}$ to use is not very clear.

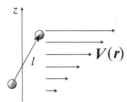

Fig. 8.6 Shear flow: We consider a macroscopic shear flow, so the gradient of V must be microscopically (especially, at the scale of the mean free path l) very small, but in the figure, it is exaggerated.

To understand the time decay of this velocity gradient we study the transport of the x-component of the momentum. Due to exchange of particles between positions with different z-coordinates, larger V_x (or larger momentum density) and smaller V_x layers mix and the gradient in the z-direction decays. This is the effect of *shear viscosity*.

To apply the general formula (8.26) to the quantity we are interested in, we must identify what $\hat{x}$ is in the shear viscosity case. In this case, it must be the x-component of the momentum density

$$\hat{x}(\boldsymbol{r}) = \sum_{\boldsymbol{r}_i(t) \in d\tau(\boldsymbol{r})} m v_{ix}/d\tau(\boldsymbol{r}), \tag{8.29}$$

where the summation in the numerator means the summation of all the x-components of the momentum of the particles in the volume element $d\tau$. Therefore, (here we assume the number density n is uniform) thanks to the law of large numbers we expect only the expectation value of v_x is relevant:

$$\hat{x}(\boldsymbol{r}) = nm V_x(\boldsymbol{r}) \tag{8.30}$$

is the right density to study. Therefore, (8.26) (or its z-component) reads

$$J_V = -\frac{1}{3}\bar{v} l n m \frac{\partial V_x}{\partial z}, \tag{8.31}$$

where J_V is the z-component of the "flux of the x-component of the momentum." *Shear viscosity* η is defined by

$$J_V = -\eta \partial V_x/\partial z, \tag{8.32}$$

Comparing this with (8.31), we get the shear viscosity η:

$$\eta = \frac{1}{3} m n \bar{v} l. \tag{8.33}$$

With the already obtained estimate of l (7.2) and the mean speed $\bar{v} = \sqrt{8 k_B T/\pi m}$ (see **Q6.1**(2)), we obtain

$$\eta = \frac{2}{3d^2} \sqrt{\frac{m k_B T}{\pi^3}}. \tag{8.34}$$

This is independent of the number density n as noted by Maxwell. We generally expect that the viscosity increases with density, but in gases, higher densities imply shorter free paths or a shorter mixing distance (actually the mean free path length is $\propto 1/n$) and the expected density effect is cancelled. Also notice that the viscosity increases with

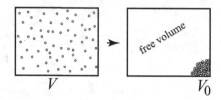

Fig. 8.7   The idea of van der Waals.

temperature. Although this is contrary to the behavior we usually encounter in liquids, it is easy to understand in the gas phase, because higher temperatures imply better mixing between layers with different V.

8.13 Maxwell Estimated the Molecular Size[7]

To establish the reality of atoms, we wish to determine the number of particles N and their size d. Even if we could determine the mean free path ℓ, we can determine only the combination Nd^2.

In 1873,[8] van der Waals (1837–1923) proposed his equation of state of imperfect gases (cf. Chapter 32)

$$P(V - V_0) = Nk_BT - \frac{\alpha}{V}(1 - V_0/V). \tag{8.35}$$

His basic idea is as follows (see Fig. 8.7): Since molecules are not point masses but have volumes, they cannot run everywhere they wish (at least they must avoid each other). However, if we collect all the volumes of the molecules at a corner of the container (its volume is V_0), then the centers of mass of the molecules could freely move around in the "free volume" $V - V_0$. Therefore, if we ignore the attractive interactions, the "hard-core" gas would look like an ideal gas with a reduced volume:

$$P(V - V_0) = Nk_BT. \tag{8.36}$$

The rest of the van der Waals equation is to take care of the attractive intermolecular forces (cf. **32.5**). Thus, from $V_0 \simeq bN\pi d^3/6$, where b is a geometrical constant of order unity, we can estimate Nd^3. We know Nd^2 from the viscosity, so we can estimate N and d. The method can give an estimate of Avogadro's constant $N_A \simeq (4 \sim 8) \times 10^{23}$ (see **Q8.1**).[9]

---

[7] Actually, his first calculation of 1873 followed the method proposed by Loschmidt in 1865, who identified $\pi d^3/6$ as the volume per molecule in the liquid phase. Therefore, $(\pi d^3/6)/(1/n) = V_L/V_G$, where V_L is the molar volume of the liquid phase and V_G that of the gas phase. Thus, we obtain $d = 6\sqrt{2}(V_L/V_G)\ell$. Loschmidt estimated this as $8(V_L/V_G)\ell$, where ℓ was estimated by (8.27). Since we get d, we can count the number of molecules in V_L. Maxwell estimated $N_A \sim 4.3 \times 10^{23}$.

[8] [In 1873 Maxwell's *A Treatise on Electricity and Magnetism* was published; Bruckner composed his symphony No. 3.]

[9] **Definition of Avogadro's constant** Avogadro's constant is defined as the number of atoms in 0.012 kg of ^{12}C. For example, a recent estimate of Avogadro's constant value, $6.02214078(18) \times 10^{23}$, is due to B. Andreas et al. (2011) "Determination of the Avogadro constant by counting the atoms in a ^{28}Si crystal," *Phys. Rev. Lett.*, **106**, 030801.

8.14 Thermal Conductivity

The thermal conductivity λ is defined as (called *Fourier's law*)

$$\boldsymbol{J}_H = -\lambda \operatorname{grad} T, \tag{8.37}$$

where $\boldsymbol{J}_H$ is the heat flux (the thermal energy flux). The transported density $\hat{x}$ must be the thermal energy contained in the unit volume. Let us assume that the gas is a monatomic gas:

$$\hat{x}(\boldsymbol{r}) = \frac{\sum_{\boldsymbol{r}_i(t) \in d\tau(\boldsymbol{r})} m v_i^2/2}{d\tau(\boldsymbol{r})}, \tag{8.38}$$

so

$$\hat{x}(\boldsymbol{r}) = \frac{3}{2} n k_B T(\boldsymbol{r}), \tag{8.39}$$

where $T(\boldsymbol{r})$ is the temperature field thanks to the law of large numbers.

Equation (8.26) reads

$$\boldsymbol{J}_H(\boldsymbol{r}) = -\frac{1}{3}\bar{v}l \operatorname{grad}\left(\frac{3}{2} n k_B T(\boldsymbol{r})\right) = -\frac{1}{2} n k_B \bar{v}l \operatorname{grad} T(\boldsymbol{r}). \tag{8.40}$$

Comparing this with (8.37), we obtain

$$\lambda = \frac{1}{2} n k_B \ell \bar{v}. \tag{8.41}$$

Notice that $\eta = nmD$, $\eta/\lambda = 2m/3k_B$, and $\lambda/D = 3nk_B/2$. The last relation tells us $\lambda = c_V D$, where c_V is the specific heat per volume of gas under constant volume.[10] Again, we should note that these relations do *not* tell us anything about the microscopic properties of the gas particles.

8.15 Dimensional Analysis of Transport Coefficients[11]

The dimension of a quantity X is usually denoted by $[X]$. The basic dimensions are represented by the following symbols: length L, mass M, and time T. For example, $[d] = L$. To obtain the dimension of a quantity, go back to its definition. For example, $[D]$ is obtained from $\boldsymbol{J} = -D \operatorname{grad} n$ as follows. The particle number flux is the number of particles going through a unit area in unit time, so $[J] = 1/L^2 T$, because the number of particles is dimensionless (a pure number), $[n] = 1/L^3$. Gradient is essentially differentiation with length, so $[\operatorname{grad}] = 1/L$ (differentiation is something like division). Therefore, $[J] = 1/L^2 T = [D]/L^4$, so $[D] = L^2/T$.

---

[10] Here c_V is the energy required to raise the temperature of the unit volume of the gas by 1 K under constant volume.

[11] **Introduction to dimensional analysis** See, e.g., Appendix 3.5A of Oono, Y. (2013). *The Nonlinear World*. Tokyo: Springer.

For $[\eta]$ let us go back to its definition: $J_p = -\eta \, \mathrm{grad} \, v$, where J_p is the momentum flux, and v is the velocity. Since the dimension of momentum is ML/T, $[J_p] = (ML/T)/L^2 T = M/LT^2$, $[v] = L/T$, so $[\eta] = [J_p]L/[v] = M/LT$.

For $[\lambda]$ again let us go back to its definition $J_H = -\lambda \, \mathrm{grad} \, T$ (in this formula T is temperature, so $k_B T = E$ is energy). Therefore, $[E]/L^2 T = [\lambda][E/k_B]/L$, so we obtain $[\lambda/k_B] = 1/LT$.

Thus, we obtain $[D/\eta] = L^3/M$, so $\eta/D \sim mn$ is concluded. We get $[k_B D/\lambda] = L^3 = 1/[n]$, which gives $\lambda \sim nk_B D \sim c_V D$. We also get $[k_B\eta/\lambda] = M$, which implies $\eta/\lambda \sim m/k_B$. These are the relations mentioned earlier.

8.16 Significance of $J \propto \mathrm{grad} \, \hat{x}$

Due to collisions the particles cannot go straight for a long distance (actually, it is a zigzag random walk as we saw in Chapter 7). If there were no collision, the particles could move along their straight "ballistic" trajectories, so the amount of "X" transported between two locations must be proportional to the difference of $\hat{x}$ between these two locations (not to the slope of $\hat{x}$ as we learned for linear transport phenomena) irrespective of the distance over which transportation occurs. Thus, the flux proportional to the gradient of $\hat{x}$ is actually a clear sign of molecular collisions occurring at the microscopic scale.

Problem

Q8.1 Estimating Avogadro's Constant from Diffusion

There is a pure gas which roughly obeys a van der Waals equation of state with the excluded volume $V_0 = 5.1 \times 10^{-5} \, \mathrm{m^3/mole}$. Note that

$$V_0 = N_A \frac{1}{2} \frac{4\pi}{3} d^3 = \frac{2\pi}{3} N_A d^3, \tag{8.42}$$

where N_A is Avogadro's constant and d is the diameter of the gas particle (atom or molecule spherically approximated).[12]

(1) This gas has a density of $5.894 \, \mathrm{kg/m^3}$ under 1 atm at $T = 273 \, \mathrm{K}$. What is the root-mean-square velocity of the gas particles for this gas?

(2) The diffusion coefficient was observed to be $D = 4.8 \times 10^{-6} \, \mathrm{m^2/s}$. Using the simple gas kinetic estimate of D (i.e., $D = lv/3$), obtain the mean free path l. Here, the reader may identify v with the root-mean-square velocity just computed in (1).

---

[12] Here $4\pi d^3/3$ is the excluded volume by a spherical particle of diameter d. This is due to mutual overlap of excluded volumes of particles, so to avoid double counting $1/2$ is multiplied to get V_0 per mole.

(3) Try to estimate Avogadro's constant from the data given here.[13] Assume that the volume at 1 atm and 273 K is 22.4 ℓ/mole.

Solution

(1) $\langle v^2 \rangle = 3P/\rho$, so $\sqrt{3 \times 1.013 \times 10^5/5.89} = 227$ m/s. We could use the mean speed $\bar{v}$ as in the text: $\bar{v} = 1.63\sqrt{\langle v^2 \rangle}$. The estimate of N_A will be altered, but the order does not change.

(2) $\ell = 3D/v = 6.34 \times 10^{-8}$ m.

(3) Let $V = 22.4 \times 10^{-3}$ m^3 be the volume of this gas at 1 atm:

$$N_A d^2 = V/\sqrt{2}\pi\ell, \quad N_A d^3 = 3V_0/2\pi. \tag{8.43}$$

Therefore,

$$d = \frac{3V_0}{2\pi} \frac{\sqrt{2}\pi\ell}{V} = \frac{3\ell V_0}{\sqrt{2}V}, \tag{8.44}$$

which is 3.067×10^{-10} m $= 3.1$ Å. A reasonable value (van der Waals radius $= 2.2$ Å for xenon) and

$$N_A = V/\sqrt{2}\pi\ell d^2 = \frac{V}{\sqrt{2}\pi\ell} \frac{2V^2}{9\ell^2 V_0^2} = \frac{\sqrt{2}V^3}{9\pi\ell^3 V_0^2} = 8.466 \times 10^{23}. \tag{8.45}$$

---

[13] The data here are for xenon.

9

Brownian Motion

Summary

* Mesoscopic particles suspended in a fluid move around vigorously and randomly. Brown recognized the universality of this motion called Brownian motion.
* Brownian motion is caused by thermal noise; a quantitative mesoscopic theory due to Einstein enables us to estimate k_B and N_A.
* The general features of Brownian motion may be understood in terms of mechanics with a noise term (a Langevin equation).
* The trajectory of a Brownian particle is closely related to random walks: $\langle r^2 \rangle = 2dDt$.

Key words

Brownian motion, Langevin equation, mesoscopic, Einstein relation, diffusion equation, Einstein–Stokes relation

The reader should be able to:

* Explain the key idea of the mesoscopic approach using Einstein's Brownian motion theory as an example.
* Derive the Einstein relation.
* Estimate the span of a random walk or a random chain polymer.

9.1 How Mesoscopic Particles Behave

At the microscopic scale molecules are colliding with fellow molecules and are recoiling forever. What happens to a particle much bigger than the molecules surrounding it? The particles we can observe optically are about a thousand times linearly as large as typical molecules (Fig. 9.1(a)). This means that the collision cross section ratio is $\sim 10^6$. Thus, numerous small impulses are imparted to the big particle from the surrounding molecules. The law of large numbers tells us that the motion of the big particle must be extremely slow compared with the gas particles, and its motion is due to the "$o[N]$" part of (4.4) (i.e., due to the deviation from the law of large numbers). That is, we observe a typical

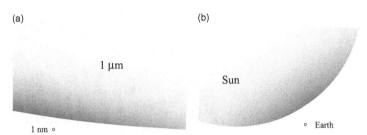

Fig. 9.1 Brownian particle (1 μm radius) vs molecule (1 nm radius); we say the Sun is much bigger than the Earth, but the Sun/Earth ratio is only 109.

mesoscopic scale motion, which we now call *Brownian motion* and what we observe through microscopes is the large deviation behavior of the mesoscopic particles.

9.2 Brown Discovered a Universal Motion (now Called Brownian Motion)

Brownian motion was discovered in the summer of 1827 by Robert Brown.[1,2,3] We are usually given an impression that he simply observed "Brownian motion." However, he conducted very careful and thorough research to establish the *universal nature* of the motion; there were many people who saw Brownian motion before him, but none did any careful study.

Since the particles for which Brown first observed the motion came from living cells (pollen tubes), initially he thought that it was a vital phenomenon (see Fig. 9.2). Removing the effects of advection, evaporation, etc., carefully, he tested many flowers. Then, he tested old pollens in the British Museum (he was the (founding) director of the botanical division), and still found active particles. He conjectured that this was an organic effect, testing even

---

[1] Brown published in the following year a pamphlet entitled "A brief account of microscopical observations made in the months of June, July and August, 1827, on the particles contained in the pollen of plants; and on the general existence of active molecules in organic and inorganic bodies." (available from Google Books). See Brush, S. G. (1968b). A history of random processes: I Brownian movement from Brown to Perrin, *Arch. Hist. Exact Sci.*, **5**, 1–36.

[2] [In 1827: The first transatlantic crossing by steam; Laplace and Beethoven died in March; Schubert composed *Winterreise*.]

[3] **Who was R. Brown?** Robert Brown (1773–1858) was perhaps the greatest botanist (and a great microscopist; Alexander von Humboldt (1769–1859) called him "the glory of Great Britain") in the first half of the nineteenth century. He wrote (1810) a classic describing the Australian flora, following his expedition (1801–1805). He was the first to recognize the two major divisions (angiosperms and gymnosperms) of seed plants (1827). He recognized the nucleus of the cell and so named it (1831).

Before departing for his Beagle expedition (Dec., 1831 to Oct., 1836), Charles Darwin (1809–1882) asked for Brown's advice. The participants of the now historical Linnean Society meeting of July 1, 1858, where the theory of natural selection was first read, were there to listen to Lyell reading the eulogy for Brown who died on June 10. Cf. an authoritative biography of Charles Darwin: Browne, J. (1995). *Charles Darwin: Voyaging*. New York: Knopf; Browne, J. (2002). *Charles Darwin: The Power of Place*. New York: Knopf.

Fig. 9.2   The pollen tube Brown observed first was from *Clarkia pulchella* (flower reddish purple, Oenotheraceae, northwest US; the genus name commemorates Clark of the Lewis and Clark expedition (1804–1806)). He observed 1/4000–1/5000 in (0.5–0.6 μm) particles in the pollen tube. From the quoted booklet: "the first plant examined proved in some respects remarkably well adapted to the object in view. This plant was *Clarkia pulchella*, of which the grains of pollen, taken from antherae full grown, but before bursting, were filled with particles of granules of unusually large size, perhaps slightly flattened, and having rounded and equal extremities. While examining the form of these particles immersed in water, I observed many of them very evidently in motion . . ." (USDA photo.).

coal with no exception found. This suggested to him that not only the vital but the organic nature of the specimens were irrelevant. He then tested numerous inorganic specimens (including a piece of the Sphinx; he also roasted his specimens).[4]

There is no doubt that his report attracted considerable attention. For example, Faraday presented it in a Friday Evening Discourse in February 1829. However, most scientists thought that this would be explained in terms of molecular motion in the future, but that there were more important questions than this, so it was not worth studying.

9.3 General Properties of Brownian Motion

There was no work published about Brownian motion between 1831 and 1857, but the phenomenon was well known. From the 1850s new experimental studies began. The established facts included:

(1) The trajectory is quite erratic without any tangent lines anywhere.
(2) Two Brownian particles are statistically independent even when they come within their diameters.
(3) Smaller particles move more vigorously.
(4) The higher the temperature, the more vigorous the motion.
(5) The smaller the viscosity of the fluid medium, the more vigorous the motion.
(6) The motion never dies out.

---

[4] According to Darwin's autobiography "His knowledge was extraordinarily great, and much died with him, owing to his excessive fear of ever making a mistake" *Charles Darwin, Thomas Huxley Autobiographies*, Edited with an Introduction by Gavin de Beer, Oxford: Oxford University Press, pp. 60–61.

In the 1860s there were experimentalists who clearly recognized that the motion was due to the impact of water molecules. Even Poincaré (1854–1912) mentioned this motion in 1900, but somehow no founders of kinetic theory and statistical mechanics paid any attention to Brownian motion.[5]

9.4 Langevin's Explanation of Brownian Motion

Due to the bombardment of water molecules, the Brownian particle executes a zigzag motion, and eventually, the displacement Δr of the particle during time span t follows $\langle \Delta r^2 \rangle \propto t$. Closely following Langevin's argument,[6] we can demonstrate this.

Let r be the position vector of the Brownian particle, and m its mass. Newton's equation of motion requires the forces acting on the particle. Since the particle is being hit "randomly," we expect a random force w (whose direction and magnitude change incessantly and erratically) acting upon the particle. If the Brownian particle moves at a constant velocity v, it would be hit by more particles of the medium on its front than on its back (imagine running in the rain). Therefore, it is natural to expect a force opposing the motion (i.e., drag) whose magnitude is proportional to the speed. Therefore, the equation of motion reads

$$m\frac{d^2 r}{dt^2} = -\zeta \frac{dr}{dt} + w, \tag{9.1}$$

where ζ is a positive constant describing the relation between the particle velocity and the resistive or frictional force the particle feels from the medium. This type of equation with a stochastic noise term is generally called a *Langevin equation*. Since

$$r \cdot \frac{d^2 r}{dt^2} = \frac{d}{dt}\left(r\frac{dr}{dt}\right) - \left(\frac{dr}{dt}\right)^2 = \frac{d}{dt}\left(\frac{1}{2}\frac{dr^2}{dt}\right) - \left(\frac{dr}{dt}\right)^2, \tag{9.2}$$

we have

$$\frac{m}{2}\frac{d^2 r^2}{dt^2} - m\left(\frac{dr}{dt}\right)^2 = -\frac{\zeta}{2}\frac{dr^2}{dt} + w \cdot r. \tag{9.3}$$

Let us "ensemble-average" this equation. That is, we prepare many such Brownian particles and average the equations for them. Denote this averaging procedure by $\langle \ \rangle$. Since

---

[5] **Why no founders?** According to Hiroshi Ezawa, they never expected the particle fluctuations large enough to be observable. If all molecular bombardments are statistically independent, this conclusion seems inescapable. In reality, molecules are correlated and the fluid medium in which Brownian particles float must be handled as a continuum fluid with fluctuations (i.e., handled at least by fluctuating hydrodynamics). Thus, the Langevin equation (9.1) is not a true mechanical description of the Brownian particle, but of its (extremely) time-coarse-grained picture. Microscopically (even mesoscopically), the real "random force" acting on the Brownian particle is very complicated.

[6] Langevin, P. (1908). Sur la théorie du mouvement Brownien, *C. R. Acad. Sci.*, **146**, 530–533. A translation may be found in Lemons, D. S. and Gythiel, A. (1997). Paul Langevin's 1908 paper "On the theory of Brownian motion" ["Sur la théorie du mouvement brownien," *C. R. Acad. Sci.* (Paris) **146**, 530–533 (1908)], *Am. J. Phys.*, **65**, 1079–1081.

the averaging procedure is linear and time-independent, we can exchange the order of differentiation and averaging:

$$\frac{m}{2}\frac{d^2 \langle r^2 \rangle}{dt^2} - m\left\langle \left(\frac{dr}{dt}\right)^2 \right\rangle = -\frac{\zeta}{2}\frac{d\langle r^2 \rangle}{dt} + \langle w \cdot r \rangle. \tag{9.4}$$

Langevin said, "The average value of the term $w \cdot r$ is evidently null by reason of the irregularity of the complementary forces w." Also, thanks to the equipartition of kinetic energy in equilibrium (Chapter 2), the second term on the left-hand side is known:

$$\frac{1}{2}m\left\langle \left(\frac{dr}{dt}\right)^2 \right\rangle = \frac{3}{2}k_B T, \tag{9.5}$$

where k_B is the Boltzmann constant and T is the absolute temperature of the system.

If we introduce

$$z = \frac{d\langle r^2 \rangle}{dt}, \tag{9.6}$$

(9.4) reads

$$\frac{m}{2}\frac{dz}{dt} + \frac{\zeta}{2}z = 3k_B T. \tag{9.7}$$

Notice that this "3" is the spatial dimensionality $\eth$. Equation (9.7) implies that, after a sufficiently long time,[7] the time derivative dz/dt should vanish and $z = 6k_B T/\zeta$ (note that 6 here is $2\eth$), or

$$\langle r^2 \rangle = \frac{6k_B T}{\zeta}t = \frac{2\eth k_B T}{\zeta}t. \tag{9.8}$$

That is, the absolute value of the displacement during (a macroscopic) time t is proportional to $\sqrt{t}$.

9.5 Relation to Random Walk

As we have seen, due to thermal noise a Brownian particle executes an erratic motion. Let Δr_i be the total displacement between time $(i-1)\tau$ and $i\tau$, where τ is a mesoscopic time scale which is very small from our point of view (say, 1 ms). Then, we may model the movement of the particle by a *random walk* (cf. Fig. 7.2). After n steps (after $t = n\tau$), the total displacement of the Brownian particle is given by

$$r(t) = \Delta r_1 + \Delta r_2 + \cdots + \Delta r_n. \tag{9.9}$$

Let us compute the mean square displacement. This is a repetition of what we did in **7.3**:

$$\langle r^2 \rangle = \sum_i \langle \Delta r_i^2 \rangle + 2\sum_{i<j} \langle \Delta r_i \cdot \Delta r_j \rangle. \tag{9.10}$$

---

[7] This is actually the order of the mesoscopic scale relaxation time: $\tau \simeq m/\zeta$. This is very short for a macroscopic observer like us.

Since the movement of Brownian particles is uniform (e.g., throughout the duration of the motion the displacements are statistically the same), we may expect $\langle \Delta r_1^2 \rangle = \langle \Delta r_2^2 \rangle = \cdots$. There must not be any systematic direction to go: $\langle \Delta r_i \rangle = 0$. We expect Δr_i are totally random (statistically independent): $\langle \Delta r_i \cdot \Delta r_j \rangle = \langle \Delta r_i \rangle \cdot \langle \Delta r_j \rangle = 0$ for $i \neq j$. Therefore, (9.10) implies

$$\langle r^2 \rangle = n \langle \Delta r_1^2 \rangle \propto t, \tag{9.11}$$

which is consistent with (9.8).

9.6 Einstein's Theory of Brownian Particle Flux

Einstein quantitatively demonstrated, in 1905,[8] that the cause of Brownian motion is thermal motion of molecules, three years before Langevin's work discussed previously.

Einstein considered the diffusion process of a collection of Brownian particles in a fluid. The diffusion flux J may be written as

$$J = -D \operatorname{grad} n, \tag{9.12}$$

where n is the number density of the Brownian particles, D is defined by this equation (Fick's law; see **8.8**).

Einstein's key idea was that a Brownian particle may be treated both as a large molecule and, at the same time, as a tiny macroscopic particle (i.e., he consciously introduced the "*mesoscopic scale*" description of Nature, although he did not invent the word "mesoscopic").

Since we regard Brownian particles as molecules, we may apply Dalton's law of partial pressures. We assume the number density n of the Brownian particles is very small, so they do not interact with each other; we may regard the collection as an ideal gas (the particles are treated microscopically):

$$P = n k_B T. \tag{9.13}$$

The average of mesoscopic quantities must be understandable macroscopically (i.e., in terms of macroscopic laws); Einstein did not explicitly say this, but as emphasized repeatedly, this is the key mesoscopic feature. Let f be the average force acting on each particle (see Fig. 9.3).

The total force acting on the slice in the figure is $nf A dr$. This force is due to the partial pressure difference:

$$nf A \cdot dr = -AP(r + dr) + AP(r) = -A \operatorname{grad} P \cdot dr. \tag{9.14}$$

---

[8] **1905** This was Einstein's annus mirabilis; he published four papers in *Annalen der Physik* (on the photoelectric effect, Brownian motion, special relativity, and $E = mc^2$) (every not-crazy paper submitted to this journal seems to have been accepted for publication; no peer review was needed for sound science). Einstein, A. (1905). Über die von der molekularkinetischen Theorie der Wärme geforderten Bewegung von in ruhenden Flüssigkeiten suspendierten Teilchen, *Ann. Phys.*, **17**, 549–560. (On the motion of suspended particles in stationary fluid required by the molecular kinetic theory of heat.)

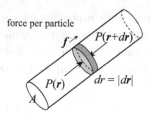

The total force acting on the thin slice of thickness $dr = |dr|$ may be understood as the force due to the pressure difference, so $nfA \cdot dr = -A[P(r + dr) - P(r)] = -A\,\text{grad}\,P \cdot dr$, which gives (9.15).

That is,[9]

$$nf = -\text{grad}\,P. \tag{9.15}$$

Recall that this is exactly the "entropic force" discussed at the end of Chapter 7.

On average a Brownian particle behaves as a macroscopic particle moving slowly, so its (average) velocity v due to pushing by f must obey

$$\zeta v = f, \tag{9.16}$$

where ζ is the friction constant between the particle and the surrounding fluid (drag coefficient) (which we used in Langevin's approach **9.4**).

The diffusion flux J is (cf. **8.3**)

$$J = nv = nf/\zeta, \tag{9.17}$$

so (9.13) and (9.15) tell us that (assuming the temperature is uniform)

$$J = -\frac{1}{\zeta}\text{grad}\,P = -\frac{k_B T}{\zeta}\text{grad}\,n. \tag{9.18}$$

9.7 Einstein's Formula

Comparing this with Fick's law (9.12), we obtain

$$D = k_B T/\zeta, \tag{9.19}$$

which is called the *Einstein relation*. This equation allows us to obtain k_B or, since the gas constant R is known, to calculate Avogadro's constant N_A.

Einstein's original paper used $\zeta = 6\pi a\eta$, where a is the radius of the Brownian particle and η the shear viscosity of the fluid:

$$D = \frac{k_B T}{6\pi a\eta}. \tag{9.20}$$

---

[9] In Chapter 5 we used a similar argument to study the force balance. Here, we are *not* discussing the force balance. Here f is the force due to the fluid. To reproduce the argument in **5.7**, we must discuss the force balance between f and gravity: $nf - nmg\mathbf{e}_z = 0$ (if the z-axis is taken upward).

where $\zeta = 6\pi a\eta$ is Stokes' law for the drag coefficient for a sphere of radius a in a fluid of shear viscosity η.[10] Thus, the original Einstein's relation (9.20) is often called the *Einstein–Stokes formula*.

9.8 The Einstein–Stokes Relation: Dimensional Analysis

We already know $[D] = L^2/T$ (see **8.15**). In dimensional analysis, first we must itemize all the quantities we believe relevant. In the present example, diffusion should be slow with large a or large η, and also it is related to thermal motion, so T should matter; T always appears with k_B, so we may conclude that D should depend on a, η and $k_B T$. $k_B T$ has the dimension of energy, $[k_B T] = M(L/T)^2$. We already know $[\eta] = M/LT$. $[D]$ does not contain M, so we should get rid of M: $[k_B T/\eta] = (ML^2/T^2)/(M/LT) = L^3/T$. Therefore, $k_B T/a\eta$ must have the same dimension as D. Thus, $D \propto k_B T/a\eta$.

9.9 Displacement of Particles by Diffusion

Einstein's relation (9.19) with (9.8) due to Langevin implies

$$\langle r^2 \rangle = 2\partial Dt. \tag{9.21}$$

Einstein, before Langevin, derived this equation in a different way (as discussed later), studying the time evolution of the number density $n(t, r)$ of the Brownian particles that obeys the diffusion equation (recall **8.8**):

$$\frac{\partial n}{\partial t} = D\Delta n, \tag{9.22}$$

where Δ is the Laplacian. The easiest method to solve (9.22) is to use the Fourier transformation (see Appendix 9A), but here we simply quote the result. If n is normalized by the total number of particles, we get the probability density distribution function of the Brownian particles $f(t, r)$. Needless to say, f obeys the same diffusion equation. Let us assume at $t = 0$, all the probability is concentrated at the origin (i.e., $f(0, r) = \delta(r)$; recall Chapter 6 for δ). Then,

$$f(t, r) = \left(\frac{1}{4\pi Dt}\right)^{\partial/2} e^{-r^2/4Dt}. \tag{9.23}$$

---

[10] **Relevant hydrodynamics** The derivation of Stokes' law is not very trivial; see, for example, Landau, L. D. and Lifshitz, E. M. (1987) *Fluid Mechanics*. 2nd edition. Oxford: Butterworth-Heinemann.

As noted in footnote 5, p. 97 the actual random force acting on the Brownian particle is hydrodynamic, and complicated. However, as can be guessed from the random walk picture (9.9), after sufficient coarse-graining, simple randomness can quantitatively explain the average behaviors of the Brownian particle.

Therefore, after t, the mean square displacement must be[11]*

$$\langle r(t)^2 \rangle = 2 \eth D t. \tag{9.24}$$

That is, if we observe the mean square displacement of a particle, then the diffusion constant D of a collection of such particles may be measured. Perrin (1870–1942) implemented this measurement and obtained Avogadro's constant (see **Q9.1**).[12]

9.10 Einstein's Fundamental Idea: Summary

Let us summarize Einstein's fundamental idea, which is the key idea of nonequilibrium statistical mechanics as we may learn later (briefly in Appendix 26A):

> Mesoscopic dynamics dominated by fluctuations, if averaged, obeys macroscopic phenomenological laws.

This was later more clearly stated by Onsager as the *regression hypothesis*: when driving is turned off, the average of the mesoscopic observable returns to equilibrium, following the macroscopic phenomenological law. From our point of view what Onsager wished to tell us is that the mesoscopic observables obey the large deviation principle around the macroscopic law due to the law of large numbers (as is explicit in Appendix 26A).

9.11 Summary of the Boltzmann Constant, k_B[13]

It would be practical to have some sense of the magnitude of the Boltzmann constant.

$$k_B = 1.3806503 \times 10^{-23} \, \text{J/K}$$

$$= 1.3806503 \times 10^{-2} \, \text{pN} \cdot \text{nm/K}$$

$$= 8.617343 \times 10^{-5} \, \text{eV/K}.$$

The *gas constant R* is defined by

$$R \equiv N_A k_B = 8.314462 \, \text{J/mol} \cdot \text{K} = 1.986 \, \text{cal/mol} \cdot \text{K}.$$

---

[11]*Do not forget that (9.23) is a Gaussian density distribution in $\eth$-space. Since $r^2 = x_1^2 + x_2^2 + \cdots + x_\eth^2$, (9.23) is actually a product of $\eth$ independent Gaussian density distributions for each orthogonal component x_i: $P(t, x_i) = (4\pi D t)^{-1/2} e^{-x_i^2/4Dt}$. We immediately see $\langle x_i^2 \rangle = 2Dt$. Consequently, we have $\langle r^2 \rangle = \eth \langle x_1^2 \rangle = 2\eth D t$.

[12] Perrin, J. (1916). *Atoms*. London: Constable, translated by D. L. Hammick. Available online: https://archive.org/details/atomsper00perruoft.

[13] **Representative energy scales** 5∼10 pN is a typical force felt or exerted by molecular machines; a few nm is a typical displacement of molecular motors. Cf., the diameter of DNA is 2 nm (its pitch is 3.4 nm); the α-helix pitch is 3.6 amino acid residues = 0.54 nm. To ionize an atom, a few electronvolts are needed, so, if T is the room temperature (300 K), it is about $100 \, k_B T$. Note that even on the surface of the Sun (with the temperature corresponding to the black-body radiation of about 6000 K, cf. **30.6**), hydrogen atoms are not significantly ionized.

Here, $N_A = 6.02214078(18) \times 10^{23}$/mol is *Avogadro's constant* and 1 cal = 4.18605 J.

It is convenient to remember that at room temperature (300 K):[14]

$$k_B T = 4.14 \, \text{pN} \cdot \text{nm}$$
$$= 0.026 \, \text{eV},$$
$$RT = 2.49 \, \text{kJ/mol} = 0.6 \, \text{kcal/mol}.$$

Problems

Q9.1 Perrin Reproduced

One experiment replicating Perrin's experiment in a modern setting uses polystyrene particles of diameter (i.e., $2a$) $0.5 \, \mu\text{m}$ suspended in a buffer solution of viscosity $\eta = 1.03 \times 10^{-3} \, \text{Pa·s}$ at $T = 300 \, \text{K}$. A two-dimensional stage was recorded by a microscope with a CCD camera. The mean square average displacement in x is observed as $\langle x^2 \rangle = 15.6 \times 10^{-13} \, t \, \text{m}^2$ after t seconds. Assuming that the gas constant $R = 8.31 \, \text{J/mol·K}$ is known, estimate Avogadro's constant N_A.

Solution

The relation we use is $\langle x^2 \rangle = 2Dt$ and the Einstein–Stokes relation $D = k_B T / 6\pi a\eta$. Therefore, $k_B = (3\pi a\eta/T) \times 15.6 \times 10^{-13} = 1.26 \times 10^{-23}$, or $N_A = R/k_B = 6.58 \times 10^{23}$.

Q9.2 Two-Dimensional Lattice Random Walk

On a triangular lattice or a honeycomb lattice (see Fig. 9.4) with the same edge (i.e., lattice bond) length ℓ is a random walker. The walker starts from the origin O and walks along the edges. At every second they choose randomly any of the edges connected to their current position and move to the nearest-neighbor lattice point along the chosen edge. If their ith step displacement is denoted by vector $\boldsymbol{a}_i$, their total displacement during time t seconds is given by

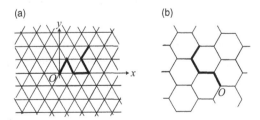

Fig. 9.4   Triangular lattice (a) and honeycomb lattice (b).

---

[14] Under physiological condition, hydrolysis of a single ATP molecule provides about $20 \, k_B T$.

$$R = a_1 + a_2 + \cdots + a_t. \tag{9.25}$$

Here, all the step vectors a_i are lattice bond vectors.

(1) After t seconds on which lattice (T = triangular or H = honeycomb) can they be further away from the origin on average?

(2) Now, on the triangular lattice due to a strong wind blowing constantly in the $+x$ direction, the walker tends to choose $+x$ direction with probability 0.5, but still chooses the remaining five directions randomly (with probability 0.1 for each).

   (a) What is the average position (x and y coordinates) of the walker after t seconds?
   (b) What is the variance of the y-coordinate after t seconds?
   (c) What is the mean square displacement $\langle R^2 \rangle$ of the walker after t seconds?

Solution

(1) Thanks to the statistical independence of steps and the average step displacement being zero (i.e., $\langle a_i \rangle = 0$, so $\langle a_i \cdot a_j \rangle = \ell^2 \delta_{ij}$), we obtain

$$\langle R^2 \rangle = \sum_{i=1}^{t} \langle a_i^2 \rangle. \tag{9.26}$$

Here, $\langle \ \rangle$ is an ensemble average. Obviously, $\langle a_i^2 \rangle = \ell^2$, so $\langle R^2 \rangle = t\ell^2$. Does this calculation depend on spatial dimension or the lattice structure?

We have H = T. This might be slightly counterintuitive, because the honeycomb lattice walk seems less "zigzag" than the other case. Do not forget that for H there is a significant probability (1/3) to retrace the immediate-past step to return to the same position the walker was 2 sec ago.

(2) (a) The position after t seconds is given by (9.25). Therefore, the average position is $\langle R \rangle = t \langle a_1 \rangle$.

$$\langle a_i \rangle = 0.5(\ell, 0) + 0.1\ell \sum_{k=1}^{5} \left(\cos \frac{k\pi}{3}, \sin \frac{k\pi}{3} \right), \tag{9.27}$$

but from the symmetry without actual calculation

$$\langle a_i \rangle = 0.5(\ell, 0) + 0.1(-\ell, 0) = (0.4\ell, 0). \tag{9.28}$$

Therefore, $\langle R \rangle = (0.4\ell t, 0)$.

(b) Let us write $R = (X, Y)$. Then, $Y = \sum_{i=1}^{t} y_i$, where y_i is the y-component of the ith step vector. We know $\langle Y \rangle = 0$, so using the statistical independence of steps, we have

$$V(Y) = \langle Y^2 \rangle = \sum_{i=1}^{t} \langle y_i^2 \rangle, \tag{9.29}$$

where

$$\langle y_1^2 \rangle = 0.6 \times 0 + 0.4 \left(\ell \sin \frac{\pi}{3} \right)^2 = 0.3\ell^2 \tag{9.30}$$

Therefore, $V(Y) = 0.3\ell^2 t$.

(c) We need

$$\langle R^2 \rangle = \sum_{i=1}^{t} \langle a_i^2 \rangle + \sum_{i \neq j} \langle a_i \cdot a_j \rangle. \tag{9.31}$$

Although each step is statistically independent (so we may write $\langle a_i \cdot a_j \rangle = \langle a_i \rangle \cdot \langle a_j \rangle$), its average is not zero in this case, so we cannot ignore the cross terms. There are $t(t-1)$ cross terms, but they are all the same: if $i \neq j$,

$$\langle a_i \cdot a_j \rangle = \langle a_i \rangle \cdot \langle a_j \rangle = \langle a_1 \rangle^2. \tag{9.32}$$

We have already computed $\langle a_1 \rangle = (0.4\ell, 0)$. Obviously, $\langle a_i^2 \rangle = \ell^2$

$$\langle R^2 \rangle = \ell^2 t + 0.16\ell^2 t(t-1). \tag{9.33}$$

What is the variance of R?

Appendix 9A How to Obtain Equation (9.23)

Use of Fourier transformation (see **A.3–A.5**) is the best.[15] Fourier transformation $\mathcal{F}$ is defined as follows (in 3-space):

$$[\mathcal{F}f](k) \equiv \hat{f}(k) = \left(\frac{1}{2\pi}\right)^3 \int_{\mathbb{R}^3} d^3x\, f(x)\, e^{ik \cdot x}. \tag{9.34}$$

Notice that differentiation becomes multiplication:

$$[\mathcal{F}(\nabla f)](k) = -ik\hat{f}(k). \tag{9.35}$$

This can be demonstrated by (essentially an integration by parts)

$$\begin{aligned}
[\mathcal{F}(\nabla f)](k) &= \left(\frac{1}{2\pi}\right)^3 \int_{\mathbb{R}^3} d^3x\, \nabla f(x)\, e^{ik \cdot x} \\
&= \left(\frac{1}{2\pi}\right)^3 \int_{\mathbb{R}^3} d^3x\, \left[\nabla\left(f(x)\, e^{ik \cdot x}\right) - ik f(x)\, e^{ik \cdot x}\right].
\end{aligned} \tag{9.36}$$

The first term in the second line of (9.36) vanishes (assuming f vanishes at infinity), and we get (9.35). From $\hat{f}$ we can recover f by the inverse transformation:

$$f(x) = [\mathcal{F}^{-1}\hat{f}](x) = \int_{\mathbb{R}^3} d^3k\, \hat{f}(k)\, e^{-ik \cdot x}. \tag{9.37}$$

Let us Fourier transform the diffusion equation (9.22). We get

$$\frac{d\hat{n}(t,k)}{dt} = -Dk^2 \hat{n}(t,k). \tag{9.38}$$

---

[15] Körner, T. W. (1988). *Fourier Analysis*. Cambridge: Cambridge University Press, is a wonderful book everyone should read.

This is an ordinary differential equation (k is a mere parameter). The initial condition is $n(0, x) = \delta(x)$ (i.e., initially all the particles are at the origin). Its Fourier transform is (recall **6.6**)

$$\hat{n}(0, k) = 1/8\pi^3. \tag{9.39}$$

Therefore, the solution to (9.38) is

$$\hat{n}(t, k) = \frac{1}{8\pi^3} e^{-Dk^2 t}. \tag{9.40}$$

Now, we inverse-transform this to get

$$n(t, r) = \int_{\mathbb{R}^3} d^3 k \, \frac{1}{8\pi^3} e^{-Dk^2 t - ik \cdot r} \tag{9.41}$$

$$= \frac{1}{8\pi^3} \int_{\mathbb{R}^3} d^3 k \, e^{-tD(k + ir/2Dt)^2 - r^2/4Dt} \tag{9.42}$$

$$= \frac{1}{8\pi^3} \left(\frac{\pi}{tD} \right)^{3/2} e^{-r^2/4Dt} = \left(\frac{1}{4\pi tD} \right)^{3/2} e^{-r^2/4Dt}. \tag{9.43}$$

The procedure from (9.41) to (9.42) is the completion of square in the exponent (see **5.4**), and then the calculation from (9.42) to (9.43) is the usual 3D Gaussian integral (**5.3**).

10 The Langevin Equation and Large Deviation

Summary

* Noise in the Langevin equation must have a proper magnitude not to disrupt equilibrium (the fluctuation–dissipation relation).
* The conservation of Brownian particles implies the Smoluchowski equation governing their density distribution function.
* Noise in the Langevin equation may be naturally understood with the aid of large deviation.

Key words

fluctuation–dissipation relation, Langevin noise, Smoluchowski equation, large deviation principle

The reader should be able to:

* Reproduce the argument leading to the fluctuation–dissipation relation.
* Set up a Langevin equation with an appropriate noise.
* Get the equilibrium distribution from the Smoluchowski equation.
* Intuitively understand the various time scales in describing Nature.
* Be accustomed to the large deviation point of view.

Now that we can count the number of molecules, the reader may wish to go to the mainstream thermal physics. You can go to the next part, skipping this section. This chapter outlines in **Unit 10.8** the large deviation principle, one of the three pillars of probability theory. This unit may be understood without reading the rest of the chapter.

Thermal noise is of superb importance to Brownian motion. However, Einstein ingeniously avoided discussing noise. Langevin, although he explicitly wrote it in his equation, never discussed it. We noted that, truly microscopically, noises are complicated; only after coarse-graining they exhibit certain general features. Let us look at noises in equilibrium as deviations from averages.

10.1 Overdamped Langevin Equation

The basic equation for a Brownian particle in an external potential U reads

$$m\frac{d^2x}{dt^2} = -\zeta\frac{dx}{dt} - \nabla U + w. \tag{10.1}$$

An ordinary Brownian particle is overdamped (ζ is very large and m is very small; or the time scale m/ζ is quite short from our macroscopic point of view), so let us discuss this case:

$$0 = -\zeta\frac{dx}{dt} - \nabla U + w \tag{10.2}$$

or

$$\frac{dx}{dt} = -\frac{1}{\zeta}\nabla U + v, \tag{10.3}$$

where the random velocity $v = w/\zeta$. When we observe a Brownian particle, we observe x, so we usually "sense" the overdamped equation (10.3).

10.2 Langevin Noise: Qualitative Features

We will see how the following two properties naturally characterize the statistical features of the thermal noise v appearing in the overdamped Langevin equation (10.3) which describes the particle position we directly observe:

 (i) v is Gaussian and isotropic,
(ii) its amplitude must be related to the ambient temperature T and ζ.

Qualitatively, (ii) may be understood as follows (see Fig. 10.1). Brownian motion occurs even in an equilibrium state in which no macroscopic changes nor flow/transport

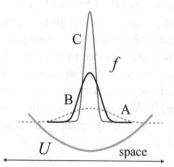

Fig. 10.1   Illustration of the fluctuation–dissipation relation for a Brownian particle in a potential U. Here f is the probability density distribution function to find the particle at the given location with: A, too large noise; B, just right noise; C, too small noise for a given T and ζ. To reproduce the correct equilibrium state, the noise must be carefully chosen.

phenomena occur. Therefore, the spatial distribution of the particle in a potential U after a time long enough to forget its initial condition should be consistent with the Boltzmann factor $\propto e^{-\beta U}$, where $\beta = 1/k_B T$. For example, if the viscosity of the suspending liquid is large, ζ is large, and as can be seen from (10.3), the effect of the systematic force due to the potential energy becomes relatively small. If the amplitude of the noise ν for the overdamped Langevin equation is not reduced appropriately, then obviously the distribution would spread too much;[1] when the ambient liquid is viscous, large noise often pushes the particle away from the potential minimum. Before the particle reaches the minimum, another noise displaces the particle further away from the potential minimum. Thus, a larger damping effect must be balanced with a smaller noise amplitude for the (overdamped) Langevin equation to describe the equilibrium state correctly. Such relations between noise (fluctuation) and damping (dissipation) imposed by the stability of equilibrium is called the *fluctuation–dissipation relation*. The item (5) in **9.3** is a qualitative expression of this relation.

10.3 Langevin Noise

Now, we wish to model the noise ν quantitatively. We assume that each of its orthogonal components is statistically independent, so let us study its x-component ν_x as a representative. As a function of time $\nu_x(t)$ changes quite rapidly and erratically, so we assume its ensemble average satisfies

$$\langle \nu_x(t) \rangle = 0, \tag{10.4}$$

and, using the δ-function (Chapter 6),

$$\langle \nu_x(t)\nu_x(s) \rangle = A\delta(t - s), \tag{10.5}$$

where A is a positive numerical constant. This implies that there is no memory in noise (different times are uncorrelated).[2] Since the three spatial components are uncorrelated, we could write

$$\langle \nu_i(t)\nu_j(s) \rangle = A\delta_{ij}\delta(t - s). \tag{10.6}$$

10.4 Smoluchowski Equation

The fluctuation–dissipation relation is usually derived from the condition that the equilibrium distribution of the particles is correctly described by the Boltzmann factor. Our strategy is as follows:

---

[1] **Warning** Do not confuse the scaled ν (random velocity) and the original noise w (random force). The original noise must maintain the equipartition of energy despite ζ, so larger ζ requires larger amplitude of w.

[2] As the reader will see later, this depends on the time scale we observe the system. Microscopically, this cannot be true.

(i) Find the equation governing the particle number density (called the Smoluchowski equation).

(ii) The required parameter in the equation is related to A in (10.6).

(iii) The compatibility condition between the stationary distribution and the Boltzmann factor determines A (fluctuation–dissipation relation).

Let us derive the transport equation for the number density (step (i)). Since our system is close to equilibrium, driving forces of the number density field must not be large. Thus, we may linearly superpose the flux due to the gradient of n (due to the entropic force) and the one due to the actual systematic force.

The entropic force portion must have the form (see (9.18) with (9.19))

$$J = -D\nabla n. \tag{10.7}$$

The systematic force-driven portion reads (recall (9.16); see also Fig. 8.2)

$$nv = -n\frac{1}{\zeta}\nabla U, \tag{10.8}$$

where U is the potential for the driving force. Therefore, the total flux reads

$$J = -D\nabla n - \frac{1}{\zeta}n\nabla U. \tag{10.9}$$

The conservation of particles (recall **8.7**) reads

$$\frac{\partial n}{\partial t} = -\text{div}\,J = \nabla \cdot \left(n\frac{1}{\zeta}\nabla U + D\nabla n\right). \tag{10.10}$$

This equation is called the *Smoluchowski equation*.[3]

10.5 Relation Between the Langevin Noise and D

Step (ii)

Since the superposition applies, we can find the relation between D and A using the case without U. Let us solve (10.3) without U, assuming that the particle starts from the origin:

$$r(t) = \int_0^t ds\, v(s). \tag{10.11}$$

From this we obtain

$$\langle r^2(t)\rangle = \int_0^t ds \int_0^t ds'\, \langle v(s) \cdot v(s')\rangle = \partial A \int_0^t ds \int_0^t ds'\, \delta(s-s') \tag{10.12}$$

$$= \partial A \int_0^t ds = \partial At. \tag{10.13}$$

---

[3] Or called the *Fokker–Planck equation*, or *Kolmogorov's forward equation*.

Einstein just solved (10.10) without U and found $\langle r^2(t) \rangle = 2\partial Dt$ (see (9.24)). Therefore, we must conclude

$$A = 2D. \tag{10.14}$$

Einstein already gave $D = k_B T / \zeta$ (**9.7**), so we must conclude $A = 2k_B T / \zeta$.

Here, however, we wish to determine A using the just derived $A = 2D$ and the consistency of the noise and the equilibrium distribution. The starting point is the Smoluchowski equation governing the number density function of the particle governed by the (overdamped) Langevin equation (10.3):

$$\frac{\partial n}{\partial t} = \nabla \cdot \left(n \frac{1}{\zeta} \nabla U + \frac{A}{2} \nabla n \right). \tag{10.15}$$

10.6 Smoluchowski Equation and Fluctuation–Dissipation Relation

Step (iii)

In equilibrium, the time derivative must vanish, so (10.15) gives

$$\nabla \cdot \left(n \frac{1}{\zeta} \nabla U + \frac{A}{2} \nabla n \right) = 0. \tag{10.16}$$

This means

$$n \frac{1}{\zeta} \nabla U + \frac{A}{2} \nabla n = \text{ const.} \tag{10.17}$$

Very far away from the potential minimum n should vanish, so the integration constant must vanish. Therefore, the equation (10.17) reads

$$\frac{2}{A\zeta} \nabla U + \nabla \log n = 0. \tag{10.18}$$

This implies that

$$n \propto e^{-2U/A\zeta}, \tag{10.19}$$

which must be proportional to the Boltzmann factor. Consequently, we must conclude that $2/A\zeta = \beta$ or

$$A = 2k_B T / \zeta, \tag{10.20}$$

which is the *fluctuation–dissipation relation*. Since $D = A/2$, we have recovered the Einstein relation (9.19).

Thus, the (overdamped) Langevin noise that does not disrupt the equilibrium state is given by

$$\langle v_i(t) v_j(s) \rangle = 2 \frac{k_B T}{\zeta} \delta_{ij} \delta(t - s). \tag{10.21}$$

10.7 Significance of the Fluctuation–Dissipation Relation

The importance of the fluctuation–dissipation relation is that it allows us to set up an equation governing the dynamics at the mesoscopic scale without referring to microscopic dynamics (i.e., mechanics).

Since the mesoscopic world is described in terms of deviations from the law of large numbers, if the mesoscopic descriptive results are averaged, we should recover macroscopic descriptions due to the law of large numbers. The noise part is the deviation from the law of large numbers. The large deviation principle and the fluctuation–dissipation relation dictate the mesoscopic description of the world close to equilibrium. To understand this picture of the mesoscopic world, we need the large deviation theory. The remaining portion of this chapter will discuss the large deviation theoretical framework to understand Langevin equations

First, let us look at the general theoretical framework called the *large deviation theory* to study deviations from the law of large numbers.

10.8 Mesoscopics and Large Deviation Principle

The law of large numbers tells us that in the sample number $N \to \infty$ limit (here, we stick to the same notation as in Chapter 4 to minimize complication)

$$P\left(\left|\frac{1}{N}\sum_{k=1}^{N}X_k - E(X_1)\right| > \varepsilon\right) \to 0. \tag{10.22}$$

If N is not sufficiently large, this asymptotic relation cannot be used for empirical studies. We must refine this asymptotic law. Perhaps the most natural refinement is to try to actually evaluate how small or large this probability is. For iid[4] stochastic variables $\{X_n\}$ with $V(X_1) < \infty$, it is known that the decay rate of the above probability is exponential:

$$P\left(\left|\frac{1}{N}\sum_{k=1}^{N}X_k - E(X_1)\right| > \varepsilon\right) \approx e^{-NI_\varepsilon}, \tag{10.23}$$

where $\approx$ implies that the ratio of the logarithms of the both sides converges to unity in the large N limit, and I_ε is a (ε-dependent) positive constant.

For physicists, the following form is convenient (this is explained statistical-mechanically in **18.15**):

$$P\left(\frac{1}{N}\sum_{k=1}^{N}X_k \in v(y)\right) \approx e^{-NI(y)}, \tag{10.24}$$

---

[4] Identically and independently distributed.

where $v(y)$ is a volume element around y, and $I(y)$ is called the *rate function* (or *large deviation function*), satisfying

$$I(y) \begin{cases} > 0 & \text{if } y \neq \langle X_1 \rangle, \\ = 0 & \text{if } y = \langle X_1 \rangle. \end{cases} \tag{10.25}$$

The second equality is the law of large numbers. (10.24) + (10.25) is called the *large deviation principle*.[5] The rate function often behaves as

$$I(y) \simeq \frac{1}{2A}(y - \langle X \rangle)^2 \tag{10.26}$$

for not too large y, where A is a positive constant.

Notice that $e^{-NI(y)}$ gives an estimate of the probability density of fluctuation Ny (a large fluctuation). That is, when we study $S_N = \sum_{k=1}^{N} X_k$, it is on average $N\langle X_1 \rangle$, but there are fluctuations of significance, whose distribution can be inferred from $I(y)$.[6]

We will use this framework to understand fluctuations in equilibrium system later (Chapter 26). Here, we apply this framework to a much more nontrivial situation: time averaging.

10.9 Three "Infinitesimal Times," dt, δt, Δt

Let us go back to the basic observation that our world often allows descriptions at three levels with distinct space-time scales. The three time scales are illustrated in Fig. 10.2.

We already discussed the transport phenomena which describe the time evolution at our macroscopic time scale (Chapter 8). The change occurring at the ms scale may be regarded as "the change during an infinitesimal time" (for transport phenomena), so dt in

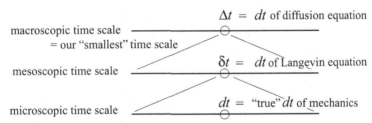

Fig. 10.2 Here dt is the microscopically infinitesimal time scale (perhaps 10^{-15} s or less); Δt is the "infinitesimal time scale" for us macroorganisms (perhaps, 10^{-3} s), which is usually written as dt from our point of view; δt is the "infinitesimal time scale" in the mesoscopic world, and dt in the Langevin equation may be this time scale.

---

[5] A further mathematical requirement is that the level sets of the rate function must be convex.

[6] **Rate function summary** $I(y)$ has a unique minimum at $y = \langle X_1 \rangle$, its level set is convex, and if X_1 has a finite variance, I is differentiable near the global minimum at $y = \langle X_1 \rangle$. There is no book suitable for physicists, but two reviews may be accessible: a relatively new one: Tourchette, H. (2009). The large deviation approach to statistical mechanics, *Phys. Rep.*, **478**, 1–69 and an old one: Oono, Y. (1989). Large deviation and statistical physics, *Prog. Theor. Phys. Suppl.*, **99**, 165–205.

the transport equation is actually Δt (perhaps about 1 ms). To describe the true molecular dynamics dt is of order fs ($= 10^{-15}$ s) or less, so from this "true dt" point of view, Δt is almost eternal.[7] The macroscopic and microscopic time scales are very disparate. Connecting them are mesoscopic phenomena characterized by a time scale δt of the order, perhaps, 1 ns.[8]

10.10   Time Averaging to Get Mesoscopic Results

At the macroscopic scale, if the deviation from equilibrium is gentle, we see transport phenomena.[9] The macro time derivative dX/dt is actually the ratio of the change ΔX during Δt and Δt:

$$\text{macro derivative} \quad \frac{dX}{dt} = \frac{X(t + \Delta t) - X(t)}{\Delta t}, \tag{10.27}$$

but the fundamental theorem of calculus tells us that this is the time average during Δt of the microscopic derivative (usually identified with the true mathematical derivative):

$$\frac{\Delta X}{\Delta t} = \frac{1}{\Delta t} \int_t^{t+\Delta t} \frac{dX}{dt} dt. \tag{10.28}$$

The relation between the mesoscale derivative and the microscopic true derivative is analogous:

$$\frac{\delta X}{\delta t} = \frac{1}{\delta t} \int_t^{t+\delta t} \frac{dX}{dt} dt. \tag{10.29}$$

Note, furthermore,

$$\frac{\Delta X}{\Delta t} = \frac{1}{\Delta t} \int_t^{t+\Delta t} \frac{\delta X}{\delta t} \delta t, \tag{10.30}$$

which gives (10.28) with an appropriate succession of the δt time intervals.

What is the relation between the above time derivatives on various scales and the law of large numbers? The microscopic time derivative is deterministic but unpredictable, and it has only a microscopic time-scale memory (extremely forgetful from our time scale). That is, after, say, 0.1 ps we may expect that dX/dt behaves statistically independently. Thus, the ratio appearing on the right-hand-side of both (10.28) and (10.29) may be understood as empirical expectation values just as S_N/N in the law of large numbers. Since $\Delta t/dt \gg 1$ is really large, for $\Delta X/\Delta t$ the law of large numbers holds, and we may ignore its fluctuations. This is the macroscopic phenomenological law. A typical example is the transport theory (Chapter 8).

---

[7] **Time scale ratio: we are almost eternal** Note that 10^9 s is about 31.7 a (= years). If dt is 1 s, Δt corresponds to 32 ka; If dt is 1 min, Δt corresponds to 1.9 Ma; If dt is 1 d, Δt corresponds to 2.74 Ga; cf., the Earth was born 4.56 Ga ago.

[8] If δt is 1 d, Δt is about 2.7 ka, dt is 0.86 sec. If δt is 1 s, Δt is 11.6 d and dt is 1 μs.

[9] Here, "gentle" means that the macroscopic observables change sufficiently slowly that the law of large numbers holds on the space-time scale where the macrovariable changes are infinitesimal.

In contrast, $\delta t/dt$ is large but not huge, so we must worry about fluctuations (deviations from the long-time average). Thus, at the mesoscopic time scale we see Brownian motion and the equations describing this time scale are Langevin equations (Chapter 9).

10.11 Langevin Equation As a Result of Large Deviation Theory

Let us study the overdamped Langevin equation (10.3) that describes the motion of a Brownian particle. If we respect the time scale, it should actually be written as (here the conserved force is replaced with a general force $\boldsymbol{F}$):

$$\frac{\delta \boldsymbol{x}}{\delta t} = \frac{1}{\zeta}\boldsymbol{F} + \boldsymbol{v}. \tag{10.31}$$

If we average this over a macroscopic time scale, the result should yield the macroscopic law: $\boldsymbol{v} = \boldsymbol{F}/\zeta$, so

$$\frac{\Delta \boldsymbol{x}}{\Delta t} = \frac{1}{\zeta}\boldsymbol{F}. \tag{10.32}$$

This is due to the law of large numbers: Since $\Delta t/dt \gg 1$,

$$P\left(\left|\frac{\Delta \boldsymbol{x}}{\Delta t} - \frac{\boldsymbol{F}}{\zeta}\right| > \varepsilon\right) \simeq 0. \tag{10.33}$$

It is natural to expect that the mesoscopic deviation from the macroscopic behavior must be described by the large deviation principle. After all, this is the essence of Onsager's regression hypothesis (see Appendix 26A). What form of large deviation principle should we expect? Since δt is a $\delta t/dt$ collection of dt, N in (10.24) must be proportional to $\delta t/dt$. However, we do not know precisely what dt is, so let us write

$$P\left(\frac{\delta \boldsymbol{x}}{\delta t} - \frac{\boldsymbol{F}}{\zeta} \in v(\boldsymbol{v})\right) \approx \exp\left[-\delta t\, I(\boldsymbol{v})\right], \tag{10.34}$$

where v denotes the volume element around $\boldsymbol{v}$, and the large deviation function I should read

$$I(\boldsymbol{v}) = \frac{1}{2A}\boldsymbol{v}^2, \tag{10.35}$$

where A is a positive constant to be determined (but, as we will learn, it turns out to be exactly the same A introduced in (10.5)). If we use this for (10.34), we obtain the density distribution function $f(\boldsymbol{v})$ for the noise:

$$f(\boldsymbol{v}) \propto \exp\left\{-\frac{\delta t}{2A}\boldsymbol{v}^2\right\}. \tag{10.36}$$

Thus, fluctuations are Gaussian, and[10*]

$$\langle \boldsymbol{v}^2 \rangle = \eth A/\delta t, \tag{10.37}$$

where $\eth$ is the spatial dimensionality.

[10*] $\boldsymbol{v}$ is a $\eth$-dimensional (column) vector $(v_x, v_y, \ldots)^T$ (^{T} implies transposition) and each component satisfies $\langle v_x^2 \rangle = A/\delta t$. Therefore, $\langle \boldsymbol{v}^2 \rangle = \eth A/\delta t$.

The variance looks unpleasant with the mesoscopic infinitesimal δt appearing downstairs, but we already know what its proper interpretation should be from Chapter 6: the δ-function:

$$\delta(t - s) = 0 \text{ for } t \neq s, \tag{10.38}$$

$$\delta(t - s)\, dt = 1 \text{ for } t = s. \tag{10.39}$$

In short, basically, the "needle" of length $1/dt$ located at $t = s$ is $\delta(t - s)$. Thus, the real meaning of (10.37) is:[11]

$$\langle \boldsymbol{v}(t) \cdot \boldsymbol{v}(s) \rangle = \eth A \delta(t - s), \tag{10.40}$$

or

$$\langle \boldsymbol{v}(t)\boldsymbol{v}^T(s) \rangle = AI\delta(t - s), \tag{10.41}$$

where I is the $\eth \times \eth$ unit matrix (do not forget that our vectors are column vectors; ^{T} implies transposition: "column $\leftrightarrow$ row"). If we demand the fluctuation–dissipation relation **10.6**, we must impose

$$A = 2k_B T/\zeta. \tag{10.42}$$

10.12 Langevin Equation: Practical Summary

Let us write down the Langevin equation governing an overdamped Brownian particle with a friction coefficient ζ in the potential U satisfying the fluctuation–dissipation relation (e.g., (10.42)) at temperature T as a summary. The equation reads (here δt is written as dt)

$$\frac{d\boldsymbol{x}}{dt} = -\frac{1}{\zeta}\frac{\partial U}{\partial \boldsymbol{x}} + \boldsymbol{v}(t), \tag{10.43}$$

with the Gaussian noise satisfying $\langle \boldsymbol{v}(t) \rangle = 0$ and

$$\langle \boldsymbol{v}(t)\boldsymbol{v}(s)^T \rangle = \frac{2k_B T}{\zeta}I\delta(t - s), \tag{10.44}$$

where I is the $\eth \times \eth$ unit matrix. (10.44) tells us that the memory duration of the noise is almost instantaneous at the mesoscopic time scale.

Problems

Q10.1 Brownian Motion of a Harmonic Oscillator: Basic Questions

There is a 1D harmonic oscillator in a viscous fluid, obeying the following (overdamped) Langevin equation (in 1-space):

---

[11] Here "dt" is really δt, so the $\delta(t - s)$ is the delta function for the mesoscopic time scale. That is, from the microscopic point of view it has a width of order δt. For simplicity, we use only one symbol for δ-functions.

$$\frac{dx}{dt} = -\frac{1}{\zeta}kx + v, \tag{10.45}$$

where ζ and k are positive constants (the viscous damping factor and the spring constant, if the reader wishes to interpret them), v is an appropriate equilibrium thermal noise.

(1) What is the magnitude A of the noise v

$$\langle v(t)v(s)\rangle = A\delta(t-s), \tag{10.46}$$

if the correct Boltzmann factor $e^{-kx^2/2k_BT}$ governs the equilibrium distribution of the oscillator position along the x-axis at temperature T? (The reader can read off the answer from the text)

(2) The root-mean-square displacement $\sqrt{\langle x^2\rangle}$ of the oscillator is 1.2 nm at $T = 295$ K. What is the spring constant k?

(3) Suppose a 1D harmonic oscillator in a viscous fluid obeys the following Langevin equation

$$\frac{dx}{dt} = -\frac{1}{z}x + v. \tag{10.47}$$

The equilibrium mean square displacement is given by $\langle x^2\rangle = V$. What is the noise amplitude A as defined in (10.46)? Assume that the temperature of the system is T.

Solution

(1) As the hint says, we can simply read off the answer: $\langle v^2\rangle dt = 2k_BT/\zeta$ or $\langle v(t)v(s)\rangle = (2k_BT/\zeta)\delta(t-s)$, because we know the distribution is just $\propto e^{-\beta U}$, where $U = kx^2/2$ in our case.

(2) The Boltzmann factor gives just the Gaussian distribution for x. By inspection, we can read $\sigma^2 = k_BT/k$ off. Hence, $\langle kx^2\rangle = k_BT$ or $k = k_BT/\langle x^2\rangle$ ($k_B = 1.38 \times 10^{-23}$ J/K).

$$k = 1.38 \times 10^{-23} \times 295/(1.2 \times 10^{-9})^2 = 2.83 \times 10^{-3} \text{ N/m}. \tag{10.48}$$

That is, $k = 2.83$ pN/nm (pico newton/nanometer is just the right size for biomolecules).

(3) $\langle x^2\rangle = V$ implies the harmonic potential of $U = \frac{1}{2}(k_BT/V)x^2$. That is, the spring constant is $k = k_BT/V$ (just as we saw in (2)), $z = \zeta/k$. Therefore, $\zeta = zk = k_BTz/V$, so $\langle v^2\rangle dt = 2k_BT/\zeta = 2V/z$ or $\langle v(t)v(s)\rangle = (2V/z)\delta(t-s)$.

To review the key point of the fluctuation–dissipation relation, let us solve this question from a more basic level, constructing the Smoluchowski equation governing the particle density n. We follow **10.4**. The particle flux consists of two parts, the part due to the force $(-x/z)n$ and the part due to the entropic force $-(A/2)\nabla n$, where we used the relation between the diffusion constant and the noise amplitude $D = A/2$. The number of particles is conserved, so we have a conservation law.

$$\frac{\partial n}{\partial t} = -\text{div}\left(-\frac{x}{z}n - \frac{A}{2}\nabla n\right). \tag{10.49}$$

In equilibrium, there is no time-dependence, and also we assume there are no particles far away. Thus, we conclude

$$-\frac{x}{z}n - \frac{A}{2}\nabla n = 0 \tag{10.50}$$

or

$$\nabla \log n = -\frac{2x}{Az} \tag{10.51}$$

Hence, we get $n \propto e^{-x^2/Az}$. This gives the variance of x to be $\langle x^2 \rangle = Az/2 = V$, so $A = 2V/z$ just as we saw earlier.

Q10.2 Diffusing Globular Proteins

There are two proteins of mass m and M. Let us assume that the protein molecules are spherical and their average densities are the same. We know $M/m = 100$. For the smaller molecule to diffuse across a fixed length L in a cell it takes 0.23 s on average. What is the best guess of the time needed for the larger protein to diffuse across the same distance L?

Solution

$L^2 = \langle r^2 \rangle = 2\partial Dt$, and $D = k_B T/6\pi a\eta$. This means Dt is the same, so t/a is constant. Since we assume that the proteins are spherical and with the same density, $a \propto M^{1/3}$. That is, $t/M^{1/3}$ is constant. Hence, $t = 0.23(M/m)^{1/3} = 1.07$ s.

Do not use the gas phase formula obtained in **8.11** to calculate the diffusion constant in liquids.

Q10.3 Large Deviation for Binomial Distribution

Let $X_n = 1$ or 0 (head or tail) with equal probability. Find the large deviation function $I(y)$ for the empirical average: $(1/N)\sum_{n=1}^{N} X_n$.

Solution

$$-\log P\left(\frac{1}{N}\sum_{n=1}^{N} X_n \simeq y\right) = -\log \binom{N}{Ny} 2^{-N} \tag{10.52}$$

$$= N \log N - (Ny)\log(Ny) - N(1-y)\log[N(1-y)] - N\log 2 \tag{10.53}$$

$$= -N[y \log y + (1-y)\log(1-y) + \log 2]. \tag{10.54}$$

This implies

$$I(y) = y \log y + (1-y)\log(1-y) + \log 2. \tag{10.55}$$

As expected $I(1/2) = 0$. Let $x = y - 1/2$. Then,

$$I(y) = (x + 1/2)\log(2x + 1) + (1/2 - x)\log(1 - 2x) \simeq 2x^2 = 2\left(y - \frac{1}{2}\right)^2. \quad (10.56)$$

That is, as we wished,

$$P\left(\frac{1}{N}\sum_{n=1}^{N} X_n \simeq y\right) \approx e^{-2N(y-1/2)^2}. \quad (10.57)$$

This is essentially the same as (7.19).

PART II

STATISTICAL THERMODYNAMICS: BASICS

Macrosystems in Equilibrium

Summary

* Important observables for macrosystems are extensive (additive) or intensive.
* Even though underlying mechanics is reversible, macroscopic systems exhibit irreversibility.

Key words

additivity, extensive, intensive, fourth law, emergence of irreversibility

The reader should be able to:

* Explain why the total mechanical energy of a macroscopic system is additive.
* Explain why irreversibility naturally occurs in systems with many particles.

11.1 How to Describe a Macroscopic System in Mechanics

We do not need any special way to describe a macroscopic system, if we wish to describe it purely mechanically. Mechanical entities are atoms and molecules, so a system is mechanically described by the system Hamiltonian governing these particles.

The Hamiltonian of a system consisting of N point particles of mass m interacting with a potential energy $U(\boldsymbol{q}_1, \dots, \boldsymbol{q}_N)$ has the following form

$$H = \sum_{i=1}^{N} \frac{1}{2m} \boldsymbol{p}_i^2 + U(\boldsymbol{q}_1, \dots, \boldsymbol{q}_N). \tag{11.1}$$

☐ In quantum mechanics $\{\boldsymbol{q}_i\}_{i \in \{1,\dots,N\}}$ are the position operators of individual particles, $\{\boldsymbol{p}_i\}_{i \in \{1,\dots,N\}}$ are the momentum operators, and H is the Hamiltonian or the energy operator. In classical mechanics (in the classical approximation of mechanics) $\{\boldsymbol{q}_i\}_{i \in \{1,\dots,N\}}$ are position coordinates and $\{\boldsymbol{p}_i\}_{i \in \{1,\dots,N\}}$ are the momentum coordinates of the particles. They are jointly called the *phase coordinates*. See **A.14**, **A.35**.

The first term in (11.1) describes the kinetic energy K (here, K is the total kinetic energy). Usually, we will assume no external forces, and so U depends on the mutual relative, not absolute, particle positions, and H is the total mechanical energy.

Since we are interested in the "intrinsic" properties of the system, we are not interested in the translation and rotation of the system as a whole. Thus, we are interested in the Hamiltonian of the system observed from the coordinate system relative to which the system does not exhibit any overall translational and rotational motion (the co-moving coordinate system). The total mechanical energy of the system observed by the co-moving observer is understood as the "intrinsic" mechanical energy of the system and is called the *internal energy*[1] in thermodynamics.

11.2 Conservation of Mechanical Energy and the First Law of Thermodynamics

Unless there is a net exchange of energy with the external world, the total "intrinsic" mechanical energy (11.1) of a system stationary relative to the observer should be conserved according to the conservation of mechanical energy. Thus, the internal energy of a system must be a conserved quantity due to the conservation of mechanical energy. This is the essence of the *first law of thermodynamics* (see **13.1**). This was recognized by Carnot, Mayer, Joule, Helmholtz,[2] and others, but Helmholtz most clearly recognized the first law as a consequence of the conservation of mechanical energy, especially due to the fact that intermolecular interactions have potential functions.[3]

11.3 What Mechanics?

Quantum mechanics is the fundamental mechanics governing the world of atoms and molecules.[4] Therefore, for an overview of thermal physics, the microscopic world must be

---

[1] If there are electromagnetic effects, we include the electromagnetic energy in the internal energy.

[2] **Mayer, Joule, and Helmholtz** J. R. von Mayer (1814–1878) was a physician who realized that there had to be some relation among chemical energy (nutrients), work done by labor, etc., and conceived the conservation of "something." In 1842, using Mayer's cycle (see **15.1**), he determined the work equivalent of heat and demonstrated the quantitative conversion of work into heat (cf. **Q15.1**(1)) before Joule.

As we will see in **13.6** Joule reported his result in 1847, but he had already demonstrated the quantitative conversion of electric energy into heat in 1840. Interconversions of various physical and chemical quantities were discovered at that time by Volta, Oersted, Seebeck, Faraday, and others. Joule wanted to verify the quantitative equivalence of all these "energies" (although the concept became fully recognized only in the 1850s). To establish the concept of "energy" Helmholtz's work in 1847 was crucial. For a book on Helmholtz's life (1821–1894) and work see Meulders, M. (2010). *Helmholtz from Enlightenment to Neuroscience.* Boston: MIT Press.

[3] **Is Helmholtz's mechanische Weltanschauung empirical?** From the empiricist's point of view, however, a precise statement is that the empirically established first law of thermodynamics can be explained simply, *if we postulate* that the internal energy of thermodynamics is the total mechanical energy of the microscopic model of a system that obeys a conserved mechanics. We must not forget that Helmholtz's mechanical world view (mechanische Weltanschauung) was aprioristic (i.e., mechanics is an inevitable – in contrast to an empirical – truth of the world) as Mach pointed out.

Remember that thermodynamics was intact when classical mechanics was dethroned. Conservation of energy transcends mechanics.

[4] However, as empiricists, we must clearly recognize that there is no experimental demonstration that a macroscopic collection of atoms obey many-body mechanics. See a discussion in **12.1**.

described in terms of quantum mechanics. However, there is no (especially introductory) exposition that tries to do this squarely. We discussed Bernoulli's kinetic theory and derived the equation of state of an ideal gas in Chapter 2. Furthermore, in Chapter 5, we discussed the Maxwell distribution, but there was no quantum mechanics anywhere. Still, the obtained results are in agreement with the empirical world for a considerably wide range of phenomena. There may be many reasons for this, but the following two seem crucial.

(i) In many cases, our empirical experiences do not depend on microscopic details. For example, although it carries "mechanics" in its name, statistical mechanics does not actually use mechanics very much (as we will see later).

(ii) Due to the interaction between the systems and their environments, classical approximations may often become very accurate.

We casually introduce walls into thermodynamics and statistical mechanics, but we must clearly recognize that walls considerably spoil the intrinsic quantum mechanics of the system inside. Thus, macroscopic systems governed by pure quantum mechanics must be very rare. In statistical mechanics very often a system + heat bath is treated as an isolated macroscopic quantum mechanical system. However, since no unitary time evolution is realistic for any actual macrosystems, the point of view that the foundation of thermodynamic must be furnished by pure mechanics deserves critical scrutiny.

Conservation of energy has nothing to do with the number of particles in the system. What are the conspicuous features of a system consisting of numerous particles? Two features come to our mind immediately: additivity of energy and irreversibility of time evolution.

11.4 Additivity of Energy

The usual intermolecular interaction decays spatially sufficiently quickly. A general form of a two-body (or binary) interaction potential is illustrated in Fig. 11.1. The very popular *Lennard-Jones potential* has the following form

$$\varphi(r) = 4\varepsilon \left[\left(\frac{\sigma}{r} \right)^{12} - \left(\frac{\sigma}{r} \right)^{6} \right], \tag{11.2}$$

where ε is an energy scale and σ is the core size (a length scale) (Fig. 11.1).

Let us divide a macroscopic system in $\eth$-space into two halves with a clean cutting surface.[5] Then, the sum of the internal energies of the halves is almost (in the large system size limit) indistinguishable from the total internal energy of the system before dividing, if the two-body interaction potential decays spatially faster than $r^{-\eth}$.[6] Therefore, if the interaction potential decays faster than $r^{-\eth}$, the internal energy is proportional to the volume of the system.

---

[5] In order to avoid, e.g., a fractal surface. Such cutting is said to divide the system into *van Hove volumes*.

[6] The molecular potential is, generally speaking, not two-body, so, precisely speaking, U cannot be written as a sum of two-body interactions. Here, for simplicity, we assume that we can majorize $|U|$ by a sum of two-body interactions of the form mentioned in the text.

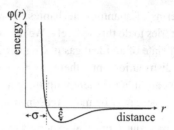

Fig. 11.1 The two-body intermolecular force potential $\varphi(r)$. The repulsive portion is very steep (any steep function will do to describe it, say, $1/r^{12}$), and is due to Pauli's exclusion principle restricting the electron-cloud overlap. The attractive portion $1/r^6$ is caused by the induced dipole–dipole interaction (the London force). Roughly speaking, the two-body intermolecular force is characterized by the repulsive core (or hardcore) diameter σ (the representative length scale) and the depth of the potential well ε (the representative energy scale).

☐ The Coulomb and gravitational interaction energies decay as $1/r$. For the Coulomb interaction, since the opposite charges gather around a given charge, the raw or bare Coulomb interaction between charges is shielded as $e^{-\alpha r}/r$ for some positive α, if the system is charge-neutral, so we need not worry about the Coulomb force. For gravity, there is no way to shield it, but the size L of the usual macroscopic object we encounter in thermal physics is macroscopic but is not cosmically huge. That is, there is a very broad "intermediate zone" for L satisfying "microscopic scale (e.g., σ) $\ll L \ll$ the cosmic scale," where we may ignore gravity. If the system is very heavy or cosmically large, then our ordinary statistical thermodynamics does not hold.

The charge or magnetic dipole–dipole interaction potential behaves just as $1/r^0$. This marginal case is delicate. For example, the sample shape matters to the internal energy as the reader may know well for magnetic energy of a magnet.

11.5 Why We Are Interested in Additive Quantities

To study a macroscopic system we often study additive observables that are proportional to the system size if the system is uniform (think of internal energy). An important reason for this is that the amount of material (e.g., number of particles) is additive. It is also natural to consider additive quantities, because they are big for big systems. If an observable becomes (relatively) smaller for larger systems, we need not pay much attention to such observables to understand the bulk properties of a macroscopic system. Thus, we are interested in observables that are independent of the system size (*intensive quantities*) and those proportional to the amount of materials (number of particles)[7] in the system (*extensive quantities*).[8] Strictly speaking, "additivity" and "extensivity" must be distinguished, but we will use the term "extensivity" to mean "additivity" as well.

---

[7] Precisely speaking, quantities proportional to conserved quantities related to the numbers of particles such as the baryon number or the electric charge.

[8] **Superextensive quantities?** Are there any "superextensive" quantities we must pay attention to that grow faster than the system size? Empirically, we do not have such quantities (we have already ignored the most important such effect: gravity, which is fortunately very weak) in our world. Perhaps, this might be a prerequisite for organisms with some logical capability to emerge.

We could say if we halve a macrosystem, the resultant halves are also macrosystems. We cannot continue this halving process indefinitely unless the initial system is "infinitely large." This means that if we wish to make a statistical mechanics of a macrosystem, logically we are required to take the huge system limit while "keeping the properties of the system." If the system size becomes infinite, almost all extensive quantities must diverge. Thus, strictly speaking, we can discuss only the densities of extensive quantities called *thermodynamic densities*. The system size infinite limit keeping all the densities constant is called the *thermodynamic limit*. We must be able to show the existence of the thermodynamic limit. A prototype is discussed in **Q11.1**.

11.6 The Fourth Law of Thermodynamics[9]

As discussed previously, macroscopically important observables are extensive or intensive. All the thermodynamic observables relevant to the macroscopic description of a system (i.e., the thermodynamics of the system) are either extensive or intensive. This is called the *fourth law of thermodynamics*.

11.7 Time-Reversal Symmetry of Mechanics

The world of mechanics is time-reversal symmetric. In the case of classical mechanics, Newton's equation of motion of a closed (isolated) system is an autonomous differential equation of second order without any first-order derivative: For an N-particle system, generally we have

$$\frac{d^2 q_i}{dt^2} = F_i(q_1, \ldots, q_N) \tag{11.3}$$

for $i = 1, \ldots, N$. Since the forces F_i are time t independent for a closed system, $t \to -t$ does not change the equations.

The Schrödinger equation (derived in **A.14**) for an isolated system reads

$$i\hbar \frac{\partial \psi}{\partial t} = H\psi, \tag{11.4}$$

where H is a self-adjoint operator independent of time. In this case $t \to -t$ might seem to alter physics, but what we observe is real, so the physics must be intact under complex conjugation.[10] Thus, quantum physics is also intact under time reversal.

---

[9] **The Fourth Law** This was emphasized and named by P. T. Landsberg (1922–2010). According to him, "There is nothing very startling about it since this law is always implicitly adopted, and people know about it. It is rarely displayed, however, as a 'law'." The real significance is that the law seems to be valid for nonequilibrium macrosystems as well. See Landsberg, P. T. (1972). The fourth law of thermodynamics, *Nature*, **238**, 229–231. This law will be explicitly invoked at various junctures in this book.

[10] Hermitian conjugation, more precisely. Recall that when Schrödinger first wrote his equation down, he could choose $-i$ instead of i as the reader can find in his original logic (see **A.14**).

In the long run, however, we are all dead and will never be resurrected (see Lucretius, Chapter 1). The world we live in is definitely irreversible. How can we reconcile this with mechanics?

11.8 Irreversibility from Mechanics?

All the ambitious young men (Boltzmann 1844–1906, Einstein 1879–1955, ...) tried to explain irreversibility from mechanics.[11]

Boltzmann derived in 1872,[12] from a pure microscopic mechanical description of atoms, an equation (called the Boltzmann equation, see **11.9**) that governs the irreversible time evolution of dilute gases. His colleague Loschmidt (1821–1895) asked why Boltzmann could derive an irreversible equation from a reversible equation. This question made Boltzmann realize that his derivation included a sort of coarse-graining, discarding subtle correlation effects, and that a statistical plausibility argument also sneaked in. Boltzmann also realized that the initial "nonequilibrium" states always contained more order, so the time evolution always drives the system in the direction to be less ordered.

In 1896,[13] Zermelo (1871–1953; who was from Dec. 1894 to Sept. 1897 an assistant to Planck (1858–1947)), utilizing Poincaré's recurrence theorem,[14] pointed out that Boltzmann's argument was logically flawed: sooner or later the state of a closed system returns to a state indefinitely close to the starting state, so no irreversibility occurs. Boltzmann admitted the flaw, but since he was a theoretical physicist, he responded, basically, that Zermelo knew mathematics but did not understand physics; think how long it takes such recurrence to happen. It would take far longer than the age of the universe even for a very small system.[15]

[11] **Physicists were philosophical in those days** "Boltzmann had an exceptional preparation in philosophy, compared with a physicist of today (especially if young and/or American), but certainly at a level that was not impossible to find in the *Mitteleuropa* of his days ..." Cercignani, C. (1998). *Ludwig Boltzmann: The Man who Trusted Atoms*. Oxford: Oxford University Press. p. 192.

[12] [In 1872 the world's first national park (Yellowstone NP) was established; Claude Monet painted *Impression, Sunrise*; B. Russell was born.]

[13] [In 1896 Röntgen discovered X-rays; Becquerel discovered radioactivity; Puccini composed *La Bohéme*; Battle of Adwa, in the first Italo-Ethiopian War.]

[14] **Poincaré's theorem on recurrence** Irrespective of the nature of dynamics (time-reversible symmetric or not), a measure-theoretical dynamical system (roughly, a mechanical system preserving a probability measure) can return to a state indefinitely close to its initial condition. A more precise version (for discrete time dynamical systems) is: Let μ be an invariant measure of a map f. If a set E satisfies $\mu(E) > 0$, then the trajectory $f^n(\omega)$ starting from almost all $\omega \in E$ returns to E infinitely many times. This "cyclic" return of a system to its initial state is called the *Poincaré cycle*. There is a quantum version as well.

[15] See section 1.4 Boltzmann Controversy in Ebbinghaus, H.-D., (2007). *Ernst Zermelo: An Approach to his Life and Work*. Berlin: Springer.

Zermelo and statistical mechanics Needless to say, Zermelo's greatest cultural contribution is the axiomatic set theory and the formulation of Axiom of Choice, but he never lost interest in the foundation of statistical mechanics; he was the translator of Gibbs' *Elementary Principle in Statistical Mechanics* (1902) to German. Although Zermelo appreciated Gibbs much more than Boltzmann, his criticism of Gibbs' theoretical framework is also serious.

11.9* The Boltzmann Equation[16]

The Boltzmann equation governs the number density in the position-momentum space $n(\gamma, t)$ with $\gamma = (\boldsymbol{q}, \boldsymbol{p})$, where $\boldsymbol{q} \in \mathbb{R}^3$ is the position vector and $\boldsymbol{p} \in \mathbb{R}^3$ the momentum vector for a single particle of mass m. It has the form

$$\frac{\partial n}{\partial t} + \frac{\boldsymbol{p}}{m} \cdot \frac{\partial n}{\partial \boldsymbol{r}} + \boldsymbol{F} \cdot \frac{\partial n}{\partial \boldsymbol{p}} = \int d^3 \boldsymbol{p}_1 \int d\sigma \, v_r [n(\gamma') n(\gamma_1') - n(\gamma) n(\gamma_1)], \qquad (11.5)$$

where the right-hand side describes the effect of two-body collisions.[17] We will not discuss this equation any further.[18]

11.10 What is the Lesson?

What we have learned from these debates is that:

(1) Very often the initial state is special (away from equilibrium) so (even following the pure mechanics) for a long time irreversible behaviors are observed.
(2) However, if we can wait for long enough, any finite dynamical system (even after coarse-graining) almost comes back to its initial special state, but the required time is enormous, and we never experience this reversibility for macrosystems.

A toy model can illustrate (1) and (2). Suppose a point is going around a unit circle with period 1. The point is under the influence of a noise that makes its angular speed fluctuate (see Fig. 11.2).

If there are only two such oscillators, their average position may become close to the origin, but then the average recovers its original amplitude fairly easily. If we have many ($N \gg 1$) such oscillators, after the average becomes close to zero, it stays small for an enormously long time, and then will return close to the original value. The waiting time for this recovery is likely to be of order e^{cN}, where c is a positive constant of order 1.

Thus, as long as the system is finite, Zermelo is right, but as to the waiting time Boltzmann is right.

---

[16] Boltzmann, L. (1872). Weitere Studien über das Wärmegleichgewicht unter Gasmolekülen, *Wiener Berichte*, **66**, 275–370. Notice that Boltzmann's equilibrium statistical mechanics (1878; see **Q16.2**) originated from his efforts to answer Loschmidt's criticism against the explanation of irreversibility using this equation.

[17] Here $d\sigma$ is the differential scattering cross section and v_r is the relative speed of colliding particles.

[18] **We do not dwell on the Boltzmann equation** The derivation of hydrodynamics and computing transport coefficients from the Boltzmann equation as well as the study of its general properties were quite important topics in the traditional nonequilibrium studies, but we will not discuss this equation, because it is applicable only to fairly dilute gases. Thus, the derivation of hydrodynamics via this equation is quite unsatisfactory. Furthermore, unless we use a complicated two-body interaction potential, we must use different two-body potentials (e.g., the Lennard-Jones potential (11.2) with different parameters σ and ε) for different transport coefficients (say, shear viscosity and heat conductivity) for the *same* gas at the *same* temperature and density.

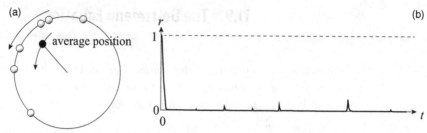

Fig. 11.2 (a): Ensemble of points with angular speeds slightly fluctuating around 2π rad/s. The averaged position spirals toward the center. (b): A schematic plot of the radial position of the center of mass.

11.11* Purely Mechanical Toy Model and Arrow of Time

Some readers might question the previous argument, because "noise" is introduced from outside; we could easily make a deterministic example almost indistinguishable from the aforementioned model. Let the coordinates of the ith particle be

$$x_i(t) = \cos 2\pi t + \eta_{ix}(t), \quad y_i(t) = \sin 2\pi t + \eta_{iy}(t), \tag{11.6}$$

where η_{ix}, η_{iy} are the coordinates of the (scaled) Sinai billiard model (Fig. 1.1) on the square $(-\varepsilon, \varepsilon] \times (-\varepsilon, \varepsilon]$ (take $\varepsilon > 0$ appropriately small). Each particle has its own billiard, so subscript i is attached. The resultant n-rotating particle system is a perfectly time-reversal symmetric and deterministic system. Even for this system **11.10** (1) holds.[19]

If we reverse time in the middle of the decay of the initial peak in Fig. 11.2(b), the system perfectly climbs back to the initial state. The *reader might* say after all the arrow of time is inserted by hand with a special initial condition and studying only $t > 0$; from the same "initial" condition we could study $t < 0$ to observe a picture very similar to the mirror image of Fig. 11.2(b) with respect to the vertical axis at $t = 0$.

In reality, we always have some external effect (actually, no macrosystem can be really isolated), so even if the time is reversed, the decay likely continues for large systems. One might say this is the ultimate explanation of the direction of the time arrow.

To describe the time evolution of the world, we need an observer. How can we, the observers, judge which direction of time is future and which is past as organisms? Notice that the passing of time t is never felt directly; one might say time is felt only through the accumulation of regrets. The time direction we experience while living in the world is the direction we call future. If the world is observed by an intelligent organism, there is no other direction of time. Thus, the question of the direction of the time arrow is a pseudo question.

---

[19] The reader might still say the toy model lacks many-body dynamics, because each particle obeys its own dynamics. It is, however, easy to revise the model so the noise η_i is constructed by the projection of a high-dimensional Sinai-like billiard model.

11.12 The Law of Large Numbers Wins

The macrostate we are likely to observe in the toy model discussed here for large N corresponds to the almost flat bottom portion in Fig. 11.2(b). The compatible microstates are "most generic" without any particular nonuniform distribution of the points along the circle. This macrostate is the most probable in the sense that it is compatible with the maximum number of particle configurations (i.e., the microstates), so, as we discussed in Chapter 7, it is the equilibrium state. The equilibrium state can be reached purely mechanically without any external effects. Although, for a finite system, as illustrated by the spikes in Fig. 11.2(b), the equilibrium state does not last forever, if the system is large enough (N is large), the duration of the equilibrium state is very long; the system is almost always in equilibrium, once it reaches the equilibrium state. This is believed to be the general feature of the usual macroscopic systems, classical or quantum.

Observe that this equilibrium state is compatible with the law of large numbers with respect to the microstates: the average position of all the particles is almost at the origin (and forever so in the $N \to \infty$ limit). If a system contains a huge number of particles, the system eventually reaches a state compatible with the law of large numbers and would stay there forever. This is the essence of the irreversible tendency of macroscopic systems.

We can claim that thermodynamics (Chapter 12 and subsequent chapters) describes the features of macrosystems that survive the filter of the law of large numbers (i.e., the features that are with an overwhelming probability (almost sure) in the $N \to \infty$ limit). We may say the only reliable conclusions of statistical mechanics (which starts at Chapter 17) are those compatible with the law of large numbers (and deviations from them understandable with the aid of the large deviation principle).

Problem

Q11.1 Existence of Ground State Energy Density

Let us assume that the system Hamiltonian satisfies the following two conditions:

(i) If two macroscopic cubes are juxtaposed with the spacing between the two parallel surfaces larger than d, there are only attractive interactions between the two cubes.

(ii) The ground state energy $E(N)$ of the cube with N particles may be bounded from below as

$$E(N) \geq -NB, \tag{11.7}$$

where B is a positive constant. This is a stability condition for the system not to collapse indefinitely. If the interaction potential has a sufficiently hard core, and the attractive interaction is not too long-ranged, this stability condition holds.

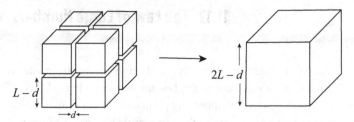

Fig. 11.3   Let us juxtapose 2^D cubes of edge $L - d$ with spacing d as illustrated to make a cube of edge $2L - d$. Within this cube after joining smaller cubes, the particles can go anywhere within the cube. This rearrangement can reduce the total energy. Repeating this procedure n times, we can make a cube of edge $2^n L - d$.

Using the procedure explained in Fig. 11.3, we join 2^D cubes of edge $L - d$ containing N particles with spacing d. In the new cube of edge $2L-d$, $2^D N$ particles rearrange themselves and a new ground state with energy $E(2^D N)$ is realized. Show that $E(N)/N$ converges in the large N limit. This means $E(N)/V$ converges in the large volume limit.

Solution

With the aid of the usual variational principle for the ground state we have

$$2^D E(N) \geq E(2^D N). \tag{11.8}$$

Repeating this, we have

$$E(N)/N \geq 2^{-D} E(2^D N)/N \geq 2^{-2D} E(2^{2D} N)/N \geq \cdots \geq 2^{-nD} E(2^{nD} N)/N \geq \cdots . \tag{11.9}$$

Thanks to (ii) this is bounded from below by $-B$, so this monotone decreasing sequence converges. The total volume divided by 2^{nD} also converges to L^D, so total internal energy/total volume converges in the large volume limit. Thus, the ground-state energy density exists.

Strictly speaking, this demonstration is for a special sequence of cubes. The key ideas to remove this limitation are:

(a) In the large volume limit, the surface effect can be ignored.
(b) Any shape may be approximated by a collection of cubes.

For physicists, these should be easily acceptable assertions.

12 Thermodynamic Description

Summary

* A phenomenological description of macroscopic systems is a description solely in terms of quantities observable and controllable at the macroscopic scale.
* A macrosystem in a certain constant environment can be in a time-independent state called the equilibrium state.
* An equilibrium state is described in terms of thermodynamic coordinates consisting of internal energy and work coordinates.

Key words

phenomenology, equilibrium state, thermodynamic coordinates, work coordinates, thermodynamic space, quasistatic process, state function

The reader should be able to:

* Explain what an equilibrium state is.
* Explain why the thermodynamic coordinates are privileged coordinates.
* Explain what the thermodynamic space is and what the state functions are.

Can we have a theoretical framework to systematize empirical facts of macrosystems without referring to microscopic details?

12.1 What is Phenomenology?

A phenomenological description of macrosystems is a description solely in terms of quantities that may be observed, controlled, and described at the macroscopic scale. If we could make a closed (complete) system of theory in terms of such quantities, the result is called *phenomenology*.[1]

---

[1] For a general discussion of phenomenology, see Oono, Y. (2013). *The Nonlinear World*. Tokyo: Springer, chapter 3 Phenomenology.

The phenomenology of macroscopic systems in equilibrium (see **12.2**) we now have is called *thermodynamics*. The reader must clearly recognize that, in contrast to (the supposedly more fundamental) statistical mechanics, thermodynamics survived the quantum revolution unscathed; and it actually helped launch the revolution. Quantum mechanics has no problem with thermodynamics (for now), but even if quantum mechanics is replaced by something else in the future, thermodynamics will remain intact.

In physics, "phenomenology" has not necessarily been respected; it could almost be a pejorative (e.g., in high-energy physics). However, notice that when underlying microscopic descriptions are impossible or only approximate, phenomenology may be the only realistic rational means for human beings to grasp the world. As we will see soon, thermodynamics correctly captures a certain universal feature of the macroscopic world which is quite insensitive to the microscopic details. A phenomenology is not an approximate way to understand the world nor a crude version of something more accurate; thermodynamics is not an approximation to a certain more advanced and accurate theory.[2] Perhaps, intelligent beings could emerge only in such worlds that allow phenomenological descriptions.

However, one may well claim that as an empirical science, quantum mechanics is much more precisely validated than thermodynamics. Then, a phenomenology such as thermodynamics should be regarded as a secondary non-fundamental part of physics, even from the purely empirical science point of view. Rigorously speaking, however, there is no firm empirical verification that a macroscopic object is governed by pure quantum mechanics. It is usually argued that we can check the hypothesis that all the particles in macrosystems are governed by quantum mechanics by comparing its consequences with empirical facts. However, as already noted and as we will see later, statistical mechanics actually uses only a small portion of mechanics, so it is hardly possible that phenomenology can be used to confirm mechanics.[3]

As we will learn soon, thermodynamics is a theoretical system distilled from empirical facts, and its fundamental laws dictate the basic characteristics of the world. Precise empirical verifications of its principles tend to be regarded as technological feats; if verification fails, the usual reaction is to suspect that the experiment is flawed (and always that has been the case).

Since every macroscopic system consists of numerous microscopic constituents, any change at any space-time scale is possible. It is hard to expect that violent changes could be phenomenologically described, because such changes could easily involve short wavelength and high frequency changes. We must select reasonably "calm" space-time systems/states to construct a macroscopic phenomenology.

---

[2] Hal Tasaki's summary.

[3] **Implication of pure mechanics** There are theoretical works that demonstrate thermodynamic results purely in terms of mechanics. Such demonstrations (if valid) should be interpreted as the verification of the consistency of pure mechanics to thermodynamics, because pure mechanics holds only under unrealistic conditions for macroscopic systems in general.

12.2 Equilibrium States

A macroscopic system (a system with extremely many particles[4]) can take a special state called a *thermodynamic equilibrium state* (called an equilibrium state for simplicity). To specify the state, we introduce the concept, *generalized isolation*: a system is subjected under a generalized isolation condition if it is isolated but may be coupled to uniform field(s) conjugate to work coordinates (to be defined shortly). A macrosystem is in an *equilibrium state*, if it is left under a generalized isolated condition and eventually reaches a time-independent state.[5]

12.3* External Disturbance: Mechanics vs. Thermodynamics

According to empirical facts, if efforts are made sufficiently to remove external uncontrolled effects (let us call this "in the $R \to 0$ limit," R for external random perturbations), then thermodynamic observation results are reproducibly obtained. However, it is certain that this limit $R \to 0$ is far from the rigor of removing external disturbances required by mechanics experiments; notice, for example, that an appropriate heat bath may be attached to a system without affecting its thermodynamic observables. The empirical fact that foundational experiments on quantum mechanics are extremely hard eloquently attests to the difficulty of maintaining really isolated mechanical systems.[6]

If a system contains only a few particles, we can take a time span where the $R \to 0$ limit is essentially realized, so critical experiments on quantum mechanics have been accomplished. However, for a larger number of particles and for a longer time span this is prohibitively difficult. That is, the limit $R \to 0$ and neither the long time $t \to \infty$ nor the large system limit $N \to \infty$ is commutative in mechanics. Pure mechanics demands the

---

[4] **What is macroscopic?** From the strictly macroscopic phenomenological point of view, since whether atoms and molecules exist or not is an irrelevant issue, it is more logical to say, "a system around us that we can see by our naked eyes" (as explained in Chapter 1, this characterization is actually scientific), but this book should be practical as well, so anything useful will be used to understand thermal physics.

[5] **Equilibrium: another possible characterization** Equivalently, a macrosystem is in an equilibrium state, if we can devise a (macroscopically) constant and spatially uniform environment (without any dissipative currents) into which we can embed the system (with appropriate boundary conditions) and still the state does not change in time. Total isolation is a possible environment. However, not all the equilibrium states of a given system may be realized under the total isolation condition. Notice that this characterization of equilibrium never requires the isolation of a system from the external world. Also it never asks how the equilibrium state is realized. The state may be prepared in contact with a heat bath, for example.

[6] **Purely mechanical macrosystems are fictitious** "In essence, macroscopic systems are open, and their evolution is almost never unitary." (Zurek, W. H. (2003). Decoherence, einselection, and the quantum origins of the classical. *Rev. Mod. Phys.*, **75**, 715–775).

For classical systems, everyone including Maxwell clearly recognizes the difficulty in isolating mechanical systems. It is said quantum systems are much more stable than classical (esp. chaotic) systems, but as noted in the text from our macroscopic point of view quantum systems are sufficiently fragile.

$R \rightarrow 0$ limit first, but thermodynamics does not require $R \rightarrow 0$ rigorously (may not be required even at the end as noted in the preceding paragraph).

It should be clear by now that the regime that mechanics can handle and the regime where thermodynamics lives hardly overlap each other (since $N \gg 1$) under realistic conditions. The reader might still say that mechanics can tell us something reasonable about macroscopic systems as discussed in Chapter 11. However, first of all, we never used dynamics in the extensivity study. Dynamics is used in the explanation of irreversibility, but in this case, we need not stick to mechanics. The point is simply that pure mechanics being compatible with irreversibility does not imply it is required for irreversibility.

Therefore, the pure mechanical study of the foundation of macrophenomenology is, though a highly interesting mathematical question, not really a natural science issue. The situation is parallel to ergodic theory. As a part of mathematics it is extremely important, but for the foundation of statistical mechanics it has been a red herring.

12.4 Thermodynamic Coordinates: a Privileged Set of Variables

Since thermodynamics must be a phenomenology at the macroscopic scale, to construct a closed (self-contained) theoretical framework, we must carefully choose the physical quantities we deal with. The most fundamental of them are the *thermodynamic coordinates*. They are extensive quantities and consist of internal energy E and other variables (called *work coordinates*) that we can observe and control mechanically (and electromagnetically) macroscopically.[7] The system volume V is often among them. For a magnetic system, magnetization M is included.

Thermodynamic coordinates are a very special set of variables to describe equilibrium states that are privileged in the following sense:

---

[7] **How can we choose thermodynamic coordinates?** The reader may complain that the explanation in the text never clearly discusses how to choose these coordinates for a given system. They must include the internal energy E. The work coordinates are so chosen that they can be independently controlled and that the first and the second laws of thermodynamics (Chapter 13) must hold.

The equilibrium state is an equivalence class (for a mathematical definition of "equivalence class" see footnote 1 of Chapter 17) of macroscopically distinguishable states (notice that not all the macroscopic observables may be thermodynamically meaningful observables) by the values of the thermodynamic coordinates. Therefore, logically speaking, the uniqueness of the equilibrium state specified by thermodynamic coordinates is merely by definition (i.e., tautological). However, if an equivalence class includes easily distinguishable macrostates, it is very likely that the chosen work coordinate set is not complete, because our capabilities to observe (to see) and to control (to do) have evolved hand in hand, so usually there is a close correspondence between them macroscopically; where we can detect differences, we can apply different actions.

Suppose, for example, for a gas system E and V are chosen. If the gas particles have magnetic moments, but we fail to include M in the work coordinate set, what happens? Consequently, the magnetic field B is out of our sight, but perhaps the environmental B changes and, e.g., the second law might be violated apparently. Then, at least we recognize that our thermodynamic coordinate system is incomplete and we must look for the cause. As can be seen from this explanation, if we do not touch some variables at all (as independent variables) and are confident that there is no natural cause to change them, we can drop them from the list of the work coordinates.

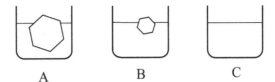

A   B   C

Fig. 12.1 A–C contain the same amount of water (liquid + floating ice) at 0 °C under 1 atmospheric pressure. However, their internal energies are distinct; A has the least internal energy and C the most. In elementary thermodynamics, often the temperature T appears as a key variable instead of internal energy E, but these examples clearly tell us that T cannot distinguish equilibrium states that are distinct. Analogously, in the liquid–gas phase transition under constant pressure P and T, E, and V can change. These examples indicate that thermodynamic coordinates are the fundamental "privileged" set of variables to describe thermodynamic equilibrium state; generally speaking, intensive variables such as T and P fail to describe macrostates uniquely.

(i) To understand thermodynamic coordinates we need only (macroscopic) mechanics and electromagnetism. We accept these macromechanics (perhaps relativistically corrected) as our firm empirical ground. The work that a system does or that a system receives can be described through the changes in these coordinates and is quantified solely mechanically and electromagnetically. Notice that we need not understand what "heat" or "temperature" is. Thus, as long as the thermodynamic properties are concerned, a macroscopic system is regarded as a black box with mechanically controllable handles.

(ii) The thermodynamic coordinates uniquely specify the equilibrium state. The significance of this statement is illustrated by the example in Fig. 12.1. Notice that temperature T or pressure P is *not* qualified to be included in the thermodynamic coordinates.

12.5 Thermodynamic Space

The space spanned by the thermodynamic coordinates is called the *thermodynamic space* or thermodynamic state space. For a given macroscopic system, each equilibrium state uniquely corresponds to a point in its thermodynamic space.

Very Important Warning However, the converse is not true. That is, although any equilibrium state has its unique representative point in the thermodynamic space, even the same point may correspond to a nonequilibrium state. The crucial point is that the point in the thermodynamic space itself cannot tell us how it is changing (reversibly or irreversibly). For example, a hot coffee in a high-quality thermos very gradually cools toward the room temperature. The process could be indefinitely slow, so its state is always infinitesimally close to a certain equilibrium state for a certain length of time. However, it is patently a state undergoing an irreversible process. Whether a point in the thermodynamic space is in equilibrium or not depends on the context (and the time scale).

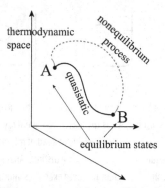

Fig. 12.2 A and B are equilibrium states. A quasistatic process connecting A and B is in the thermodynamic space (however, see the warning in the text). From A to B a process need not be quasistatic. Then, most such processes are outside the thermodynamic space (broken curve).

12.6 Quasistatic Processes

In thermodynamics we wish to study the change of one equilibrium state to another through various processes. Most processes allowed to the system cannot be described in the thermodynamic space, because most processes are via various nonequilibrium states. A *quasistatic process* is a process along which both the system and its environment are (infinitesimally) close to equilibrium and can retrace their evolution precisely, i.e., "reverse their footsteps." Thus, a quasistatic process is also called a *retraceable process*.[8] Here "retraceable" means that, after retracing the process, the world returns exactly to the original (macro)state. Most processes are not retraceable and so cannot be described by the thermodynamic coordinates alone. In contrast, in a retraceable process, the system is always close enough to equilibrium that the thermodynamic coordinates provide a sufficient description. In practice, retraceability often requires that the system evolve slowly (hence the term "quasistatic").[9]

As is clear from our definition, a quasistatic process is reversible. A process from state A to another state B is a *reversible process*, if there is a means to go from B to A with no trace of the process left in the world. Logically speaking, a reversible process need not be retraceable, but usually, a reversible process is synonymous to a quasistatic process just defined.

---

[8] In this book, "quasistatic process" is used in this strict sense. The reader must be careful, since different definitions of "quasistatic process" may be used in other books.

[9] **Why can we often realize a reversible process by slowing down?** Just as the slowly cooling coffee or slow leakage of gas exemplifies, slowing down is not enough to be reversible. However, if the driving force of the change is small, then the energy dissipation rate is a higher order small quantity (recall, e.g., the electrical power loss V^2/R, where V is the electric potential difference = the driving force across the resistor of resistance R). Thus slowing down by reducing driving forces allows us to realize quasistatic processes. We could reduce the changing rate by choking the flux instead of reducing the driving force. In this case slowing down the changing rate does not help us to realize reversible change. The two examples at the beginning of this footnote are the examples. Thus, for example, heat flow is reversible only if the flow is between two infinitesimally different temperatures (see **15.6**).

Warning (corresponding to the earlier "very important warning"). A quasistatic process is described as a continuous curve in the thermodynamic space, but the converse is not always true just as the case of the correspondence between equilibrium states and the points in the thermodynamic space. That is, a continuous path in the thermodynamic space does not unconditionally describe a quasistatic process. Again, think of gradually cooling coffee in a thermos (see also **Q15.4**: ultraslow leakage of gas from a container).

12.7 State Quantities/Functions

If the value of a macroscopic quantity of an equilibrium state is uniquely specified by the state, the quantity is called a *state quantity* (or *state function*). In particular, any observable that is a function defined on the thermodynamic space is a state function. Its value is indifferent to how the state is realized. For example, the equilibrium volume of a system is a state function; temperature is another example.

12.8 Simple System

An equilibrium system need not be spatially homogeneous at the macroscopic scale. If a system is spatially homogeneous, we call it a *simple system*.[10]

12.9 Compound System

Two or more simple systems considered as a single system with or without certain interactions among them are called a *compound system*. Even if the component simple systems are in equilibrium, the compound system as a whole may not be in equilibrium. The thermodynamic space of the compound system is the direct product of the thermodynamic spaces of the constituent simple systems. Just as in the thermodynamic space of a simple system, a point in the thermodynamic space of a compound system may correspond to a nonequilibrium state. We have to specify carefully the interactions among the constituent subsystems.

12.10 Why Thermodynamics Can Be Useful

When an initial equilibrium state and a final equilibrium state are given, the change of a state function between these two states does not depend on the actual process but only on

---

[10] If we need spatially inhomogeneous states, the system will be partitioned into sufficiently homogeneous macroscopic subsystems; if this is impossible, we will not discuss it thermodynamically in this book.

these initial and final equilibrium states. Even if the actual process connecting these two states is violent and irreversible, devising an appropriate quasistatic process connecting the same end points, we can thermodynamically compute the change of any state function. This makes thermodynamics extremely useful in practice.[11]

Traditional thermodynamics often includes the zeroth law of thermodynamics about the existence of temperature (see **12.13** for completeness), but there are much more important general properties of the systems in thermodynamic equilibrium.

If one wishes to have a formal axiomatic system, the existence of the objects of the theoretical system must be guaranteed. Thus, it is often stated at the beginning that if a macrosystem is left isolated for a sufficiently long time, an equilibrium state is realized. Unfortunately, as noted already, this cannot guarantee the existence of all the equilibrium states we are interested in. Furthermore, it is not our concern how equilibrium states are realized. Besides, isolation may not be really meaningful in practice. Empirically, it is much more important that we can check experimentally whether a system is in equilibrium or not, although this is not very easy in practice; we must check that all the basic principles are satisfied.

Next, an important property is mentioned to make the fourth law (**11.6**) operationally meaningful.

12.11 Partitioning–Rejoining Invariance

A (simple) macroscopic system in equilibrium is partitioning–rejoining invariant: if a macroscopic system in equilibrium is divided into two halves (of about the same sizes[12]), the halves are themselves in equilibrium. If they are rejoined, the result is indistinguishable from the original system as far as the thermodynamic observables are concerned (see Fig. 12.3).[13]

---

[11] **Existence of quasistatic process** The astute reader may wonder whether we can always devise a quasistatic process connecting any two equilibrium states that can be connected by a certain process. Strictly speaking, this is assumed, but if the two states belong to the same connected component of a realizable equilibrium state set in the thermodynamic space, this is a very natural assumption.

[12] Needless to say, the dividing surface should not be fractal. That is the resultant halves must be van Hove volumes. Let us call such partition *van Hove partition* (see **11.4**).

[13] **How to partition** A homogeneous equilibrium state may be anisotropic (like an ordered magnetic system) or even with movement (superfluidity). In such cases, careful matching of the parts is required.

   Isn't partitioning necessarily a violent process destroying equilibrium? An official answer is, no. Suppose we wish to divide an ice cube into halves. We could imagine a huge ice block and saw it into halves; the damage would be confined to the neighborhood of the separation surface, and the effects are not extensive. Actually, we can avoid such a brute-force separation. Since we are interested in the thermodynamic states, the actual process does not matter. Thus, we could reversibly melt the ice cube, divide the resultant liquid water into halves, and then make half ice cubes (controlling the crystal axes at least in principle). It should always be possible to invent a quasistatic process to realize partitioning.

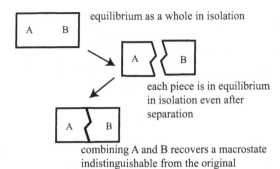

equilibrium as a whole in isolation

each piece is in equilibrium
in isolation even after
separation

combining A and B recovers a macrostate
indistinguishable from the original

Fig. 12.3 Partitioning–rejoining invariance of equilibrium states: this could be understood as a characterizing feature of macroscopic systems within macroscopic phenomenology without referring to molecules.

12.12 Thermal Contact

Empirically we know that even if there is no exchange of work or matter, two systems in contact can exchange their energies. Such a special contact is called *thermal contact*. If two systems A and B are in thermal contact and are in equilibrium as a compound system **12.9**, we say A and B are in *thermal equilibrium*.

12.13* The Zeroth Law of Thermodynamics

If systems A and B are in thermal equilibrium, and if B and C are in thermal equilibrium, then so are systems A and C, if in thermal contact. That is, the thermal equilibrium relation is an equivalence relation.[14] This is often called the *zeroth law of thermodynamics*. Then, usually, a demonstration is given of the statement that thermal equilibrium is temperature equilibrium: there is a scalar quantity called *temperature* (or more precisely, an empirical temperature) that takes identical values for two systems if and only if they are in thermal equilibrium.

We will not show this (= the existence of temperature), because we do not need this traditional zeroth law to develop thermodynamics and because many more assumptions are needed than the zeroth law stating that the thermal equilibrium relation is an equivalence relation.[15]

[14] For equivalence relation see footnote 1 of Chapter 17.
[15] See Lenker, T. D. (1979). Caratheodory's concept of temperature, *Synthese*, **42**, 167–171.

13 Basic Thermodynamics

Summary

* The first law is the conservation of energy (but for sufficiently slow processes).
* The internal energy does not decrease adiabatically, if there is no net change of work coordinates (the second law).
* The second law implies that the thermodynamic space is foliated into adiabats equal to constant-entropy surfaces.
* Entropy is adiabatic reversibility invariant, and $dS = d'Q/T$ for reversible import of heat $d'Q$.

Key words

conjugate pair, Clausius' law, Kelvin's law, Planck's law, adiabatic process, adiabat, entropy, fundamental equation, Gibbs relation

The reader should be able to:

* Recognize that the first law is essentially the law of conservation of energy, with some extra requirements by thermodynamics.
* Understand various expressions of the second law.
* Understand how to demonstrate the existence of entropy, starting with empirical observations.
* Write down the Gibbs relation, recognizing its general structure.

13.1 The First Law of Thermodynamics

The *first law of thermodynamics* is essentially the conservation of mechanical energy (internal energy) of the system as already discussed in **11.2**. Mayer, Joule, Helmholtz, and others established the existence of a state function E called *internal energy*. Its change ΔE cannot be explained solely in terms of the total work W supplied to the system, and the discrepancy Q is understood as the energy transferred in the form of heat to the system:[1]

---

[1] **Macrokinetic vs. thermal energy** Take one liter (1 kg) of water. We need 2 J of energy to accelerate it from rest to 2 m/s, but this same amount of energy cannot even raise the temperature of the same amount of water by 1 mK. This tells us how revolutionary the steam engine was to convert thermal energy even partially into macrokinetic energy.

$$\Delta E = W + Q. \tag{13.1}$$

Thus, in terms of thermodynamic coordinates that can be understood and quantified solely mechanically and electromagnetically, something called heat (whose true nature is not very clear) is macroscopically quantified.[2]

Notice that although E is a state function, neither W nor Q is a state function; W and Q are distinguished by how energy is transferred, so they depend explicitly on the path connecting the initial and the final equilibrium states (the path need not be in the thermodynamic space). They are not the differences of some state functions, so never write them as ΔW or ΔQ. Their differentials are not exact (or perfect), so inifinitesimal work and heat transfered are expressed as $d'W$ and $d'Q$ instead of dW and dQ.

13.2 Sign Convention

Let us make the *sign convention* for W, Q, etc., explicit. The sign is seen from the system's point of view: everything imported to the system is positive, and exported negative. For example, if we do some work to the system in the ordinary sense, $W > 0$. If we get some useful work, in the ordinary sense, from the system, $W < 0$.

When the change is quasistatic, W is determined by the equilibrium states of the system along the quasistatic path. Let us look at typical examples.

13.3 Volume Work

The work $d'W$ required to change the system volume from V to $V + dV$ is given by (see Fig. 13.1):

$$d'W = -Fdl, \tag{13.2}$$

if the infinitesimal displacement of the piston is dl and the external force is F. Here, the differential expressing the infinitesimal work is written as $d'W$ instead of dW to indicate clearly that the infinitesimal is not the differential of a state function (not a perfect differential[3]). According to our sign convention, if the system is doing work, then $d'W > 0$, but this happens when the volume shrinks, that is, when $d\ell < 0$. Therefore, (13.2) has a minus sign. If the volume change is quasistatic, the system pressure P and the external force F are always in balance,

$$F = A \times P, \tag{13.3}$$

---

[2] **Work–heat distinction** We will assume that work and heat may be clearly distinguished throughout this exposition. If the processes we consider are slow enough and macroscopically controllable, then this is true, but if a process is violent, the distinction could be meaningless.

[3] Or not an exact 1-form.

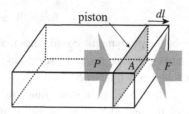

Fig. 13.1   Work done by volume change.

where A is the cross section of the piston. Thus, we have arrived at the formula applicable to the quasistatic process:

$$d'W = -PAdl = -PdV. \tag{13.4}$$

If the process is fast, there would not be sufficient time for the system to equilibrate. For example, when we compress the system rapidly, the force necessary and the force given by (13.3) can be different. Even worse, the pressure P may not be well defined if the change is too rapid. Consequently, (13.4) does not hold.

Thus, although the first law is essentially the conservation of energy, to express it in terms of a small number of variables, the change must be gentle (quasistatic).

13.4 Magnetic Work

As an example of electromagnetic work let us derive the expression for the work required to increase the magnetization $\boldsymbol{M}$ in the magnetic field $\boldsymbol{B}$:

$$d'W = \boldsymbol{B} \cdot d\boldsymbol{M}. \tag{13.5}$$

Put a bar magnet of length L in a solenoid with n turns of wire per unit length (Fig. 13.2). If a current I flows through the coil of the solenoid, a magnetic field $\boldsymbol{B}$ of intensity $B = \mu_0 nI$ is created, where μ_0 is the magnetic permittivity of vacuum. With the bar magnet whose magnetization density is $\boldsymbol{m}$, the magnetic field $\boldsymbol{B}'$ of the place occupied by the magnet is given by (see **13.5** for the convention)

$$\boldsymbol{B}' = \boldsymbol{B} + \mu_0 \boldsymbol{m}. \tag{13.6}$$

Let us vary the current: $I \rightarrow I + \delta I$. The work needed for this change is, if no Ohmic loss exists, the total change of $\mathcal{E} \times I$, where $\mathcal{E}$ is the induced electromotive force due to Faraday's law. Here $\mathcal{E}$ is zero before changing I, so we have

$$\delta W = -\int dt\,(I + \delta I(t))\delta\mathcal{E}(t) = -I\int dt\,\delta\mathcal{E}(t) \tag{13.7}$$

to the first order. Here, the minus sign is required, because the work is done against the electric potential drop. Let A be the cross section of the solenoid. Then, the magnetic flux

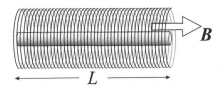

L

A bar magnet in a solenoid.

without the bar magnet is $\Phi = \mu_0 nAI$, and the contribution of the bar magnet is ma, where a is its cross section. Faraday's law tells us, since there are nL turns,

$$\delta \mathcal{E}(t) = -nL \left(\mu_0 n \frac{dI}{dt} A + \mu_0 \frac{dm}{dt} a \right). \tag{13.8}$$

Thus, (13.7) reads (recall $\mu_0 nI = B$)

$$\delta W = \mu_0 nLI(n\delta IA + a\delta m) = \frac{1}{\mu_0} B\delta BLA + B\delta M = LA\delta \left(\frac{1}{2\mu_0} B^2 \right) + B\delta M, \tag{13.9}$$

where $La\delta m = \delta M$ (the total magnetization change). The first term in this expression is the change of the vacuum magnetic energy independent of the presence of the magnet. Therefore, only the second term increases the energy of the magnet. Thus, we have (13.5). Another derivation can be found in **Q13.4**.

Following the derivation similar to **Q13.4**, the work done by the electric field is analogously obtained as

$$d'W = \boldsymbol{E} \cdot d\boldsymbol{P}, \tag{13.10}$$

where $\boldsymbol{E}$ is the external electric field and $\boldsymbol{P}$ is the polarization.

13.5* Convention for Electromagnetic Field

In this book we adopt the IS units. That is, the magnetic field is induced by current, and Maxwell's equations in a vacuum read

$$\operatorname{div} \boldsymbol{E} = 0, \quad \operatorname{curl} \boldsymbol{E} = -\frac{\partial \boldsymbol{B}}{\partial t}, \tag{13.11}$$

$$\operatorname{div} \boldsymbol{B} = 0, \quad \operatorname{curl} \boldsymbol{B} = \frac{1}{c^2} \frac{\partial \boldsymbol{E}}{\partial t}. \tag{13.12}$$

If there is a material, two auxiliary fields, the electric displacement field $\boldsymbol{D}$ and the magnetizing field $\boldsymbol{H}$, are introduced:

$$\boldsymbol{D} = \varepsilon_0 \boldsymbol{E} + \boldsymbol{P}, \; \boldsymbol{H} = \frac{1}{\mu_0} \boldsymbol{B} - \boldsymbol{M} \tag{13.13}$$

Here, $\boldsymbol{P}$ is polarization and $\boldsymbol{M}$ magnetization.

13.6 Prehistory of the Second Law

Joule (1818–1889) quantitatively demonstrated (in 1847) that work can be converted into heat, and believed that work and heat were equivalent, but long before him Carnot (1796–1832) had already established (published in 1824) the impossibility of complete conversion of heat into work.[4,5,6,7] Thomson's brother told him (in 1844) to pay attention to the work of Carnot.[8,9] Thomson recognized the importance of Joule's work, but, since he also accepted Carnot, he did not believe work and heat were equivalent (Joule, however, rejected Carnot, because the latter used the caloric theory). Thomson realized that Carnot's work could establish a universal scale of temperature, and introduced the concept of absolute temperature (see Chapter 14), but he failed to grasp the real relation between heat and work.

Carnot clearly recognized that only when there are hotter and colder heat baths can we produce work; there is a fundamental asymmetry between heat and work. Clausius (1822–1888) understood this, in essence and in plain terms, as follows: Temperature is the "value" of energy. No one can simply raise the "value" of energy (i.e., transfer energy from a colder to a hotter bath) without any other change; work corresponds to heat at $T = \infty$. Thus, thermodynamics was established by Clausius.[10]

---

[4] **Industry was far ahead** The reader should compare the years in this footnote and those in the main text of this unit. Watt's steam engines were developed during 1760–1770; Trevithick's steam locomotive (Puffing Devil) was introduced in 1804; and Stephenson's *Locomotion No. 1* was built to run on the Stockton and Darlington Railway in 1825. Steamboats were operating earlier than this. Robert Fulton's boat with a Watt steam engine in 1807 travelled between New York and Albany, a distance of 240 km in 32 hrs.

[5] As noted in **11.2** Mayer had already determined the work equivalent of heat in 1842. [In 1847 Helmholtz published, "On the conservation of force; a physical memoir"; C. Bronte published *Jane Eyre*.]

[6] **Joule and Thomson meet** When Joule presented his work, he was asked to be brief and was not given any time for discussion, but Thomson was in the audience. In those days Thomson subscribed to the caloric theory (which regarded heat as a substance called caloric), so he was rather skeptical, but he accepted the idea of conversion of work into heat.

[7] [In 1824 Beethoven's Symphony No. 9 premiered; The Battle of Ayacucho in the Peruvian War of Independence.]

[8] [In 1844 The first electrical telegram was sent by Morse; Goodyear patented vulcanization.]

[9] **Carnot was totally forgotten** The work was analytically rewritten in 1834 by Clapeyron (see Fig. 15.3), and Thomson knew (in 1844 or 1845) Carnot's work through Clapeyron's exposition. Thomson searched the used book stores throughout Paris for the original and got it only in 1848. Clausius was able to know Carnot only through Clapeyron and Thomson in 1850.

Clapeyron B. P. E. Clapeyron (1799–1864) was an engineer who supervised the construction of the first railway line connecting Paris to Versailles and Saint-Germain. He introduced the diagram representation of the Carnot cycle and considerably simplified Carnot's original proof of his theorem; see **15.4**. He also clarified the concept of reversibility and stated the definitive version of Carnot's theorem.

[10] **Thermodynamics is established by Clausius** Now, it seems generally accepted that Clausius and Thomson (independently) constructed thermodynamics, but this is largely due to the British propaganda by Tate. Neglecting Mayer may be in the same vein. Notice that in the original formulation Clausius' principle included Kelvin's as well.

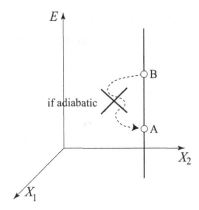

Thermodynamic space and Planck's principle: The vertical axis denotes internal energy E and X_1, X_2 represent work coordinates. A vertical move implies a purely thermal process. Adiabatically, there is no way to move from a state B to another state A that is vertically below it according to Planck's principle.

13.7 The Second Law of Thermodynamics

The *second law of thermodynamics* summarizes what Carnot and Clausius clearly understood as follows. Two famous expressions are:

Clausius' principle: Heat cannot be transferred from a colder to a hotter body without any other effects.[11]
Kelvin's principle: A process cannot occur whose only effect is the complete conversion of heat into work.[12]

We use the second law in the following form:

Planck's principle: For any adiabatic process with all the work coordinates returning to their original values, $\Delta E \geq 0$ (see Fig. 13.3).

An "*adiabatic system*" is a process that occurs in a system without any exchange of heat with its surroundings (a process occurring in a Dewar jar; a system always $\Delta E = W$ holds).[13] Intuitively, what Planck's principle implies is that due to dissipation that usually accompanies any macroscopic process heat is generated, so adiabatically the system heats up (and its internal energy inevitably increases).

---

[11] The principle expressed faithfully to Clausius' original idea may be read again in plain terms: "It is impossible to raise the 'value' of energy without any other change in the world."

[12] Notice that Clausius' principle contains Kelvin's principle, if we understand work as the heat energy at $T = \infty$ as Clausius recognized.

[13] **Adiabatic wall and adiabatic system** There is a special wall called an *adiabatic wall* such that for a system surrounded by this wall (= adiabatic system) the necessary work to bring the system from a given initial equilibrium state to a specified final equilibrium state is independent of the actual process but is dependent only on these two end states of the process.

   The first law implies adiabatically and quasistatically $dE = \sum_i x_i dX_i$, where (x_i, X_i) are conjugate pairs (such as $\{-P, V\}$, $\{B, M\}$) for work coordinates.

13.8* Planck's Principle, Kelvin's Principle, and Clausius' Principle are Equivalent

None is more fundamental than the other two. Let us sketch the general logic.[14]

If Planck's principle is violated, then a cyclic change of the work coordinates can reduce the system energy adiabatically. That is, work can be produced by using the system itself as a single heat bath. Therefore, Kelvin's principle is violated. If Kelvin's principle is violated, then we can get work from a cold bath and do work on a hotter bath to increase its energy. Thus, Clausius' principle (his original statement in footnote 11) is violated. If Clausius' principle is violated, then we can convert a uniform adiabatic system into colder and hotter halves and operate a heat engine between them to produce work by a cyclic change of the work coordinates. Thus, Planck's principle is violated.

We have demonstrated, symbolically, not P $\Rightarrow$ not K $\Rightarrow$ not C $\Rightarrow$ not P. That is, P $\Rightarrow$ C $\Rightarrow$ K $\Rightarrow$ P; all the principles mentioned here are equivalent.

13.9 Clausius Recognizes Entropy

Now, we wish to demonstrate that the second law implies the existence of a state function called "entropy," which was introduced by Clausius.[15] Clausius recognized: The constant entropy surfaces (called adiabats or isentropic surfaces) foliate the thermodynamic space. If state A has a larger entropy than state B, we can never go from A to B without cooling the system.

13.10 Entropy Exists[16]

In Fig. 13.4 we choose an arbitrary point P and a vertical line L in the realizable portion of the thermodynamic space. This line is parallel to the energy axis (along which all the work coordinates are kept constant), along which we can change the states by exchanging heat with the external world. Let us find a quasistatic adiabatic path connecting P and L.[17]

---

[14] Since we are not interested in making any axiomatic framework for thermodynamics, the demonstration of the equivalence of the principles is only a rough sketch.

[15] **Entropy and Gibbs** In introductory thermodynamics it seems unanimously recognized that entropy is a difficult concept to grasp. This may also be a misconception or misunderstanding spread by British physicists, who resisted recognizing entropy for a while. It is said even Maxwell, who used entropy correctly for the first time in Britain, misunderstood it initially. In reality, English-speaking scientists were rescued by Gibbs who correctly understood thermodynamics.

[16] The argument here is not a watertight argument. We proceed intuitively, because the gist of the argument must be there.

[17] Thus, it is retraceable. Recall **12.6**.

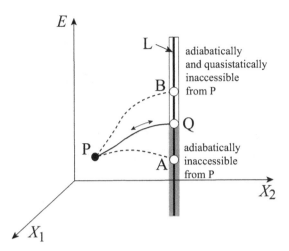

Let Q be a state on a vertical line L (along which we can move with heat exchange alone) that can be reached from state P adiabatically and quasistatically. If state A can be reached by an adiabatic process from P, then adiabatically we can go from Q to A via P, violating Planck's principle. Thus, the shaded portion is inaccessible from P adiabatically. If state B can be reached by an adiabatic and quasistatic process from P, then adiabatically we can go from B to Q via P, violating Planck's principle, again. Thus, Q is unique: there is only one point on L that can be reached from P adiabatically and quasistatically. (We can adiabatically go from P to B, but it is an irreversible process.)

Suppose the path lands on L at point Q. Can we also reach other points on L in the same fashion? Planck tells us state A below Q is inaccessible; if accessible, we can adiabatically go to A from Q, contradicting Planck's principle. If we could go to B above Q adiabatically and reversibly from P, then we can go to Q adiabatically via P from B, again contradicting Planck's principle. Thus, we have learned that the point on L we can reach from P adiabatically and quasistatically is only Q.

Now, moving the stick L throughout the space keeping it parallel to the energy axis, we can construct a hypersurface consisting of points adiabatically and quasistatically (retraceably) accessible from point P. This is an *adiabat* containing P. Moving the point P vertically, and repeating the above argument, we can construct different adiabats.[18]

13.11 Entropy Stratifies the Thermodynamic Space

Adiabats stratify the thermodynamic space. That is, no two different adiabatic surfaces cross each other. See Fig. 13.5 to understand that these sheets, the adiabats, cannot cross; crossing means Planck's law is violated. The adiabats do not have any "overhangs." As we can see from Fig. 13.6, the reason is the same as that for no crossing.

---

[18] We may need to move P horizontally as well to cover the whole physically realizable region in the thermodynamic space.

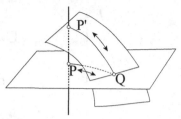

Fig. 13.5 If two adiabats cross or touch, then we can make a cycle that can be traced in any direction, because PQ and P'Q can be traced in either directions, so Planck's principle is violated.

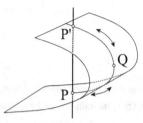

Fig. 13.6 Just as in Fig. 13.5, an overhang violates Planck's principle.

We have seen that the thermodynamic space is stratified (or foliated) into layers vertically stacked and respecting their order. Thus, we can define a state function by appropriately assigning real numbers continuously according to their heights along a vertical line; in other words, we can define a state function S which is an increasing continuous function of E under constant work coordinates.

How can we change the value of this function S? Obviously, we can change S by going up or down along L while the work coordinates are kept fixed; that is, we can change S by adding or subtracting heat Q. Since we have assumed that S increases with E, for $d'Q > 0$ we must have $dS > 0$. Therefore, we may define S so that $dS \propto d'Q$ holds (at each point of the thermodynamic space; the proportionality factor may be a function of the thermodynamic coordinates). Thus, S may be defined as a function at least once differentiable with respect to E. Also S can be defined as an extensive quantity, because Q is an extensive quantity.

13.12 Entropy and Heat

Suppose two systems are in contact through a wall that allows only the exchange of heat (that is, the two systems are in thermal contact **12.12**), and they are in thermal equilibrium. Exchange of heat $d'Q$ between the systems in thermal equilibrium is a reversible process (say, system I gains $d'Q_I = d'Q$ and II $d'Q_{II} = -d'Q$), so this process occurs within a single adiabat of the compound system (i.e., the two systems considered together as a single system). If we write $d'Q_X = \theta_X dS_X$ (X = I or II), with the aid of the additivity of S,

$$0 = dS_I + dS_{II} = d'Q \left(\frac{1}{\theta_I} - \frac{1}{\theta_{II}} \right). \tag{13.14}$$

This implies $\theta_I = \theta_{II}$. That is, when two systems are in thermal equilibrium, the proportionality constants are the same.[19] Hence, we may interpret the proportionality factor as an empirical temperature (see **12.13** for empirical temperature).

The introduced temperature can be chosen as a universal temperature T called the *absolute temperature*. Hence, in the quasistatic process we write:[20]

$$d'Q = TdS. \tag{13.15}$$

13.13 Entropy Principle

Entropy can only be reduced by cooling (by depriving heat). That is, there is no adiabatic process to reduce the system entropy.

13.14 The Gibbs Relation

Since entropy S is a once differentiable monotone increasing function of E, we can express E as a function of S and work coordinates (collectively denoted as) X: $E = E(S, X)$. This relation was called the *fundamental equation* by Gibbs. Its differential form is an infinitesimal version of the first law of thermodynamics for the quasistatic process:

$$dE = TdS - PdV + \mu dN + \boldsymbol{B} \cdot d\boldsymbol{M} + \cdots. \tag{13.16}$$

This is called the *Gibbs relation*. Its entropic form reads

$$dS = \frac{1}{T}dE + \frac{P}{T}dV - \frac{\mu}{T}dN - \frac{\boldsymbol{B}}{T}d\boldsymbol{M} + \cdots. \tag{13.17}$$

Note that all ds denote exact differential in the above, because differentiated quantities are all state functions (recall **12.7**).

The chief concern of thermodynamics up to Gibbs was to formulate the second law and to prove the existence of entropy. Gibbs then utilized entropy and reformulated thermodynamics as a system that was also of practical use. The very starting point of this new formulation was this relation (13.16), which Gibbs wrote down for the first time.

---

[19] We must also show that $\theta_1 \neq \theta_2$ if the two systems are not in thermal equilibrium. We see this from $dS_1 + dS_2 \neq 0$ in such cases, because the compound system cannot stay in the original adiabat due to irreversibility.

[20] **Identification of absolute temperature** We must show, in the flow of the logic of this book, that this T is identical to the T appearing in the ideal gas law. To this end we have only to consider the Carnot cycle **15.3**, or to compute the efficiency η of an ideal engine (the fundamental idea of Thomson on the absolute temperature); we will obtain $\eta = 1 - \theta_1/\theta_2$ (assuming that $\theta_1 < \theta_2$). If we use an ideal gas we obtain $\eta = 1 - T_1/T_2$, so θ and T are identical (up to the choice of units).

On the use of the ideal gas As we will see (in Chapter 23) ideal gases contradict the third law of thermodynamics, so there are people who assert that ideal gases are unphysical and should not be used to develop the basic theoretical framework. However, if pressure is sufficiently reduced, then any real gas becomes an ideal gas, however low its temperature may be. Therefore, as long as we clearly recognize this condition, there is no fundamental difficulty in using ideal gases to develop fundamental theories.

Notice that each term consists of a product of a conjugate pair:[21] an intensive quantity and d[the corresponding (i.e., conjugate) extensive quantity]. Do not forget the minus sign in front of PdV (recall **13.3**).

Problems

Q13.1 Equivalence of Heat and Work (Very Elementary)

This is an extremely elementary question. A block of mass $M = 1\,g$ is at rest in space (vacuum). Another block of mass $2M = 2\,g$ with velocity $V = 1.5\,km/s$ collides with the first block and the two blocks stick to each other. Both blocks are initially at 200 K.

(1) Assuming that there is no radiation loss of energy and that there is no rotation of the resultant single block, obtain the temperature of the resultant single block after equilibration. Assume that the specific heat of the material making the blocks is 2.1 J/g K.

(2) If rotation of the resultant body is allowed, what can be said about its final temperature? In particular, is it possible not to raise the temperature of the resultant single block?

Solution

(1) The total initial macroscopic kinetic energy is $(2M)V^2/2 = 2.25 \times 10^3\,J$. The final total kinetic energy of macroscopic motion is (the necessary speed is determined by the conservation of linear momentum: $2MV/3M$)

$$\frac{1}{2}(3M)(2V/3)^2 = \frac{2}{3}MV^2 = 1500\,J. \tag{13.18}$$

Therefore, 750 J should have become the energy of thermal motion. Thus, $750/6.3 = 119\,K$ is the increase in temperature, so the final temperature is 319 K.

(2) Rotational motion can be excited, so the temperature increase is reduced. However, this rotation is due to the nonzero angular momentum around the center of mass of the initial system. Now, the question is whether the rotational kinetic energy can preserve the kinetic energy of relative motion. If the second body has an extremely long thin rod to connect it to the other body to become a single block, then we can reduce the loss of rotational kinetic energy indefinitely (compute the final rotational kinetic energy and compare it with the relative kinetic energy). That is, the temperature increase can be made indefinitely small.

---

[21] **Conjugate pairs with respect to energy and with respect to entropy** As we have seen in the text the Gibbs relation reads $dE = TdS + \sum xdX$, where X are the work coordinates. We say (T, S) or (x, X) is a conjugate pair with respect to energy. We can rewrite this as $dS = (1/T)dE - \sum (x/T)dX$. We call $(1/T, E)$ or $(-x/T, X)$ a conjugate pair with respect to entropy. If we say simply "conjugate pair," usually it is with respect to energy.

Q13.2 Newcomen vs. Watt

Newcomen (1664–1729) invented his engine in 1712,[22] and Watt (1736–1819) his engine in 1763–75 (sporadically developed). What is the fundamental difference between these two engines?

Solution

The Newcomen engine first fills its cylinder with steam, and, by injecting cool water in it, creates a vacuum. The atmospheric pressure pushes the piston into this vacuum and this is the main source of the work. The steam is used to create vacuum, and does not produce work by itself. That is why it is called the atmospheric engine. The heat is wasted to reheat the cylinder. In contrast, Watt's engine uses the pressure produced by steam to do work. Furthermore, later Watt realized that without supplying steam any further, the pressure of the steam can do work while decreasing its temperature (adiabatically). See Fig. 15.3. Thus, it is clear that heat engine in the modern sense was invented by Watt.

Q13.3 Free Expansion and Joule–Thomson Processes are Irreversible[23]

Using the Gibbs relation for E or for $H = E + PV$ (called *enthalpy*[24]), show

$$(1)\ \left(\frac{\partial S}{\partial V}\right)_E > 0 \ \text{ and (2) } \left(\frac{\partial S}{\partial P}\right)_H < 0. \tag{13.19}$$

Solution

(1) $dE = TdS - PdV = 0$ implies

$$\left(\frac{\partial S}{\partial V}\right)_E = \frac{P}{T} > 0. \tag{13.20}$$

Therefore, the free adiabatic expansion ($\Delta V > 0$) of a gas is an irreversible process ($\Delta S > 0$).

(2) $dH = d(E + PV) = TdS + VdP = 0$ implies

$$\left(\frac{\partial S}{\partial P}\right)_H = -\frac{V}{T} < 0. \tag{13.21}$$

This implies the Joule–Thomson process (cf. **Q16.2**), which is an adiabatic process, is irreversible, since $\Delta P < 0$ for the process.

Q13.4 Magnetic Work Using Magnetic Dipoles

The work required to increase the magnetization $\boldsymbol{M}$ of a block in the external magnetic field $\boldsymbol{B}$ is written as $d'W = \boldsymbol{B} \cdot d\boldsymbol{M}$. We have shown this with the aid of induction. Here, we place

---

[22] [In 1712: Peter the Great move the Russian capital to Saint Petersburg; Frederick the Great became the King of Prussia; Cromwell died; Jean-Jacques Rousseau was born.]

[23] If the reader needs a partial derivative review, read **24.6**: Partial derivative review and **24.8**: Remarks on the notation of partial derivatives.

[24] Discussed in Section 16 but here we need only the definition.

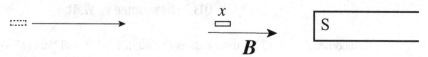

Fig. 13.7 The magnetic field **B** is prepared by a large constant bar magnet, and the magnetic dipole initially at infinity is brought to position x along the x-axis. The field is parallel to the axis.

a magnetic dipole in a magnetic field and calculate the work done to the system. This is not very trivial, because not all the energy is stored in the block under study; some portion is stored as the potential energy in the "relation" between the block and the device creating the external magnetic field. We know that this potential energy is $-\boldsymbol{B} \cdot \boldsymbol{M}$ (probably, the reader remembers that the energy of a magnetic dipole $\boldsymbol{\mu}$ is minimum, when $\boldsymbol{B}$ and $\boldsymbol{\mu}$ are parallel: $E = -\boldsymbol{\mu} \cdot \boldsymbol{B}$).

Let us assume that a block contains numerous (but not interacting) magnetic dipoles $\boldsymbol{\mu}_i$ and $\boldsymbol{M} = \sum_i \boldsymbol{\mu}_i$, where the summation is over all the magnetic dipoles in the block. We have only to study individual magnetic moments, so take a representative $\boldsymbol{\mu}$. We assume that the (size of the) magnetic dipole changes due to the magnetic field, so the field dependence is explicitly written as $\mu(\boldsymbol{B})$. At position x the (x-component of the) force acting on the small magnetic dipole parallel to the x-axis (see Fig. 13.7) is given by ($+$ in the positive x-direction; from now on we assume the field is in the x-direction)

$$F = \mu(B(x))\frac{dB(x)}{dx}. \tag{13.22}$$

Since we are doing thermodynamics, we must bring the magnetic dipole from infinity gently to the present position x. To perform such an experiment, we must apply a force opposing the above force (i.e., $-F$) while displacing the magnetic dipole.

(1) What is the work W we do to the whole system (the dipole plus the bar magnet) while dragging the dipole from $-\infty$ to x? (Since the force we exert and the displacement are both given, it is an elementary question.)

(2) However, this energy W is stored not only in the block containing the magnetic dipoles, but also between the bar magnet and the dipole as the potential energy at x (as given above). Show that the energy stored in the dipole is

$$W + \mu(B(x))B(x) = \int_{\mu(0)}^{\mu(B(x))} B(x')d\mu(B(x')). \tag{13.23}$$

Therefore, $dE = BdM$, if only $\boldsymbol{M} (= \sum \mu)$ is changed among the work coordinates.

Solution

(1) The force we exert is $-F$ (not F; without our application of brake, the "block" would fly to the bar magnet)

$$W = -\int_{-\infty}^{x} Fdx' = -\int_{0}^{B(x)} \mu(B(x'))dB(x'). \tag{13.24}$$

Here, the dependence of μ on B is explicitly written. This implies that the total work done to the system consisting of the block (containing dipoles) and the bar magnet reads

$$W = - \int_0^{B(x)} M(B)dB. \tag{13.25}$$

(2) The total energy supplied, W, must be equal to the sum of the potential energy, $(-\mu B)$ and the energy stored in the dipole. Therefore, $W - (-\mu B)$, i.e., (13.23) is the energy stored in the dipole itself. Let us honestly compute the sum (13.23).

$$- \int_0^{B(x)} \mu(B(x'))dB(x') + \mu(B(x))B(x)$$

$$= - \int_0^{B(x)} \mu(B(x'))dB(x') + \int_0^{B(x)} d[\mu(B(x'))B(x')]$$

$$= - \int_0^{B(x)} \mu dB + \int_0^{B(x)} d[\mu B] = \int_0^{\mu(B(x))} B(x')d\mu(B(x')).$$

If we sum this over all the dipoles in the block, we get

$$\int_{M(0)}^{M(B(x))} B(x')dM(B(x')). \tag{13.26}$$

Therefore, $dE = BdM$.

14

Thermodynamics: General Consequences

Summary

* Under an adiabatic condition, spontaneous changes imply increase of entropy.
* Under an adiabatic condition, equilibrium is reached when the entropy is maximum (entropy maximization principle).
* Entropy is a concave (convex upward) function of thermodynamic coordinates.
* There is a general strategy to extend theories under adiabatic conditions to those under more general conditions. Thus, Clausius' inequality $\Delta S \geq Q/T_e$ is derived.

Key words

Clausius' inequality, entropy maximization principle, internal energy minimization principle, convexity, equilibrium conditions

The reader should be able to:

* Explain why entropy is concave.
* Understand the logic to show Clausius' inequality.
* Be familiar with convex functions.

14.1 Summary of Basic Principles[1]

The basic laws of thermodynamics are the summary of the experiences of us macroscopic organisms ([N] indicates the closest traditional 'Nth law"):

[O] There is a state called an equilibrium state. Equilibrium states of a system may be described in terms of thermodynamic coordinates (E, X_i), where E is the internal

---

[1] **Nernst's joke on the three principles** Kurt Mendelssohn writes, "When lecturing on 'his' heat theorem, Nernst was careful to point to an interesting numerical phenomenon concerning the discovery of the three fundamental laws of thermodynamics. The first one had three authors, Mayer, Joule and Helmholtz; the second had two, Carnot and Clausius; whereas the third was the work of one man only, Nernst. This showed conclusively that thermodynamics was now complete since the authorship of a hypothetical fourth law would have to be zero" *The World of Walther Nernst: The Rise and Fall of German Science 1864–1941* (ebook form from Plunket Lake Press, 2015; the original 1973) Chapter 4.

energy and X_i are work coordinates. The equilibrium state exhibits partitioning–rejoining invariance (see Chapter 12).

[I] The conservation of energy: $\Delta E = Q + W$, or for infinitesimal changes $dE = d'Q - PdV + BdM + xdX$;[2] the variables appear in "conjugate pairs": $(-P, V), (B, M), (x, X)$ (for a generic pair), etc. (see Chapter 13).

[II] The thermodynamic space is foliated into $S = $ constant (hyper)surfaces. With adiabatic quasistatic (thus reversible) processes we cannot get out of a given $S = $ const. surface. With work only, $\Delta S < 0$ never happens; to reduce entropy we definitely need cooling. For a quasistatic process the Gibbs relation holds: $dE = TdS + \sum_i x_i dX_i$. Often $dS = \frac{1}{T}dE - \sum \frac{x}{T}i dX_i + \cdots$ is convenient (see Chapter 13).

[III] $S = 0$ in the limit $T \to 0$. This is the third law we will encounter in Chapter 23.

[IV] Thermodynamic variables are either extensive or intensive. The total amount of an extensive quantity of a compound system is the sum of the extensive quantities of the subsystems (additivity) (see Chapter 11).

Practically Speaking:

(i) Thermodynamics can be used to compute the state function change caused by any process connecting an initial equilibrium state A and a final equilibrium state B.

(ii) To this end we devise a convenient quasistatic path from A to B in the thermodynamic space along which we can use the Gibbs relation mentioned in [II].

14.2 Entropy Maximization Principle

Entropy cannot be reduced by any adiabatic process. Therefore, if an equilibrium state changes into another equilibrium state through modification of only the work coordinates under an adiabatic condition,[3] the entropy of the system generally increases.

Suppose the initial system is in equilibrium but with some constraints (say, compartmentalized with internal walls). If we remove the constraints, the system would evolve to a new equilibrium state (see Fig. 14.1). Since the change is spontaneous, generally, this final state has a larger entropy. This is the *principle of increasing entropy*.[4]

A spontaneous change in an adiabatic system increases its entropy, so if the system entropy reaches the maximum under a given constraint, the system reaches its equilibrium

---

[2] **Standard state function symbols** We stick to the standard notations:
E: internal energy, S: entropy, T: temperature, P: pressure, V: volume, $\boldsymbol{B}$ (B or h): magnetic field, $\boldsymbol{M}$ (M): magnetization, μ: chemical potential, N: number of particles. We use X for a generic work coordinate (extensive quantity) and x for its intensive conjugate (with respect to energy).

[3] "Adiabatic" implies no exchange of heat. Then, the reader may infer that thermal contact with a single heat bath is admissible if there is no net heat exchange. This is correct. Notice that "adiabatic condition" does not mean $R \to 0$ (external noise zero) limit (see Chapter 12).

[4] However, this does not claim the system entropy increases at every intermediate time during the evolution process, because thermodynamic entropy is defined only for equilibrium states.

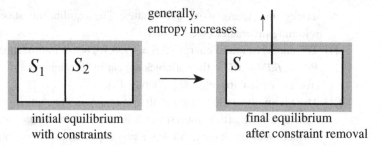

generally, entropy increases

S_1 S_2 → S

initial equilibrium
with constraints

final equilibrium
after constraint removal

Fig. 14.1 Initially, assume that the system is in equilibrium in the presence of a wall, which may be understood as a symbol of a certain constraint. The total entropy of this system as a whole (i.e., as a compound system **12.9**) is $S_1 + S_2$. The whole system is under an adiabatic condition. When the wall is removed (i.e., the constraint is removed), the system evolves to a new equilibrium state with a larger entropy spontaneously (irreversibly). That is, the final entropy S must satisfy $S \geq S_1 + S_2$ (the principle of increasing entropy).

state under the constraint. This is the *entropy maximization principle*.[5] Thus, the change δS of the system entropy due to any virtual change (perturbation) of the system tells us that (*stability and evolution criteria*):

$$\delta S < 0 \iff \text{the state is thermodynamically stable,} \tag{14.1}$$

$$\delta S > 0 \iff \text{the state spontaneously evolves.} \tag{14.2}$$

Therefore, the second law gives us a *variational principle* in terms of entropy to find a stable equilibrium state for an adiabatic system.

In this description of the stability criterion, we mentioned "virtual changes or perturbations."[6] In reality, however, they are not virtual in most cases, but are actually produced spontaneously by thermal fluctuations. Thus, as long as thermal fluctuations are not suppressed, whenever the system entropy can increase, the system evolves to maximize its entropy; behind any variational principle are fluctuations to substantiate it.

14.3 Entropy is Concave

Let us join two systems made of the same substances to make a single system. The entropy maximization principle tells us that the entropy of the resultant compound system is given by

---

[5] **Remark on entropy maximum principle** Astute readers would say that under an adiabatic condition, if entropy is maximum, then the state is in equilibrium, but the converse – if equilibrium, its entropy is maximum – is not demonstrated. This is true. However, in the usual thermodynamics, this converse is postulated.

Generally speaking, even if thermodynamics tells us that a process is not forbidden, whether the system actually spontaneously realizes the process or not is a matter of kinetics or dynamics, and, logically speaking, thermodynamics cannot say anything about it. Still, in the overwhelming majority of cases thanks to thermal fluctuations, such a process actually occurs spontaneously. Therefore, we may assume that the entropy max condition is equivalent to the equilibrium condition under adiabatic conditions.

[6] Up to this point δS is due to perturbations that are uniform throughout the system. However, as will be noted in Chapter 25, the perturbations can be spatially nonuniform (can be localized in small regions).

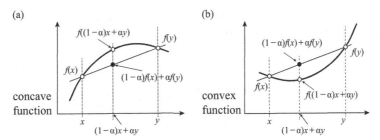

Fig. 14.2 An example of the concave function (a) and that of the convex function (b) are illustrated; the resultant inequality for (b) is called Jensen's inequality. The black dots correspond to the right-hand sides of (14.4) and (14.8), respectively.

$$S(E, X) = \max[S(E_1, X_1) + S(E_2, X_2)], \tag{14.3}$$

where the maximum is taken over all the partitions of E and X between the two systems as $E = E_1 + E_2$ and $X = X_1 + X_2$ (X collectively denotes work coordinates). This implies with the aid of the extensivity of S (i.e., $S(\alpha E, \alpha X) = \alpha S(E, X)$):

$$S\left((1 - \alpha)E + \alpha E', (1 - \alpha)X + \alpha X'\right) \geq (1 - \alpha)S(E, X) + \alpha S(E', X') \tag{14.4}$$

for any $\alpha \in [0, 1]$. That is, S is a concave function (its graph is convex upward) of all the thermodynamic coordinates (see Fig. 14.2(a)). This implies that the local stability criterion (14.1) holds globally as well.

14.4 Internal Energy Minimization Principle

The entropy maximization principle implies for any deviation ΔX of X from the equilibrium value:[7]

$$S\left(E, X_{\text{eq}}\right) - S\left(E, X_{\text{eq}} + \Delta X\right) \geq 0. \tag{14.5}$$

Therefore, since S is an increasing function of energy, we can increase the internal energy E in the second term to $E' \geq E$ under the $X = X_{\text{eq}} + \Delta X$ condition to satisfy

$$S\left(E, X_{\text{eq}}\right) - S\left(E', X_{\text{eq}} + \Delta X\right) = 0. \tag{14.6}$$

This implies that under the constant entropy condition, if an extra constraint to fix X at $X_{\text{eq}} + \Delta X$ is removed, then the internal energy surely decreases in equilibrium, since

---

[7] Since entropy is defined only for equilibrium states, this means, precisely speaking, that if, with some constraints, we make a new equilibrium state with $X_{\text{eq}} + \Delta X$ and E, then (14.5) holds.

$E \leq E'$. That is, if the internal energy is minimized under a constant entropy condition, the system must be in equilibrium.

14.5 Internal Energy is Convex

Let us join two systems made of the same substances to make a single system. The energy minimization principle tells us that the internal energy of the resultant compound system is given by

$$E(S_1 + S_2, X) = \min\left[E(S_1, X_1) + E(S_2, X_2)\right], \tag{14.7}$$

where the minimum is taken over all the partitions of S and X between the two systems as $S = S_1 + S_2$ and $X = X_1 + X_2$. This implies with the aid of the extensivity of E

$$E\big((1 - \alpha)S + \alpha S', (1 - \alpha)X + \alpha X'\big) \leq (1 - \alpha)E(S, X) + \alpha E\big(S', X'\big) \tag{14.8}$$

for any $\alpha \in [0, 1]$. That is, E is a convex function (its graph is convex downward) of all the variables (i.e., entropy and work coordinates) (see Fig. 14.2(b)).

14.6 Extension to Non-Adiabatic Systems

To consider a system which is not isolated, that is, a system which is interacting with its environment, we construct a bigger isolated system composed of the system itself (I) and its interacting environment (II) (see Fig. 14.3). We assume that both the systems are macroscopic, so we may safely ignore the surface effect.

The environment is always in equilibrium with its intensive thermodynamic variables (e.g., temperature) kept constant. To realize this we take a sufficiently big system (called a *reservoir* like a thermostat or a chemostat) as the environmental system II. We assume any change in system II caused by the changes in system I is quasistatic.

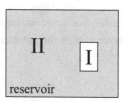

Fig. 14.3 The system II is the environment for the system I we are interested in. System II is called a reservoir if it is always infinitesimally close to equilibrium with constant intensive parameters.

14.7 Clausius' Inequality

The entropy change of the compound system $I + II$ is given by the sum of the entropy change ΔS_I of the system I and that ΔS_{II} of the environment II.[8] Since the whole system $I + II$ is adiabatic, a natural process occurring in the whole system must satisfy (after reaching a new equilibrium state of the compound system; entropy maximization principle)

$$\Delta S_I + \Delta S_{II} \geq 0. \tag{14.9}$$

Let Q be the heat transferred to the system I from the environment II. From our assumption, we have

$$\Delta S_{II} = -Q/T_e, \tag{14.10}$$

where T_e is the temperature of the environment (system II). The minus sign is because II is losing heat to I.[9] Combining (14.9) and (14.10) yields the following inequality:

$$\Delta S_I \geq Q/T_e. \tag{14.11}$$

This is *Clausius' inequality* for non-adiabatic systems. If the change in I is reversible, then $\Delta S_I = Q/T_e$ should hold; (14.11) implies that "excess entropy" has been produced in I by the irreversibility of the process. For adiabatic systems we recover the principle of increasing entropy (in **14.2**), because $Q = 0$.

☐ In the above discussion the reader might have assumed that $Q > 0$ (heat is imported from the environment), but notice that the sign of Q does not matter. That is, if the system exports heat $|Q|$ (i.e., import $Q = -|Q| < 0$), still (14.11) holds. In this case usually $\Delta S_I < 0$, so we see $|\Delta S_I| \leq |Q|/T_e = -Q/T_e$: due to irreversibility (intuitively due to produced entropy) ΔS_I is less negative than the reversible case Q/T, so $|\Delta S_I| \leq |Q|/T$.

14.8 Equilibrium Conditions between Two Systems

As an application of the entropy maximization principle, let us study the equilibrium conditions for two systems I and II interacting through various walls (see Fig. 14.4).

(i) Consider a rigid impermeable wall (no matter exchange allowed) which is diathermal (allowing heat exchange). The two systems are in thermal contact and exchange internal energy in the form of heat. The total entropy of the system S is the sum of

Fig. 14.4 The thick vertical segment denotes the wall that selectively allows the exchange of certain extensive quantities.

---

[8] Remember that in thermodynamics the change is always between two equilibrium states.
[9] This is an informal expression of: II is losing energy in the form of heat.

the entropy of each system S_I and S_{II}. The total internal energy E is also the sum of subsystem internal energies E_I and E_{II} (additivity). We isolate the compound system and study the equilibrium condition for the system. We should maximize the total entropy with respect to the variation of E_I and E_{II}:

$$\delta S = \frac{\partial S_I}{\partial E_I}\delta E_I + \frac{\partial S_{II}}{\partial E_{II}}\delta E_{II} = \left(\frac{\partial S_I}{\partial E_I} - \frac{\partial S_{II}}{\partial E_{II}}\right)\delta E_I = 0, \tag{14.12}$$

where we have used that $\delta E = 0$ or $\delta E_I = -\delta E_{II}$. Hence, the equilibrium condition is

$$\frac{\partial S_I}{\partial E_I} = \frac{\partial S_{II}}{\partial E_{II}}, \tag{14.13}$$

or $T_I = T_{II}$. That is, thermal equilibrium is equivalent to the equality of the temperatures.

(ii) Consider a diathermal impermeable wall which is movable. In this case the two systems can exchange energy (in the form of heat) and volume. If we assume that the total volume of the system is kept constant ($\delta V = \delta V_1 + \delta V_2 = 0$), the equilibrium condition should be

$$\delta S = \frac{\partial S_I}{\partial V_I}\delta V_I + \frac{\partial S_{II}}{\partial V_{II}}\delta V_{II} = \left(\frac{\partial S_I}{\partial V_I} - \frac{\partial S_{II}}{\partial V_{II}}\right)\delta V_I = 0, \tag{14.14}$$

and $T_I = T_{II}$, that is,

$$\frac{\partial S_I}{\partial V_I} = \frac{\partial S_{II}}{\partial V_{II}} \tag{14.15}$$

and $T_I = T_{II}$. Therefore, $P_I = P_{II}$ is also required.

☐ **Remark** If the movable wall is adiabatic, heat exchange is impossible, so I and II are adiabatic and their entropies are separately maximized. However, we cannot use these maximizations, because heat may be generated by the motion of the wall and is distributed to I and II in a certain way. In our situation, the total energy is still constrained, so the Gibbs relation (13.16) requires $P_I = P_{II}$. We cannot say anything about the temperatures. Notice that this nonuniqueness is a prediction of thermodynamics: it tells us that under the given condition and constraints, different temperatures are realizable, depending on the details of the process.

Problem

Q14.1 The Impossibility of Macroscopic Relative Motion in Equilibrium

Entropy S is a function of the internal energy E seen from the center-of-mass co-rotating coordinates of the system:

$$E = U - P^2/2M - LI^{-1}L/2, \tag{14.16}$$

where U is the total energy, P the total momentum, L the total angular momentum, M the total mass of the system, and I the inertial moment tensor around the center of mass. That is, the internal energy E of the system is the total energy U minus the total kinetic energy of the macroscopic translational and rotational motions.

Let us divide the system into several subsystems. We put subscript a to macroobservables of the subsystem a. We have $\boldsymbol{P} = \sum \boldsymbol{P}_a$ and $\boldsymbol{L} = \sum_a \boldsymbol{L}_a$. The angular momentum $\boldsymbol{L}_a$ of the subsystem a around the center of mass of the whole system may be written as $\boldsymbol{L}_a = \boldsymbol{\ell}_a + \boldsymbol{r}_a \times \boldsymbol{P}_a$, where $\boldsymbol{P}_a$ is the (linear) momentum of subsystem a, $\boldsymbol{r}_a$ is the relative position vector of the center of mass of subsystem a with respect to the center of mass of the whole system, and $\boldsymbol{\ell}_a$ is the angular momentum of subsystem a around its own center of mass. The Gibbs relation for subsystem a reads (for simplicity, other extensive variables than entropy are suppressed in $dE = TdS + \cdots$; we write U as a function of S, momentum and angular momentum):

$$TdS_a + \boldsymbol{v}_a \cdot d\boldsymbol{P}_a + \boldsymbol{\omega}_a \cdot d\boldsymbol{\ell}_a - dU_a = 0, \tag{14.17}$$

where $\boldsymbol{v}_a$ is the center of mass velocity of subsystem a, and $\boldsymbol{\omega}_a$ is its angular velocity around its center of mass.

The additivity of entropy implies (we use the fact that the total energy is preserved: $dU = \sum_a dU_a = 0$)

$$TdS + \sum_a \boldsymbol{v}_a \cdot d\boldsymbol{P}_a + \sum_a \boldsymbol{\omega}_a \cdot d\boldsymbol{\ell}_a = 0. \tag{14.18}$$

This is the Gibbs relation for an isolated macrosystem with relative macroscopic motions.[10]

(1) Demonstrate that the only macroscopic motion allowed to the system compatible with thermodynamic equilibrium is a rigid-translation plus rotation as a whole. This might be obvious, because "thermodynamical equilibrium" implies that all the frictions and dissipations are allowed to occur to increase the system entropy.

(2) **[$T \geq 0$]** Show that negative absolute temperature is not allowed for ordinary macroobjects.

Solution

Equation (14.18) can be rewritten with the aid of the expression of $\boldsymbol{L}_a$ as:[11*]

$$TdS + \sum_a \left(\boldsymbol{v}_a - \boldsymbol{\omega}_a \times \boldsymbol{r}_a \right) \cdot d\boldsymbol{P}_a + \sum_a \boldsymbol{\omega}_a \cdot d\boldsymbol{L}_a = 0. \tag{14.19}$$

Now, let us write down the condition to maximize the total entropy of the system under the constant total momentum, and total angular momentum. Using Lagrange's multipliers $\boldsymbol{v}$ and $\boldsymbol{\omega}$, we obtain

$$\frac{\partial S}{\partial \boldsymbol{P}_a} = \frac{\boldsymbol{v} + \boldsymbol{\omega}_a \times \boldsymbol{r}_a - \boldsymbol{v}_a}{T}, \tag{14.20}$$

$$\frac{\partial S}{\partial \boldsymbol{L}_a} = \frac{\boldsymbol{\omega} - \boldsymbol{\omega}_a}{T}. \tag{14.21}$$

---

[10] We have assumed the temperature equilibrium has been attained.
[11*] We have used $\boldsymbol{A} \cdot \boldsymbol{B} \times \boldsymbol{C} = \boldsymbol{B} \cdot \boldsymbol{C} \times \boldsymbol{A} = \boldsymbol{C} \cdot \boldsymbol{A} \times \boldsymbol{B}$, where vectors are 3-vectors.

Consequently,

$$v_a = v + \omega_a \times r_a, \quad \omega_a = \omega. \tag{14.22}$$

That is, in equilibrium, a macroscopic system can move only as a rigid body: no macroscopic relative motion of the parts is allowed.

In the above consideration we have assumed that the momenta are exchanged between the subsystems. Suppose a macroscopic system may be divided into two subsystems that can move relatively without friction. Then, note that this argument does not work (i.e., without dissipation relative macroscopic motion is allowed even in equilibrium as should be physically obvious).[12]

(2) Almost as a byproduct of (1) we can conclude that the absolute temperature is non-negative. Suppose T is negative. Then, S is a decreasing function of internal energy. Therefore, if $T < 0$, the system entropy can be increased by pumping energy into the macroscopic degrees of freedom (see (14.16)).[13] Consequently, the system explodes. Therefore, T cannot be negative.

However, if the system internal energy is bounded from above (needless to say, no spatial degrees of freedom are allowed), then it cannot increase indefinitely and this argument does not apply (as we will see later (in **Q24.2**) for the spin system not interacting with the lattice degrees of freedom).

---

[12] **No death, no birth** However, no local velocity fluctuation can be generated thermally without dissipation. Nothing that cannot decay can be produced spontaneously in a stable world. Recall the fluctuation–dissipation relation in Chapter 10. Recall Helmholtz's theorems on vortices, if the reader knows fluid dynamics.

[13] Small-scale organization of microscopic thermal noise into mesoscopic-scale collective motions will always occur as fluctuations, but this does not have a serious effect on ordinary thermodynamically stable systems.

15

Entropy Through Examples

Summary

* The crux of thermodynamic computation is to devise a quasistatic process.
* There is a universal bound for the efficiency to convert heat into work (Carnot's theorem).
* $\Delta S > 0$ implies molecules have more diverse behaviors due to the change, so we must ask more questions to determine their states than before the change.
* Units of entropy: $1\,\text{J/(K·mol)} = 0.173$ bits/molecule.

Key words

reversible engine, Carnot's theorem, heat pump, entropy of mixing, bit

The reader should be able to:

* Demonstrate that the reversible engine is the best engine.
* Estimate the entropy change due to various irreversible processes.
* Intuitively understand entropy change in terms of the number of yes–no questions (in bits).

Let us become familiar with thermodynamics, especially entropy, through elementary examples.[1]

15.1 Mayer's Relation

Demonstrate *Mayer's relation*: $C_P = C_V + R$ for an ideal gas, where C_P is the constant pressure molar specific heat and C_V the constant volume molar specific heat of the gas.

First of all, we must identify the quantities in terms of thermodynamic variables. The specific heats under constant V and constant P are defined, respectively, as

$$C_V = \left(\frac{\partial Q}{\partial T}\right)_V, \quad C_P = \left(\frac{\partial Q}{\partial T}\right)_P. \tag{15.1}$$

---

[1] If the reader needs a partial derivative review, read **24.6** Partial derivative review and **24.8** Remarks on the notation of partial derivatives.

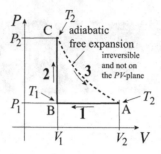

Fig. 15.1 Mayer's cycle consists of isobaric compression **1**, constant volume heating **2**, and adiabatic free expansion **3** (recall the law of constant temperature due to Gay-Lussac in Chapter 2). **1** and **2** require quasistatic heat exchange with environment.

The first law tells us $dE = d'Q - PdV$, so

$$C_V = \left(\frac{\partial E}{\partial T}\right)_V, \quad C_P = \left(\frac{\partial E}{\partial T}\right)_P + P\left(\frac{\partial V}{\partial T}\right)_P. \tag{15.2}$$

For an ideal gas E is dependent only on T (recall that E is the kinetic energy of unhindered molecular motion for ideal gases), so $dE = C_V dT$ (under any condition). We know $V = RT/P$, so

$$C_P = \left(\frac{\partial E}{\partial T}\right)_P + P\left(\frac{\partial V}{\partial T}\right)_P = C_V + R. \tag{15.3}$$

Mayer obtained this relation with the aid of *Mayer's cycle* (see Fig. 15.1). Notice that **3** in Fig. 15.1 does not change E, so for the ideal gas A and C are at the same temperature, say, T_2. Let the temperature at B be T_1. The work supplied by the isobaric compression process **1** is $W = -P_1(V_1 - V_2)$. The heat is discarded during this process simultaneously: $Q_1 = C_P(T_1 - T_2)$ (< 0). During the process **2** heat $Q_2 = C_V(T_2 - T_1)$ is absorbed. Therefore, for the cycle as a whole, we have $W + Q_1 + Q_2 = 0$. $R - C_P + C_V = 0$ is obtained, because

$$0 = P_1(V_2 - V_1) + C_P(T_1 - T_2) + C_V(T_2 - T_1) = R(T_2 - T_1) + C_P(T_1 - T_2) + C_V(T_2 - T_1). \tag{15.4}$$

Mayer used his cycle to determine the work equivalence of heat in 1842 as mentioned in **11.2** (see **Q15.1**(1)), five years before Joule.

15.2 Poisson's Relation

Show for an ideal gas that along an adiabatic quasistatic path $PV^\gamma = $ const., where $\gamma = C_P/C_V$. This is called *Poisson's relation*.

The first law implies $dE = -PdV$ (adiabatic and quasistatic). Also $dE = C_V dT$ (ideal gas). Therefore,

$$0 = C_V dT + PdV = C_V d(PV/R) + PdV = (C_V/R + 1)PdV + (C_V/R)VdP. \tag{15.5}$$

That is, $\gamma d \log V + d \log P = 0$ with the aid of Mayer's relation **15.1**.

15.3 Reversible Engine: Carnot's Theorem

Obtain the efficiency η (see (15.7) for the definition) of a reversible heat engine, and demonstrate that there is no engine more efficient than the reversible engine (Carnot's theorem).

A heat engine is a device that absorbs heat from a high temperature heat bath (temperature T_H) and converts a portion into work. The remaining energy is discarded to a low temperature heat bath (temperature T_L). See Fig. 15.2. Let Q_H and Q_L be the heats the engine absorbs from the high and low temperature heat baths, respectively, per one cycle, and W the work the engine obtains per one cycle (we expect $Q_H > 0$, $Q_L < 0$, and $W < 0$). The first law implies (since the engine does not produce energy)

$$W + Q_H + Q_L = 0. \tag{15.6}$$

The *efficiency* of an engine is the ratio of the work the engine produces (the benefit we get) to the heat it absorbs from the high temperature reservoir (the expenditure we pay). Therefore, we define the engine efficiency as (be careful with the sign convention):

$$\eta \equiv \frac{|W|}{Q_H} = -\frac{W}{Q_H} = 1 + \frac{Q_L}{Q_H}. \tag{15.7}$$

Let ΔS_H be the entropy increase of the engine in a single cycle due to the import of heat from the high temperature bath, and ΔS_L due to the import of heat from the low-temperature bath. Clausius' inequality (14.11) tells us that

$$\Delta S_H \geq \frac{Q_H}{T_H}, \quad \Delta S_L \geq \frac{Q_L}{T_L}. \tag{15.8}$$

Since the engine returns to the original state after a single cycle, $\Delta S = \Delta S_H + \Delta S_L = 0$:

$$0 = \Delta S_H + \Delta S_L \geq \frac{Q_H}{T_H} + \frac{Q_L}{T_L} \implies \frac{Q_H}{T_H} \leq -\frac{Q_L}{T_L}, \tag{15.9}$$

which implies $T_L/T_H \leq -Q_L/Q_H$. Thus, we have

$$\eta = 1 + \frac{Q_L}{Q_H} \leq 1 - \frac{T_L}{T_H}. \tag{15.10}$$

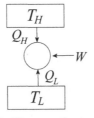

A heat engine operating between two heat baths. We assume the standard sign convention seen from the engine (the circle in the figure). Thus, $Q_H > 0$, $W < 0$, and $Q_L < 0$.

If the engine is reversible, it attains the maximum efficiency $\eta = 1 - T_L/T_H$. This statement is called *Carnot's theorem*.[2]

Thomson saw here a possibility of introducing the universal temperature scale based solely on the thermodynamic principles free from any materials; he reached the concept of the *absolute temperature* in terms of the maximum efficiency.

15.4*   The Original Carnot's Argument using the Carnot Cycle of an Ideal Gas

Carnot conceived the following engine (the *Carnot engine*)[3] which utilizes an ideal gas (in this exposition, 1 mole of it) as its working substance (see Fig. 15.3). This original demonstration of Carnot's theorem is much harder than the one we just saw, but may be a good elementary thermodynamics exercise:

(i) The engine does work through expansion while absorbing heat from the high temperature heat source (at T_H) (A $\to$ B in Fig. 15.3).

(ii) Then, it continues to expand while doing work and cools from T_H to T_L (B $\to$ C). Notice that this portion was Watt's novelty in his engine (**Q13.2**).

(iii) Next, the engine volume isothermally shrinks (i.e., some positive work is done to the engine) while discarding heat to the low temperature heat source at T_L (C $\to$ D).

(iv) Finally, the system is compressed adiabatically (again some positive work is done to the engine) and its temperature goes up from T_L to the original T_H (D $\to$ A).

The work added to the system (engine) is

$$W = -\oint_{\text{ABCDA}} PdV, \tag{15.11}$$

so it is equal to $(-1)\times$ the area surrounded by the warped quadrangle ABCD in Fig. 15.3. That is, the work done by the engine during its one cycle is the area of ABCD.

During the isothermal process A $\to$ B the engine does some work to the environment, but its internal energy is constant, because this is an isothermal process for an ideal gas; the work must be paid by the heat Q_H absorbed from the high temperature heat source at T_H. Therefore, (notice $dE = d'Q - PdV = 0$)

$$Q_H = \int_{A \to B} PdV = \int_{A \to B} \frac{RT_H}{V}dV = RT_H \log \frac{V_B}{V_A} > 0. \tag{15.12}$$

---

[2] If $T_H = +\infty$, then the reversible efficiency is 1. Recall **13.7**, according to Clausius, that work is heat from a bath at $T = \infty$.

[3] **Carnot's original used the caloric theory** The actual original argument due to Carnot relied on the caloric theory (which regarded heat as a substance called caloric), so the exposition given here cannot literally be his original argument, but a correct transliteration was done by Clausius. We need this demonstration to identify the absolute temperature introduced by the ideal gas law and with θ we introduced to relate heat and entropy change in Chapter 13.

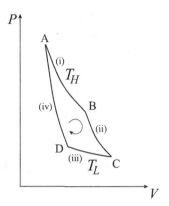

Fig. 15.3 The Carnot cycle: AB and CD are quasistatic isothermal processes, and BC and DA are quasistatic adiabatic processes. BC is the key element of Watt's engine as discussed in **Q13**.2. The working substance is an ideal gas, so during the isothermal process its internal energy is constant. This implies that during isothermal processes the work the system does (or is done to the system) and the heat it absorbs (or it discards) must be identical. Understanding the Carnot engine with the *PV* diagram was originally due to Clapeyron (1834; thus this diagram is called Clapeyron's graph), who advocated Carnot's work.

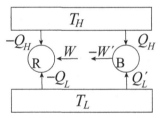

Fig. 15.4 The reversible engine R (left) is now used as a heat pump, and is driven by a (imaginary) better engine B (right) that can produce work $|W'| > |W|$ using the identical heat sources.

Similarly, during the isothermal process C→D the heat $|Q_L|$ discarded (i.e., Q_L (< 0) absorbed) by the system to the low temperature heat source at T_L must be identical to the work done to the system, so we have

$$|Q_L| = -\int_{C\to D} PdV = -\int_{C\to D}\frac{RT_L}{V}dV = RT_L\log\frac{V_C}{V_D}. \qquad (15.13)$$

To relate these two formulas, we need the ratios of the volumes related by quasistatic adiabatic processes. Poisson's relation, **15.2**, $PV^\gamma = $ const. implies $TV^{\gamma-1} = $ const. Consequently, $T_H V_A^{\gamma-1} = T_L V_D^{\gamma-1}$ and $T_H V_B^{\gamma-1} = T_L V_C^{\gamma-1}$ hold. This implies that $T_H/T_L = V_D^{\gamma-1}/V_A^{\gamma-1} = V_C^{\gamma-1}/V_B^{\gamma-1}$, or $V_B/V_A = V_C/V_D$. Using this relation in (15.12) and (15.13), we obtain the equality in (15.9). The rest is identical to the argument above, and we get $\eta = 1 - T_L/T_H$.

Carnot's original proof of his theorem went as follows. Suppose we have an engine B better (more efficient) than the reversible engine R, which can be driven backward by supplying work. Let us drive the reversible engine R backward with engine B and use R as a "heat pump" (see Fig. 15.4).

Here $|W'| > |W|$, so if we use the output of the "better engine" to drive the reversible engine, we can still utilize the work $|W'| - |W|$. Since the net heat imported to the two engines from the hotter bath is zero, inevitably, $|Q'_L| < |Q_L|$. That is, $|Q_L| - |Q'_L|$ is absorbed from the colder bath. This implies that work has been extracted from a single heat bath, violating Kelvin's principle. Hence, there cannot be any better engine than the reversible engine.

15.5 Fundamental Equation of Ideal Gas

The thermodynamic space of an ideal gas is spanned by internal energy E and volume V. Compute the entropy difference between the initial equilibrium state (E_1, V_1) and the final equilibrium state (E_2, V_2) for a 1 mole of ideal gas.

Since entropy is a state function, to compute its change between two equilibrium states, we may invent a convenient quasistatic process connecting these two states.

The Gibbs relation tells us along a quasistatic path

$$dS = \frac{1}{T}dE + \frac{P}{T}dV. \tag{15.14}$$

Since for a (1 mole) ideal gas $E = C_V T$ and $PV = RT$,

$$dS = \frac{C_V}{E}dE + \frac{R}{V}dV = C_V d\log E + R d\log V. \tag{15.15}$$

Here dS is a perfect differential, so we have only to integrate this along a convenient path (this is the meaning of inventing a convenient process):

$$S(E_2, V_2) = S(E_1, V_1) + C_V \log \frac{E_2}{E_1} + R \log \frac{V_2}{V_1}. \tag{15.16}$$

In contrast to the usual equation of state $PV = RT$, the relation (which Gibbs called the *fundamental equation*) $S = S(E, V)$ gives us "everything" we wish to know about the ideal gas:

$$\frac{P}{T} = \left(\frac{\partial S}{\partial V}\right)_E = \frac{R}{V}, \quad \frac{1}{T} = \left(\frac{\partial S}{\partial E}\right)_V = \frac{C_V}{E}. \tag{15.17}$$

☐ Poisson's relation $PV^\gamma = $ const. **15.2** must imply $\Delta S = 0$, since we derived Poisson's relation assuming $\Delta S = 0$, but as an exercise, let us explicitly check this. Since $R = C_P - C_V$, (15.16) reads

$$S(E_2, V_2) = S(E_1, V_1) + C_V \left[\log \frac{E_2}{E_1} + (\gamma - 1)\log \frac{V_2}{V_1}\right]. \tag{15.18}$$

For an ideal gas $P = RT/V = RE/C_V V$, so Poisson's relation implies $EV^{\gamma-1} = $ const. If this relation holds, the above equation certainly gives $S(E_2, V_2) = S(E_1.V_1)$.

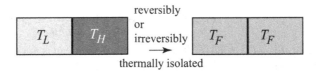

Fig. 15.5 Initially, two blocks have different temperatures. What is the common temperature T_F when the blocks reach a thermal equilibrium under an adiabatic condition?

15.6 Adiabatic Heat Exchange between Two Blocks at Different Temperatures

There are two blocks with the same heat capacity C at temperatures T_L and T_H ($> T_L$), respectively. If we bring these blocks to thermal equilibrium reversibly or irreversibly, what is the final common temperature T_F (see Fig. 15.5)?

Irreversible Case

If we make a thermal contact between them (assume that the system as a whole is thermally isolated), a "perfectly" irreversible process occurs, and the final temperature is $T_F = (T_L + T_H)/2$, because no work is involved under thermal isolation condition (the first law). Needless to say, the final entropy of this system must be larger than the initial one, i.e., $\Delta S > 0$ (Δ always means "final" − "initial"). To use thermodynamics, we must invent a quasistatic process connecting the initial and the final equilibrium states. We bring one block from T_L to T_F, and the other from T_H to T_F quasistatically, and then join these two. This last step does not change anything. Let us study each block separately.

An important observation is that if the heat exchange is across infinitesimal temperature difference dT, then equilibration due to heat transfer is a quasistatic process (no increase of entropy[4]). Therefore, we may prepare numerous heat baths at various temperatures, and use them appropriately in turn to change the temperature of the block gradually (quasistatically). Along this process, we may use thermodynamics. Since $dQ = CdT$, $dS = CdT/T$:

$$\Delta S_1 = \int_{T_L}^{T_F} \frac{CdT}{T} = C \log \frac{T_F}{T_L}. \tag{15.19}$$

We can perform an analogous calculation for the other box, so combining the answers, we get

---

[4] "No increase of entropy" here means that the entropy of the system *and* the heat bath stays constant (as explained here). Needless to say, the entropy of each component changes.

Heat transfer, however slow, is reversible only across infinitesimal temperature difference The entropy change due to thermal contact between the block of heat capacity C at $T + dT$ and a heat bath at T is $CdT/T - C \log[(T + dT)/T] = O[dT^2]$, so it is a higher-oder infinitesimal, and may be ignored. Here, CdT/T is the entropy increase of the bath due to the import of heat CdT and $\int_{T+dT}^{T}(C/T')dT' = -C \log[(T + dT)/T]$ is the decrease of the entropy of the block due to cooling.

$$\Delta S = \Delta S_1 + \Delta S_2 = C \log \frac{T_F^2}{T_L T_H} = 2C \log \frac{T_F}{\sqrt{T_L T_H}}. \tag{15.20}$$

Here, $T_F = (T_L + T_H)/2$. Since the arithmetic mean is not smaller than the geometric mean (or due to the concavity of log; recall Fig. 14.2(a)),

$$\Delta S = 2C \left[\log \frac{T_L + T_H}{2} - \frac{\log T_L + \log T_H}{2} \right] > 0. \tag{15.21}$$

Reversible Case

Equation (15.20) implies that if $T_F = \sqrt{T_L T_H}$, then $\Delta S = 0$. There must be a reversible adiabatic process to realize this. Notice that the internal energy of the system is not conserved:

$$\Delta E = 2CT_F - (CT_L + CT_H) = 2C \left(\sqrt{T_L T_H} - (T_L + T_H)/2 \right) < 0. \tag{15.22}$$

To realize this reversible process and to extract work we can set up a reversible engine between the two blocks and operate it until there is no temperature difference. Let us assume that T_H' is the temperature of the hotter block and T_L' that of the colder block at a certain instant. Since by operation of the engine, the block temperatures change, we analyze the engine working when the hotter block temperature is between T_H' and $T_H' + dT_H'$ (notice that $dT_H' < 0$). The entropy change must be zero (a reversible engine) during this temperature change:

$$dS = \frac{dQ_H}{T_H'} + \frac{dQ_L}{T_L'} = C \frac{dT_H'}{T_H'} + C \frac{dT_L'}{T_L'} = 0. \tag{15.23}$$

This implies $d \log(T_H' T_L') = 0$ or $T_L' T_H' = $ constant. That is, $T_F^2 = T_L T_H$, or, as we know, the final temperature must be $T_F = \sqrt{T_L T_H}$.

We know the efficiency of the reversible engine:

$$-\frac{d'W}{d'Q_H} = 1 - \frac{T_L'}{T_H'}, \tag{15.24}$$

or

$$-d'W = \left(1 - \frac{T_L'}{T_H'} \right) d'Q_H. \tag{15.25}$$

Here, $d'Q_H = -CdT_H'$, because decrease of T_H' implies positive $d'Q_H$ seen from the engine. Therefore,

$$-d'W = - \left(1 - \frac{T_L'}{T_H'} \right) CdT_H' = \left(1 - \frac{T_L' T_H'}{T_H'^2} \right) C(-dT_H'). \tag{15.26}$$

That is, the work we can take out from the system is (note that $T_F = \sqrt{T_L T_H}$):

$$-W = C(T_H - T_F) - CT_F^2 \left(\frac{1}{T_F} - \frac{1}{T_H} \right) = 2C \left(\frac{T_L + T_H}{2} - \sqrt{T_L T_H} \right). \tag{15.27}$$

This is positive as shown before (cf. (15.22)). Of course, this is a stupid way to compute W; the answer is obvious from the first law. Trust thermodynamics.

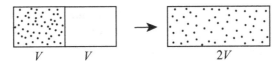

Fig. 15.6 If the volume is doubled, in order to locate a molecule as accurately as before expansion we need to know which V (left half or right half) it is in.

Let us try to understand entropy intuitively in terms of "information." The rest of this chapter is a preview of Chapter 21.

15.7 Sudden Doubling of Volume

We know adiabatic free expansion is irreversible. Let us double the volume of an ideal gas from $V_I = V$ to $V_F = 2V$ by adiabatic sudden expansion (say, by removing the middle wall in Fig. 15.6). There is no supply of heat nor work, so the internal energy must be conserved. Since the process does not change the internal energy of the gas, we can compute the entropy change, using the formula (15.16), or integrating $dS = (P/T)dV$; this latter approach is equivalent to devising an appropriate quasistatic process. For one mole of the gas

$$\Delta S = R \log \frac{V_F}{V_I} = R \log 2. \tag{15.28}$$

$R = 8.31\,\text{J/(K·mol)}$, and $\log 2 = 0.693$, so $\Delta S = 5.75\,\text{J/(K·mol)}$.

15.8 Entropy and Information: Preview

Entropy change and gain or loss of "information" are closely related as we will see in Chapter 21. Here is a preview. Suppose the volume is doubled while keeping the temperature constant (not to alter the motional states; here, we discuss the ideal gas as a classic mechanical object) as in the previous example. To locate a molecule as accurately as before the expansion (with the same device) we have only to know additionally which V (left or right) it is in. This knowledge is obtained from a question answered by a single (equally likely) Yes or No ("is it in the left box?"). We express this as follows: after the volume expansion, to describe the state microscopically (i.e., to describe the microstate), we need one *bit* more information for each molecule compared to the state before expansion.[5]

---

[5] **Bit** The "bit" is the unit of information we can obtain from an answer of a single yes–no question whose answer we cannot guess at all; our expectation of Y or N is with probability 1/2. We will discuss this in Chapter 21. Here, simply accept this intuitively.

Thus, we may naturally expect that the entropy 5.75 J/(K·mol) corresponds to the information 1 bit/molecule (i.e., 1 J/(K·mol) = 0.174 bit/molecule). In other words, 1 bit of information corresponds to $k_B \log 2 = 0.691 k_B = 9.54 \times 10^{-24}$ J/K of entropy.[6]

15.9 Mixing Entropy

Mixing of two different substances also causes an increase in entropy (even without any interactions between them). Suppose there are two kinds of ideal gas A and B (N particles each in a separate container of the same volume V). They are at the same T and P, and can be mixed at constant T and P (due to Dalton's law of partial pressures) by removing the separating wall at the midpoint of the box (see Fig. 15.7).

If we use the information–entropy relation above, we can guess the mixing entropy. After mixing, if we pick up a single molecule, we must know whether it is A or B. Before mixing, this information was given "for free" by the particle position. That is, the mixing process prepares a state that requires one more bit (one extra yes–no question, say, "is it A?") to specify the state of its individual molecules. Therefore, $\Delta S = 2N k_B \log 2$ is our guess (because there are $2N$ particles).

The mixing process may be decomposed into the processes illustrated in Fig. 15.7. First, we expand each gas separately to prepare the state at temperature T and pressure $P/2$ (we can do so by adiabatic "free" expansion as just discussed), and then superpose these two gases to make the final mixture.[7] Therefore the final superposition step does not cause any thermodynamic change. Therefore, ΔS must be just the sum of the "$V \to 2V$" expansion entropy changes, and our guess is correct.

☐ A more general case is that the amount of A and B are different; T and P are the same but the volumes are V_A and V_B, respectively. Then, the final volume is $V_A + V_B$, so the free expansion entropies for A and B are

$$\Delta S_A = N_A k_B \log \frac{V_A + V_B}{V_A}, \quad \Delta S_B = N_B k_B \log \frac{V_A + V_B}{V_B}. \tag{15.29}$$

That is, the mixing entropy is given by (notice that P, T constant $\Rightarrow N \propto V$)

$$\Delta S = N_A k_B \log \frac{N_A + N_B}{N_A} + N_B k_B \log \frac{N_A + N_B}{N_B}. \tag{15.30}$$

If we introduce the mole fractions $x_A = N_A/(N_A + N_B)$ and $x_B = N_B/(N_A + N_B)$, we can rewrite (15.30) as

$$\Delta S = (N_A + N_B) k_B (-x_A \log x_A - x_B \log x_B). \tag{15.31}$$

We have learned that if we mix distinct gases A and B, entropy increases. Suppose we have two gases C and D, and wish to know whether they are distinct gases or not. Let us measure the mixing entropy to know this. Is this feasible?

---

[6] When the ambient temperature is T, roughly speaking, "memorizing" 1 bit of information costs energy of $0.691 k_B T$ (or 1 bit of "knowledge" may be converted to work of $0.691 k_B T$. See Chapter 22).

[7] This is realizable with the aid of semipermeable membranes (walls that can allow only A or B to go through) as illustrated in Fig. 15.7.

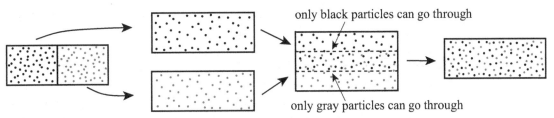

only black particles can go through

only gray particles can go through

Fig. 15.7 The mixing may be considered as two irreversible volume doublings and subsequent superposition of the expanded gases. The last superposition step may be reversibly realized with the aid of semipermeable walls as described in the figure. Thus the last step does not change the system entropy.

15.10 Entropy Changes due to Phase Transition

Another way to change the system entropy is through a phase transition, e.g., melting or evaporation.[8] When a solid melts, a latent heat Q_m is absorbed at a constant temperature (the melting temperature T_m), so the system entropy changes by (under a given pressure)

$$\Delta S_m = Q_m/T_m. \tag{15.32}$$

We have a similar formula for boiling:

$$\Delta S_b = Q_b/T_b, \tag{15.33}$$

where Q_b is the latent heat of evaporation (boiling heat) and T_b is the boiling temperature.

☐ For water $\Delta S_m = 21.9\,\text{J/(K·mol)} = 3.7$ bits/molecule and $\Delta S_b = 109\,\text{J/(K·mol)} = 18$ bits/molecules at 1 atm. Can we understand these entropy changes intuitively? Upon melting, water molecules should rotate more freely in 3-space (3D space). If we simply specify the orientation direction by one of the octants, 3 bits/molecules may not be unreasonable. When evaporated, the volume is expanded by about 1300 times, so even specifying where a molecule is requires extra $\log_2 1300 \simeq 10$ bits. Therefore, although we cannot quantitatively explain these 18 bits by such a crude idea,[9] still we can partially understand why ΔS_b is much larger than ΔS_m.

Problems

Q15.1 Basic Questions

(1) For argon at 1 atm ($= 1.013 \times 10^5$ Pa) and 300 K whose volume is 22.4 ℓ the constant pressure specific heat is 0.125 cal/(g·K), and the heat capacity ratio $\gamma = 1.66$. Estimate the mechanical equivalence of heat, assuming that the gas is ideal. Ar = 39.948 amu. (Use $PV = RT$ and Mayer's relation to compute R; needless to say, the reader must pretend that they do not know how to convert PV in J to that in cal.) (1 cal = 3.8 J according to these data).

---

[8] We will discuss what phase transitions are statistical-mechanically in Chapter 31.
[9] Water is a very difficult fluid, because the molecules can make 3D networks thanks to hydrogen bonding.

(2) What are the needed work and heat to reversibly and isothermally double the volume of 1 mole of ideal gas at 300 K? ($W = -Q = -1729$ J)

(3) Let us assume that when air rises, it undergoes adiabatic expansion. What is the changing rate of temperature as air rises? That is, find $dT(h)/dh$, where h is the altitude. Poisson's relation or its corollary $PT^{\gamma/(1-\gamma)} = $ const. may be used. (Use the force balance condition along a vertical column $dP(h) = -n(h)mgdh$ (cf. Fig. 5.2) and the equation of state $P(h) = n(h)k_BT(h)$. The result is $dT(h)/dh = -(\gamma - 1)mg/\gamma k_B$, where m is the (average) molecular mass of air.)

Q15.2 Entropy Change due to Heating

This is a very elementary problem.

(1) 1 kg of water (specific heat $= 4.2$ kJ/(kg·K)) at $0\,°C$ is in contact with a $50\,°C$ heat bath, and eventually reaches $50\,°C$. What is the entropy change of the water? What is the increase of the entropy of this water plus the heat bath?

(2) Instead of (1) first the water at $0\,°C$ is in contact with a $25\,°C$ heat bath. Then, after reaching thermal equilibrium, the water is in contact with a $50\,°C$ heat bath to reach the final temperature $50\,°C$ as in (1). What is the increase of the entropy of the water plus the two heat baths?

(3) Show that in the two-step heating process whatever the first heat bath temperature T is between $0\,°C$ and $50\,°C$, the total change of entropy of the water plus heat baths is less than the case of (1).

Solution

(1) Entropy is a state function. Thus, the increase of the entropy of the system (water) is the entropy difference between the two equilibrium states. To get ΔS we use a quasistatic process connecting these states. Along such a process $dS = CdT/T$, so we get

$$\Delta S = \int_{273}^{323} \frac{CdT}{T} = 4.2 \log \frac{323}{273} = 0.706 \text{ kJ/K}. \tag{15.34}$$

The entropy increase of the heat bath is $-4.2 \times 50/323 = -0.650$ kJ/K. Therefore, the total increase of the entropy of the system plus the bath $= +56$ J/K.

(2) The system entropy increase is the same as (1) (it is the difference of the state function values). The entropy absorbed by two baths is

$$- 4.2 \times 25(1/298 + 1/323) = -0.677 \text{ kJ/K}. \tag{15.35}$$

Therefore, the total entropy change of the system plus the baths $= 28.6$ J/K.

(3) We have only to compute the bath entropy change, which is given by

$$-4.2 \times ((T - 273)/T + (323 - T)/323) = -4.2(2 - 273/T - T/323) \text{ kJ/K}. \tag{15.36}$$

Notice that $f(T) = A/T + T/B$ ($0 < A < B$) is smaller than $1 + A/B = f(A) = f(B)$ (its minimum is $f(\sqrt{AB})$) for $T \in (A, B)$. Therefore, for $T \in (A, B)$ the entropy discarded to the heat bath is larger than the $T = 323$ K case. This implies that the total change of entropy of the water + heat baths is less than the case of (1).

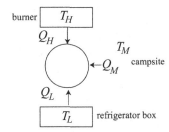

Fig. 15.8 An idealized LPG fridge $Q_H > 0$, $Q_L > 0$, and $Q_M < 0$ are required.

Q15.3 Gas Fridge

There is a refrigerator that uses an external heating process (such as used in campers using LPG). Let us imagine an ideal fridge (i.e., reversible fridge) importing heat Q_H from the high temperature heat reservoir (say, a burner) at temperature T_H. Let T_M be the temperature of the campsite. The temperature inside the cooled box is T_L ($T_H \gg T_M > T_L$ is the usual case). The energy balance of the device may be as in Fig. 15.8. For this device to work as a fridge, Q_H and Q_L must be positive (i.e., the device absorbs these heats) and Q_M must be negative (this heat must be discarded). Since Q_H is supplied by some energy source, the "goodness" of the fridge may be measured by the cost–performance ratio:

$$\eta = Q_L/Q_H. \tag{15.37}$$

(1) Write down the energy balance equation (i.e., $\Delta E = 0$ for a cycle). We strictly apply our sign convention: in $+$, out $-$.
(2) Write down the reversibility condition (i.e., $\Delta S = 0$).
(3) Using these equations, obtain η in terms of T_H, T_M, and T_L.
(4) Looking at the obtained η, the reader will realize that this "goodness measure" improves (increases) as T_H is raised: the hotter the burner, the cooler the box. Isn't it counterintuitive? Explain very briefly why it is not counterintuitive.

Solution

(1) The reader should stick to the algebraic sign convention: $Q_H + Q_M + Q_L = 0$.
(2) Since reversibility may be assumed for the ideal case, we may use $dS = d'Q/T$:

$$\frac{Q_H}{T_H} + \frac{Q_M}{T_M} + \frac{Q_L}{T_L} = 0. \tag{15.38}$$

(3) Using result (1) and (15.38) to get rid of Q_M, we obtain

$$\frac{Q_H}{T_H} + \frac{Q_L}{T_L} = \frac{Q_L + Q_H}{T_M}, \tag{15.39}$$

so

$$\frac{1}{T_H} + \eta \frac{1}{T_L} = (1 + \eta)\frac{1}{T_M}, \tag{15.40}$$

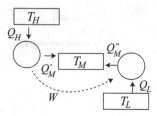

Fig. 15.9 Conceptual dissection of the LPG fridge.

or (recall $T_L < T_M < T_H$)

$$\left(\frac{1}{T_L} - \frac{1}{T_M}\right)\eta = \frac{1}{T_M} - \frac{1}{T_H}. \tag{15.41}$$

Therefore,

$$\eta = \frac{\frac{1}{T_M} - \frac{1}{T_H}}{\frac{1}{T_L} - \frac{1}{T_M}} = \left(1 - \frac{T_M}{T_H}\right)\frac{T_L}{T_M - T_L}. \tag{15.42}$$

This is just the product of the reversible heat engine efficiency working between T_H and T_M and the reversible refrigerator efficiency (Q_L/W in Fig. 15.9) working between T_M and T_L. Thus, the conceptual dissection Fig. 15.9 is quite natural. This does not mean every such fridge contains an engine. The dissection is only conceptual. This answers (4) as well. Increasing T_H makes the engine efficiency better, so the overall efficiency increases. Thus, the hotter the burner, the cooler the fridge (although we are not actually lowering T_L in this problem).

Q15.4 Explosion in a Box

Inside a thermally insulated (i.e., adiabatic) empty (i.e., vacuum) box of volume $10V$ is a small can of volume V containing one mole of an ideal gas at temperature T. The can is punctured and the gas escapes to a larger box and eventually reaches a new equilibrium state (see Fig. 15.10).

(1) What is the change of the total internal energy of the system due to puncturing the can?
(2) What is the total entropy change due to puncturing the can?

Solution

(1) No heat nor work is exchanged with the outside world, so $\Delta E = 0$. Since our gas is ideal, this implies that the temperature (when definable) is invariant.
(2) We may use the equation of state $S = S(E, V)$. Thus,

$$\Delta S = R\log 10.$$

What if the hole made by punctuation is extremely small and molecules can go through it only one by one? No change, because $\Delta E = 0$ does not change, and the final state does not change, so whatever the process is as long as the system is energetically isolated, the

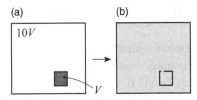

Fig. 15.10 Initially, the can is filled with a gas (a) and is inside a vacuum box of volume 10V. Then, it is punctured and the gas escapes to reach the final equilibrium state (b).

result cannot change. However, the reader might be suspicious. Let us actually compute the entropy change along the extreme-slow leak process.

This is very slow but not a quasistatic process (recall the Warning in **12.6**). Actually, if we pay attention to a small portion $\delta N'$ of the gas going out from the can into the box, the process is patently an irreversible expansion. Thus, entropy increases. This increase may be computed, and after integrating all these infinitesimal increases δS of entropy, we get exactly the same result (as demonstrated here).

The following detailed calculation is not at all recommended (except for thermodynamics exercise), but let us follow the very slow process just described. Suppose N' molecules have already leaked out from the can (assume that the leakage is very slow, so the can and the box are almost in equilibrium[10]). Then, the can pressure is $P = (N - N')k_BT/V$. Let $\delta N'$ be the further small amount of leak from the can. Before going out of the can, this portion $\delta N'$ occupies the volume (note that $N - N'$ molecules still occupy the volume V of the can):

$$\delta V_i = \frac{\delta N'}{N - N'}V \tag{15.43}$$

in the can. The pressure in the outer box is $P = N'k_BT/9V$, so the escaping molecules $\delta N'$ occupy the volume

$$\delta V_f = \frac{\delta N'}{N'}9V. \tag{15.44}$$

That is, the leaked $\delta N'$ changes its volume from δV_i to δV_f. Therefore, the entropy increase due to this escape is (notice that the amount of molecules going out is $\delta N'$)

$$\delta S = k_B \delta N' \log \frac{\delta V_f}{\delta V_i} = k_B \delta N' \log \frac{9(N - N')}{N'}. \tag{15.45}$$

We should integrate this from N to $N/10$ (= the remaining amount in the can):

$$\Delta S = \int_N^{N/10} dN' \, k_B \log \frac{9(N - N')}{N'} = Nk_B \log 10. \tag{15.46}$$

---

[10] Remember that a quasistatic process is a chain of equilibrium states. If only the can or the box is considered, the process for each system is quasistatic, but as a combined system it is never in equilibrium, so it is not a quasistatic process for the whole system. Recall **12.9**.

Q15.5 If $\Delta S = 0$, There Ought to be a Reversible Process

Suppose there are two identical containers A and B containing identical amounts of water, but their temperatures are distinct: A is at T_H and B at T_L ($< T_H$). The opposite situation in which A is at T_L and B at T_H obviously has the same entropy as the former case. There must be a reversible way to exchange the temperatures only. Can we do this only with heat transfer without using engines?[11] We can use this device to utilize the thermal energy in the used water in a bath tub to warm up clean water, which may be the usual tap water.

---

[11] We allow gentle flow to move (relocate) water.

Summary

* Helmholtz free energy A tells us about reversible work under isothermal condition: $\Delta A = W$.
* $-A$ is a convex function of T obtained from internal energy by the Legendre transformation.
* Convex analysis guarantees that internal energy can be recovered from free energies.

Key words

Helmholtz free energy, Legendre transformation, Gibbs free energy, enthalpy, free energy minimum principle, convex function

The reader should be able to:

* Understand $\Delta A \leq W$.
* Understand the (geometrical) meaning of the Legendre transformation.

In reality, the variables $S, V, X, \ldots$ for the ordinary Gibbs relation (13.16) are often hard to control or at least awkward. For example, to keep volume constant may be more difficult than to keep pressure constant. To keep temperature constant may be easier than an adiabatic condition.

16.1 Isothermal System

Under T constant (an isothermal condition) we must allow a "free" exchange of heat between the system and its ambient world to maintain the system temperature. Therefore, we wish to pay attention to the right-hand side of

$$dE - d'Q = d'W = -PdV + xdX. \tag{16.1}$$

Under a quasistatic condition, $d'Q = TdS$, so (16.1) reads isothermally

$$dE - TdS = d(E - TS) = -PdV + xdX. \tag{16.2}$$

This implies that the introduction of the quantity

$$A = E - TS, \tag{16.3}$$

called the *Helmholtz free energy*,[1] is convenient. Notice that for an isothermal process

$$dA = d'W. \tag{16.4}$$

Thus, ΔA is the work the system obtains by a reversible process under constant temperature (i.e., a reversible isothermal process).

16.2 The Free Energy Change, ΔA, by an Irreversible Process

Work, W, is always measurable with the aid of mechanics. What happens if the work exchange is not reversible under isothermal conditions?[2]

If we do work W to the system irreversibly (that allows some dissipation of work), the system must discard heat to the heat reservoir to prevent the final equilibrium state from becoming hotter (recall Planck's principle in Chapter 13). This implies that, even if we do the actual work W, effectively the system receives less energy from the supplied work. We must conclude

$$\Delta A \leq W \tag{16.5}$$

under isothermal conditions. That is, to change the system free energy, we must provide the system with at least ΔA of work (called the *minimum work principle*).

Pay attention to the sign convention: "coming in is +." Suppose the system does work of amount $|W|$ ($W < 0$) to the outside. This implies that the system is supplied with work of $-|W| = W$, so according to (16.5), $\Delta A \leq -|W|$ must hold. Since $\Delta A < 0$, and $|\Delta A|$ is the amount of decrease of the system free energy, when the system does work to the outside, (16.5) implies

$$|\Delta A| \geq |W|. \tag{16.6}$$

That is, the work produced by the system cannot exceed the amount of the free energy lost by the system. The maximum work we can gain from the system is $|\Delta A|$ (*the maximum work principle*).

16.3* Clausius' Inequality and Work Principle

The reader might have felt that the preceding argument sounds like a hand-waving argument, so let us derive (16.5) from Clausius' inequality **14.7**. Let I be the system and II the heat reservoir as in Fig. 14.1:

---

[1] Old literatures use F instead of A.

[2] **What does "isothermal" mean?** Temperature is not definable if a system is not in equilibrium, so the reader may well question what an isothermal irreversible process means. In this book any process whose initial and final states have the same temperature is called an *isothermal process*. Anything can happen in between. In particular, the temperature even if well defined need not be constant. The system need not be immersed in a constant-temperature heat bath during the process.

$$\Delta S_{\mathrm{I}} \geq Q/T. \tag{16.7}$$

Here, we assume $T_e = T$. Q is the heat given to system I, so the heat bath loses Q or gains $-Q$. Let us supply W to system I (however, there is no guarantee that this work is completely received by system I as work due to its dissipation caused by irreversibility). The first law applied to heat bath II reads

$$\Delta E_{\mathrm{II}} = -Q. \tag{16.8}$$

The definition of the Helmholtz free energy and an isothermal condition imply

$$\Delta E_{\mathrm{I}} = \Delta A_{\mathrm{I}} + T\Delta S_{\mathrm{I}}. \tag{16.9}$$

Since the total energy has been increased by W due to the supplied work,

$$W = \Delta E = \Delta E_{\mathrm{I}} + \Delta E_{\mathrm{II}} = \Delta A_{\mathrm{I}} + T\Delta S_{\mathrm{I}} - Q. \tag{16.10}$$

Clausius' inequality implies $T\Delta S_{\mathrm{I}} - Q \geq 0$, so this implies

$$\Delta A_{\mathrm{I}} - W = Q - T\Delta S_{\mathrm{I}} \leq 0. \tag{16.11}$$

This is what we wished to demonstrate.

16.4 Free Energy Minimum Principle

If there is no exchange of work, irreversibility under isothermal conditions implies

$$\delta A < 0. \tag{16.12}$$

This implies that, if there is no spontaneous change (i.e., the equilibrium state is stable), then any virtual change δA must satisfy

$$\delta A \geq 0. \tag{16.13}$$

That is, in the stable equilibrium state under constant T the Helmholtz free energy must be the global minimum (because A is a convex function for constant T). This is the *free energy minimum principle*.

16.5 Gibbs Relation for A

The Gibbs relation now reads

$$dA = -SdT - PdV + xdX, \tag{16.14}$$

so we see, as designed, the natural set of independent thermodynamic variables is (T, V, X) instead of (S, V, X). If the reader wishes to understand the real logic behind this change of independent variables, read **16.8**: Legendre transformation.

16.6 Gibbs Free Energy

It is often convenient to study systems not only under constant temperature but also under constant pressure. Then, the work due to volume change (volume work $-PdV$) must be freely exchanged between the system and the external world to keep the system pressure constant, so we should rewrite the Gibbs relation as

$$dE - TdS + PdV = xdX + \cdots, \tag{16.15}$$

but since T and P are constant, it is convenient to introduce the following *Gibbs free energy* G

$$G = E - TS + PV. \tag{16.16}$$

Quite an analogous argument as the case of the Helmholtz free energy tells us that under constant T and P, if no work other than due to volume changes exists, then

$$\delta G < 0, \iff \text{spontaneous changes can occur}, \tag{16.17}$$

$$\delta G > 0, \iff \text{the equilibrium is stable}. \tag{16.18}$$

Again, this is the principle of minimum free energy (now under constant T and P).

A process with very large $\Delta G > 0$ looks miraculous if it happens spontaneously. In biological systems such reactions are abundant. The secret of biochemistry is to couple reactions with $\Delta G > 0$ to reactions sufficiently spontaneous with $\Delta G < 0$ to make "miraculous biological processes" possible.

16.7 Enthalpy

The Gibbs free energy may be written as

$$G = H - TS, \tag{16.19}$$

where

$$H = E + PV \tag{16.20}$$

is called the *enthalpy*.

If there are only volume works, then $d'W = -PdV$, so under constant pressure the first law reads

$$dH = dE + PdV = d'Q. \tag{16.21}$$

That is, the increase of enthalpy is the heat absorbed by the system under constant pressure. For example, if a chemical reaction occurs in a system, then the change of enthalpy is the reaction heat under constant pressure ($\Delta H < 0$ implies an exothermic reaction; see Chapter 27). If a phase transition, say melting, occurs and the latent heat is Λ, it is the enthalpy change at the phase transition point (solid $\rightarrow$ liquid): $\Delta H = \Lambda$.

We have learned that $E = E(S, V, X)$, the fundamental relation, is convex (see **14.5**). Its convexity allows us to recover $E = E(S, V, X)$ from various free energies. Thus, for example, $A = A(T, V, X)$ is also called the fundamental relation. To understand this extremely important structure of thermodynamics, we need rudiments of *convex analysis*. We must understand its crucial tool, the Legendre transformation, geometrically (and intuitively).

16.8 Legendre Transformation

It is said that $E \to A = E - TS$ or $E \to G = E - TS + PV$ allows us to change the independent variables from (S, V, X) to (T, V, X) or (T, P, X). This is called (in most introductory textbooks) the *Legendre transformation*.

In the transformation $E(S, V, X) \to A(T, V, X) = E - TS$ we use $(\partial E / \partial S)_{V,X} = T$ to fix T (henceforth we suppress V and X). Therefore, to obtain A for a given T we look for a point on the curve $E = E(S)$ where its tangent has a slope T. This is equivalent to looking for a point where the difference (*signed* distance) between the curve $E = E(S)$ and the line $E = TS$ is minimized (see Fig. 16.1). Thus, we see $A = \min_S \{E - TS\}$.

Mathematically (and aesthetically), it is far better to interpret $A = \min_S \{E - TS\}$ as $-A = \max_S \{TS - E\}$.[3] The merits are:

(i) E is a convex function of S, so $-A$ is a convex function of T.[4]
(ii) The Legendre transformation is an involutive transformation;[5] Legendre transformation applied twice recovers the original object: $E = \max_T \{ST - (-A)\} = \max_T \{TS + A\}$.

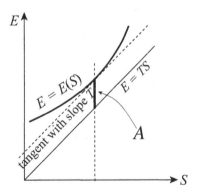

Fig. 16.1 If we fix the Xs, E is a monotone increasing convex function of S (see (14.8)). At the value of S corresponding to the vertical dotted line the difference between $E(S)$ and TS is the smallest, where the slope of the tangent to $E = E(S)$ is equal to T.

---

[3] In convex analysis max and min are always replaced by sup (supremum) and inf (infimum), respectively, but throughout this book we will not be meticulous.
[4] "convex" always implies "convex downward" in mathematics.
[5] **Involution** A map f whose inverse is itself (i.e., $f \circ f = 1$) is called an involutive map or *involution*. For example, the Hermitian conjugation † is an example of involutions.

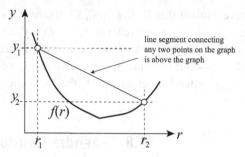

Fig. 16.2 The function f is a convex function if the line segment connecting any two points on its graph (here $(r_1, f(r_1))$ and $(r_2, f(r_2))$) is not below it. A convex function must be continuous, but need not be differentiable as illustrated here.

Thus, we can recover E from A. This is explained in the following units. A branch of analysis studying convex functions is called *convex analysis*.[6]

16.9 Convex Function

Let $f : S \to \mathbb{R}$, where S is a subset of $\mathbb{R}^n$. Let us write $y = f(r)$ ($r \in \mathbb{R}^n$). The function f is a *convex function*, if the line segment connecting $(r_1, f(r_1))$ and $(r_2, f(r_2))$ for any $r_1, r_2 \in S$ is above the graph of $y = f(r)$ (a one-dimensional case is illustrated in Fig. 16.2).[7]

A convex function is a continuous function,[8] but may not be differentiable.

16.10 Geometrical Meaning of Legendre Transformation

The mathematically standard definition of the Legendre transformation for a convex function $f(x)$ to another convex function $f^*(\alpha)$ is

$$f^*(\alpha) = \max_x \{\alpha x - f(x)\}. \tag{16.22}$$

A demonstration that f^* is also a convex function is given in **16.11**. There, it is also shown that

$$f(x) = \max_\alpha \{\alpha x - f^*(\alpha)\}. \tag{16.23}$$

---

[6] The bible of convex analysis is: Rockafellar, R. R. (1970). *Convex Analysis*. Princeton: Princeton University Press (since 1997 in the series Princeton Landmarks in Mathematics). For physicists, some patience and clever reading may be required.

[7] The domain S of the function f must be a convex set; a set is a *convex set* if the segment connecting any two points in the set is again in the set.

[8] **Convex functions are continuous** A convex function is continuous. A rough sketch of demonstration may be: Suppose $f(x)$ is convex on $[b - \varepsilon, b + \varepsilon]$. Let $A_+ = [f(b + \varepsilon) - f(b)]/\varepsilon$ and $A_- = [f(b) - f(b - \varepsilon)]/\varepsilon$ (note that $A_- \leq A_+$). Then, convexity implies that on $(b - \varepsilon, b]$ (resp., on $[b, b + \varepsilon)$) $A_+(x - b) + f(b) \leq f(x) \leq A_-(x - b) + f(b)$ (resp., $A_-(x - b) + f(b) \leq f(x) \leq A_+(x - b) + f(b)$), so $x \to b$ implies $f(x) \to f(b)$.

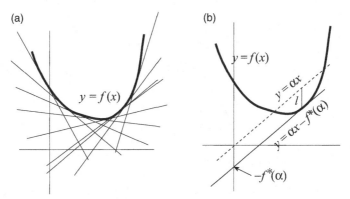

Fig. 16.3 (a): The totality of tangent lines can reconstruct a convex function; if we know them, their envelope curve is the original convex function. (b): l is the maximum gap between the dotted line $y = \alpha x$ and the convex curve $y = f(x)$ (we pay attention to its sign; maximum of $\alpha x - f(x)$). Therefore, if we choose $f^*(\alpha) = \max_x\{\alpha x - f(x)\}$, then $y = \alpha x - f^*(\alpha)$ is the tangent line in the figure. This gives a geometrical meaning of the Legendre transformation $f \to f^*$.

That is, the Legendre transformation is involutive: $f^{**} = f$. This clearly implies that Legendre transformations perfectly preserve the information in the original convex function.

That f^* preserves all the information in f may be understood intuitively from the following geometrical meaning of the Legendre transformation: a convex curve can be reconstructed from the totality of its tangent lines (see Fig. 16.3(a)), where a tangent line of a convex curve is a line sharing at least one point with the curve, and all the points on the curve are on one side of the line or on it (i.e., none on the other side).

A line with a slope α is specified by its y-section $-f^*(\alpha)$: $y = \alpha x - f^*(\alpha)$. If this line is tangent to f, $f^*(\alpha)$ is just given by the Legendre transformation of f (Fig. 16.3(b)).

16.11 The Function f^* is Also Convex and $f(x) = f^{**}(x)$

The function f^* is also convex and $f(x) = f^{**}(x) = \max_\alpha\{\alpha x - f^*(\alpha)\}$ can be illustrated by Fig. 16.4. This graphic demonstration uses the fact that any convex function is a primitive function of an increasing function g: $f(x) = \int^x g(x')dx'$.

In (a) of Fig. 16.4 the pale gray area is $f(x)$. Legendre transformation maximizes the signed area $\alpha x - f(x)$, the dark gray area, by changing x. That is, the (signed) area bounded by the α-axis, the horizontal line through α, the vertical line through x, and the graph of $g(x)$ is maximized with respect to x. When $\alpha = g(x)$, this dark gray area becomes maximum; if x goes beyond this point, a negative area would be added.

Figure 16.4(b) illustrates the case just this condition is satisfied. For this particular pair (x, α) $f^*(\alpha) + f(x) = \alpha x$ is realized; this equality is called *Fenchel's equality*. From these illustrations it is obvious that the relation between f and f^* is perfectly symmetric, so f^* is convex, and $f(x) = \max_\alpha\{\alpha x - f^*(\alpha)\}$, or $f^{**} = f$; * is involutive.

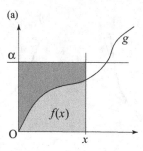 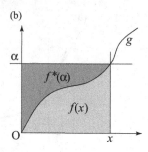

Fig. 16.4 Illustration of the relation between f and f^* in 1D.

This illustration works only for a single variable case, but the Legendre transformation and its salient features (such as $f = f^{**}$) are not confined to one dimension. Read a textbook of convex analysis.

16.12 Legendre Transformation Applied to E: Summary

Since E is a convex function of entropy S and all the work coordinates (see **14.5**), in particular it is a convex function of S. Therefore, its Legendre transform with respect to S: $\max_S\{TS - E\} = E^* = -A$ is a convex function of T (i.e., A is concave or convex *upward* as a function of T).[9] Since the Legendre transformation is an involution, we recover $E = (-A)^* = \max_T\{TS - (-A)\} = E^{**}$.

The usual formula we see in thermodynamics textbooks such as $A = E - TS$ is just an example of Fenchel's equality: $E + (-A) = TS$.

Problems

Q16.1 Basic Problems

(1) In a hydrogen fuel cell $H_2 + (1/2)O_2 \longrightarrow H_2O$ occurs. The reaction heat is $\Delta H = -286$ kJ/mole of hydrogen. The entropies of the substances per mole are: 70 J/K for H_2O, 131 J/K for H_2, and 205 J/K for O_2.[10] Under constant T and P (actually in the so-called standard state: 1 atm and 300 K), what is the electric energy that may be converted to extractible work? ($\Delta G = \Delta H - T\Delta S = -286 \times 10^3 + 300 \times 163 = -237$ kJ/mole.)

(2) 1 mole of an ideal gas at 300 K is quasistatically and isothermally compressed from 5 to 25 atm. Find ΔE, ΔS, ΔA and ΔG. ($\Delta E = 0$, $\Delta S = -8.316\log 5 = -13.4$ J/K, $\Delta A = \Delta E - T\Delta S = -T\Delta S = 4.02$ kJ, $\Delta G = \Delta A + \Delta(PV) = \Delta A$.)

[9] **Warning** A is convex with respect to V and X (when T is fixed) because E is, and $-A$ is convex with respect to T (when X and V are fixed), but $-A$ is *not* convex as a multivariate function of T, V, X.

[10] They are absolute entropies we will encounter in Chapter 23, but the reader may simply regard them as the entropy relative to some basic states.

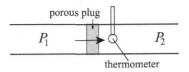

The throttle experiment by Joule and Thomson. A gas in the left side with pressure P_1 is gently squeezed into the right with pressure P_2 ($< P_1$), and the temperature change is measured. The whole system is thermally isolated.

Q16.2 Joule–Thomson Effect

In the *Joule–Thomson experiment* performed in 1852–1854 (Fig. 16.5) a gas of pressure P_1 is pushed out through a porous plug (e.g., cotton plug) to the one of pressure P_2 ($< P_1$) gently under thermal insulation (no exchange of heat). The process is called the *Joule–Thomson process* (or *throttling process*), and the temperature change due to this process is called the *Joule–Thomson effect*.

(1) Show that enthalpy H is preserved in this process.
(2) (This exercise is technically better done after Chapter 24, so come back later.) Show that the Joule–Thomson coefficient $(\partial T/\partial P)_H$ can be rewritten as

$$\left(\frac{\partial T}{\partial P}\right)_H = \frac{1}{C_P}\left(T\left(\frac{\partial V}{\partial T}\right)_P - V\right). \tag{16.24}$$

(3) Let the equation of state of a gas be

$$PV = RT + BP, \tag{16.25}$$

where B is called the second virial coefficient, and is generally a function of T. If $B < 0$, the interatomic interaction is attractive. Compute the Joule–Thomson coefficient for this gas using (16.24).

Solution

(1) Suppose the volume V_1 on the left-hand side is extruded to the volume V_2 on the right-hand side. The first law tells us $\Delta E = W = (-P_2 V_2) - (-P_1 V_1) = \Delta(-PV)$, so $\Delta H = 0$.
(2) (It is better, technically, to do this type of questions after Chapter 24.)

$$\left(\frac{\partial T}{\partial P}\right)_H = \frac{\partial(T,H)}{\partial(P,H)} = \frac{\partial(P,T)}{\partial(P,H)}\frac{\partial(T,H)}{\partial(P,T)} = \frac{1}{C_P}\left(-T\left(\frac{\partial S}{\partial P}\right)_T - V\right). \tag{16.26}$$

Notice that

$$\left(\frac{\partial S}{\partial P}\right)_T = \frac{\partial(S,T)}{\partial(P,T)} = \frac{\partial(S,T)}{\partial(P,V)}\frac{\partial(P,V)}{\partial(P,T)} = -\left(\frac{\partial V}{\partial T}\right)_P. \tag{16.27}$$

Thus,

$$\left(\frac{\partial T}{\partial P}\right)_H = \frac{1}{C_P}\left(T\left(\frac{\partial V}{\partial T}\right)_P - V\right). \tag{16.28}$$

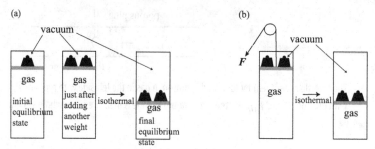

Fig. 16.6 (a) Left: the initial state; Center: just before the irreversible sinking occurs; Right: the final state. The system is kept at T. (b) By adjusting the force F, we wish to lower the weights quasistatically.

(3) Immediately we get

$$\left(\frac{\partial T}{\partial P}\right)_H = \frac{1}{C_P}\left(T\frac{dB}{dT} - B\right). \tag{16.29}$$

Thus, generally, for lower temperatures this is positive. That is, the gas cools through the Joule–Thomson process. For many gases around the room temperature, this is the case, indicating that the interparticle interaction is generally attractive, contrary to Newton's springy molecules (see Chapter 2).

Q16.3 Gas under Weights

A vertical cylinder of cross section A containing an ideal gas is equipped with a piston (with a negligibly small mass) and is placed in a room at temperature T. Initially, on the piston is a weight of mass M (ignore the ambient pressure, or we do this experiment in the vacuum as illustrated in Fig. 16.6(a)). Now we put another identical weight on the piston. The cylinder is rigid but does not isolate the content thermally. What is the percentage of the potential energy of the weights lost as heat, etc., to the environment?

Solution

Let V be the initial volume and assume the gas is n moles. The initial pressure is $Mg/A = P$. The piston moves by $V/2A$, because the volume is halved with an added weight, so $W = 2Mg(V/2A) = nRT$ is the potential energy lost from the weights between the initial and the final states. The increase of the free energy of the gas is (notice $\Delta E = 0$ for isothermal process for an ideal gas):

$$\Delta A = \Delta E - T\Delta S = -T\Delta S = -TnR\log\frac{V/2}{V} = nRT\log 2 \tag{16.30}$$

Hence, $(W - \Delta A)/W \times 100\% = 100 - 69.3 = 31\%$. We can directly obtain ΔA as well since $dA = -SdT - PdV$. Here T is constant, so

$$\Delta A = -\int_V^{V/2}\frac{nRT}{V}dV = nRT\log\frac{V}{V/2} = nRT\log 2. \tag{16.31}$$

Let us do this process gently by applying an appropriate force F (Fig. 16.6(b)). Then, the work W_{rev} (reversible work) done to the gas is a reversible work, so $\Delta A = W_{rev}$. W_{rev} is the potential energy difference − the work we do through F, so clearly $W > W_{rev} = \Delta A$. Without our assistance, it is clear that the potential energy of the weights is lost as heat, and the loss should be $W - \Delta A$.

Summary

* If we have a translation table between mechanical and thermodynamic quantities, we can calculate thermodynamic quantities with the aid of mechanics.
* The table must include not only mechanical quantities (i.e., thermodynamic coordinates) but also entropy.
* Boltzmann's principle provides the translation of entropy: $S = k_B \log w$, where w is the number of microstates compatible with the thermodynamic state of interest.
* With very natural observations as to thermodynamics and mechanics, we can understand this principle from thermodynamics.
* Thermodynamics is obtained from almost any microstate that gives the thermodynamic coordinates correctly.

Key words

phase space, microstate (classical and quantum), microcanonical ensemble, microcanonical partition function, Boltzmann's principle, typical state

The reader should be able to:

* Recognize what thermodynamics can and cannot do.
* Clearly explain the meaning of the quantities appearing in Boltzmann's principle.
* Use Boltzmann's principle for simple examples.
* Understand why statistical mechanics can give (or is consistent with) thermodynamics.

17.1 Power and Limitation of Thermodynamics

As the reader has realized, thermodynamics is very powerful when the right inputs are introduced, but it cannot tell us anything specific to a particular system. For example, the fundamental equation $S = S(E, V)$ cannot be supplied by thermodynamics; when we computed this, even for an ideal gas, we relied on $E = C_V T$ and $PV = RT$, neither of which is thermodynamically obtainable.

The reader must clearly know what thermodynamics can and cannot do. Thermodynamics can compute the changes of state functions (state variables) between two equilibrium states irrespective of the actual process that has happened, if the fundamental equation of state of the system is known. Thermodynamics cannot calculate materials-specific (or system-specific) input required for this computation.

17.2 Why Statistical Mechanics?

We know that the microscopic world underlies the world we experience daily, and that its description in terms of mechanics is much more detailed than what the macroscopic phenomenology can offer. We hope that looking at the microscopic details, we may be able to obtain the information that thermodynamics needs but cannot provide.

To describe a macrosystem in terms of particle mechanics, we need far more variables than the dimension of the thermodynamic space (recall scooping out the ocean by a tablespoon and also Chapter 11). Thus, it is a natural guess that we need some statistical means: *statistical mechanics*.

17.3 What Do We Really Need?

However, as is emphasized repeatedly, the macroscopic world is the world governed by the law of large numbers, so if we know how to get the expectation values, we do not need statistics explicitly. We need only the *translation table* of thermodynamic quantities in terms of (the expectation values of) microscopically described mechanical quantities.

We have learned that the most fundamental macroscopic description of any equilibrium state is in terms of thermodynamic coordinates (E, X), where E is the internal energy and X (collectively) are the work coordinates (Chapter 12). We know E is the system mechanical energy. X may be the volume V, magnetization M, etc., and can be described in terms of microscopic mechanical variables easily and/or naturally. Many work coordinates are expressed as sums of microscopic mechanical quantities, so the law of large numbers tells us that not even their distributions are required (recall **4.2**, especially (4.4)). Thus, we can write down the translation table for thermodynamic coordinates relatively easily.

17.4 Translation of Entropy: Boltzmann's Principle

However, thermodynamics is "thermo" dynamics. Indeed, we have learned that entropy $S = S(E, X)$ is the fundamental equation (see **13.14**) we need in order to use thermodynamics. Therefore, the translation table must include S.

The translation table was completed by Boltzmann in the following form, *Boltzmann's principle*:

$$S = k_B \log w(E, X), \tag{17.1}$$

where k_B is the Boltzmann constant, and $w(E, X)$ is the "number" of "microscopic states (*microstates*)" compatible with the macrostate (E, X). That is, a macrostate is an equivalence class of microstates by the agreement of macroobservables (thermodynamic coordinates in the present context).[1] To understand this statement precisely, we must clarify what "microstate" means.[2]

Those who are not very familiar with quantum mechanics can read an outline of the principles of quantum mechanics in the Appendix: Introduction to Mechanics at the end of this book. The bra-ket notation (the so-called Dirac notation) will be fully used. If the reader is not familiar with it, see Appendix 17A: Linear algebra in Dirac's notation.

17.5 Counting Microstates

For a quantum system (the classical case appears at the end of this unit) its state is a ket in the state space (the Hilbert space $\mathcal{H}$ on which the system Hamiltonian is self-adjoint; in Fig. 17.1 it is called the microstate space). We know a complex constant multiple of a ket expresses the same microstate as the original ket, so microstates must correspond to "directions" (rays) or normalized kets (cf. **A.12**). If two kets are not orthogonal, they are not quite distinct, so if we wish to collect distinct microstates, we should collect orthonormal kets. Thus, the totality $\tilde{w}(E, X)$ of microstates corresponding to the thermodynamic state (E, X) must be a subspace of $\mathcal{H}$, and the number of distinct microstates must be the dimension of $\tilde{w}(E, X)$: $w(E, X) = \dim \tilde{w}(E, X)$.

We have not yet specified how to choose the microstates (normalized kets) corresponding to the thermodynamics state (E, X). Let H be the system Hamiltonian, and let $\hat{X}$ collectively describe the operators corresponding to the observable X. If all these operators commute, we have simultaneous eigenkets $|E, X\rangle$ (cf. **A.25**). $\tilde{w}(E, X)$ is the totality of kets, whose energy is close to E,[3] and the work coordinate eigenvalues are close to X. $w(E, X) = \dim \tilde{w}(E, X)$, which is called the *microcanonical partition function* of the system.

---

[1] **Equivalence relation and equivalence class** A binary relation $\approx$ on a set X is an *equivalence relation*, if and only if it satisfies the following three axioms: for any a, b, and $c \in X$ (i) $a \approx a$ (reflexivity), (ii) $a \approx b \iff b \approx a$ (symmetry), and (iii) $a \approx b, b \approx c \implies a \approx c$ (transitivity). An *equivalence class* of $x \in X$ related to x by $\approx$ is a set $[x] = \{y \mid y \approx x, y \in X\}$. That is, $[x]$ is the totality of elements in X indistinguishable from x by the equivalence relation $\approx$. Note that for any $x, y \in X$, $[x] = [y]$ or $[x] \cap [y] = \emptyset$.

[2] The reader may understand that a "microstate" corresponds to an "elementary event" in probability (see Chapter 3).

[3] Precisely speaking, say, the energy is in $(E - \Delta E, E]$ for a macroscopically small leeway ΔE. We discussed a related topic for an ideal gas in **4.5**.
Why we need the leeway Empirically speaking, any macroobservable should have some leeway in their values. Thus, all other X should also have some width ΔX. However, the real reason for the necessity of some widths is not because errors are inevitable. The widths are needed for $\tilde{w}(E, X)$ to be sufficiently large so the concept of "almost sure" (or overwhelming majority) is meaningful. Thus, from the purely theoretical point of view, the leeways are not needed for certain X if $\tilde{w}(E, X)$ can still be very large (e.g., for V or N usually we do not need any widths). In some cases, even ΔE is not needed (e.g., **18.1**).
  The choice of ΔE, if needed, should be macroscopically small (say, 0.1% of E) for E to be meaningful, but otherwise ΔE can be any value as long as $(1/N) \log(\Delta E/E)$ is negligible for large N.

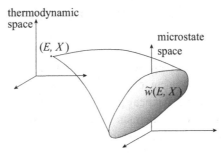

For each equilibrium state (E, X) in the thermodynamic space, we can imagine a subset $\tilde{w}(E, X)$ of the microstate space (called the phase space, classical-mechanically) consisting of all the microstates that give the same (macroscopically indistinguishable) thermodynamic coordinates.

☐ In this elementary book, we will not discuss the cases in which H and $\hat{X}$ are not commutative. For the noncommutative so-called non-Abelian case, we cannot have simultaneous eigenstates of the observables corresponding to the thermodynamic coordinates. How to construct $\tilde{w}(E, X)$ is still a research topic.

If we use the classical approximation (classical mechanics), the most detailed description of the system is in terms of the canonical coordinates. The most popular canonical coordinates for a N-particle system consist of $6N$-vector $\gamma = (r_1, \ldots, r_N, p_1, \ldots, p_N)$, where r_i is the position vector of the ith particle, and p_i its momentum vector. This $6N$-space is called the *phase space* of the system. We compute the thermodynamic coordinates for each phase point γ, and make a subset $\tilde{w}(E, X)$ of the phase space consisting of the phase points with the thermodynamic coordinate values (E, X) (allowing macroscopically small leeways). Its $6N$-volume $W(E, X)$ divided by h^{3N}, where h is Planck's constant, is $w(E, X)$, which is called the *microcanonical partition function*.[4]

17.6 Statistical Mechanics is Completed

Thus, the microcanonical partition function $w(E, X)$ is unambiguously defined by mechanics, and statistical mechanics is complete. The rest is taken care of by the Gibbs relation

$$dS = \frac{1}{T}dE + \frac{P}{T}dV - \frac{\mu}{T}dN - \frac{x}{T}dX + \cdots . \tag{17.2}$$

In practice, it is wise to use thermodynamics as much as possible with sparing use of statistical mechanics.

Two problems remain: How can we use Boltzmann's principle in practice? Why is the principle plausible? Let us first use it to become familiar with the tool.

---

[4] **Why h^{3N}** The factor h^{3N} is due to the classical-quantum correspondence. Classically, each γ is regarded as a microstate, but since the position and the momentum of a particle cannot be specified simultaneously in quantum mechanics (the uncertainty relation **A.28**), a $6N$-cube of edge length $\sqrt{h}$ should be regarded as a single microstate. We will come back to this point in Chapter 20.

17.7 Classical Approach to Classical Ideal Gas

We can use the completed translation table to compute S from mechanics. However, a quantum-mechanical evaluation is more involved than the classical approximation. Let us follow the classical procedure first to compute $w(E, X)$ for a classical ideal gas, which consists of N noninteracting point particles of mass m in a volume V. The system Hamiltonian is the pure kinetic energy:

$$H = \sum_{i=1}^{N} \frac{p_i^2}{2m}. \tag{17.3}$$

The microcanonical partition function $w(E, V)$ collects all the microstates with the volume V and the total energy in $(E - \Delta E, E]$, where ΔE is a (macroscopically small) leeway. Let us write down the phase volume $W(E, V) \equiv h^{3N} w(E, V)$:

$$W(E, V) = \int_{r_1, \dots, r_N \in V, E - \Delta E < \sum_{i=1}^{N} p_i^2 / 2m \leq E} d\Gamma, \tag{17.4}$$

where $d\Gamma = d^3 r_1 \cdots d^3 r_N d^3 p_1 \cdots d^3 p_N$ is the volume element of the $6N$-dimensional phase space. The space and momentum integrals can be decoupled totally:

$$W(E, V) - \int_V d^3 r_1 \cdots \int_V d^3 r_N \int_{\sum p_i^2 / 2m \in (E - \Delta E, E]} d^3 p_1 \cdots d^3 p_N \tag{17.5}$$

$$= V^N \int_{\sum p_i^2 / 2m \in (E - \Delta E, E]} d^3 p_1 \cdots d^3 p_N. \tag{17.6}$$

The last integral is the volume of the skin of thickness $\propto \Delta E / \sqrt{E}$ of a $3N$-ball ($3N - 1$-sphere[5]) of radius $\sqrt{2mE}$, which must be proportional to $E^{3N/2 - 1} \Delta E$, so (for an exact evaluation see **Q17.5**)

$$w(E, V) \propto V^N E^{3N/2} \Delta E. \tag{17.7}$$

Here, $N \gg 1$, so 1 is ignored in $3N/2 - 1$. Therefore, Boltzmann tells us that

$$S = k_B \log w(E, V) = N k_B \log V + \frac{3}{2} N k_B \log E + k_B \log \Delta E + \cdots, \tag{17.8}$$

where the remaining terms are $N \times$ a constant. Using thermodynamic relations, we get

$$\frac{1}{T} = \left(\frac{\partial S}{\partial E} \right)_V = \frac{3}{2} \frac{N k_B}{E} \tag{17.9}$$

or $E = (3/2) N k_B T$ and

$$\frac{P}{T} = \left(\frac{\partial S}{\partial V} \right)_E = \frac{N k_B}{V}, \tag{17.10}$$

---

[5] Notice that in mathematics, 1-sphere is the edge of a disk, 2-sphere (the ordinary sphere) is the skin of a 3-ball (the ordinary ball), etc.

which is the equation of state. Boltzmann confirmed his principle first with the aid of the classical ideal gas as the reader can see from **Q17.2**, and then confirmed it for "any" classical particle system (see **Q17.3**).

Incidentally, (17.8) does not satisfy the fourth law: if we double extensive quantities (experimentally, we have only to join two identical systems to make a compound system): $N \to 2N$, $E \to 2E$, $V \to 2V$, then we must also have $S \to 2S$, but this does not hold. When we use statistical mechanics, a wise practice is to make a shortcut with the aid of thermodynamics, since it is reliable. We should demand that the fourth law holds. Then, we are forced to accept the following form:

$$S = k_B \log w = N k_B \log \frac{V}{N} + \frac{3}{2} N k_B \log \frac{E}{N} + k_B \log \Delta E + \cdots . \tag{17.11}$$

This corresponds to replacing w with $w/N!$ (see Chapter 20). The form (17.11) is not surprising, because in the thermodynamic limit (roughly, a large system size limit) all the extensive variables diverge, so the meaningful fundamental equation must be in terms of the (thermodynamic) densities (or quantities per particle) alone (see **31.11**).

17.8 Quantum Mechanical Study of Classical Ideal Gas

Next, let us obtain the microcanonical partition function $w(E, V)$ for a classical ideal gas quantum mechanically. In quantum mechanics, a classical ideal gas is characterized by the absence of any quantum effect: here a possible quantum effect is that if particles come too close, their indistinguishability must be respected. For such effects to be ignorable, the average de Broglie wavelength of each particle must be sufficiently shorter than the average interparticle distance. The de Broglie wavelength of a particle of mass m is on average

$$\lambda \sim h/\sqrt{m k_B T}, \tag{17.12}$$

since $p^2/m \sim k_B T$ and the de Broglie wavelength $\lambda = h/p$. On the other hand, the average particle distance is $\sqrt[3]{V/N}$, so the condition for quantum effects to be ignored is $\sqrt[3]{V/N} \gg \lambda$, i.e.,

$$\frac{N}{V} \ll \left(\frac{m k_B T}{h^2} \right)^{3/2}. \tag{17.13}$$

If this condition is satisfied, a fluid is said to be classical.

Let us compute $w(E, V)$ for a noninteracting N particle quantum system. To this end, we must solve a one-particle Schrödinger equation in the cube of edge length L such that $V = L^3$:

$$-\frac{\hbar^2}{2m} \Delta \psi = E \psi. \tag{17.14}$$

Here, Δ is the Laplacian and we impose a boundary condition: $\psi = 0$ on the wall. The eigenfunctions have the following form

$$\psi_{\boldsymbol{k}} \propto \sin k_x x \sin k_y y \sin k_z z \tag{17.15}$$

with

$$\boldsymbol{k} \equiv (k_x, k_y, k_z) = \frac{\pi}{L}(n_x, n_y, n_z) \equiv \frac{\pi}{L}\boldsymbol{n}, \tag{17.16}$$

where $n_x, \dots$ are positive integers: $1, 2, \dots$. The eigenfunction $\psi_{\boldsymbol{k}}$ belongs to the 1-particle eigenenergy $\hbar^2 \boldsymbol{k}^2 / 2m$. That is, the single-particle energy reads

$$\varepsilon(\boldsymbol{n}) = \frac{h^2}{8mL^2}\boldsymbol{n}^2. \tag{17.17}$$

Consequently, the microstate of the system may be uniquely specified as $\{\boldsymbol{n}_1, \boldsymbol{n}_2, \dots, \boldsymbol{n}_N\}$, where $\boldsymbol{n}_i$ is the three-dimensional positive integer vector specifying the state of the ith particle (see (17.16)). We wish to count the microstates whose energy is between $E - \Delta E$ and E. The total energy is a sum of (17.17), so with the aid of $3N$-positive integer vector $\boldsymbol{M}$ we can write the total energy of the gas as $h^2 \boldsymbol{M}^2 / 8mL^2$, which must be in $(E - \Delta E, E]$. Thus, we have to count the integer vectors $\boldsymbol{M}$ satisfying $(L/h)\sqrt{8m(E - \Delta E)} < |\boldsymbol{M}| \leq (L/h)\sqrt{8mE}$ with all the components being positive. When a big number is counted, we may use

$$\sum_{\text{conditions on } n} 1 = \int_{\text{conditions on } n} dn. \tag{17.18}$$

We see that the problem is quite similar to the classical-mechanical computation done before:

$$w(E, V) = \frac{S_{3N-1}}{2^{3N}} \int_{(L/h)\sqrt{8m(E-\Delta E)}}^{(L/h)\sqrt{8mE}} M^{3N-1} dM \tag{17.19}$$

$$\sim \left(\frac{L}{h}\right)^{3N} E^{3N/2-1} \Delta E = \frac{V^N E^{3N/2} \Delta E}{h^{3N}}. \tag{17.20}$$

Here, $|\boldsymbol{M}|$ is written as M and the factor $1/2^{3N}$ is to count only the contribution of the "first quadrant" (because all the components of $\boldsymbol{M}$ must be positive), and S_{3N-1} is the surface area of the unit $(3N - 1)$-sphere (see **Q17.5**). The result obtained agrees with (17.6). We need not repeat the calculation beyond this point.

17.9 Projection and Dimension

Let us formalize the quantum mechanical calculation of $w(E, V)$. Our task is to compute the dimension of a subspace spanned by the totality of the state vectors satisfying a certain condition.

If a subspace $\mathcal{V}$ is, for example, a 2-space, there is an orthonormal basis consisting of two kets, say, $|1\rangle$ and $|2\rangle$. The component in the $|1\rangle$ direction of a certain ket $|A\rangle$ may be written as $\langle 1|A\rangle$. Similarly we can handle $|2\rangle$. Combining these components, the projection of $|A\rangle$ to the subspace $\mathcal{V}$ spanned by $|1\rangle$ and $|2\rangle$ is written as (for the Dirac notation see Appendix 17A):

$$|1\rangle\langle 1|A\rangle + |2\rangle\langle 2|A\rangle = (|1\rangle\langle 1| + |2\rangle\langle 2|) |A\rangle. \tag{17.21}$$

The operator in the parenthesis, $P \equiv |1\rangle\langle 1| + |2\rangle\langle 2|$, is the projection operator onto $\mathcal{V}$. Notice that

$$P^2 = P. \tag{17.22}$$

Generally speaking, a linear operator P satisfying $P^2 = P$ is called a *projection* (or a projection operator).[6] The trace of this projection is (inside a trace Tr cyclic permutations are allowed):

$$\text{Tr}P = \text{Tr}(|1\rangle\langle 1| + |2\rangle\langle 2|) = \text{Tr}(\langle 1|1\rangle + \langle 2|2\rangle) = 2, \tag{17.23}$$

which is the dimension of $\mathcal{V}$. That is, if we can construct a projection $P_\mathcal{V}$ onto a subspace $\mathcal{V}$, then $\dim \mathcal{V} = \text{Tr}(P_\mathcal{V})$.

How can we apply the projection formalism to the classical ideal gas? Let the (normalized) single-particle eigenket with the quantum number $\boldsymbol{n}_j$ be $|\boldsymbol{n}_j\rangle$. Then, an N particle energy ket is given by

$$|\boldsymbol{M}\rangle = \prod_{i=1}^{N} |\boldsymbol{n}_i\rangle. \tag{17.24}$$

The projection onto the subspace $w(E, V)$ reads

$$P_{w(E,V)} = \sum_{\substack{(L/h)\sqrt{8m(E-\Delta E)} < |\boldsymbol{M}| \leq (L/h)\sqrt{8mE,} \\ \text{all the components of } \boldsymbol{M} \text{ positive}}} |\boldsymbol{M}\rangle\langle \boldsymbol{M}|. \tag{17.25}$$

If we take its trace, we get (17.19), so we get $w(E, V)$ in agreement with the already obtained result.

17.10 Derivation of Boltzmann's Formula: Key Observations

The empirical observations we wish to rely on are:

[O] There is a macrostate called an equilibrium state that does not depend on time macroscopically.

[X] The needed observation time for thermodynamic coordinates is very short (say, 1 μs or much less if the system is large enough).

The usual picture of microscopic dynamics is that the instantaneous microstate wanders around the microstate space; in particular, if the macrostate is (E, X), it wanders around in $\tilde{w}(E, X)$ (see Fig. 17.1). Traditionally, "statistics" of "statistical mechanics" was understood as taking statistics over all the microstates in $\tilde{w}(E, X)$. Consequently, a misconception was spread that the key to statistical mechanics was the even mechanical sampling over $\tilde{w}(E, X)$ (the ergodic theoretical justification of statistical mechanics).

---

[6] In particular, if $P^* = P$ as in our case, the projection is called an orthogonal projection. Notice that $P(1-P) = 0$ for any projection P.

For the ordinary macro object containing $N \sim 10^{23}$ particles, what is the time scale required to sample $\tilde{w}$ evenly? It is roughly the time scale of the Poincaré cycle $\sim e^N$.[7] The observation time [X] implies that during one thermodynamic observation, only an extremely tiny fraction of $\tilde{w}$ is sampled.[8] However, [O] implies that if we repeat this experiment 1 billion years later, we get the same thermodynamic result. A 1 ps observation in the year 10^9 CE will cover again only very tiny portion of $\tilde{w}$ somewhere else.

What is the most natural conclusion? If we sample a microstate from $\tilde{w}(E, X)$, it almost surely corresponds to an equilibrium state with the thermodynamic coordinates (E, X). That is, if we wish to know thermodynamics, we have only to observe a single microstate; *almost every microstate sampled from a macrosystem in equilibrium gives the same thermodynamics.*[9]

17.11 † Typical States in $\tilde{w}(E, X)$ Give the Identical Equilibrium State

Astute readers may question our conclusion, saying that $\tilde{w}(E, X)$ must contain numerous microstates that cannot correspond to the macroscopic equilibrium state specified by (E, X), because the membership of $\tilde{w}(E, X)$ is solely due to the agreement of the total values E, X. For example, even if we specify E, this E might not be distributed evenly in the system. Or, in the container there may be a macroscopic flow. However, according to our experience such states are never observed. Combining [O], [X], and this empirical result, we must confirm the italicized statement in **17.10**.

17.12 Derivation of Boltzmann's Formula[10]

Let us derive Boltzmann's principle from thermodynamics (and mechanics). The micro-canonical density operator ρ is given by[11]

---

[7] The time scale of the Poincaré cycle is roughly the time scale required for a given closed dynamical system to return to its intial condition. Poincaré's recurrence theorem guarantees that any state can return to any starting neighborhood. Recall **11.8**.

[8] Notice that if the system is really isolated, then energy eigenkets are invariant; pure-mechanically, in equilibrium, there is no wandering in quantum mechanics.

[9] The critical reader may not accept this mathematically idealized conclusion. What we can conclude empirically is the homogeneity of $\tilde{w}(E, X)$ after an extremely short time averaging. Note that in the thermodynamic limit the needed observation time is infinitesimal.

[10] **Einstein's statistical mechanics** This is not the author's original; it is an adaptation of Einstein's derivation of the canonical ensemble theory to the quantum microcanonical ensemble. See Einstein, A. (1903). Eine Theorie der Grundlagen der Thermodynamik, *Ann. Phys.* **316**, 170–187. See **Q18.1** for the derivation of the canonical formalism. It is easier.

Technical remark: Here, the assumed distribution presupposes the principle of equal probability, but actually it is not needed. Besides, we will see (Chapter 19) that thermodynamics allows us to assume the principle.

[11] A density operator is a counterpart of distribution function for the ordinary statistics, and average of an observable A is given by $\mathrm{Tr}\,(\rho A)$. See Chapter 19.

$$\rho = \frac{1}{w(E)} \sum_{\varepsilon} |\varepsilon\rangle \chi_\Delta(\varepsilon - E)\langle\varepsilon|, \tag{17.26}$$

where χ_Δ is the indicator of $(-\Delta, 0]$ and the sum is over the eigenvalues of H: $H|\varepsilon\rangle = \varepsilon|\varepsilon\rangle$. The normalization condition implies

$$1 = w(E)^{-1} \sum_{\varepsilon} \chi_\Delta(\varepsilon - E). \tag{17.27}$$

We assume that the Hamiltonian H has a parameter λ that can be controlled externally to supply work to the system.

Assume that the change $\lambda \to \lambda + \delta\lambda$ causes the internal energy change $E \to E + \delta E$ as well. Since the normalization condition must be maintained, from (17.27) we obtain

$$0 = -\frac{\delta w}{w^2} \sum_{\varepsilon} \chi_\Delta(\varepsilon - E) + \frac{1}{w} \sum_{\varepsilon} \chi_\Delta'(\varepsilon - E)(\delta\varepsilon - \delta E). \tag{17.28}$$

Since most microstates give identical thermodynamics, $\delta\varepsilon$ and $\delta'W$ (work due to $\delta\lambda$) are identical for almost all the microstates. Therefore, (17.28) reads

$$\delta \log w = -(\delta E - \delta'W)\frac{1}{w} \sum_{\varepsilon} \chi_\Delta'(\varepsilon - E). \tag{17.29}$$

The first law of thermodynamics tells us $\delta E - \delta'W = \delta'Q$. Also, we have

$$-\frac{1}{w} \sum_{\varepsilon} \chi_\Delta'(\varepsilon - E) = \frac{1}{w} \frac{\partial}{\partial E} \sum_{\varepsilon} \chi_\Delta(\varepsilon - E) \tag{17.30}$$

$$= \frac{\partial}{\partial E} \left[\frac{1}{w} \sum_{\varepsilon} \chi_\Delta(\varepsilon - E)\right] + \frac{1}{w^2} \frac{\partial w}{\partial E} \sum_{\varepsilon} \chi_\Delta(\varepsilon - E) \tag{17.31}$$

$$= \frac{1}{w} \frac{\partial \log w}{\partial E} \sum_{\varepsilon} \chi_\Delta(\varepsilon - E) = \frac{\partial \log w}{\partial E}. \tag{17.32}$$

Therefore, we obtain

$$\delta \log w = \eta \delta'Q, \tag{17.33}$$

where

$$\eta \equiv \left(\frac{\partial \log w}{\partial E}\right)_X. \tag{17.34}$$

What is η? We have already computed this for a classical ideal gas in (17.9):[12]

$$k_B \frac{\partial \log w(E, V)}{\partial E} = \frac{3k_B N}{2E}, \tag{17.35}$$

which is $1/T$. Using thermodynamics $\delta S = \delta'Q/T$, and adjusting the units if needed, we get

$$\delta S = k_B \delta \log w. \tag{17.36}$$

---

[12] A more general calculation is classically done by Boltzmann himself. See **Q17.3**.

Here, the variation δ is quite a general variation, so we have obtained the exact differential equation

$$dS = k_B d \log w. \qquad (17.37)$$

Integrating this, we get Boltzmann's formula.

Problems

Q17.1 Einstein Model with Microcanonical Formalism

The total energy of the system is $E = \mathcal{M}\varepsilon$, where ε is the size of the energy quantum and $\mathcal{M}$ is the number of energy quanta (phonons originally). There are N lattice sites and each lattice site can accommodate some energy quanta.[13] Each microstate is distinguished by the distribution of energy quanta over the lattice sites.

(1) Obtain the entropy S as a function of E (or $\mathcal{M}$).
(2) Obtain the average number $\mathcal{M}/N$ of energy quanta (phonons) for each lattice site ($\mathcal{M}\varepsilon/N$ is the average energy per lattice site) as a function of T. Discuss the high and low temperature limits of $\mathcal{M}/N$ and S.
(3) Prepare two identical such systems, 1 and 2, and put them into thermal contact. Suppose the total energy is $2\mathcal{M}\varepsilon$. How sharp is the most likely energy partition (i.e., an even partition of energy between the two identical systems) in equilibrium?

Solution

In the following, the results from Chapter 23 will be quoted for later convenience, but the reader may ignore them totally. Notice that the system has only one thermodynamic coordinate, E.

(1) The problem is equivalent to distributing $\mathcal{M}$ indistinguishable objects into N distinguishable bins. See (3.47). Hence,

$$w(E) = \binom{\mathcal{M} + N - 1}{\mathcal{M}}. \qquad (17.38)$$

We may ignore 1, since $\mathcal{M}$ and N are both macroscopic. Boltzmann's principle tells us that

$$S(E)/k_B = \log w(E) = (N + \mathcal{M}) \log(N + \mathcal{M}) - \mathcal{M} \log \mathcal{M} - N \log N, \qquad (17.39)$$

where we have used (7.14) with $\mathcal{M} = E/\varepsilon$.

---

[13] The model is equivalent to the Einstein model of insulating solids. As we will discuss in Chapter 23, Einstein introduced a collection of N quantized harmonic oscillators to explain the low temperature behavior of the specific heat of solids. Each oscillator localized at a lattice point is understood as a container of phonons (energy quanta) of energy $\varepsilon = \hbar\omega$. (We ignore the zero-point energy.)

(2) To introduce T we use the Gibbs relation:

$$\frac{1}{T} = \frac{dS}{dE} = \frac{1}{\varepsilon}\frac{dS}{d\mathcal{M}} = \frac{k_B}{\varepsilon}\log\frac{N+\mathcal{M}}{\mathcal{M}}. \tag{17.40}$$

From this, we get the average energy of the site:

$$\frac{\mathcal{M}}{N}\varepsilon = \frac{\varepsilon}{e^{\beta\varepsilon}-1}. \tag{17.41}$$

If the temperature is sufficiently high, $\mathcal{M}/N \simeq k_B T/\varepsilon \gg 1$ (this actually corresponds to the equipartition of energy, see Chapter 20), then the entropy as a function of T reads

$$S(T) = (\mathcal{M}+N)\log[\mathcal{M}(1+N/\mathcal{M})] - \mathcal{M}\log\mathcal{M} - N\log N \tag{17.42}$$

$$= N\log(\mathcal{M}/N) + (\mathcal{M}+N)[N/\mathcal{M} - N^2/2\mathcal{M}^2 + \cdots] \tag{17.43}$$

$$= N\log(\mathcal{M}/N) + N + N^2/2\mathcal{M} + \cdots . \tag{17.44}$$

Since $\mathcal{M} \gg N$ in this case, $S(T) = N\log(ek_B T/\varepsilon)$, which corresponds to (23.15). In the opposite limit $T \to 0$, $\mathcal{M}/N \to 0$, so $\lim_{T\to 0} S(T) = 0$ can be concluded (see Chapter 23).

(3) We saw an analogous case of distributing gas particles in Chapter 7. To study the probability of deviation from the even partition of energy, let us define $\mathcal{M}_1 = \mathcal{M}+x$ (resp., $\mathcal{M}_2 = \mathcal{M}-x$). Then, the microcanonical partition function of the whole system $w(x) = w(\mathcal{M}_1, \mathcal{M}_2)$ is given by

$$w(x) = \binom{N+\mathcal{M}+x,}{N}\binom{N+\mathcal{M}-x,}{N} \tag{17.45}$$

Let us compute $\log[w(x)/w(0)]$ with the aid of the already calculated result just above (7.19) $(A+x)\log(A+x) = A\log A + (\log A + 1)x + x^2/2A + o[x^2]$:

$$\log[w(x)/w(0)] = -x^2\left[\frac{1}{\mathcal{M}} - \frac{1}{N+\mathcal{M}}\right] = -\frac{N}{\mathcal{M}(N+\mathcal{M})}x^2. \tag{17.46}$$

This implies that x obeys a Gaussian distribution of mean 0 and variance $\mathcal{M}(N+\mathcal{M})/2N$. The reader might think the variance is very large, but we must compare it with the equilibrium expectation $\mathcal{M}$. The ratio

$$\frac{\sqrt{\langle x^2\rangle}}{\mathcal{M}} = \frac{1}{\sqrt{2N}}\sqrt{1+N/\mathcal{M}} \tag{17.47}$$

implies, as long as T is not extremely small, the width of the central peak scales as $1/\sqrt{\text{system size}}$. That is, the internal energy is evenly partitioned very sharply between the two macrosystems.

Q17.2 How Boltzmann Reached His Principle

Let us taste the original paper.[14] Main steps are stated as questions (and answers to them). Let us consider a (classical ideal) gas consisting of N particles in a container with volume V. Let w_n be the number of particles with the (one-particle) energy between $(n-1)\varepsilon$ and $n\varepsilon$ ($\varepsilon > 0$). Thus, the set $\{w_n\}$ specifies a microstate of the gas with w_n particles in the one-particle energy bin $((n-1)\varepsilon, n\varepsilon]$.

(1) Show that maximizing the number of ways ("Komplexionszahl" Z_K) to realize a collection of microstates ("Komplexion") specified by $\{w_n\}$ is equivalent to the minimization condition for

$$M = \sum w_n \log w_n. \tag{17.48}$$

(2) Write $w_i = w(x)\epsilon$ and simultaneously take the $n \to \infty$ and $\varepsilon \to 0$ limits, maintaining $x = n\varepsilon$ constant. Show that minimizing M is equivalent to minimizing

$$M' = \int w(x) \log w(x) dx. \tag{17.49}$$

(3) We should not ignore the constraints that the total number of particles is N and the total energy is E. Under these conditions, derive Maxwell's distribution in 3-space by minimizing M'. (Regard x to be the molecular canonical coordinates $\boldsymbol{r}, \boldsymbol{p}$.)

(4) Now, Boltzmann realized that $\log Z_K \equiv \log N! - M'$ gives the entropy of the ideal gas. Based on this observation, he proposed

$$S \propto \log(\text{number of Komplexions}). \tag{17.50}$$

Compare this and the formula for S obtained thermodynamically, as Boltzmann did, to confirm his proposal.

Remark Usually, the story ends here (so did Boltzmann's original paper). However, Boltzmann later confirmed for macrosystems (fluids with interparticle interactions) described by E and V that his formula of entropy (17.1) satisfied the Gibbs relation for general classical many-body systems; in particular, $(dE + PdV)/T$ is a perfect differential (see **Q17.3**).

Solution

(1) The Komplexionszahl reads (Chapter 3 Appendix 3A)

$$Z_K = \frac{N!}{w_1! \, w_2! \cdots w_n! \cdots}, \tag{17.51}$$

so maximizing this is equivalent to minimizing the denominator or its logarithm (see (7.14)):

---

[14] Boltzmann, L. (1877), Über die Beziehung zwischen dem zweiten Hauptsatze der mechanischen Wärmetheorie und der Wahrscheinlichkeitsrechnung respective den Sätzen über Wärmegleichgewicht, *Wiener Berichte*, **76**, 373–435 ("On the relation between the second law of thermodynamics and probability calculation concerning theorems of thermal equilibrium"). [In 1877 Tchaikovsky composed *Swan Lake*; Bell's first commercial phone service; the last phase of American Indian Wars.]

$$\log(w_1! \, w_2! \cdots w_n! \cdots) = \sum_n \log w_n! = \sum (w_n \log w_n - w_n) = M - N. \quad (17.52)$$

(2) Substituting the quantities in M as indicated, we have

$$\sum_n w_n \log w_n = \sum_x w(x)\varepsilon \log[w(x)\varepsilon] = \sum_x w(x)\varepsilon \log w(x) + \sum_x w(x)\varepsilon \log \varepsilon, \quad (17.53)$$

where $\sum_x$ formally means the summation over all x. The first term in the rightmost expression is a Riemann sum, so we obtain (17.49). The second term is $N \log \varepsilon$ and is unrelated to the number of complexions, so we may ignore it.

(3) In the original paper Boltzmann regarded the variable x as the three components of velocity vector v_x, v_y, v_z separately. Here, we combine the momentum $\boldsymbol{p}$ and the position $\boldsymbol{r}$ as x: $w(x) = w(\boldsymbol{r}, \boldsymbol{p})$. The constraints are

$$N = \int d^3r d^3p \, w(\boldsymbol{r}, \boldsymbol{p}), \quad E = \int d^3r d^3p \, w(\boldsymbol{r}, \boldsymbol{p})E(\boldsymbol{p}), \quad (17.54)$$

where $E(\boldsymbol{p}) = \boldsymbol{p}^2/2m$ is the energy of a single-particle state $(\boldsymbol{r}, \boldsymbol{p})$ with m being the mass of a gas particle. Using Lagrange's technique, we should maximize (α and β are Lagrange's multipliers):

$$\int d^3r d^3p \, w(\boldsymbol{r}, \boldsymbol{p})[\log w(\boldsymbol{r}, \boldsymbol{p}) + \alpha + \beta E(\boldsymbol{p})]. \quad (17.55)$$

Hence, ($E = (3/2)Nk_BT$ is used to fix β; α is fixed by the total number of particles N):

$$w(\boldsymbol{r}, \boldsymbol{p}) = \frac{N}{V} \frac{1}{(2\pi m k_B T)^{3/2}} e^{-p^2/2k_B Tm}. \quad (17.56)$$

We have obtained the Maxwell distribution.

(4) If we compute (17.50) (i.e., $N \log N - M'$) with the aid of w in (17.56)

$$S = N \log V + N\left(\frac{3}{2} + \frac{3}{2}\log(2\pi k_B Tm)\right) = N \log VT^{3/2} + \text{const.} \quad (17.57)$$

This agrees with the entropy obtained thermodynamically (apart from $\log N!$; the formula is not extensive). See the text around (15.16).

Q17.3 Boltzmann's Principle does not Contradict Thermodynamics[15]

Let us check that Boltzmann's principle (within classical physics) is indeed consistent with thermodynamics: that is, if $S = k_B \log w(E, V)$, we get the correct Gibbs relation (Chapter 13):

$$dS = \frac{1}{T}dE + \frac{P}{T}dV, \quad (17.58)$$

where $w(E, V)$ is the number of microstates satisfying that the energy is in $(E - \delta E, E]$ and the volume is in $(V - \delta V, V]$. Here, we clearly know what E and V are that appear both in

---

[15] The following is (hopefully) an accessible version of what can be found in Gallavotti, G. (1999). *Statistical Mechanics A Short Treatise*. Berlin: Springer.

mechanics and in thermodynamics. The pressure P can be computed mechanically, and T is related to the average kinetic energy K of the system.

Using the Boltzmann formula, we can write

$$dS = k_B \frac{1}{w} \frac{\partial w}{\partial E} dE + k_B \frac{1}{w} \frac{\partial w}{\partial V} dV. \tag{17.59}$$

Therefore, if we can compute partial derivatives in the above and identify their meanings, we should accomplish what we wish. This is actually what Boltzmann did in 1884. The demonstration is not very trivial, so here we wish to use the following relation (however, this is shown in (4) after the solution to (3); (4) exists only in Solution)

$$k_B \frac{1}{w} \frac{\partial w(E, V)}{\partial V} \rightarrow \frac{P}{T} \tag{17.60}$$

(in the thermodynamic limit) and consider only the energy derivative. We can write (here we drop h^{3N}):

$$w(E, V) = \int_{[E]} d^{3N}\boldsymbol{q} d^{3N}\boldsymbol{p} - \int_{[E-\delta E]} d^{3N}\boldsymbol{q} d^{3N}\boldsymbol{p} = \delta E \frac{\partial}{\partial E} \int_{[E]} d^{3N}\boldsymbol{q} d^{3N}\boldsymbol{p}, \tag{17.61}$$

where $[E]$ denotes the phase volume with energy not larger than E, $\boldsymbol{q}$ and $\boldsymbol{p}$ collectively denote the position and momentum vectors of all the N particles and $d^{3N}\boldsymbol{q} = d^3\boldsymbol{q}_1 \cdots d^3\boldsymbol{q}_N$, etc. We assume that the gas is confined in the volume V. Let $E = K(\boldsymbol{p}) + U(\boldsymbol{q})$, where K is the total kinetic energy and U the total intermolecular potential energy. The phase space integration may be written as

$$\int_{[E]} d^{3N}\boldsymbol{q} d^{3N}\boldsymbol{p} = \int_{V^N} d^{3N}\boldsymbol{q} \int_{K(\boldsymbol{p}) \leq E - U(\boldsymbol{q})} d^{3N}\boldsymbol{p}. \tag{17.62}$$

Thus, the integration with respect to $\boldsymbol{p}$ is the calculation of the volume of the $3N$-sphere of radius $\sqrt{2m(E - U(\boldsymbol{q}))}$.

(1) Show that

$$\frac{\partial}{\partial E} \int_{[E]} d^{3N}\boldsymbol{q} d^{3N}\boldsymbol{p} = \int_{V^N} d^{3N}\boldsymbol{q} \frac{S_{3N-1}}{3N} 2m \frac{3N}{2} [2m(E - U(\boldsymbol{q}))]^{3N/2 - 1}, \tag{17.63}$$

where S_{3N-1} is the surface area (volume) of the $(3N - 1)$-dimensional unit sphere.

(2) Using this formula, we can differentiate $w(E, V)$ with E. Obtain

$$\frac{1}{w(E, V)} \frac{\partial w(E, V)}{\partial E} = \left(\frac{3N}{2} - 1\right)\left\langle \frac{1}{K(\boldsymbol{p})} \right\rangle. \tag{17.64}$$

(3) We know from the kinetic theory that the average kinetic energy of a point particle is proportional to T (precisely speaking, the average of $\boldsymbol{p}^2/2m = 3k_BT/2$). Assuming that all the kinetic energies of the particles are statistically independent,[16] demonstrate that the formula (17.64) is indeed equal to $1/k_BT$.

---

[16] This is not really a trivial statement; we need that the system is "normal." That is, the intermolecular interaction range must be very short, and the interactions are sufficiently repulsive in the very short range.

Solution

(1) The integration with respect to $\boldsymbol{p}$ in (17.62) is the calculation of the volume of the $3N$-sphere of radius $\sqrt{2m(E - U(q))}$.

$$\int_{[E]} d^{3N}\boldsymbol{q}\, d^{3N}\boldsymbol{p} = \int_{V^N} d^{3N}\boldsymbol{q}\, S_{3N-1} \int_0^{\sqrt{2m(E-U(q))}} p^{3N-1}dp$$

$$= \int_{V^N} d^{3N}\boldsymbol{q}\, \frac{S_{3N-1}}{3N}[2m(E - U(q))]^{3N/2}. \qquad (17.65)$$

From this,

$$\frac{\partial}{\partial E}\int_{[E]} d^{3N}\boldsymbol{q}\, d^{3N}\boldsymbol{p} = \int_{V^N} d^{3N}\boldsymbol{q}\, \frac{S_{3N-1}}{3N}2m\frac{3N}{2}[2m(E - U(q))]^{3N/2-1} = w(E, V)/\delta E. \qquad (17.66)$$

(2) The result of (1) gives us w, so we must differentiate this once more.

$$\frac{\partial^2}{\partial E^2}\int_{[E]} d^{3N}\boldsymbol{q}\, d^{3N}\boldsymbol{p} = \int_{V^N} d^{3N}\boldsymbol{q}\, \frac{S_{3N-1}}{3N}(2m)^2\frac{3N}{2}\left(\frac{3N}{2} - 1\right)[2m(E - U(q))]^{3N/2-2} \qquad (17.67)$$

$$= \int_{V^N} d^{3N}\boldsymbol{q}\, \frac{S_{3N-1}}{3N}(2m)\frac{3N}{2}\left(\frac{3N}{2} - 1\right)[2m(E - U(q))]^{3N/2-1}\frac{1}{K(p)}. \qquad (17.68)$$

Comparing this with (17.66), we get

$$\delta E\frac{\partial^2}{\partial E^2}\int_{[E]} d^{3N}\boldsymbol{q}\, d^{3N}\boldsymbol{p} = \left(\frac{3N}{2} - 1\right)w(E, V)\left\langle\frac{1}{K(p)}\right\rangle. \qquad (17.69)$$

That is,

$$\frac{1}{w(E, V)}\frac{\partial w(E, V)}{\partial E} = \left(\frac{3N}{2} - 1\right)\left\langle\frac{1}{K(p)}\right\rangle. \qquad (17.70)$$

(3) We wish to demonstrate in the $N \to \infty$ limit $(3N/2-1)\langle 1/K(p)\rangle = 1/k_BT$. Obviously,

$$\left(\frac{3N}{2} - 1\right)\left\langle\frac{1}{K(p)}\right\rangle = \left\langle\frac{1}{(2K(p)/3)/N}\right\rangle. \qquad (17.71)$$

The law of large numbers tells us that

$$\frac{2K(p)}{3N} = \frac{2}{3}\frac{1}{N}\sum\frac{p^2}{2m} \to \frac{2}{3}\left\langle\frac{p^2}{2m}\right\rangle = k_BT. \qquad (17.72)$$

(4) (Extra exercise) Let us show (17.60). Assume that the system is a cube and we change the volume by extending it into the x-direction as $dV = Ad\ell$ (as Fig. 13.1), where A is the cross section of the cube and $d\ell$ is the extension of the edge in the x-direction. Then,

$$w(E, V + dV) - w(E, V) = N\int_{dV} d^3\boldsymbol{q}_1 \int d^{3(N-1)}\boldsymbol{q}'\, d^{3N}\boldsymbol{p} = N\int^* d^{3N}\boldsymbol{q}\, d^{3N}\boldsymbol{p}, \qquad (17.73)$$

where $d^{3(N-1)}q'$ is the usual phase volume element without d^3q_1 and the domain of the integral $\int^*$ is such that particle 1 is kept in dV but the remaining canonical variables are in the energy shell $(E - \delta E, E]$. Thus, we have obtained

$$\frac{1}{w(E, V)} \frac{\partial}{\partial V} w(E, V) dV = \frac{N}{w(E, V)} \int^* d^{3N}q d^{3N}p. \tag{17.74}$$

The pressure reads (cf. Bernoulli's calculation in Chapter 2)

$$PdV = \frac{N}{w(E, V)} \int_{dV} d^3q_1 2mv_{1x}^2 \int \int d^{3(N-1)}q' d^{3N}p = \frac{N}{w(E, V)} \int^* 2mv_{1x}^2 d^{3N}q d^{3N}p \tag{17.75}$$

Therefore,

$$PdV = \frac{N}{w(E, V)} \int^* 2mv_{1x}^2 d^{3N}q d^{3N}p. \tag{17.76}$$

Dividing this with (17.74), we obtain

$$\frac{PdV}{\frac{\partial}{\partial V} \log w(E, V) dV} = \int^* 2mv_{1x}^2 d^{3N}q d^{3N}p \Big/ \int^* d^{3N}q d^{3N}p. \tag{17.77}$$

The right-hand side is expected to converge to $k_B T$.[17] Here, we expect that the system satisfies thermodynamics, so dV and the rest must be statistically independent. Therefore, we may use the law of large numbers to conclude this. Thus, we have arrived at $k_B \partial \log w(E, V)/\partial V = P/T$.

Q17.4 High-Dimensional Volume is Near its Skin

We should exploit the magnitude of N (high dimensionality of the microstate space). Calculation of the volume of the microcanonical partition function $w(E, V)$ is just computing the volume of a thin skin of a large dimensional ball. The most important fact to remember about a high-dimensional object is that most of its volume is in its skin. Explain this fact.

Solution

Let the linear dimension of a $\eth$-object be R and its volume be $CR^\eth$, where C is a constant dependent on the shape. We skin it. If the skin thickness is $\delta R/2 \ll R$, then the skinned object is similar to the original shape, and its linear dimension is $R - \delta R$. Thus, the ratio of the skinned volume to the original volume is

$$C(R - \delta R)^\eth / CR^\eth = (1 - \delta R/R)^\eth. \tag{17.78}$$

Therefore, for example, even if $\delta R/R = 10^{-10}$, if $\eth = 10^{20}$, this ratio is almost zero. For high-dimensional objects, its volumes is almost in its skin, and this situation does not change unless the skin is excessively thin. This consideration implies that the microcanonical partition function $w(E, X)$ may be replaced by the number of all the microstates whose energy is less than or equal to E.

---

[17] Purely statistical-mechanically, this is not unconditional; we need the system stability and "short-rangedness" of the interactions as explained in Chapter 11.

Q17.5 The Volume of a $\eth$-ball

(1) Let $S_{\eth-1}$ be the surface area of the unit $(\eth-1)$-sphere (the boundary of a unit $\eth$-ball). Show that the volume of $\eth$-ball of radius R is given by $B_{\eth}(R) = (S_{\eth-1}/\eth)R^{\eth}$.

(2) Demonstrate that

$$I_{\eth} = \int_{-\infty}^{\infty} dx_1 \cdots \int_{-\infty}^{\infty} dx_{\eth} e^{-a(x_1^2+\cdots+x_{\eth}^2)} = (\pi/a)^{\eth/2}, \tag{17.79}$$

where a is a positive constant. Compute the same integral in the polar coordinate system to show (cf. in 2-space we use $2\pi r dr$, in 3-space $4\pi r^2 dr$; that is $S_1 = 2\pi$, $S_2 = 4\pi$):

$$I_{\eth} = S_{\eth-1}\Gamma(\eth/2)a^{-\eth/2}/2. \tag{17.80}$$

(3) Comparing (17.79) and (17.80), we obtain $S_{\eth-1} = 2\pi^{\eth/2}/\Gamma(\eth/2)$, so with the aid of $\Gamma(x+1) = x\Gamma(x)$,

$$B_{\eth}(R) = \frac{\pi^{\eth/2}}{\Gamma(\eth/2+1)}R^{\eth}. \tag{17.81}$$

Using $S_{\eth-1}$, we can evaluate the classical microcanonical partition function of the ideal gas as

$$w(E, V) = \frac{2mV^N(2mE)^{3N/2-1}\pi^{3N/2}}{h^{3N}N!\,\Gamma(3N/2)}\Delta E. \tag{17.82}$$

Confirm this.

Appendix 17A   Linear Algebra in Dirac's Notation

The reader must have a reasonable familiarity with linear algebra.

17A.1 Ket and Bra

Column vectors are regarded as *ket vectors* (*kets*) and are denoted as $|a\rangle$. Their transpositions are called *bra vectors* (*bras*) and are denoted as $\langle a|$. The ordinary vector space V is called the *ket space*, and its dual space V^* is called the *bra space*.

17A.2 Bracket Product

The scalar product $a \cdot b$ of two vectors a and b is denoted as $\langle a|b\rangle$ (*bracket product*). The *norm* of a ket $|a\rangle$ is defined as $\|a\| = \sqrt{\langle a|a\rangle}$. We also write $\overline{\langle a|b\rangle} = \langle b|a\rangle$.

17A.3 Orthonormal Basis

Let V be a m-dimensional ket space. A *basis* $u = \{|e_i\rangle\}$ of V is a set of m linearly independent kets with which any ket in V may be written as their linear combination. If u is a basis satisfying $\langle e_i|e_j\rangle = \delta_{ij}$, we say u is an *orthonormal basis* (*ON basis*) of V.

17A.4 Componentwise Representation

$a_i = \langle e_i | a \rangle$ is called the ith component (with respect to the ON basis $u = \{|e_i\rangle\}$). We can write

$$|a\rangle = \sum_i |e_i\rangle \langle e_i | a \rangle. \tag{17.83}$$

This is true for any ket $|a\rangle$ in V, thus we have the resolution of unity.

17A.5 Resolution of Unity

We have the following important identity (called the *resolution of unity* with respect to the ON basis u):

$$\sum_{e_i \in u} |e_i\rangle \langle e_i | = 1. \tag{17.84}$$

For example, $\langle a | b \rangle = \langle a | \sum_i |e_i\rangle \langle e_i | b \rangle = \sum_i \bar{a}_i b_i$ as we know from elementary linear algebra.

17A.6 Changing Bases

Changing an ON basis u to $u' = \{|f_i\rangle\}$ is always easy with the aid of resolutions of unity (17.84):

$$a_i = \langle e_i | a \rangle = \langle e_i | \sum_j |f_j\rangle \langle f_j | a \rangle = \sum_j \langle e_i | f_j \rangle \langle f_j | a \rangle, \tag{17.85}$$

where $\langle f_j | a \rangle$ is the jth component of $|a\rangle$ with respect to u', and $U = \text{matr.}(\langle e_i | f_j \rangle)$ is the transformation matrix, which is unitary.

17A.7 Unitary Operator

If a linear operator U satisfies $U^* U = U U^* = 1$, U is called a *unitary operator*. The unitary operator U preserves the norm. Note that the bra corresponding to $U|a\rangle$ is $\langle a | U^*$, so $\|U|a\rangle\|^2 = \langle a | U^* U | a \rangle = \||a\rangle\|^2$.

Any norm preserving operator satisfies $U^* U = 1$. Such a U is unitary if and only if the range of U is the whole V. (Demonstration: Such a U is bijective, so it has a two-sided inverse W: $UW = WU = 1$, because W exists such that for any $|a\rangle$ and $|b\rangle$ $U|a\rangle = |b\rangle$ implies $W|b\rangle = |a\rangle$ and vice versa, so $WU|a\rangle = |a\rangle$ and $UW|b\rangle = |b\rangle$. Therefore, $U^* = U^* UW = W$. Hence, $UU^* = 1$.)

17A.8 Linear Operator on V

A linear map A from V into itself is called a *linear operator* (on V): $A(\alpha|a\rangle + \beta|b\rangle) = \alpha A|a\rangle + \beta A|b\rangle$ for any kets in V and $\forall \alpha, \beta \in \mathbb{C}$. Using a resolution of unity, we can make a matrix representation of any linear operator as

$$A = \left(\sum_i |e_i\rangle\langle e_i| \right) A \left(\sum_j |e_j\rangle\langle e_j| \right) \equiv \sum_{i,j} |e_i\rangle\langle e_i|A|e_j\rangle\langle e_j| = \sum_{i,j} |e_i\rangle A_{ij}\langle e_j|, \quad (17.86)$$

where $A_u = \text{matr.}(\langle e_i|A|e_i\rangle) = \text{matr.}(A_{ij})$ is the matrix representation of A with respect to the ON basis u. For example, the matrix expression of the product AB of two linear operators A and B reads

$$\langle e_i|AB|e_k\rangle = \langle e_i|A \left(\sum_j |e_j\rangle\langle e_j| \right) B|e_k\rangle = \sum_j \langle e_i|A\,|e_j\rangle\langle e_j|B|e_k\rangle = \sum_j A_{ij}B_{jk}. \quad (17.87)$$

17A.9 Unitary Transformation of the Linear Operator

This is just the change of the ON basis from u to $u' = \{|f_j\rangle\}$:

$$(A_{u'})_{ij} = \langle f_i|A|f_j\rangle = \sum_{m,n} \langle f_i|e_m\rangle\langle e_m|A|e_n\rangle\langle e_n|f_j\rangle. \quad (17.88)$$

Notice that $U = \text{matr.}(\langle e_i|f_j\rangle)$ is a unitary matrix; we could read this formula as $A_{u'} = U^*A_uU$ as matrix algebra.

17A.10 Eigenvalue Problem

If $A|\lambda\rangle = \lambda|\lambda\rangle$, for a nonzero ket $|\lambda\rangle$, λ is called an *eigenvalue* of A and $|\lambda\rangle$ an *eigenvector* belonging to λ.

If $A = A^*$ (*Hermitian* or *self-adjoint*), then $\lambda \in \mathbb{R}$, because $\langle\lambda|A|\lambda\rangle = \langle\lambda|A^*|\lambda\rangle = \overline{\langle\lambda|A|\lambda\rangle}$. For two eigenvalues λ and μ ($\neq \lambda$) $\langle\lambda|\mu\rangle = 0$, because $\langle\lambda|A|\mu\rangle = \lambda\langle\lambda|\mu\rangle = \mu\langle\lambda|\mu\rangle$ implies $\langle\lambda|\mu\rangle = 0$.

The eigenvalues of a unitary operator are on the unit circle: $\langle\lambda|U^*U|\lambda\rangle = \lambda^*\lambda\langle\lambda|\lambda\rangle = \langle\lambda|\lambda\rangle$, so $|\lambda| = 1$.

Suppose V is finite-dimensional with an ON basis $u = \{|e_i\rangle\}_{i=1}^N$. Then, the eigenvalue problem $A|\lambda\rangle = \lambda|\lambda\rangle$ reads

$$\sum_j \langle e_i|A|e_j\rangle\langle e_j|\lambda\rangle = \lambda\langle e_i|\lambda\rangle. \quad (17.89)$$

This means ($A_u = \text{matr.}(\langle e_i|A|e_j\rangle)$)

$$\det(A_u - \lambda I) = 0. \quad (17.90)$$

This degree N polynomial of λ is called the *characteristic equation* for A. That is, the eigenvalues of A are the zeros of (17.90)

17A.11 Spectral Decomposition of Linear Operator

Let A satisfy $A = A^*$ (formally self-adjoint[18]). Let the totality of the eigenvalues of A be $\{\mu_i\}$ (repeating the same values according to their multiplicity equals degeneracy), but

---

[18] In this Appendix the vector space is assumed to be finite-dimensional. In this case $A = A^*$ is enough to characterize the eigenvalues of A, but if the space is not finite-dimensional as in many examples in quantum

let us first assume that all the eigenvalues $\{\mu_m\}$ are distinct. The set of the corresponding normalized eigenkets $\{|\mu_m\rangle\}$ makes an ON basis of V as can be seen from **17A.10**. Thus, we have the *spectral decomposition* of A:

$$A = \sum_m |\mu_m\rangle\langle\mu_m| A \sum_n |\mu_n\rangle\langle\mu_n| = \sum_n |\mu_n\rangle\mu_n\langle\mu_n|. \tag{17.91}$$

Notice that eigenvalues and eigenkets depend on A continuously. Therefore, it is always possible to make all the eigenvalues distinct by arbitrary small modifications of A. Now, we can apply the above spectral decomposition. Then, we remove the perturbation. Because of the continuity, the spectral decomposition formula must remain in the same form. Thus, the spectral decomposition theorem must be true for any self-adjoint operator.[19]

This is equivalent to saying that there is a unitary matrix U such that U^*AU is diagonal.

17A.12 Functions of Self-Adjoint Operators

Let A be a self-adjoint operator such that $A = \sum_\mu |\mu\rangle\mu\langle\mu|$. If $f(\mu)$ are all well defined for the eigenvalues μ of A, we define $f(A)$ as:[20]

$$f(A) = \sum_\mu |\mu\rangle f(\mu)\langle\mu|. \tag{17.92}$$

17A.13 Commuting Matrices are Diagonalizable Simultaneously

Let A and B be self-adjoint and $[A, B] = 0$ (A and B commute). Let $A|\mu_i\rangle = \mu_i|\mu_i\rangle$. Then

$$\langle\mu_i|AB|\mu_j\rangle = \langle\mu_i|BA|\mu_j\rangle \Rightarrow \mu_i\langle\mu_i|B|\mu_j\rangle = \langle\mu_i|B|\mu_j\rangle\mu_j. \tag{17.93}$$

If $\mu_i \neq \mu_j$, $\langle\mu_i|B|\mu_j\rangle = 0$. If B is not diagonal yet, we may further diagonalize it in each eigenspace belonging to μ_i of A without spoiling the already diagonalized matrix expression of A.

17A.14 Normality is Equivalent to Unitary Diagonalizability

A matrix A is *normal* if $AA^* = A^*A$. A necessary and sufficient condition for A to be diagonalized by a unitary transformation is that A is normal. (Equivalently, a necessary and sufficient condition for the eigenvectors of A to be an orthogonal basis is that A is normal.) Let $A_R = (A + A^*)/2$ and $A_I = (A - A^*)/2i$. Then, $A = A_R + iA_I$ and both A_R and A_I are self-adjoint. Furthermore, if A is normal, then they commute, so they are simultaneously diagonalizable by the same unitary transformation. That is, A is diagonalizable. The converse is obvious.

mechanics, for A to be Hermitian (i.e. self-adjoint), we must also require normality $AA^* = A^*A$ or that the domain of A is the whole vector space.

[19] If V is finite dimensional, our logic is complete, but if V is infinite-dimensional and especially if A is not bounded, our naive argument does not go through. See Akhiezer N. I. and Glazman, I. M. (2013) *Theory of Linear Operator in Hilbert Space* Dover Books on Mathematics.

[20] Here, we assume the vector space is of finite dimension. For a more general case if f is integrable with respect to the spectrum of A, $f(A)$ is well defined.

Statistical Mechanics of Isothermal Systems

Summary

* Isothermal systems are handled by the canonical formalism: $A = -k_B T \log Z$, where $Z = \mathrm{Tr}\, e^{-\beta H}$ is the canonical partition function.
* Microcanonical and canonical formalisms give identical results if the system is large (if $\log N/N \ll 1$) (ensemble equivalence).

Key words

canonical partition function, Gibbs–Helmholtz formula, ensemble equivalence, Stirling's formula, Schottky defect, Schottky type specific heat

The reader should be able to:

* Compute the microcanonical partition functions and canonical partition functions for simple systems.
* Understand the ensemble equivalence.
* Recognize the relation between the energy gap and the Schottky type specific heat.

18.1 Schottky Defects

Let us consider an isolated crystal with point defects (vacancies) on the lattice sites (*Schottky defects*; see Fig 18.1). To create one such defect we must move an atom from a lattice point to the surface of the crystal. The energy cost for this is assumed to be ε. Although the number n of vacancies are macroscopic, we may still assume it to be very small compared to the number N of all lattice sites. Therefore, we may assume that the volume of the system is constant, so the (internal) energy E of the system is the only macroscopic (thermodynamic) variable which completely specifies the macrostates.

We must compute the microcanonical partition function w as a function of the total energy E, which is given by

$$E = n\varepsilon. \tag{18.1}$$

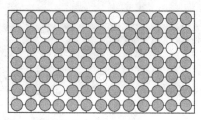

Fig. 18.1 The vacancies (white disks) are the Schottky defects.

We may interpret this as the internal energy and consider w as a function of n.[1] A microstate of this system is specified by the locations to place n vacancies. Since all the lattice points can be distinguished, the number of the ways to place n vacancies is obviously

$$w(n) = \binom{N}{n}. \tag{18.2}$$

18.2 Entropy of the Crystal with Schottky Defects

To compute the entropy with the aid of Boltzmann's principle, we use *Stirling's formula* to evaluate $\log N!$ asymptotically for large N:

$$N! \approx (N/e)^N, \tag{18.3}$$

or

$$\log N! \approx N \log N - N. \tag{18.4}$$

We know these formulas already (Section 7), but a bit more detail is given in **18.3**.

The following formula using (18.3) is worth remembering (see Chapter 21):

$$\log \binom{A}{B} = -A \left[\frac{B}{A} \log \frac{B}{A} + \left(1 - \frac{B}{A} \right) \log \left(1 - \frac{B}{A} \right) \right]. \tag{18.5}$$

Boltzmann's principle gives us

$$S = -N k_B \left[\frac{n}{N} \log \frac{n}{N} + \left(1 - \frac{n}{N} \right) \log \left(1 - \frac{n}{N} \right) \right]. \tag{18.6}$$

Using the Gibbs relation ($dE = TdS$ in the present case), we get (recall $dE = \varepsilon dn$):

$$\frac{1}{T} = \frac{dS}{dE} = \frac{1}{\varepsilon} \frac{dS}{dn} = \frac{k_B}{\varepsilon} \log \frac{N-n}{n}. \tag{18.7}$$

When (18.6) is differentiated, the derivatives of the logarithmic terms cancel each other, so essentially, we have only to differentiate the prefactors in front of the logarithms. This is very easy, and we immediately get (18.7).

---

[1] Notice that we do not specify the energy leeway ΔE, because in this example, even without ΔE, $w(n)$ can be large.

If the temperature is sufficiently low or ε is sufficiently large (i.e., $\varepsilon/k_B T \gg 1$), the above formula reduces to

$$\frac{\varepsilon}{k_B T} \simeq \log \frac{N}{n}, \tag{18.8}$$

because $N \gg n$. Hence, under this low temperature condition, the internal energy E reads

$$E = \varepsilon N e^{-\varepsilon/k_B T}, \tag{18.9}$$

which may be guessed from the Boltzmann factor.

18.3* Stirling's Formula

Let us rederive (18.4). We start from the gamma function for non-negative integer N:[2]

$$\Gamma(N+1) = \int_0^\infty dt\, t^N e^{-t} = N!. \tag{18.10}$$

To confirm this we make the following generating function $g(a)$:

$$g(a) = \sum_{N=0}^\infty \frac{a^N \Gamma(N+1)}{N!} = \int_0^\infty dt \sum_{N=0}^\infty \frac{(at)^N}{N!} e^{-t} = \int_0^\infty dt\, e^{-(1-a)t} = \frac{1}{1-a} = \sum_{N=0}^\infty a^N. \tag{18.11}$$

Since a can be anything (although $|a| < 1$ must be required), we have demonstrated (18.10). Since N is very large, if we rewrite it as

$$\Gamma(N+1) = \int_0^\infty dt\, t^N e^{-t} = \int_0^\infty dt\, e^{N(\log t - t/N)}, \tag{18.12}$$

the integrand should be concentrated around the peak of $\log t - t/N$, which is at $t = N$. Therefore,[3]

$$N! = \int_0^\infty dt\, e^{N(\log t - t/N)} \simeq e^{N\log N - N}. \tag{18.13}$$

This is what we wanted. A better formula is obtained by taking the width of the peak into account; the result is called *Stirling's formula*. A still better formula called *Gosper's formula* is known:[4]

$$N! \sim e^{-N} N^N \sqrt{(2N + 1/3)\pi}. \tag{18.14}$$

---

[2] For the gamma function and related topics, Havil, J. (2003). *Gamma: Exploring Euler's Constant* (foreword by F. J. Dyson). Princeton: Princeton University Press may be an enjoyable book.

[3] This approach is called *Laplace's method*.

[4] This formula cannot be obtained by improving the calculation in (18.13). Stirling's formula is the one with the factor $2N + 1/3$ in the formula replaced with $2N$. Gosper's formula gives an amazing $0! \sim \sqrt{\pi/3} = 1.02\cdots$. A proof can be found in Mortici, C. (2011). On Gosper's formula for the gamma function, *J. Math. Inequalities*, **5**, 611–614.

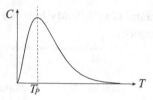

Fig. 18.2 The Schottky type specific heat, which has a peak indicating the energy gap of the order $k_B T_P (= \varepsilon/2)$.

18.4 Schottky Type Specific Heat

The specific heat C of the system can be obtained from (18.9) as

$$C = \frac{dE}{dT} = Nk_B \left(\frac{\varepsilon}{k_B T} \right)^2 e^{-\varepsilon/k_B T}. \tag{18.15}$$

Notice that C has a peak at a certain temperature T_P (see Fig. 18.2). This type of specific heat is called the *Schottky type specific heat*, which tells us the presence of an energy gap for elementary excitations in the system.

18.5 Thermostat

Let us construct a formalism for isothermal systems.

Isolated systems are not so easy to handle, compared with thermostatted systems. We have introduced the Helmholtz free energy A to study thermostatted systems (Chapter 16). We have learned that the information of any thermodynamic potential can be completely obtained from A (the equivalence of thermodynamic potentials thanks to the Legendre transformation). It would be advantageous if there is a statistical mechanical means that allows us to compute the Helmholtz free energy directly from mechanics.

Let us extend the formalism for isolated systems to thermostatted systems. The logic we use is a familiar one (recall **14.6**): we assume the system I is surrounded by a huge system II, and the total system I + II is isolated. We allow that the system I can freely exchange heat with its surroundings II.[5]

The total energy E_0 of the compound system is given by

$$E_0 = E_I + E_{II}. \tag{18.16}$$

The number of microstates for system I (resp., II) with energy E_I (resp., E_{II}) is denoted by $w_I(E_I)$ (resp., $w_{II}(E_{II})$). Thermal contact is a very weak interaction, so the two systems are

---

[5] The astute reader may ask why we have to start from the microcanonical ensemble; haven't we argued in **17.11** that a natural single microstate is enough to do statistical mechanics? Indeed, a (macroscopic) subsystem of a big whole system, if the latter is sufficiently larger than the former, obeys the canonical distribution when the whole system is in any natural microstate. The basic idea is clearly mentioned as the key estimate in Tasaki, H. (1998). From quantum dynamics to the canonical distribution: general picture and a rigorous example, *Phys. Rev. Lett.*, **80**, 1373–1376.

statistically independent. Hence, the number of microstates for the compound system with the energies E_I in I and E_{II} in II is given by

$$w_I(E_I)w_{II}(E_{II}). \qquad (18.17)$$

The total number $w(E_0)$ of microstates for the compound system must be the sum of this product over all the ways to partition energy between I and II. Therefore, we get

$$w(E_0) = \sum_{0 \leq \mathcal{E} \leq E_0} w_I(\mathcal{E})w_{II}(E_0 - \mathcal{E}). \qquad (18.18)$$

The system II is huge compared with I. Expand the entropy as follows:

$$S_{II}(E_0 - \mathcal{E}) = S_{II}(E_0) - \mathcal{E}\frac{\partial S_{II}}{\partial E_{II}} + \frac{1}{2}\mathcal{E}^2\frac{\partial^2 S_{II}}{\partial E_{II}^2} + \cdots \qquad (18.19)$$

and denote the temperature of the heat bath (i.e., system II) by T:

$$\frac{\partial S_{II}}{\partial E_{II}} = \frac{1}{T}. \qquad (18.20)$$

We wish to use (18.19) in equilibrium, so $\mathcal{E}$ should be close to the internal energy of system I. Therefore, due to the extensivity of internal energy this should be of order N_I, the total number of particles in system I. Therefore,

$$\mathcal{E}\frac{\partial S_{II}}{\partial E_{II}} = O[N_I]. \qquad (18.21)$$

The second derivative in (18.19) is proportional to $\partial T/\partial E_{II} = 1/C_{II}$, where C_{II} is the specific heat of II, which is $O[N_{II}]$:

$$\mathcal{E}^2\frac{\partial^2 S_{II}}{\partial E_{II}^2} = -\frac{\mathcal{E}^2}{T^2 C_{II}} = \frac{O[N_I]^2}{O[N_{II}]} = O[N_I]\frac{O[N_I]}{O[N_{II}]} \ll O[N_I]. \qquad (18.22)$$

Therefore, the ratio of the second term and the third term in (18.19) is of order N_I/N_{II}, which is negligibly small. Thus, we have:[6]

$$w_{II}(F_0 - \mathcal{E}) = \exp\left\{\left[S_{II}(F_0) - \mathcal{E}/T + O\left[N_I/N_{II}\right]\right]/k_B\right\}. \qquad (18.23)$$

Therefore, (18.18) reads (assuming the bath is large enough)

$$w(E_0)e^{-S_{II}(E_0)/k_B} \simeq \sum_{\mathcal{E}} w_I(\mathcal{E})e^{-\beta\mathcal{E}}, \qquad (18.24)$$

where the standard notation $\beta = 1/k_BT$ is used.

With the aid of Boltzmann's principle, we have

$$k_B\log w(E_0) = S_I(E_I) + S_{II}(E_{II}) = S_I(E_I) + S_{II}(E_0 - E_I), \qquad (18.25)$$

---

[6] A critical reader might question whether the argument around here is mathematically legitimate. This will not be discussed here, because we can generally prove that the microcanonical and the canonical formalism are equivalent (ensemble equivalence). The argument here may be regarded as a mere heuristics to motivate the canonical ensemble approach.

so with the same approximation to discard $O[N_I/N_{II}]$ as (18.24) (from now on let us drop the subscript I to denote the system):

$$k_B \log \left[w(E_0) e^{-S_{II}(E_0)/k_B} \right] = S(E) + S_{II}(E_0 - E) - S_{II}(E_0) = S(E) - E/T = -A/T. \quad (18.26)$$

That is, (18.24) reads

$$e^{-\beta A} = \sum_{\mathcal{E}} w_I(\mathcal{E}) e^{-\beta \mathcal{E}}. \quad (18.27)$$

18.6 Canonical Formalism

Thus, we have arrived at our desired formalism, the *canonical formalism*: Let

$$Z = \sum_{\mathcal{E}} w(\mathcal{E}) e^{-\beta \mathcal{E}}. \quad (18.28)$$

Then,

$$A = -k_B T \log Z. \quad (18.29)$$

Here Z is called the *canonical partition function*, and this method to compute the free energy is called the *canonical formalism*.

A more microscopic expression is possible:

$$Z = \sum_{\text{all microstates}} e^{-\beta \mathcal{E}} = \text{Tr}\, e^{-\beta H}. \quad (18.30)$$

Here, the sum over all the microstates implies a summation over all the eigenvalues of the Hamiltonian H, so quantum mechanically, we may use the trace to compute the partition function. We can easily understand (18.28):

$$\sum_{\text{all microstates}} e^{-\beta \mathcal{E}} = \sum_{\mathcal{E}} \left[\sum_{\text{all microstates with energy} \sim \mathcal{E}} e^{-\beta \mathcal{E}} \right] = \sum_{\mathcal{E}} w(\mathcal{E}) e^{-\beta \mathcal{E}}. \quad (18.31)$$

18.7 The Gibbs–Helmholtz Equation

Once the canonical partition function is known, we know the Helmholtz free energy of the system. Since $A = E - TS$ or $E = A + TS$, if we can compute the entropy S, we can get the internal energy. We know the Gibbs relation from **16.5**

$$dA = -SdT - PdV + xdX + \cdots, \quad (18.32)$$

so the internal energy can be obtained as

$$E = A - T \left(\frac{\partial A}{\partial T} \right)_{V,X} = A + \frac{1}{T} \left(\frac{\partial A}{\partial (1/T)} \right)_{V,X} = \left(\frac{\partial (A/T)}{\partial (1/T)} \right)_{V,X}. \quad (18.33)$$

Notice that this can be obtained more directly from the entropy form of the Gibbs relation (with a Legendre transformation $S \to S - E/T = -A/T$)

$$-d\left(\frac{A}{T}\right) = d\left(S - \frac{E}{T}\right) = -Ed\left(\frac{1}{T}\right) + \frac{P}{T}dV + \cdots. \tag{18.34}$$

The formula is called the *Gibbs–Helmholtz equation*, which may be rewritten with the aid of the canonical partition function as

$$E = \left(\frac{\partial \beta A}{\partial \beta}\right)_{V,X} = -\frac{\partial \log Z(\beta)}{\partial \beta}. \tag{18.35}$$

Here, it is explicitly denoted that the canonical partition function Z is a function of β.

18.8   Schottky Defects Revisited

Let us revisit the Schottky defects with the canonical formalism. With $w(n)$ known (see (18.2)), it is easy to compute Z:

$$Z = \sum_{n=0}^{N} w(n)e^{-\beta n \varepsilon} = \left(1 + e^{-\beta \varepsilon}\right)^{N}. \tag{18.36}$$

Here, we have used the binomial theorem (if the reader is uncomfortable, review Appendix 3A):

$$\sum_{n=0}^{N} w(n)e^{-\beta n \varepsilon} = \sum_{n=0}^{N} \binom{N}{n}\left(e^{-\beta \varepsilon}\right)^{n} 1^{N-n} = \left(1 + e^{-\beta \varepsilon}\right)^{N}. \tag{18.37}$$

However, we could probably write down the right-most formula immediately: the canonical partition function is a sum over all the possible microstates

$$Z = \sum_{\varepsilon(1) \in \{0,\varepsilon\}, \cdots, \varepsilon(N) \in \{0,\varepsilon\}} e^{-\beta \sum_{i=1}^{N} \varepsilon(i)}, \tag{18.38}$$

where $\varepsilon(i)$ is the energy of the ith lattice point (occupied 0 or empty ε). Here, do not forget that a "microstate" is a microscopically described state of the whole macro system; in our case $(\varepsilon(1), \varepsilon(2), \ldots, \varepsilon(N))$ is a microstate. Do not confuse the microstate and the elementary states $\varepsilon(i)$ of individual microscopic entities. Notice that all the combinations of the lattice states show up, so the following factorization should work; we should get (18.36) immediately.

18.9   Factorization of Canonical Partition Function

Since this factorization is the key that makes the canonical formalism often easier than the microcanonical formalism, a detailed illustration follows.

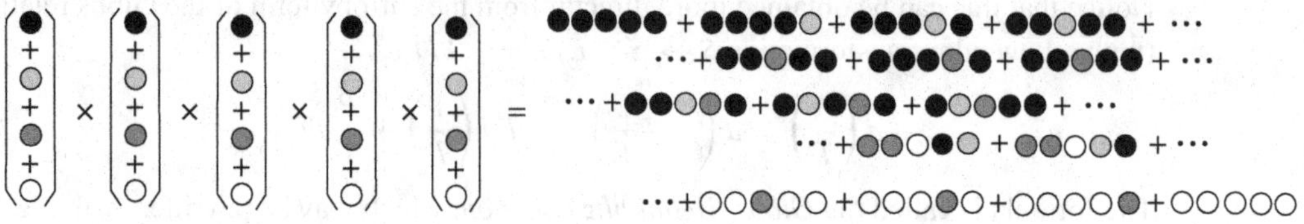

Fig. 18.3 Illustration using a five-lattice point toy model with four states (disks with different "colors") for each lattice site. Each column on the left-hand side corresponds to the sum over all states at each lattice point (i.e., Z_1 in (18.42)). Thus, the left-hand side illustrates the expression of Z corresponding to (18.41). The right-hand side illustrates the partition function Z of the system corresponding to the expression (18.40); 5-ball strings correspond to microstates. All the possible microstates appear once and only once on the right-hand side.

Suppose there are N lattice points. Each lattice point has several states, $a, b, c, \ldots$, with the corresponding "excitation energies," $\varepsilon(a)$, $\varepsilon(b)$, etc. Since the total energy of the system, that is, the energy of the microstate $\{a_1, \ldots, a_N\}$, where a_i is the state of the ith lattice point, reads

$$\mathcal{E} = \varepsilon(a_1) + \varepsilon(a_2) + \cdots + \varepsilon(a_N), \tag{18.39}$$

the canonical partition function is computed as

$$Z = \sum_{a_1, a_2, \ldots, a_N \in \{a,b,c,\ldots\}} e^{-\beta[\varepsilon(a_1) + \varepsilon(a_2) + \cdots]}. \tag{18.40}$$

Here, the summation is over all the possible combinations of the states of individual particles. Since all the combinations appear once and only once, we can rewrite this as (see Fig. 18.3):

$$Z = \left(\sum_{a_1 \in \{a,b,c,\ldots\}} e^{-\beta\varepsilon(a_1)} \right) \left(\sum_{a_2 \in \{a,b,c,\ldots\}} e^{-\beta\varepsilon(a_2)} \right) \cdots \left(\sum_{a_N \in \{a,b,c,\ldots\}} e^{-\beta\varepsilon(a_N)} \right) = Z_1^N, \tag{18.41}$$

where

$$Z_1 = \sum_{a_1 \in \{a,b,c,\ldots\}} e^{-\beta\varepsilon(a_1)} \tag{18.42}$$

is the "canonical partition function" of a single lattice point about its (internal) states. Notice that we cannot usually do this for the microcanonical approach, because not all the microstates appear in the computation of the microcanonical partition function w (i.e., only those with total energy around E).

Thus, if "particles" or "lattice points" do not interact with each other, we can guess

$$\sum_{\text{microstates}} = \left(\sum_{\text{single particle/site states}} \right)^N. \tag{18.43}$$

Thus we can get (18.36) immediately.

As noted just above (18.43), if two systems are not interacting (except for heat exchange), then the canonical formalism allows us to handle them as statistically independent. This very convenient feature does not exist in the microcanonical formalism.

18.10 Schottky Defect by Canonical Formalism (a continuation)

From (18.36) the Helmholtz free energy of the lattice with Schottky defects immediately follows:

$$A = -Nk_B T \log \left(1 + e^{-\beta\varepsilon}\right). \tag{18.44}$$

We can get the entropy by differentiation (although using $TS = E - A$ is recommended):

$$S = -\frac{\partial A}{\partial T} = Nk_B \log \left(1 + e^{-\beta\varepsilon}\right) + N\frac{\varepsilon}{T}\frac{e^{-\beta\varepsilon}}{(1 + e^{-\beta\varepsilon})}. \tag{18.45}$$

Other partition functions (other formalisms) will be introduced later. If we wish to study the thermodynamics of a system, any formalism will do, as long as the system is large enough (roughly speaking, if $\log N/N \ll 1$). We have so far discussed the microcanonical and the canonical formalisms. Let us check that the canonical result for S agrees with the microcanonical result for this simple example. The microcanonical approach gives us

$$S = -Nk_B \left[\frac{n}{N}\log\frac{n}{N} + \left(1 - \frac{n}{N}\right)\log\left(1 - \frac{n}{N}\right)\right] \tag{18.46}$$

and from (18.7)

$$\frac{n}{N} = \frac{1}{1 + e^{\beta\varepsilon}}. \tag{18.47}$$

Combining both, we get

$$S = -Nk_B \left[\frac{1}{1 + e^{\beta\varepsilon}}\log\frac{1}{1 + e^{\beta\varepsilon}} + \frac{e^{\beta\varepsilon}}{1 + e^{\beta\varepsilon}}\log\frac{e^{\beta\varepsilon}}{1 + e^{\beta\varepsilon}}\right]. \tag{18.48}$$

This indeed agrees with (18.45).

18.11 Frenkel Defects and Ensemble Equivalence

Let us study one more example: the *Frenkel defects* to illustrate the ensemble equivalence. As illustrated in Fig. 18.4 particles leave their original lattice points and wander into non-lattice positions (interstitial positions). There are N lattice points and M interstitial points. There are N particles in total and n particles leave their lattice points and move into

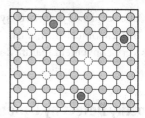

Fig. 18.4 Frenkel defects: particles with a darker color are interstitial particles (excited particles), which leave vacancies (dotted circles) behind.

interstitial sites. There is an energy cost of ε for a particle to move from a lattice point to an interstitial site. The system energy is $E = n\varepsilon$. Thus, (we write $w(E)$ as $w(n)$)

$$w(n) = \binom{N}{n}\binom{M}{n}, \tag{18.49}$$

and the canonical partition function reads

$$Z(T) = \sum_{n=0}^{\min\{N,M\}} w(n)e^{-\beta n\varepsilon} = \sum_{n=0}^{\min\{N,M\}} \binom{N}{n}\binom{M}{n}e^{-\beta n\varepsilon}. \tag{18.50}$$

Unfortunately, this cannot be summed in a closed form. However, in this case, it is easy to prove the ensemble equivalence.

18.12 Equivalence of Microcanonical and Canonical Ensembles

Since all the summands are positive in (18.50), the following inequalities are obvious:[7]

$$\max_n \left[w(n)e^{-\beta n\varepsilon} \right] \le Z(T) \le (N+1)\max_n \left[w(n)e^{-\beta n\varepsilon} \right]. \tag{18.51}$$

On the other hand, we have

$$\max_n \left[w(n)e^{-\beta n\varepsilon} \right] = \exp\left[\frac{1}{k_B}\max_S(S - E/T) \right], \tag{18.52}$$

but the "honest" definition of the Helmholtz free energy is (recall the Legendre transformation as explained in Chapter 14)

$$-A = \max_S[TS - E]. \tag{18.53}$$

Therefore, (18.51) reads ($N + 1$ is replaced with N)

$$-A/k_BT \le \log Z(T) \le -A/k_BT + \log N. \tag{18.54}$$

Here A is an extensive quantity, so it is of order N. Therefore, if we can ignore $\log N/N$ as very small, $-k_BT \log Z(T)$ (the free energy directly obtained by the canonical

---

[7] Here n must be no more than $\min\{N, M\}$, but we impose a simplifying condition $M = N$; n cannot be too large for realistic cases, so this simplification causes no physical effect, since N and M are usually of the same order.

formalism of statistical mechanics) and the free energy obtained (using thermodynamics, i.e., (18.53)) from entropy (which is computed statistical-mechanically with the aid of the microcanonical approach) are indistinguishable.

Although our demonstration of the ensemble equivalence, here the equivalence of the canonical and microcanonical formalisms, relies on a particular example, the logic we have employed is identical to that is required to demonstrate the ensemble equivalence generally and rigorously: Z is sandwiched between the maximum term and the maximum term multiplied by something proportional to N (cf. **Q28.5**).

The reader must have recognized clearly that the estimations used in statistical mechanics are very crude (not delicate at all). That is why the results are general, robust, and reliable

18.13 Frenkel Defect: Microcanonical Approach

Let us continue the Frenkel defect problem. Let us compute entropy using Boltzmann's principle.

$$S = k_B \log \binom{N}{n}\binom{M}{n}. \tag{18.55}$$

Thus, we have

$$S/k_B = -N\left[\frac{n}{N}\log\frac{n}{N} + \left(1 - \frac{n}{N}\right)\log\left(1 - \frac{n}{N}\right)\right]$$
$$- M\left[\frac{n}{M}\log\frac{n}{M} + \left(1 - \frac{n}{M}\right)\log\left(1 - \frac{n}{M}\right)\right]. \tag{18.56}$$

We need temperature, so we use the Gibbs relation $dS = (1/T)dE$ (for the present example):

$$\frac{1}{T} = \frac{dS}{dE} \tag{18.57}$$

with $dE = \varepsilon dn$:[8]

$$\frac{\varepsilon}{k_B T} = \frac{1}{k_B}\frac{dS}{dn} = \log\frac{1 - n/N}{n/N} + \log\frac{1 - n/M}{n/M} = \log\frac{(N-n)(M-n)}{n^2}. \tag{18.58}$$

Notice that when we differentiate $\log\binom{N}{n}$ with respect to n, virtually we have only to differentiate the factors outside log. Usually n is small, so we obtain

$$e^{-\beta\varepsilon} = \frac{n^2}{NM}. \tag{18.59}$$

From this we can write S in terms of T.

---

[8] Here S is now a function of E, but we do not write it explicitly.

18.14 Review: Ensemble Equivalence

We have learned two formalisms to do equilibrium statistical mechanics:

$$S = k_B \log w(E, X) \quad \text{microcanonical formalism,} \tag{18.60}$$

$$A = -k_B T \log Z(T, X) \quad \text{canonical formalism.} \tag{18.61}$$

Here, the expression of the canonical partition functions can be obtained from the Legendre transformation relation $-A = \max_S[ST - E]$ as (recall **18.12**):

$$-A = k_B T \log e^{\beta \max_S[ST-E]} = k_B T \log \operatorname{Tr} e^{-\beta H}. \tag{18.62}$$

These formalisms are equivalent if $\log N / N \ll 1$. The meaning of "equivalence" is:

A computed with the aid of thermodynamics from S according to (18.60) agrees with A directly computed statistical-mechanically according to (18.61).

S computed with the aid of thermodynamics from A according to (18.61) agrees with S directly computed statistical-mechanically according to (18.60).

In short, the reader may use any "ensemble." Here, *ensemble* means the collection of microstates with a definite summation rule (or probability assignment, if we use the principle of equal probability; see Chapter 19).

18.15 Analogy between Equilibrium Statistical Mechanics and Large Deviation Theory

The rate function I in the large deviation (see Chapter 10) is of central importance to understanding fluctuations. Its calculation may be facilitated with the following analogy to the canonical formalism. Let us define the partition function which is the moment generating function of the distribution of the empirical expectation value:[9*]

$$Z(t) = \int P\left(\frac{1}{N}\sum_{k=1}^{N} X_k \in dy\right) e^{Nty} = \left\langle e^{t\sum_{k=1}^{N} X_k}\right\rangle_P, \tag{18.63}$$

where the average sign denotes the average over the underlying probability law P. From the definition of I, we can also write Z as

$$Z(t) = \int dy\, e^{N[ty - I(y)]}. \tag{18.64}$$

[9*]Note that

$$P\left(\frac{1}{N}\sum_{k=1}^{N} X_k \in dy\right) = \left\langle \delta\left(\frac{1}{N}\sum_{k=1}^{N} X_k - y\right)\right\rangle_P dy,$$

where the average sign denotes the average over the underlying probability law P. Exchanging the order of y-integration and averaging $\langle\,\rangle_P$ gives (18.63).

Since N is not small, Laplace's method (in **18.3**) will give a good approximation. Thus it is natural to introduce an analog $a(t)$ of the free energy per particle, which is a Legendre transform of the rate function:

$$\frac{1}{N} \log Z(t) \equiv a(t) = \max_y [ty - I(y)]. \tag{18.65}$$

Therefore, with the aid of the Legendre transformation (recall **16.10**) we get the rate function as

$$I(y) = \max_t [ty - a(t)]. \tag{18.66}$$

The last relation is called the *Gärtner–Ellis theorem*.[10]

Problems

Q18.1 Einstein's Derivation of Canonical Formalism from Thermodynamics

Let us look at Einstein's approach to statistical mechanics. The correspondence rule for the thermodynamic coordinates is the same as usual. Einstein chose $P = (1/Z)e^{-\beta H}$ for his statistical principle; actually, he derived this distribution from the microcanonical distribution just as in **18.5**. We assume that the system Hamiltonian depends on parameters λ_i that may be controlled externally. Let us write the change of the Hamiltonian due to the change of these parameters as δH. Since the control parameters can be changed adiabatically, Einstein interpreted δH (averaged over the canonical distribution) as work $d'W$. Derive $A = -k_B T \log Z$.

Solution

Let $P = e^{-\log Z - \beta H}$ be the canonical distribution. Let us change the temperature (i.e., β) and the mechanical parameters $\{\lambda_i\}$:

$$0 = d \int P dy = \int dy \left[-d \log Z - H d\beta - \beta \delta H \right] P. \tag{18.67}$$

Here d is the total derivative with respect to the variables β and $\{\lambda_i\}$. The average of H is the internal energy E and that of δH is the work $d'W$, so

$$0 = -d \log Z - E d\beta - \beta d'W = -d \log Z - d(\beta E) + \beta dE - \beta d'W. \tag{18.68}$$

The first law implies $dE - d'W = d'Q$. Therefore, we get

$$\beta d'Q = d \log Z + d(\beta E). \tag{18.69}$$

Using the thermodynamic definition of entropy, we get

$$dS = \frac{1}{T} d'Q = d \left(\frac{E + k_B T \log Z}{T} \right) \tag{18.70}$$

---

[10] As we saw in Chapter 15, we need $a(t)$ to be convex, but the moment generating function is well known to be convex.

or $-k_B d \log Z = d(A/T)$. Integrating this and ignoring the arbitrary integration constant,[11] Einstein arrived at

$$A = -k_B T \log Z. \tag{18.71}$$

Q18.2 Elementary Quizzes

When the reader solves "elementary questions," it is a good habit to:

 (i) Check that the obtained result is intuitively plausible.
 (ii) Check the general conclusions (e.g., consistency with thermodynamic laws).
(iii) Take various limits (e.g., $T \to \infty$).
(iv) Find approximate calculations to obtain certain limiting results (e.g., high/low temperature analytic formulas).

(1) N lattice sites have quantum spins of $S = 1$. Associated with each spin is a magnetic moment and in a uniform magnetic field in the z-direction the energy of each spin reads $-\mu B m$, where μ is the magnetic moment and m the z-component of the spin angular momentum quantum number $(0, \pm 1)$. Compute the entropy and the magnetization M of the system.

(2) A typical spin-system problem:

Due to the ligand field the degeneracy of the d-orbit of the chromium ion Cr^{3+} is partially lifted, and the spin Hamiltonian has the following form

$$H = D\left(S_z^2 - S(S+1)/2\right), \tag{18.72}$$

where $D > 0$ is a constant with $S = 3/2$ (the cation has the electronic state described by the term $^4F_{3/2}$).

 (i) Why can we apply statistical mechanics to this "single" ion?
 (ii) Obtain the occupation probability of each energy level.
(iii) Calculate the entropy and the specific heat. Then, show the specific heat behaves as $\propto T^{-2}$ at higher temperatures.
(iv) Suppose $C = k_B(T_0/T)^2$ with $T_0 = 0.18\,\mathrm{K}$ at higher temperatures. Determine the energy spacing.

Solution

(1) The canonical partition function Z^{12} of the whole system can be written as $Z = z^N$ with the single lattice point partition function z defined as

$$z = \mathrm{Tr}\, e^{\beta \mu B m} = \sum_{m \in \{-1,0,1\}} \langle 1, m| e^{\beta \mu B m} |1, m \rangle = e^{\beta \mu B} + 1 + e^{-\beta \mu B}, \tag{18.73}$$

---

[11] This is allowed, because we have only to shift the origin of the energy.

[12] **Correct thermodynamic potential** More correctly, this is not the usual canonical partition function for which T and M are independent variables, but a kind of generalized canonical partition function which gives a Legendre transformed result $A' = -k_B T \log Z$ with $A' = A - MB$ (needless to say, this Z is the Z defined in our solution, not the true canonical partition function; Most textbooks are quite indifferent to the distinction, but the Gibbs relations are different).

where $|S, m\rangle$ is the spin angular momentum state ket (S being the total spin angular momentum). Therefore, the (generalized) Helmholtz free energy is (as noted in the footnote above $dA' = -TdS - MdB$)

$$A' = -Nk_BT \log \left(e^{\beta\mu B} + 1 + e^{-\beta\mu B} \right). \tag{18.74}$$

The internal energy is:[13]

$$E' = -\frac{\partial \log Z}{\partial \beta} = N\mu B \frac{\left(e^{-\beta\mu B} - e^{\beta\mu B} \right)}{e^{\beta\mu B} + 1 + e^{-\beta\mu B}} < 0. \tag{18.75}$$

This should be negative because the lower energy states should be occupied more. The formula $S = (E' - A')/T$ gives the system entropy

$$S = Nk_B \log(e^{\beta\mu B} + 1 + e^{-\beta\mu B}) + \frac{N\mu B}{T} \frac{\left(e^{-\beta\mu B} - e^{+\beta\mu B} \right)}{e^{\beta\mu B} + 1 + e^{-\beta\mu B}}. \tag{18.76}$$

Since $E' = -MB$,[14]

$$M = N\mu \frac{\left(e^{\beta\mu B} - e^{-\beta\mu B} \right)}{e^{\beta\mu B} + 1 + e^{-\beta\mu B}} > 0. \tag{18.77}$$

In the $T \to 0$ limit S goes to 0 (see the third law; Chapter 23), if $B \neq 0$, and $M \to N\mu$, if $B > 0$. If $B = 0$, S is independent of T, and all the states are equally probable, so $Nk_B \log 3$ as can be expected from the information point of view (Chapter 21).

In the high temperature limit all the spin states are evenly occupied, $M \to 0$, and $S \to Nk_B \log 3$ as expected (Chapter 21).

(2) (i) A state in which a particular microscopic entity assumes a particular state may be interpreted as a macroscopic state, since it is a collection of the microstates compatible with this single entity specification; needless to say the collection contains macroscopic number of microstates. Thus the marginal distribution for this entity may be obtained with the aid of statistical mechanics. If all the microscopic entities are non-interacting, as in this problem, the system partition function reads as $Z = z^N$, and the single entity marginal distribution can be obtained from z. Thus, we can apply statistical mechanics to this "single" ion. See also Chapter 19 text around (19.8).

(ii) There are four states but there are only two energy levels with $E = 3D/8$ and $-13D/8$. Therefore, $S_z = \pm 3/2$ is with $p = 1/2(1 + e^{2\beta D})$ and $S_z = \pm 1/2$ is with $p = 1/2(1 + e^{-2\beta D})$.

---

[13] More correctly, it is something like a generalized enthalpy $E' = E - MB$. Cf., our familiar enthalpy: $H = E - (-PV)$. Notice that $A' = E - MB - ST = E' - ST$. Cf., our familiar Gibbs free energy: $G = E + PV - ST = H - ST$.

[14] Or better, we can use $-(\partial A'/\partial B)_T = M$. Notice that the distinction between A and A' matters, because $dA = -SdT + BdM$. To use thermodynamics correctly we must not forget what our ensemble is. In our case, the ensemble is the (T, B)-ensemble, not (T, M). We fix T and B, not T and M.

(iii) The easiest method is to use the Shannon formula (Chapter 21).[15] The microcanonical way is probably the least useful in practice. When the reader computes S from the canonical ensemble, use $S = (E - A)/T$ with E being calculated by the Gibbs–Helmholtz equation $\partial(A/T)/\partial(1/T) = -\partial \log Z/\partial \beta = E$.[16]

$$S = -2k_B \left[\frac{1}{2(1+x)} \log \frac{1}{2(1+x)} + \frac{x}{2(1+x)} \log \frac{x}{2(1+x)} \right]$$

$$= k_B \left\{ \log[2(1+x)] - \frac{x}{1+x} \log x \right\}, \tag{18.78}$$

where $x = e^{2\beta D}$.

(iv) Setting x as previously, we have

$$C = T\frac{dS}{dT} = -(2D\beta)\frac{dS}{d2D\beta} = -2D\beta\frac{dx}{d2D\beta}\frac{dS}{dx} = -2D\beta x\frac{dS}{dx} = k_B(2D\beta)^2\frac{x}{(1+x)^2}. \tag{18.79}$$

Therefore, for large T, x is almost 1 and we get

$$C = k_B(D/k_B)^2/T^2. \tag{18.80}$$

Here $D = k_B T_0$ is estimated. The level spacing is given by $2D$ (cf. Fig. 18.2).

Q18.3 Elementary Lattice Problem

There is a 2D square lattice with M lattice points. On each lattice point is a magnetic moment that can point only in the lattice bond directions (four directions as illustrated in Fig. 18.5), but the $\pm y$ directions are the easy directions: if the dipole is along the y-axis, it is stable, that is, the energy of the dipole along the x-axis is ε (> 0 more energy) and that along the y-axis is zero. We do not pay attention to the kinetic energy of the system. Ignore the interactions among dipoles.

(1) What is the canonical partition function of the system (the temperature is T)?
(2) What is the average energy per dipole?

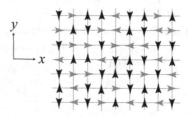

Fig. 18.5   Each dipole can point only four directions along the lattice bonds.

---

[15] There are several ways to compute entropy. If the reader knows probability explicitly, the Shannon formula (21.9) may be useful. In this case, do not forget that the sum is over the *elementary events*.
[16] As noted already in (1), these A and E are not the genuine Helmholtz free energy and internal energy in this problem.

(3) Compute the entropy $S(T)$ per dipole. What is the difference $S(\infty) - S(0)$? How many bits is this? Is this consistent with the intuitive interpretation of entropy per molecule as the number of yes–no questions?

(4) Compute the "microcanonical partition function" $w(N\varepsilon)$ $(0 \le N \le M)$.

(5) Show that the entropy computed from the microcanonical scheme (Boltzmann's principle) and the result (3) agree.

Solution

(1) Needless to say, we can start from the very definition of the canonical partition function, BUT notice that if we collect all the microstates (= microscopic states = mechanically describable whole-system states), all the states of microscopic entities (molecules, etc.; in our case dipoles sitting on the lattice points) appear once and only once (recall Fig. 18.3). Therefore, to construct the partition function for the whole lattice, we study all the states of each microscopic entity to make their individual canonical partition functions (in our case $1 + 1 + e^{-\beta\varepsilon} + e^{-\beta\varepsilon}$) and multiply them over the whole lattice:

$$Z(T) = \left(2 + 2e^{-\beta\varepsilon}\right)^M. \tag{18.81}$$

(2) Using the Gibbs relation, we get

$$\frac{E}{M} = -\frac{1}{M}\frac{\partial \log Z}{\partial \beta} = \frac{\varepsilon e^{-\beta\varepsilon}}{1 + e^{-\beta\varepsilon}}. \tag{18.82}$$

(3) Since $A = E - TS$, $S = (E - A)/T$

$$\frac{S(T)}{M} = k_B \log(2 + 2e^{-\beta\varepsilon}) + \frac{\varepsilon}{T}\frac{e^{-\beta\varepsilon}}{1 + e^{-\beta\varepsilon}}. \tag{18.83}$$

Here $S(0) = k_B \log 2$ (notice that for any n, $x^n e^{-x} \to 0$ in the $x \to \infty$ limit) and $S(\infty) = k_B \log 4$, so ΔS is just 1 bit. At $T = 0$ the dipoles are always along y (two directions), but at $T = \infty$ they can evenly assume four directions (i.e., along x- and y-directions). Thus, if we ask one yes–no question ("Is it along y?") we can reduce the uncertainty in the equilibrium state at $T = \infty$ to that at $T = 0$. In other words, to identify the state of a dipole in the $T \to \infty$ limit we need 2 bits (two questions), because we must find one particular state out of four possibilities. In the $T \to 0$ limit all the dipoles are along the easy direction, so there are only two choices for each dipole. Therefore, we need only one question to pinpoint the state of a dipole. ΔS just corresponds to the difference in the numbers of questions we must ask.

(4) We immediately obtain

$$w(N\varepsilon) = \binom{M}{N}2^M, \tag{18.84}$$

because two choices along x and along y can be selected without affecting the system energy.

A more elementary way (which is not recommended at all) is to introduce n_1, n_2, n_3, and n_4 pointing, respectively, $+x$, $-x$, $+y$, and $-y$. $N = n_1 + n_2$ (x-direction) and $M - N = n_3 + n_4$ (y-direction):

$$w(N\varepsilon) = \sum_{n_1=0}^{N} \sum_{n_3=0}^{M-N} \frac{M!}{n_1!\,(N-n_1)!\,n_3!\,(M-N-n_3)!}, \tag{18.85}$$

but an easy reorganization is: (i) choose N parallel x dipoles, and then (ii) count the number of ways to point $+$ and $-$ directions:

$$w(N\varepsilon) = \binom{M}{N} \sum_{n_1=0}^{N} \sum_{n_3=0}^{M-N} \binom{N}{n_1}\binom{M-N}{n_3} = \binom{M}{N} 2^N 2^{M-N}. \tag{18.86}$$

This is just the answer above.

(5) Thanks to Boltzmann

$$\frac{S}{M} = -k_B\left[\frac{N}{M}\log\frac{N}{M} + \left(1 - \frac{N}{M}\right)\log\left(1 - \frac{N}{M}\right)\right] + k_B\log 2. \tag{18.87}$$

Since

$$\frac{1}{T} = \frac{k_B}{\varepsilon}\frac{\partial S}{\partial N} = -\frac{k_B}{\varepsilon}\log\frac{N}{M-N}, \tag{18.88}$$

we have

$$N = \frac{M}{1 + e^{-\beta\varepsilon}}. \tag{18.89}$$

Computing N/M and using it in the entropy formula above, we again get

$$\frac{S}{M} = -k_B\left[\frac{N}{M}\log\frac{1}{1+e^{-\beta\varepsilon}} + \left(1 - \frac{N}{M}\right)\log\left(\frac{e^{-\beta\varepsilon}}{1+e^{-\beta\varepsilon}}\right)\right] + k_B\log 2 \tag{18.90}$$

$$= k_B\left[\frac{N}{M}\log(1 + e^{-\beta\varepsilon}) + \left(1 - \frac{N}{M}\right)\log(1 + e^{-\beta\varepsilon})\right]$$

$$+ k_B\log 2 - k_B\frac{e^{-\beta\varepsilon}}{1+e^{-\beta\varepsilon}}\log\left(e^{-\beta\varepsilon}\right)$$

$$= k_B\log(2 + 2e^{-\beta\varepsilon}) + \frac{\varepsilon}{T}\frac{e^{-\beta\varepsilon}}{1+e^{-\beta\varepsilon}}. \tag{18.91}$$

19   Canonical Density Operator

Summary

∗ The principle of equal probability tells us that the probability of a macrostate in an isolated system is proportional to the number of microstates compatible with the macrostate.

∗ The canonical density operator $e^{-\beta H}/Z$ allows us to compute various probabilities of events in isothermal systems.

Key words

principle of equal probability, density operator, asymptotic equipartition, Liouville's theorem, Jarzynski's equality

The reader should be able to:

∗ Understand what the density operator is and how to use it.

∗ Understand why we may use the canonical approach to small systems as well.

19.1   Principle of Equal Probability: Do We Need It?

The conventional more or less standard statistical mechanics textbooks assume a principle called the *principle of equal probability*: if we sample a microstate from $\tilde{w}(E,X)$, every microstate is equally probable. We have seen, however, that to compute thermodynamic functions statistical-mechanically, we do *not* need such an assumption. Instead, we only need the translation table in terms of mechanics of the thermodynamic coordinates plus entropy, especially the Boltzmann principle.

We will critically discuss the nature of the principle of equal probability at the end of this Chapter.

19.2   What Can We Do if We Assume Principle of Equal Probability?

If we accept the principle of equal probability, then in each microcanonical ensemble $\tilde{w}(E,X)$ the probability to sample any subset (subspace in quantum cases) $u \subset \tilde{w}(E,X)$

is proportional to its phase volume (classically) or its dimension (quantum mechanically). The principle allows us to compute, e.g., the equilibrium fluctuations of macroscopic variables (see Chapter 26, especially the justification of the interpretation of the fluctuation–response relation **26.4**).

19.3 Density Operator

Quantum mechanically, the expectation value of an observable O over a certain set of states with a statistical law is generally expressed with the aid of the *density operator* ρ as

$$\langle O \rangle = \mathrm{Tr}(\rho\, O). \tag{19.1}$$

The density operator, which is a counterpart of the density distribution function in the ordinary statistics, is a self-adjoint operator satisfying the following two conditions:

(i) normalization condition: $\mathrm{Tr}\,\rho = 1$.
(ii) positivity: for any positive definite operator A, $\mathrm{Tr}(\rho A) \geq 0$.[1]

The probability $P(A)$ for a subspace A of the microstate space (A may be understood as a quantum event) is expressed as

$$P(A) = \mathrm{Tr}(\rho P_A), \tag{19.2}$$

where P_A is the projection onto the subspace A as shown next.

☐ Let the subspace A be spanned by an orthonormal set $\{|a\rangle\}$: $P_A = \sum_a |a\rangle\langle a|$. Notice that ρ is self-adjoint, so it has an ON basis $\{|e_i\rangle\}$ to diagonalize it as $\rho = \sum_i |e_i\rangle p_i \langle e_i|$, where $\sum p_i = 1$ with $p_i \geq 0$ thanks to (i) and (ii) above.

$$\mathrm{Tr}(\rho P_A) = \mathrm{Tr}\left[\left(\sum_i |e_i\rangle p_i \langle e_i|\right)\left(\sum_a |a\rangle\langle a|\right)\right] = \sum_i p_i \langle e_i|\left(\sum_a |a\rangle\langle a|\right)|e_i\rangle. \tag{19.3}$$

Notice that $|\langle e_i|a\rangle|^2$ may be interpreted as the probability that the state $|a\rangle$ is observed when the actual state is $|e_i\rangle$ (cf. **A.26**). Then, $p_a = \sum_i p_i |\langle e_i|a\rangle|^2$ is the probability to encounter the state a, if the statistical property of the system is described by ρ. Therefore, (19.3) implies $\mathrm{Tr}(\rho P_A) = \sum_{a \in A} p_a$, which is indeed the probability to observe an event A.

19.4 Microcanonical Density Operator

If we assume the principle of equal probability, any state in a microcanonical ensemble of a system is equally probable (formally with probability $1/w(E)$[2]). Therefore, the microcanonical density operator for the energy shell $(E - \Delta E, E]$ may be written as

---

[1] **Positive definite operator** An operator A is positive definite, if $\langle *|A|* \rangle \geq 0$ for any nonzero ket $|*\rangle$.

[2] Here, for simplicity, work coordinates are suppressed. It is always easy to include them classically, and quantum mechanically if the observables corresponding to the densities are commutative. Otherwise, it is not easy, as noted in the construction of the microcanonical ensemble.

$$\rho = \sum_{\varepsilon} |\varepsilon\rangle \frac{\chi_{\Delta E}(\varepsilon)}{w(E)} \langle\varepsilon|, \tag{19.4}$$

where $\chi_{\Delta E}$ is the indicator of $(E - \Delta E, E]$. The summation is over all the eigenvalues of the system Hamiltonian.

19.5 Canonical Density Operator

Define for a Hamiltonian H

$$\rho = \frac{1}{Z} e^{-\beta H}, \tag{19.5}$$

where $Z = \mathrm{Tr}\, e^{-\beta H}$ is the canonical partition function. Then,

$$\mathrm{Tr}\,(\rho P_E) = \frac{1}{Z} w(E) e^{-\beta E}, \tag{19.6}$$

where P_E is the projection onto $\tilde{w}(E)$, the totality of the energy eigenstates of the Hamiltonian, H, belonging to $(E - \Delta E, E]$. If we assume the principle of equal probability, the probability for the system to have energy E is proportional to $w(E) w_{\mathrm{II}}(E_0 - E) \propto w(E) e^{-\beta E}$ according to (18.18). Therefore, (19.6) is the correct probability to find microstates with energy around E, so (19.5) is the correct density operator for a canonical ensemble consistent with the principle of equal probability.

19.6 Classical Case: Canonical Distribution

Just as noted in the quantum case, if we use the principle of equal probability to assign a probability measure, the probability for the system to have energy E is proportional to $w(E) w_{\mathrm{II}}(E_0 - E) \propto w(E) e^{-\beta E}$ according to (18.18). That is, the overall statistical weight is proportional to the phase volume of a set of microstates with the identical energy E times the corresponding Boltzmann factor $e^{-\beta E}$. Therefore, we can introduce the *canonical probability measure* (*canonical distribution*) μ_{can} defined on the whole phase space

$$\mu_{\mathrm{can}}(B) = \frac{1}{Z} \int_B d\Gamma e^{-\beta E(\gamma)}, \tag{19.7}$$

where Z is the canonical partition function (the normalization constant) and γ is a phase point with energy $E(\gamma)$. The integration with respect to the phase volume is over the subset B of the phase space. See the next chapter for more details.

19.7 Maxwell's Distribution Revisited

If we literally believe that the principle of equal probability holds, then the probability of a microstate γ may be obtained as

$$P(\gamma) = \text{Tr}\,(\rho|\gamma\rangle\langle\gamma|) = \frac{1}{Z}e^{-\beta E_\gamma}, \tag{19.8}$$

where E_γ is the energy of the microstate γ. Remember that a "microstate" is a microscopically described state of the whole macroscopic many-body system. Therefore, there is no experimental way to single it out, so there is no way to verify (19.8). Generally, there is no empirical justification to use statistical mechanical density operators to study individual microstates.

However, if the system consists of noninteracting microscopic entities, as the classical ideal gas, with the Hamiltonian $H(\gamma) = \sum_i \varepsilon_i$ ($\varepsilon_i = mv_i^2/2$ for a monatomic ideal gas), (19.8) implies

$$P(\gamma) \propto e^{-\beta \sum \varepsilon_i}, \tag{19.9}$$

but the totality of the microstates for a single micro-entity to have energy ε is a collection of numerous microstates, so we can rely on the principle of equal probability to obtain the distribution of ε from (19.9) as a marginal probability:

$$P(\varepsilon) \propto e^{-\beta \varepsilon}. \tag{19.10}$$

The best example is Maxwell's distribution obtained from the ideal gas canonical distribution.

Equilibrium distributions should be time-independent. Next, let us look at some of its consequences.

19.8 Equilibrium Ensemble and Time-Independence: Classical Case

We know even in equilibrium, molecules continue to move around according to mechanics. In contrast, the equilibrium distribution must be time-independent under (or in spite of) the underlying mechanics.[3] Here, we discuss classical mechanics first.

Consider a phase space. Let γ_0 be a point from which the system dynamics begins and after time, t, the system reaches γ_t in the phase space. Now, consider a collection of such intial points A_0 and set $A_t = \{\gamma_t : \gamma_0 \in A_0\}$ (see Fig. 19.1). That is, A_0 evolves to A_t according to the natural dynamics of the system. Since the number of systems in this ensemble does not change during the time evolution, the volume of A_t should not depend on t (the phase volume is invariant. See **19.9** Liouville's Theorem).

If μ is an equilibrium probability (measure) defined on the phase space $(\boldsymbol{q}, \boldsymbol{p})$, it must be time-independent. If we assume it has a density, f, defined as $\mu(d\Gamma) = f(\boldsymbol{q}, \boldsymbol{p})d\Gamma$, where $d\Gamma = d^{3N}\boldsymbol{q}d^{3N}\boldsymbol{p}$ is the phase volume element, the invariance of μ implies the invariance of f: $[f, H]_{PB} = 0$, since the phase volume is incompressible (invariant), where $[\ ,\ ]_{PB}$ is

---

[3] A critical reader might say that macroscopic systems under pure mechanics are fictitious, so the invariance question should not be crucial. However, if one wishes to claim mechanics is the basis of everything, statistical mechanics should be compatible with pure mechanics.

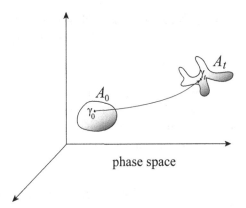

phase space

Fig. 19.1 Classical evolution of phase points. The initial condition set A_0 evolves into A_t after time t, keeping the phase volume of the sets.

the Poisson bracket (see **A.37**). If f depends only on H, certainly it is invariant. Thus, we have seen that microcanonical and canonical distribution functions are indeed invariant.

☐ **Warning**. We must note that usually there are infinitely many distinct invariant measures allowed for a given mechanical system. Which is the right equilibrium distribution? Classical mechanics cannot answer this question, because it is a matter of initial condition and nothing to do with the laws of mechanics.

19.9 Liouville's Theorem

Analytically, we can show the invariance of the phase volume as follows. Let us write the $6N$ phase coordinates of N particles as $(\boldsymbol{q}, \boldsymbol{p}) = (q_1, \ldots, q_{3N}, p_1, \ldots, p_{3N})$ (thus, e.g., "$q_{3(i-1)+1,2,3}$" describes q_{ix}, q_{iy}, and q_{iz}, respectively). The phase velocity is its time derivative $(\dot{q}_1, \ldots, \dot{q}_{3N}, \dot{p}_1, \ldots, \dot{p}_{3N})$. We can show that this flow is incompressible:[4]

$$\sum_{i=1}^{3N} \left[\frac{\partial \dot{q}_i}{\partial q_i} + \frac{\partial \dot{p}_i}{\partial p_i} \right] = 0, \tag{19.11}$$

because the equations of motion read (cf. **A.36**)

$$\dot{q}_i = \frac{\partial H}{\partial p_i}, \ \dot{p}_i = -\frac{\partial H}{\partial q_i}, \tag{19.12}$$

so

$$\sum_{i=1}^{3N} \left\{ \frac{\partial \dot{q}_i}{\partial q_i} + \frac{\partial \dot{p}_i}{\partial p_i} \right\} = \sum_{i=1}^{3N} \left\{ \frac{\partial^2 H}{\partial q_i \partial p_i} - \frac{\partial^2 H}{\partial p_i \partial q_i} \right\} = 0. \tag{19.13}$$

---

[4] Note that the following formula is the $6N$-dimensional divergence (for divergence see **8.6**) of $6N$-dimensional flow velocity; div $\boldsymbol{v} = 0$ implies that for any fixed volume, the net efflux is zero.

19.10 Equilibrium Ensemble and Time-Independence: Quantum Case

Time evolution due to quantum mechanics is a unitary transformation ($U = e^{-itH/\hbar}$; see **A.16**), so the dimension of any invariant subspace is a constant of motion. Thus, the microcanonical density operator is invariant under the natural dynamics of the system. Since H is invariant, the canonical density operator is also invariant.

19.11 Jarzynski's Equality[5]

Suppose a thermostatted system has a Hamiltonian, $H(x, \lambda(t))$, where x represents the phase coordinates, and $\lambda(t)$ is a parameter that can be changed externally from $\lambda(0)$ to $\lambda(1)$. Let W be the work needed to perform this change. Then, the free energy change ΔA from the equilibrium state with the Hamiltonian $H(x, \lambda(0))$ to that with $H(x, \lambda(1))$ may be obtained by

$$\langle e^{-\beta W} \rangle = e^{-\beta \Delta A}, \tag{19.14}$$

where the average is over many experiments with the same protocol specified by $\lambda(t)$. This is called *Jarzynski's equality*.

Precise conditions required are as follows:

(i) The system is defined on a fixed phase space and its Hamiltonian is $H(x, \lambda(t))$, where x is the phase coordinates and $\lambda(t)$ a time-dependent parameter, which may describe external control.

(ii) The initial microstate is distributed canonically at temperature T and with the Hamiltonian $H(x, \lambda_0)$.

(iii) We fix the protocol to change the system by fixing the function $\lambda(t)$ ($\lambda(0) = \lambda_0$ and $\lambda(t_f) = \lambda_1$, where t_f is the last time to modify the parameter).

We embed this system in a large thermostat that can be described by a Hamiltonian, H_0. Let us write the Hamiltonian of the whole system as $H_w(z, \lambda(t))$, where z denotes the phase coordinates of the whole system collectively. This Hamiltonian includes the interaction Hamiltonian between the system and the thermostat. We assume this effect may be ignored in equilibrium (as in Chapter 18). Here, W is the work done on the system during the change under the condition that the whole system is isolated.[6]

---

[5] Jarzynski, C. (1997). Nonequilibrium equality for free energy differences, *Phys. Rev. Lett.*, **78**, 2690–2693; Jarzynski, C. (2004). Nonequilibrium work theorem for a system strongly coupled to a thermal environment, *J. Stat. Mech.*, **2004**, P09005.

[6] The final state actually reached by the protocol at time t_f is usually very different from the final equilibrium state we are to arrive at. Despite this fact the computed free energy difference is between the correct equilibrium states at T.

19.12 Demonstration of Jarzynski's Equality

The work done on the system is given by $W = H_w(z(t_f), \lambda_1) - H_w(z(0), \lambda_0)$, because the system is, as a whole, isolated. Therefore, the average over the initial canonical distribution is given by

$$\langle e^{-\beta W} \rangle = \frac{1}{Y_0} \int dz(0) e^{-\beta H_w(z(0), \lambda_0) - \beta W} \tag{19.15}$$

$$= \frac{1}{Y_0} \int dz(0) e^{-\beta H_w(z(t_f), \lambda_1)}. \tag{19.16}$$

Here, Y_0 is the partition function for the system with the Hamiltonian $H(x, \lambda_0)$ plus the thermostat. According to Liouville's theorem the Jacobian for the variable change from $z(0)$ to $z(t_f)$ is unity.[7] Therefore, (19.16) can be rewritten as

$$\langle e^{-\beta W} \rangle = \frac{1}{Y_0} \int dz(t_f) e^{-\beta H_w(z(t_f), \lambda_1)} = \frac{Y_1}{Y_0}, \tag{19.17}$$

where Y_1 is the partition function for the system with the Hamiltonian $H(x, \lambda_1)$ plus the thermostat. Since the Hamiltonian of the thermostat is unaffected by the process, and we ignore the system–thermostat interaction Hamiltonian, $Y_1/Y_0 = Z_1/Z_0$, where Z_c is the canonical partition function for the system with the Hamiltonian $H(x, \lambda_c)$. Thus, we have demonstrated (19.14).

Since e^{-x} is convex, $\langle e^{-x} \rangle \geq e^{-\langle x \rangle}$ (a Jensen's inequality, see Fig. 14.2(b)). Therefore, (19.14) implies $e^{-\langle W \rangle} \leq e^{-\Delta A}$, or

$$\langle W \rangle \geq \Delta A. \tag{19.18}$$

Thus, apparently free energy minimum principle or (16.5) has been derived from mechanics.[8] Notice that classical mechanics cannot single out any distribution, so (19.18) implies the consistency of classical mechanics to thermodynamics.[9]

19.13* Demonstration of Jarzynski's Equality: Quantum Case[10]

Next, let us consider the quantum case of Jarzynski's equality. Suppose a quantum system has a Hamiltonian $H(\lambda)$ dependent on a parameter λ.[11] The initial parameter value is

---

[7] This requires the initial and the final phase spaces to be identical, so Jarzynski's equality does not hold, e.g., for free expansion discussed in **15.7**.

[8] **Planck's principle from quantum mechanics** Lenard proved Planck's principle from quantum mechanics (if we interpret "adiabaticity" as in mechanics): Lenard, A. (1978). Thermodynamical proof of the Gibbs formula for elementary quantum systems, *J. Statist. Phys.*, **19**, 575–586. (The paper contains much more.)

[9] A more basic problem is that a macrosystem cannot be purely (quantum or classical) mechanical (recall the $R \to 0$ discussion in **12.3**). Thus, we cannot rely on mechanics to discuss very fundamental issues about macroscopic systems. Note that analog of Jarzynski's equality holds for general Markov processes, so mechanics is actually not required.

[10] Mukamel, S. (2003). Quantum extension of the Jarzynski relation: Analogy with stochastic dephasing, *Phys. Rev. Lett.*, **90**, 170604.

[11] Here, we consider a system not imbedded in a bath (in contrast to the classical case above), but such generalization is easy.

$\lambda = 0$ and the final one is $\lambda = 1$. The eigenvalues of $H(\lambda)$ are denoted as $E_n(\lambda)$ and the corresponding eigenket is $|n; \lambda\rangle$. The initial density operator is the canonical density operator $\rho(0) = e^{\beta A(0) - \beta H(0)}$, where $A(0)$ is the initial free energy.

Let us identify W as the difference between the observed final and the initial energies $E_n(1) - E_m(0)$, we now show that the Jarzynski equality in the following form holds:

$$\left\langle e^{-\beta(E_n(1) - E_m(0))} \right\rangle = e^{-\beta \Delta A}, \tag{19.19}$$

where the average is over the initial states m and final states n (reached from state m) and $\Delta A = A(1) - A(0)$. To write this explicitly, we need the transition probability between the two observation results $E_m(0)$ initially and $E_n(1)$ finally. This can be written as $P_{n \leftarrow m} = |\langle n; 1|U|m; 0\rangle|^2$, where U is the time evolution operator corresponding to the parameter change from $\lambda = 0$ to 1 under the condition that the system is isolated. Notice that $P_{n \leftarrow m}$ is doubly stochastic: $\sum_n P_{n \leftarrow m} = \sum_m P_{n \leftarrow m} = 1.$[12] Explicitly, the left-hand side reads

$$\sum_{n,m} e^{-\beta(E_n(1) - E_m(0))} P_{n \leftarrow m} e^{\beta(A(0) - E_m(0))}$$

$$= \sum_{n,m} e^{-\beta E_n(1)} P_{n \leftarrow m} e^{\beta A(0)} \tag{19.20}$$

$$= e^{\beta A(0)} \sum_n e^{-\beta E_n(1)} = e^{\beta A(0)} e^{-\beta A(1)} = e^{-\beta \Delta A}. \tag{19.21}$$

19.14 Jarzynski's Equality and Rare Fluctuations

We know that the difference $W - \Delta A$ can be made large without bound. Still, Jarzynski's equality (19.14) holds. What does this mean?

Let us consider the simplest case; we make a cycle,[13] and the initial and the final equilibria are identical, and $\Delta A = 0$, so

$$\langle e^{-\beta W} \rangle = 1. \tag{19.22}$$

Notice that the average is over numerous runs of duration t_f (the final time $\lambda(t)$ is nonzero). If we make a rapid change, $W\ (> 0)$ can be made very large. Still, the equality must hold, so we need fairly large negative W occasionally. That is, we must wait for a fairly large-scale "violation" of the second law. This implies that we must repeat the runs extremely many times to use (19.14) for a macroscopic system or for a process with large dissipation even if a system under study is not macroscopic.

Recall how Boltzmann dismissed Zermelo's argument that quoted Poincaré's recurrence theorem (see **11.8**; see also the toy model in **11.10**). If we dismiss Zermelo, to be consistent, we might have to dismiss Jarzynski's equality as a proposition relevant to not-so-small systems.

---

[12] **(Doubly) stochastic matrix** The matrix $\{P_{i \leftarrow j}\}$ is a *stochastic matrix*, if $P_{i \leftarrow j} \geq 0$ and $\sum_i P_{i \leftarrow j} = 1$. It is called *doubly stochastic*, if $\sum_j P_{i \leftarrow j} = 1$ also holds.

[13] This only means $\lambda(t_f) = \lambda(0)$ in **19.12**. If this cyclic parameter change is done violently, surely we can supply as large W as we wish.

Finally, let us critically consider the relation between the principle of equal probability and thermodynamics.

19.15 Key Thermodynamic Facts

We first summarize fundamental facts of thermodynamic equilibrium states we wish to use. We have already noted:

[O] An equilibrium system is partitioning–rejoining invariant (Chapter 12):
 (i) each piece made by partition in isolation is in equilibrium, and
 (ii) if these pieces are rejoined, the joined result cannot be (thermodynamically) distinguished from the state before the partition (see Fig. 12.3).

[IV] Thermodynamic observables are obtainable from the partitioned system.

 Although usually not stated clearly, we know one more fact:

[T] *Invariance under thermal contact* of equilibrium states: Any equilibrium state of a thermally isolated system has a heat bath (*individual heat bath*) (or a private heat bath, so to speak) such that thermal contact with it does not alter its thermodynamic state. This is obvious, because for an equilibrium state there is a well-defined temperature.

19.16 Equilibrium State is Equivalent to a Collection of Statistically Independent Subsystems

From **19.15** [O], [IV], and [T] imply that the following procedure does not alter thermodynamic observables of the system (see Fig. 19.2).

(i) Partition a thermally isolated equilibrium system into (macroscopic van Hove; see **12.11**) pieces.
(ii) Attach each piece to its "private heat bath" for a while, and then again thermally isolate it.
(iii) Rejoin all the pieces as before to reconstruct the whole piece.[14]

Let us say that (ii) prepares *thermally independent* submacrosystems that are statistically independent. How many thermally independent submacrosystems can we find in an ordinary macroobject? One mm^3 is big enough from the molecules' point of view, so in a cube with 10 cm edge, we can easily expect far more than 10^6 thermally independent macroscopic subsystems;[15] we may safely use the law of large numbers, so thermodynamic quantities obtained from the partitioned system and the actual values are quite close.

---

[14] Incidentally, recognize that thermodynamic observables are physical quantities that are quite insensitive to subtle correlations between subcomponents (such as found in, e.g., quantum systems).
[15] How about a 1 mm cube? It contains about $\sim 10^{18}$ molecules, so notice that from any computer-simulation point of view, even if it is chopped up into millions of pieces, they are still extremely big.

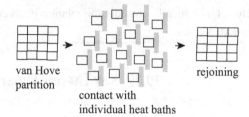

van Hove
partition

contact with
individual heat baths

rejoining

Fig. 19.2 An equilibrium system may be replaced with statistically independent pieces to obtain thermodynamic quantities.

19.17 Law of Large Numbers Implies Principle of Equal Probability

Let us decompose an isolated macrosystem into M thermally independent submacrosystems. Let γ_i represent the microstate of the ith subsystem. Let us also denote the probability for this state for this subsystem as $P(\gamma_i)$. The probability for a microstate $\{\gamma_1, \ldots, \gamma_M\}$ of the whole macrosystem is given by $P(\{\gamma_i\}) = P(\gamma_1) \cdots P(\gamma_M)$. Since $M \gg 1$ we can apply the law of large numbers to obtain

$$\frac{1}{M} \log P(\{\gamma_i\}) = \frac{1}{M} \sum_i \log P(\gamma_i) \to -s. \qquad (19.23)$$

Here, s is a constant for the given total macrosystem. That is

$$P(\{\gamma_i\}) = e^{-Ms + o(M)} \qquad (19.24)$$

independent of $\{\gamma_i\}$. This is a well-known result called the *asymptotic equipartition*.

19.18 Summary: When We Can Use the Principle of Equal Probability

We have demonstrated that for a macrosystem the microcanonical ensemble is thermodynamically (i.e., to compute thermodynamic observables) equivalent to the distribution compatible with the principle of equal probability. The demonstration clearly indicates that the use of the principle can be justified for macroobservables. Thus, the canonical distribution (density matrix) may be used to study probabilities of macroscopic events. The derivation also demonstrates that if all the subsystems are independent, even if they are not macroscopic, we may apply the principle. This demonstrates again the logic to apply the canonical distribution to small systems (e.g., to obtain Maxwell's distribution).

Problems

Except for **Q19.3** and **Q19.6** the example problems may require considerable quantum mechanics skill. The level is a bit too high for this book. Representative quantum inequalities such as Klein's and Peierls' inequalities are demonstrated. Thus, the following may be of use in advanced courses.

Q19.1 Density Matrix for a Spin System: Elementary Spin Review

Let ρ be the (canonical)[16] density operator of a single $1/2$ quantum spin system whose Hamiltonian is given by $H = -\gamma \boldsymbol{\sigma} \cdot \boldsymbol{B}$, where $\boldsymbol{\sigma}$ is $(\sigma_x, \sigma_y, \sigma_z)$ in terms of the Pauli matrices (see **A.30**).

(1) Obtain the matrix representation of ρ with respect to the basis that diagonalizes σ_z.
(2) Find the average of σ_y.
(3) Obtain the matrix representation of ρ with respect to the basis that diagonalizes σ_x.

Solution

(1) We take the direction of $\boldsymbol{B}$ to be the z axis.

$$\rho = \frac{1}{C} \begin{pmatrix} e^{\beta\gamma B} & 0 \\ 0 & e^{-\beta\gamma B} \end{pmatrix}, \tag{19.25}$$

where C is the normalization constant: the trace of the matrix in the above formula, so $C = 2\cosh \beta\gamma B$.

If the reader wishes to do the problem without specifying the coordinates, they need the following calculation. Notice that $(\boldsymbol{n} \cdot \boldsymbol{\sigma})^2 = I$, where $\boldsymbol{n}$ is a unit vector.

$$e^{\gamma \boldsymbol{B} \cdot \boldsymbol{\sigma}} = \sum_{n=0}^{\infty} \frac{1}{(2n)!} (\gamma B)^{2n} + \sum_{n=0}^{\infty} \frac{1}{(2n+1)!} (\gamma B)^{2n+1} \frac{\boldsymbol{B}}{B} \cdot \boldsymbol{\sigma} = \cosh \gamma B + \frac{\boldsymbol{B}}{B} \cdot \boldsymbol{\sigma} \sinh \gamma B. \tag{19.26}$$

(2)

$$\langle \sigma_y \rangle = \operatorname{Tr} \sigma_y \rho = \operatorname{Tr} \begin{pmatrix} 0 & -ie^{-\beta\gamma B} \\ ie^{\beta\gamma B} & 0 \end{pmatrix} = 0. \tag{19.27}$$

This should be obvious without any calculation.

(3) With the basis that diagonalizes σ_z we have

$$\sigma_x = \begin{pmatrix} 0 & 1 \\ 1 & 0 \end{pmatrix}. \tag{19.28}$$

---

[16] Strictly speaking, this is a generalized canonical density matrix, because its natural thermodynamic variables are T and B instead of T and M. Thus the corresponding thermodynamic potential is a generalized Gibbs (or Helmholtz?) free energy: $A' = E - TS - BM$ (not $A = E - TS$).

With respect to the basis the eigenvector of σ_x belonging to ± 1 is given by $(1, \pm 1)^T / \sqrt{2}$. Therefore

$$\sigma_x \begin{pmatrix} 1/\sqrt{2} & 1/\sqrt{2} \\ 1/\sqrt{2} & -1/\sqrt{2} \end{pmatrix} = \begin{pmatrix} 1/\sqrt{2} & 1/\sqrt{2} \\ 1/\sqrt{2} & -1/\sqrt{2} \end{pmatrix} \begin{pmatrix} 1 & 0 \\ 0 & -1 \end{pmatrix}. \tag{19.29}$$

That is, the following orthogonal (actually, unitary) matrix:

$$U = \begin{pmatrix} 1/\sqrt{2} & 1/\sqrt{2} \\ 1/\sqrt{2} & -1/\sqrt{2} \end{pmatrix} \tag{19.30}$$

diagonalizes σ_x as $U^* \sigma_x U$. Therefore, ρ with respect to this basis reads[17*]

$$U^* \rho U = \frac{1}{2C} \begin{pmatrix} 1 & 1 \\ 1 & -1 \end{pmatrix} \begin{pmatrix} e^{\beta \gamma B} & 0 \\ 0 & e^{-\beta \gamma B} \end{pmatrix} \begin{pmatrix} 1 & 1 \\ 1 & -1 \end{pmatrix}$$

$$= \frac{1}{C} \begin{pmatrix} \cosh \beta \gamma B & \sinh \beta \gamma B \\ \sinh \beta \gamma B & \cosh \beta \gamma B \end{pmatrix}. \tag{19.31}$$

Q19.2 Density Operator for Free Particles: Perhaps an Elementary QM Review

This problem is in two parts. The canonical density operator is given by

$$\rho = \frac{1}{Z} e^{-\beta H}, \tag{19.32}$$

where H is the system Hamiltonian and Z is the canonical partition function. Let us consider a single particle confined in a 3D cube of edge length L ($L = \infty$ may be assumed, if convenient). We wish to compute the position representation of the density operator $\langle x | \rho | x' \rangle$.

Let $U(\beta) = e^{-\beta H}$ and $H = p^2 / 2m$. There are two ways to compute $\langle x | U(\beta) | x' \rangle$ (x and x' are 3D position vectors, and bras and kets are normalized: $\langle x | x' \rangle = \delta(x - x')$).

Part A

(1) Show that

$$\frac{\partial}{\partial \beta} \langle x | U(\beta) | x' \rangle = \frac{\hbar^2}{2m} \Delta_x \langle x | U(\beta) | x' \rangle, \tag{19.33}$$

where Δ_x is the Laplacian with respect to the coordinates x.

(2) What is the initial condition (i.e., $\langle x | U(0) | x' \rangle$)?

[17*]To make the algebra quite explicit let us use the bra-ket notation (Appendix 17A). Let $\{|a\rangle\}$ (with Roman alphabet) denotes the basis diagonalizing σ_z, and $\{|\alpha\rangle\}$ (with Greek alphabet) that diagonalizing σ_x. Then, $\langle a | \sigma_z | b \rangle$ is the diagonal matrix Λ with diagonal elements ± 1. $\langle a | \sigma_x | b \rangle$ is the matrix given by (19.28). Matrix $\langle \alpha | \sigma_x | \beta \rangle = \Lambda$. Therefore, $\Lambda = U^* \sigma_x U$ reads

$$\Lambda_{\alpha\beta} = \langle \alpha | \sum_a |a\rangle\langle a | \sigma_x \sum_b |b\rangle\langle b | \beta \rangle = \sum_{a,b} \langle \alpha | a \rangle \langle a | \sigma_x | b \rangle \langle b | \beta \rangle.$$

Thus, we have $U = \text{matr.}(\langle a | \alpha \rangle)$. Since $|\alpha\rangle$ is an eigenket of σ_x, $\langle a | \alpha \rangle$ is just the (componentwisely expressed) eigenvector we obtain from the eigenvalue equation $\sum_a \langle b | \sigma_x | a \rangle \langle a | \alpha \rangle = \alpha \langle b | \alpha \rangle$. To have ρ with respect to the new basis, we write $\langle \alpha | \rho | \beta \rangle = \sum_{a,b} \langle \alpha | a \rangle \langle a | \rho | b \rangle \langle b | \beta \rangle$. Therefore, $U^* \rho U$ is what we want.

(3) Solve the equation in (1) with the correct initial condition. We may use a simple boundary condition, assuming the volume is very large (and temperature is not too low).

(4) Compute Z, using the result in (3). We may use (3) to study the finite volume system as long as the temperature is not too low.

Part B

We can directly compute $\langle x|U(\beta)|x'\rangle$ with the aid of the momentum representation of $U(\beta)$:

$$\langle p|U(\beta)|p'\rangle = e^{-\beta p^2/2m}\delta(p-p'). \tag{19.34}$$

(5) We use

$$\langle x|U(\beta)|x'\rangle = \int d^3p\, d^3p'\, \langle x|p\rangle\langle p|U(\beta)|p'\rangle\langle p'|x'\rangle. \tag{19.35}$$

What is $\langle x|p\rangle$? We may assume the infinite volume normalization (i.e., the δ-function normalization: $\langle p|p'\rangle = \delta(p-p')$).

(6) Perform the integral in (5).

Solution

(1) We immediately obtain

$$-\frac{d}{d\beta}U = HU, \tag{19.36}$$

so its position representation is obtained as given. Notice that

$$\langle x|H|x'\rangle = -\frac{\hbar^2}{2m}\Delta_x\delta(x-x'). \tag{19.37}$$

(2) $U(0) = 1$, so $\langle x|U(0)|x'\rangle = \delta(x-x')$ (if we use the continuum approximation) or $= \delta_{x,x'}$ (if we honestly treat the finiteness of the system; henceforth, we assume an infinite space for simplicity).

(3) This is a diffusion equation, so the solution may be obtained by looking up any standard textbook (or see Appendix 9A); it is essentially the Green's function of the diffusion equation with the vanishing boundary condition at infinity

$$\langle x|U(\beta)|x'\rangle = \left(\frac{mk_BT}{2\pi\hbar^2}\right)^{3/2} e^{-mk_BT(x-x')^2/2\hbar^2}. \tag{19.38}$$

This clearly exhibits that quantum effects become important at low temperatures (as can easily be guessed from the thermal wavelength proportional to $1/\sqrt{T}$).

(4) $Z = \operatorname{Tr} U(\beta)$, so (respecting the finiteness of L, i.e., the total volume $V = L^3$)

$$Z = \int d^3x\, \langle x|U(\beta)|x\rangle = \int d^3x \left(\frac{mk_BT}{2\pi\hbar^2}\right)^{3/2} = V\left(\frac{mk_BT}{2\pi\hbar^2}\right)^{3/2}. \tag{19.39}$$

(This is identical to (20.2).)

(5) $|p\rangle$ is an eigenket of H belonging to the eigenvalue $p^2/2m$: $H|p\rangle = (p^2/2m)|p\rangle$. Therefore,

$$-\frac{\hbar^2}{2m}\Delta_x\langle x|p\rangle = \frac{p^2}{2m}\langle x|p\rangle. \tag{19.40}$$

The solutions are plane waves $e^{ipx/\hbar}$ with normalization condition. Since

$$\int d^3x\, e^{i(p-p')x/\hbar} = h^3\delta(p-p'), \tag{19.41}$$

we obtain

$$\langle x|p\rangle = \frac{1}{h^{3/2}}e^{ipx/\hbar}. \tag{19.42}$$

(6)

$$\langle x|U(\beta)|x'\rangle = \int d^3p\, \langle x|p\rangle e^{-\beta p^2/2m}\langle p|x'\rangle = \int d^3p\, e^{-\beta p^2/2m+i(x-x')p/\hbar}. \tag{19.43}$$

This is a simple Gaussian integral, so indeed the answer agrees with (3).

Q19.3 Variational Principle for Free Energy (Classical Case)[18]

Let $H = H_0 + V$ be a system Hamiltonian, where H_0 is its certain portion and V the rest (usually an added potential interactions).

(1) Show that

$$A \le A_0 + \langle V\rangle_0, \tag{19.44}$$

where A is the free energy of the system with the Hamiltonian H and A_0 that with the Hamiltonian H_0. Here $\langle\ \rangle_0$ is the average over the canonical distribution of the system with the Hamiltonian H_0. The inequality is called the *Gibbs–Bogoliubov inequality*.[19]

(2) We can use the inequality to estimate A. If we can compute A_0 and $\langle V\rangle_0$, then we can estimate the upper bound of A. Its minimum may be a good approximation to A. This is the idea of the *variational approximation*. Let us study an anharmonic oscillator with the Hamiltonian

$$H = \frac{1}{2m}p^2 + \frac{1}{4}\alpha x^4, \tag{19.45}$$

where m and α are positive constants.[20] Let us define

$$H_0 = \frac{1}{2m}p^2 + \frac{1}{2}Kx^2. \tag{19.46}$$

Choose K to obtain the best estimate of A (the reader need not compute the estimate of A; it is easy but ugly).

[18] See, for example, Girardeau, M. D. and Mazo, R. M. (1973). Variational methods in statistical mechanics, In *Advances in Chemical Physics* **24**, eds. I. Prigogine and S. A. Rice, New York: Academic Press, pp. 187–255.

[19] This holds quantum mechanically as well, but the proof is not this simple. See **Q19.4**.

[20] Its canonical partition function can be calculated exactly, but let us use it to practice the variational approach.

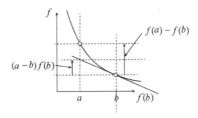

Fig. 19.3 Let f be a convex function.
Illustration of the inequality $f(a) - f(b) - (a - b)f'(b) \geq 0$, if f is also differentiable (note in the figure both $a - b$ and $f'(b)$ are negative).

Solution

(1)

$$\langle e^{-\beta V}\rangle_0 = \frac{1}{Z_0} \int d\Gamma e^{-\beta V} e^{-\beta H_0} = \frac{Z}{Z_0} = e^{-\beta(A-A_0)}. \tag{19.47}$$

Therefore, with the aid of *Jensen's inequality* (cf. Fig. 19.3):

$$e^{-\beta\langle V\rangle_0} \leq e^{-\beta(A-A_0)}. \tag{19.48}$$

That is, we are done.

(2) We know

$$A_0 = k_B T \log\left[\frac{\hbar\sqrt{K/m}}{k_B T}\right] \tag{19.49}$$

and (with the aid of $\langle x^4\rangle_0 = 3\langle x^2\rangle_0^2$ that holds for Gaussian distributions and equipartition of energy: $K\langle x^2\rangle_0/2 = k_B T/2$):

$$\left\langle\frac{1}{4}\alpha x^4\right\rangle_0 = \frac{3\alpha}{4K^2}(k_B T)^2. \tag{19.50}$$

That is,

$$A \leq k_B T \log\left[\frac{\hbar\sqrt{K/m}}{k_B T}\right] - \frac{1}{2}k_B T + \frac{3\alpha}{4K^2}(k_B T)^2. \tag{19.51}$$

Minimizing the right-hand side with respect to K, we obtain

$$\frac{1}{2K} - \frac{3\alpha}{2K^3}k_B T = 0. \tag{19.52}$$

Thus, we obtain $K = \sqrt{3\alpha k_B T}$ as the "best" harmonic approximation. This implies that

$$A \simeq k_B T\left[-\frac{3}{4}\log(k_B T) + \text{const.}\right]. \tag{19.53}$$

From this we get the molar specific heat $3R/4$, which happens to be correct as we will see in **Q20.3**.

Q19.4　Gibbs–Bogoliubov Inequality (Quantum Case)

The Gibbs–Bogoliubov's inequality

$$A \leq A_0 + \langle H - H_0 \rangle_0 \tag{19.54}$$

holds in quantum statistical mechanics as well. We first demonstrate a related inequality.

(1)　Demonstrate *Peierls' theorem*:

$$\mathrm{Tr}\, e^{-\beta H} \geq \sum_i e^{-\beta \langle i|H|i \rangle}, \tag{19.55}$$

where $\{|i\rangle\}$ is an arbitrary orthonormal basis.

(2)　Let $\{|i\rangle\}$ be the orthonormal basis consisting of the eigenstates of H_0. Then,

$$e^{-\beta A} \geq \sum_i e^{-\beta \langle i|H|i \rangle} = e^{-\beta A_0} \sum_i e^{\beta(A_0 - \langle i|H_0|i \rangle)} e^{-\beta \langle i|(H-H_0)|i \rangle}. \tag{19.56}$$

Show the Gibbs–Bogoliubov inequality with the aid of Jensen's inequality.

Solution

(1)　For a convex function f

$$\langle f(X) \rangle \geq f(\langle X \rangle), \tag{19.57}$$

where $\langle\ \rangle$ implies a (weighted) average of the variables with an appropriate sampling (Jensen's inequality; it is illustrated in Fig. 14.2(b)).

　　In the following all the bras and kets are normalized. Let A be a self-adjoint operator with normalized eigenkets $|a_i\rangle$ such as $A|a_i\rangle = a_i|a_i\rangle$. Let $|x\rangle$ be an arbitrary normalized ket. With the aid of the spectral representation (see **17A.12**), we have $f(A) = \sum |a_i\rangle f(a_i)\langle a_i|$. Therefore,

$$\langle x|f(A)|x \rangle = \sum_i |\langle a_i|x \rangle|^2 f(a_i), \tag{19.58}$$

which is an average over the probability $p_i = |\langle a_i|x \rangle|^2$ of $f(a_i)$.[21*] Therefore, Jensen tells us

$$\langle x|f(A)|x \rangle = \sum_i |\langle a_i|x \rangle|^2 f(a_i) \geq f\left(\sum_i |\langle a_i|x \rangle|^2 a_i \right) = f(\langle x|A|x \rangle) \tag{19.59}$$

Now, take any ON basis $\{|i\rangle\}$ to compute $\mathrm{Tr} f(A)$:

$$\mathrm{Tr} f(A) = \sum_i \langle i|f(A)|i \rangle \geq \sum_i f(\langle i|A|i \rangle). \tag{19.60}$$

In particular $f(x) = e^{-\beta x}$ is a convex function, so we get for any ON basis $\{|i\rangle\}$

$$\mathrm{Tr} e^{-\beta H} \geq \sum_i e^{-\beta \langle i|H|i \rangle}. \tag{19.61}$$

[21*] $\sum_i p_i = \langle x| \sum_i |a_i\rangle \langle a_i|x \rangle = \langle x|x \rangle = 1$.

(2) Notice that the inequality in (19.56) is due to Peierls' theorem, and that the equality in the same formula is easy to confirm (simply sum the exponents). Using Jensen's inequality (in the present case $\langle e^{-x} \rangle \geq e^{-\langle x \rangle}$; see Fig. 14.28),[22*]

$$\sum_i e^{\beta(A_0 - \langle i|H_0|i\rangle)} e^{-\beta\langle i|(H-H_0)|i\rangle} \geq e^{-\beta\langle H-H_0\rangle_0}, \tag{19.62}$$

where $\langle \ \rangle_0$ implies the canonical average over the ensemble defined by the Hamiltonian H_0. Thus, we have obtained

$$A - A_0 \leq \langle H - H_0\rangle_0. \tag{19.63}$$

Q19.5 Variational Principle for Density Operators

(1) For any density operator ρ

$$A \leq \mathrm{Tr}\,\rho(H + k_B T \log\rho), \tag{19.64}$$

where A is the free energy for the system whose Hamiltonian is H.
(2) Suppose ρ is the canonical density operator $\rho_c = e^{\beta(A_0 - H_0)}$ for a system with the Hamiltonian H_0. Show that the above inequality is just the Gibbs–Bogoliubov inequality (19.54).

Solution

(1) Notice that for density operators ρ and σ, if $\mathrm{Tr}\,\sigma A = 0 \Rightarrow \mathrm{Tr}\,\rho A = 0$ for any non-negative operator A, then

$$\mathrm{Tr}\,\rho\log\rho - \mathrm{Tr}\rho\log\sigma \geq 0. \tag{19.65}$$

Set $\sigma = \rho_c = e^{-\beta H}/Z$. Then we get the desired inequality. To prove (19.65) we need Klein's inequality (19.67).
(2) The answer is obvious.

Demonstration of (19.65)

Using Klein's inequality (19.67) for $f(t) = t\log t$ ($f'(t) = \log t + 1$), we get

$$0 \leq \mathrm{Tr}[f(\rho) - f(\sigma) - (\rho - \sigma)f'(\sigma)] = \mathrm{Tr}[\rho\log\rho - \sigma\log\sigma - (\rho - \sigma)(\log\sigma + 1)]$$
$$= \mathrm{Tr}[\rho\log\rho - \rho\log\sigma - (\rho - \sigma)] = \mathrm{Tr}[\rho\log\rho - \rho\log\sigma]. \tag{19.66}$$

Demonstration of Klein's inequality

Let A and B be self-adjoint operators and f be a convex differentiable function. Then, we have *Klein's inequality*:

$$\mathrm{Tr}[f(A) - f(B) - (A - B)f'(B)] \geq 0. \tag{19.67}$$

[22*] Since $|i\rangle$ are eigenkets of H_0 $\sum_i e^{\beta(A_0 - \langle i|H_0|i\rangle)} e^{-\beta\langle i|(H-H_0)|i\rangle}$ has the form $\sum_i p_i e^{-a_i}$, where p_i is the canonical probability governed by H_0 and $a_i = \beta\langle i|(H - H_0)|i\rangle$. Therefore, Jensen's inequality tells us

$$\sum_i p_i e^{-a_i} \geq e^{-\sum_i p_i a_i},$$

but $\sum_i p_i a_i = \mathrm{Tr}[\rho\beta(H - H_0)] = \beta\langle H - H_0\rangle_0$.

Let $\{|a\rangle\}$ (resp., $\{|b\rangle\}$) be orthonormal bases consisting of the eigenkets of A (resp., B). Then, the spectral representation gives us

$$\mathrm{Tr}[f(A) - f(B) - (A - B)f'(B)]$$

$$= \sum_a \langle a|[f(A) - f(B) - (A - B)f'(B)]|a\rangle \qquad (19.68)$$

$$= \sum_a \langle a| \left[f(a) - \sum_b |b\rangle f(b)\langle b| - \left(a - \sum_b |b\rangle b\langle b|\right)f'(b) \right] |a\rangle. \qquad (19.69)$$

Notice that $\sum_b \langle a|b\rangle \langle b|a\rangle = 1$, so

$$\mathrm{Tr}[f(A) - f(B) - (A - B)f'(B)] = \sum_{a,b} |\langle a|b\rangle|^2 [f(a) - f(b) - (a-b)f'(b)] \geq 0. \qquad (19.70)$$

The inequality is explained in Fig. 19.3.

Q19.6 Jarzynski's Equality[23]

A single stranded DNA with a certain binding protein is stretched slowly until the protein dissociates from the DNA. Then, the length of the DNA is returned slowly to the original relaxed state without bound proteins. The work W dissipated during the process is measured at 300 K and the experimental results were as follows:

W in pN·nm	number of times	βW	$e^{-\beta W}$
78–82	4	19.3	4.04×10^{-9}
83–87	15	20.5	1.21×10^{-9}
88–92	7	21.74	3.62×10^{-10}
93–97	4	22.94	1.082×10^{-10}
98–102	1	24.15	3.23×10^{-11}

What is the best estimate of the (Gibbs) free energy change due to binding of the protein in the relaxed state of the single stranded DNA? How is your estimate different from the simple average $\langle W \rangle$?

Solution

Notice that $k_B T = 4.14$ pN·nm and $e^{-\beta W}$ is written in the above table. Thus,

$$\sum e^{-\beta W} = 373.1 \times 10^{-10} \implies \langle e^{-\beta W} \rangle = 1.2 \times 10^{-9}. \qquad (19.71)$$

That is, our estimate of ΔA is 85.0 pN·nm. If we directly average the result, we obtain 87.4 pN·nm. Of course, we have "confirmed" the second law $\langle W \rangle \geq \Delta A$. Although we wrote A in the above, its definition is complicated.

---

[23] Inspired by R. Khafizov and Y. Chemla's optical tweezers pulling experiment on SSB (single stranded binding proteins). The numbers are only fictitious, although the magnitudes are realistic. See Suksombat, S. et al. (2015). Structural dynamics of *E.coli* single-standard DNA binding protein reveal DNA wrapping and unwrapping pathways, *Elife*, **25**: 4, 1–53.

20 Classical Canonical Ensemble

Summary

* Molecules of the same chemical species are indistinguishable.
* The canonical partition function (in the classical statistics) of a fluid system reads

$$Z = \frac{1}{N! \, h^{3N}} \int d\Gamma \, e^{-\beta H}.$$

Key words

Gibbs paradox, indistinguishability, equipartition of energy

The reader should be able to:

* Recognize the indistinguishability of molecules of the same chemical species.
* Use equipartition of energy for simple cases.
* Explain why the specific heat of diatomic molecules is not as large as the classical calculation suggests.

20.1 Classical Ideal Gas via Canonical Approach: Single Particle

In Chapter 17, we discussed the classical ideal gas quantum mechanically, but the precise calculation (although not needed to obtain thermodynamic results) was not extremely easy because of the condition imposed on the total energy. We have learned (in Chapter 18) that this restriction is lifted for the canonical approach, often resulting in an easier calculation. Let us redo the classical ideal gas with the aid of the canonical ensemble approach. We have seen that for noninteracting particle systems obtaining the one-particle (or single microentity) canonical partition function was almost all that was needed to calculate the partition function of the system (recall **18.9**). Therefore, let us compute the one-particle canonical partition function for a gas particle in a cube of edge length L ($V = L^3$).

According to the calculation in Chapter 17, each particle has eigenstates characterized by a 3-vector $\boldsymbol{n}$ whose components are non-negative integers with the eigenenergies given by (17.17), so

$$Z_1 = \mathrm{Tr}\, e^{-\beta p^2/2m} = \sum_{n_x>0, n_y>0, n_z>0} e^{-\beta h^2 n^2/8mL^2} = \int_{n_x>0, n_y>0, n_z>0} e^{-\beta h^2 n^2/8mL^2}\, d^3\boldsymbol{n}.$$

$$(20.1)$$

To calculate (20.1) use of the Cartesian coordinate system is the easiest. We get

$$Z_1(V) = V\left(\frac{2\pi m k_B T}{h^2}\right)^{3/2}. \tag{20.2}$$

We calculated the trace in (20.1) quantum mechanically in **Q19.2**(4) to get the same result. The most important feature of this formula is $Z_1 \propto V$.

Now that we have obtained the one-particle partition function, according to the discussion around (18.43), the system partition function should read

$$Z = Z_1^N. \tag{20.3}$$

20.2 Classical Ideal Gas: The Gibbs Paradox

Equation (20.3) implies

$$A(N, V) = -N k_B T \log Z_1(V). \tag{20.4}$$

Prepare two identical systems each of volume V with N particles. The free energy of each system is given by $A(N, V)$. Next, combine these two systems to make a single system. The resultant system has $2N$ particles and volume $2V$, so its free energy should be $A(2N, 2V)$. The fourth law of thermodynamics requires that

$$A(2N, 2V) = 2A(N, V). \tag{20.5}$$

Unfortunately, as we can easily check, this is not satisfied by (20.4). The key feature is $Z_1 \propto V$, so let us write $Z_1 = cV$ with a positive constant c. We may write $Z = (cV)^N$ and get

$$\log(c2V)^{2N} = 2\log(cV)^N + \log 2^{2N} \neq 2\log(cV)^N. \tag{20.6}$$

Thus we must conclude (20.3) is wrong. This is the *Gibbs paradox*.[1]

Since the fourth law is an empirical fact, we must correct (20.3) as

$$Z = f(N)Z_1^N = f(N)(cV)^N, \tag{20.7}$$

where $f(N)$ is as yet an unspecified function of N to remove the N-dependent discrepancy in (20.6). The fourth law demands (20.5):

$$\log f(2N) + 2N\log(c2V) = 2\log f(N) + 2N\log(cV). \tag{20.8}$$

---

[1] **Paradox or not?** Many authors these days claim this is not a paradox at all. We must note that Gibbs did not regard this as a paradox, because he thought he did not know what the microscopic world really was like. A *paradox* in natural science is caused by the discrepancy between what we expect according to what we supposedly know and what is real. That is, a contradiction between the reality and our world picture causes a paradox. In this sense, Gibbs' paradox is a genuine paradox. Assuming the distinguishability of individual particles (i.e., we can name the particles of the same chemical species to track them individually) is the source of the paradox (see **20.3**).

That is,

$$\log f(2N) + 2N \log 2 = 2 \log f(N) \tag{20.9}$$

or

$$f(N)^2 = 2^{2N} f(2N) \text{ (more generally, } f(N)^\alpha = \alpha^{\alpha N} f(\alpha N)). \tag{20.10}$$

The general solution to this functional equation is (set $\alpha = 1/N$; recall Stirling's formula $(N/e)^N \approx N!$):

$$f(N) = (f(1)/N)^N \propto 1/N!. \tag{20.11}$$

Consequently, thermodynamics forces us to write

$$Z = \frac{1}{N!} Z_1^N, \tag{20.12}$$

where we have discarded the unimportant multiplicative factor. We already guessed this before (see (17.11)).

Therefore, the canonical partition function for a classical ideal gas reads

$$Z_{\text{ideal}} = \frac{V^N}{N!} \left(\frac{2\pi m k_B T}{h^2} \right)^{3N/2} = \left[\frac{Ve}{N} \left(\frac{2\pi m k_B T}{h^2} \right)^{3/2} \right]^N. \tag{20.13}$$

From this we get the ideal gas entropy as $ST = E - A$ with $E = 3Nk_B T/2 = Nk_B T \log e^{3/2}$:

$$ST = Nk_B T \log \left[\frac{Ve^{5/2}}{N} \left(\frac{2\pi m k_B T}{h^2} \right)^{3/2} \right] \tag{20.14}$$

or

$$S = Nk_B \left\{ \log \frac{V}{N} + \frac{3}{2} \log T + \frac{5}{2} + \frac{3}{2} \log \frac{2\pi m k_B}{h^2} \right\}. \tag{20.15}$$

20.3 Origin of $N!$

Why does a gas require $1/N!$, but a lattice system does not? For the lattice systems we considered in Chapter 18, a spatial pattern consisting of states at individual lattice points is identified as a microstate (see Fig. 20.1(a)). In this case, the particles sitting at the lattice points are identical chemical species (atom or molecule) and we never pay any attention to the arrangement of individual particles on the lattice (we have seen a similar situation in **7.5**).

Even for a gas system, a microstate corresponds to a pattern of particle position and momentum vectors (even classically; see Fig. 20.1(b) (3) or (4)), since identical particles (molecules, elementary particles, etc.) are indistinguishable combinatorially in contradistinction to marbles. However, if we use classical mechanics to describe a gas we must name particles to describe them separately.[2] Consequently, identical patterns

---

[2] There is no other way to track particles in classical mechanics. Thus, the definition of microstates is generally incompatible with classical mechanics. However, in quantum mechanics the mechanics is decoupled from the particle distinguishability, so with quantum mechanics we can avoid the Gibbs paradox (if we wish).

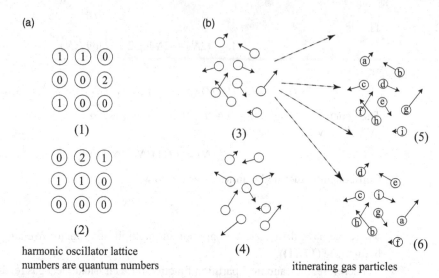

(a)

(1)

(2)

harmonic oscillator lattice
numbers are quantum numbers

(b)

(3)

(4)

(5)

(6)

itinerating gas particles

Fig. 20.1 (a) A lattice system. For the lattice system, a spatial pattern consisting of microscopic states at individual lattice points is identified as a microstate. Thus, (1) and (2) are distinct microstates since they have distinct spatial patterns. Notice that the particles sitting at lattice points are already indistinguishable.

(b) The situation does not vary very much for gasses. A pattern consisting of positions and momentum vectors of the particles corresponds to a single microstate (even classically). Thus, (3) and (4) are distinct microstates. Here, what particle is assigned to what position does not matter. In contrast to the lattice system, however, when the classical partition function is written down, particles have definite names as a, . . ., i and are distinguished. Consequently, a certain single microstate of a collection of N ($= 9$ in the figure) particles are distinguished into $N!$ apparently distinct microstates due only to the naming of the particles. For example, due to the distinct names of the particle variables required in the classical description, the microstate (3) may be written as (5), (6), or other $N! - 2$ distinct configurations.

with differently named particles classically look as if they are distinct microstates. In Fig. 20.1(b) (5) and (6) have an identical pattern to (3). Therefore, they must represent a single identical microstate, but due to the different names of the particles they are handled as distinct microstates. If there are N gas particles, the number of microstates are multiplied by $N!$ due to particle naming, so to correct this overcounting $1/N!$ must be multiplied (see **Q28.1**).[3]

---

[3] **Mesoscopic particles** Although the origin of $1/N!$ is explained in terms of particle indistinguishability in **(20.3)**, its derivation in **(20.2)** uses the fourth law: if we may assume that the two fluid systems are in the same macrostate and that thermodynamics holds, then we must have $1/N!$. Notice that the introduction of this factor removes the contribution of mixing entropy $2Nk_BT \log 2$ that disrupts extensivity. Mixing entropy can be observed only when we can undo the mixing process. Mesoscopic particles (e.g., colloidal particles) are generally distinguishable, but unless experiments to distinguish particles are performed a colloidal particle system (even polydisperse) is not regarded as a mixture. Thus, to explain thermodynamic observations we must include $1/N!$, even if particles are distinguishable. Can we then conclude that the factor $1/N!$ is not due to the indistinguishability of the particles? This is a hasty conclusion. In the case of mesoscopic particles, we do not observe distinguishability. Thus, we may still conclude that the factor $1/N!$ comes from the (operational) indistinguishability.

20.4 Classical Partition Function of Particle Systems

Let us summarize our conclusion. The canonical partition function for an ideal gas with N particles in terms of the phase integral reads

$$Z = \frac{1}{h^{3N}N!} \int d\Gamma \, e^{-\beta \sum_i p_i^2/2m},$$ (20.16)

where $d\Gamma = d^3 q_1 d^3 q_2 \cdots d^3 q_N d^3 p_1 d^3 p_2 \cdots d^3 p_N$ is the phase volume element.

The prefactors in front of the phase integral are determined while considering a particular system (the classical ideal gas), so the reader may think they are rather ad hoc. However, the factor $1/N!$ comes from the indistinguishability of the particles, so as long as identical particles are distinguished in the description of the system, this should not be peculiar to ideal gases. How about h^{3N}?

The canonical partition function Z appears in log, so it must be dimensionless (if not, the free energy shifts, instead of being simply scaled, according to the choice of units, for example). Therefore, in front of the integral whose dimension is (action)^{3N} (dimensional analytically, $[pq] = M(L/T)L = M(L/T)^2 \times T$,[4] we must have a factor cancelling this dimension. The most fundamental quantity in physics that has the dimension of action is h. Since we do not expect that the factor is peculiar to the ideal gas, it is natural to expect $1/h^{3N}$ to appear. Therefore, we define the canonical partition function of a classical fluid system in terms of the phase integral as follows:

$$Z = \frac{1}{h^{3N}N!} \int d\Gamma \, e^{-\beta H}.$$ (20.17)

This relation can be demonstrated by semi-classical analysis of the quantum statistical version of Z.

20.5 Classical Microcanonical Partition Function for a Particle System

We have introduced the classical microcanonical partition function in Chapter 17 as

$$w(E, X) = h^{-3N} W(E, X),$$ (20.18)

where $W(E, X)$ is the $6N$-dimensional volume of the subset of the phase space compatible with the thermodynamic coordinates (E, X) (with macroscopically small leeways). However, we have not taken the particle indistinguishability into account. Obviously, the correct microcanonical partition function for a classical particle system reads

$$w(E, X) = \frac{W(E, X)}{h^{3N}N!}.$$ (20.19)

---

[4] Action is energy times time. Recall h has the dimension of action; $h\nu$ is energy and $[\nu] = 1/T$.

20.6 Generalization of Equipartition of Kinetic Energy

We already know the equipartition of kinetic energy for an ideal gas with the aid of the kinetic theory of gases (Chapter 2), e.g.,

$$\left\langle \frac{1}{2}mv_x^2 \right\rangle = \frac{1}{2}k_B T. \tag{20.20}$$

Let us demonstrate, with the aid of the canonical formalism, a general theorem that implies this formula and that is applicable to any classical system.

Let x_i and x_j be two components of canonical coordinates (say, the x-component of the position vector of particle 1 and z-component of the momentum of particle 2). Then, for a classical system with Hamiltonian H we have

$$\left\langle x_i \frac{\partial H}{\partial x_j} \right\rangle = k_B T \delta_{ij}, \tag{20.21}$$

where the average is over the canonical distribution. Indeed,

$$\left\langle x_i \frac{\partial H}{\partial x_j} \right\rangle = \frac{1}{Z} \int d\Gamma \, x_i \left[-k_B T \frac{\partial}{\partial x_j} e^{-\beta H} \right] \tag{20.22}$$

$$= -\frac{1}{Z}k_B T x_i e^{-\beta H} \Big|_{|x| \to \infty} + \frac{1}{Z}k_B T \int d\Gamma \, \frac{\partial x_i}{\partial x_j} e^{-\beta H}. \tag{20.23}$$

Here, the first term due to an integration by parts must vanish, so H must increase sufficiently fast in the large variable limit. For example, if a system is spatially confined (by a potential well), certainly this is true for the spatial coordinates.

Taking x_i to be momentum coordinates, we obtain the *law of equipartition of energy* for classical kinetic energy such as

$$\left\langle p_i^2/2m \right\rangle = k_B T/2, \tag{20.24}$$

or

$$\left\langle L_i^2/2I_i \right\rangle = k_B T/2, \tag{20.25}$$

where m is the mass, I_i is the ith principal moment of inertia (ith eigenvalue of the inertial moment tensor) and L_i is the corresponding component of the angular momentum.

If the spatial position of a particle is governed by a harmonic potential with a spring constant k (i.e., the harmonic potential energy $U = kx^2/2$), we obtain, identifying x_i in (20.21) as the spatial coordinates,

$$\left\langle kx^2/2 \right\rangle = k_B T/2. \tag{20.26}$$

This implies that a single 1D harmonic oscillator has the total energy $k_B T$ on average (i.e., the average potential energy plus the average kinetic energy of the oscillator) in equilibrium, if we may handle it classically.

Since the classical kinetic energy K is quadratic in momenta (here, K is the total kinetic energy of the system):[5]*

$$\sum_i p_i \frac{\partial K}{\partial p_i} = 2K. \tag{20.27}$$

If there are N particles, then there are $3N$ variables ($3N$ momentum Cartesian components), so

$$\langle K \rangle = \frac{1}{2} \sum_i \left\langle p_i \frac{\partial K}{\partial p_i} \right\rangle = \frac{1}{2} 3N k_B T. \tag{20.28}$$

If a system is described as coupled harmonic oscillators, then the potential energy U is a quadratic function of the position (displacement) coordinates. Therefore, quite analogously the average total potential energy is

$$\langle U \rangle = \frac{1}{2} n_v k_B T, \tag{20.29}$$

if there are n_v modes (the number of independent oscillations[6]). Thus, if there are n_v modes, the harmonic oscillators contribute $n_v k_B T$ to the total internal energy in equilibrium, if we may handle them classically.

20.7 Specific Heat of Gases, Computed Classically

A direct application of the equipartition of energy is the high temperature (constant volume) specific heat per particle of a multiatomic molecular ideal gas. Let us assume that each molecule contains M atoms.[7] The Hamiltonian of each molecule can be written as

$$H = K_{CM} + K_{rot} + K_{vib} + U_{vib}, \tag{20.30}$$

where K_X is the kinetic energy associated with the motion X: CM denotes the center of mass translational motion; "rot" implies rotational motion around its center of mass; "vib" means the vibrational motion. Here U_{vib} is the potential energy for the vibrational motion.

---

[5]*If $K = \sum_{i,j=1}^{n} A_{ij} p_i p_j$, where A_{ij} is a constant, then

$$\sum_{k=1}^{n} p_k \frac{\partial}{\partial p_k} K = \sum_{k=1}^{n} p_k \sum_{i,j=1}^{n} A_{ij}(\delta_{ik}p_j + p_i\delta_{kj}) = \sum_k p_k \left(\sum_j A_{kj}p_j + \sum_i A_{ik}p_i \right) = 2K.$$

This is an example of Euler's theorem about homogeneous functions (cf **Q26.1**), so we can actually get the conclusion without any calculation.

[6] **Modes** A system with a quadratic potential may be described in terms of canonical coordinates (or corresponding operators) that makes the Hamiltonian diagonal. In other words, the system may be described as a collection of independent harmonic oscillators. The motion corresponding to each such harmonic oscillator is called a *mode*. If two or more modes have an identical angular frequency, modes cannot be uniquely chosen. However, this does not cause any problem to us because partition functions need the system energies and their degrees of degeneracy (determined by the number of the modes) only.

[7] Possible contributions of electrons and (nuclear) spins will be briefly discussed in Chapter 30.

Usually, we may assume that the molecular internal vibrations are harmonic, so all these terms are quadratic terms. Therefore, to obtain the internal energy, we need only count the number of degrees of freedom. Notice that the total number of (angular) momenta is always $3M$ for an M-atomic molecule, so, obviously

$$\langle K_{\text{CM}} + K_{\text{rot}} + K_{\text{vib}} \rangle = 3Mk_BT/2. \tag{20.31}$$

Thus, the remaining task is to count the number of vibrational modes.

For a not-linear molecule there are three translational degrees, and three rotational degrees, so there are $3M - 6$ harmonic modes. Thus, $\langle U_{\text{vib}} \rangle = (3M - 6)k_BT/2$. That is, the internal energy is $E = (3M - 3)k_BT$ per molecule, so the constant volume heat capacity $C_V = (3M - 3)R$ per mole, where R is the gas constant.

For a molecule whose shape is linear there are 3 translational degrees, and 2 rotational degrees, so there are $3M - 5$ harmonic modes.[8] Thus, $\langle U_{\text{vib}} \rangle = (3M - 5)k_BT/2$. That is, $E = (3M - 5/2)k_BT$ per molecule, so $C_V = (3M - 5/2)R$ per mole.

It is a well-known fact that these specific heat values grossly overestimate the actual specific heats of molecular gases and were regarded as a paradox before the advent of quantum mechanics.

For a diatomic gas $M = 2$ (which is linear) $C_V = (7/2)R$ according to our formula just derived. However, around the room temperature it is actually $(5/2)R$. That is, it is less by R. This is because the vibrational mode is frozen and its contribution to the kinetic and potential energies, $R/2 + R/2$, does not show up. To excite vibrations the heat bath must pay a big sum of energy (equal to the vibrational energy quantum) all at once to the molecule, so if its temperature is low, the heat bath cannot afford it. In classical mechanics the environment is allowed to pay the big sum by "installments," however low the bath temperature is, so even a high frequency vibration could be excited. However, in the real quantized world, this is virtually forbidden. Thus, the specific heat becomes small.

Problems

Q20.1 Gas Under a Weight: Pressure Ensemble

Suppose there is a vertical cylindrical container of cross section s whose top wall is a movable piston of cross section s with mass M. The piston is assumed to move only in the vertical direction (z-direction) and feels gravity (ignore gravity acting on the gas particles for simplicity). The container contains $N \, (\gg 1)$ classical noninteracting particles with mass m.

(1) Write down the Hamiltonian of the gas plus the piston system (write the piston vertical momentum as p_M).
(2) Obtain the pressure P of the gas, and write the Hamiltonian in terms of P and the volume of the gas $V = sz$, where z is the height of the piston from the bottom of the container.

---

[8] When a molecule is straight, the reader must be able to explain into what vibrational mode(s) the rotational degree is converted, comparing, e.g., water and carbon dioxide.

(3) The mechanical variables are the phase variables of the gas particles and the piston momentum p_M and vertical position $z = V/s$. Compute the canonical partition function Z_{total} of the whole system.

(4) We have already considered the system under constant T and P, and we know the appropriate thermodynamic potential is the Gibbs free energy G (Chapter 16). Then, we should have a statistical formalism that allows us to calculate G directly from mechanics. It is the *pressure ensemble* with the partition function Q defined as

$$Q = \int dV\, Z(T, V) e^{-\beta PV} \tag{20.32}$$

that corresponds to the Legendre transformation $A \to G = A + PV$, where $Z(T, V)$ is the usual canonical partition function with T and V explicitly written. Compute Q for the ideal gas in the container of the present problem. Find the relation between Z_{total} computed in (3) and Q. Obtain the equation of state (of the ideal gas) from Q.

Solution

(1)

$$H = \sum_{i=1}^{N} \frac{p_i^2}{2m} + \frac{p_M^2}{2M} + Mgz. \tag{20.33}$$

(2) From the force balance, we have

$$Ps = Mg \implies PV = Mgz. \tag{20.34}$$

Therefore, (20.33) can be rewritten as

$$H = \sum_{i=1}^{N} \frac{p_i^2}{2m} + \frac{p_M^2}{2M} + PV. \tag{20.35}$$

(3) Use $\int_0^\infty x^N e^{-ax} dx = N!/a^{N+1}$.

$$Z_{\text{total}} = \frac{1}{N!\, h^{3N+1}} \int d^N p\, d^N q\, dp_M dz\, e^{-\beta H}, \tag{20.36}$$

$$= \frac{1}{N!} \left(\frac{2\pi m k_B T}{h^2}\right)^{3N/2} \left(\frac{2\pi M k_B T}{h^2}\right)^{1/2} \int V^N e^{-\beta PV} \frac{dV}{s} \tag{20.37}$$

$$= \left(\frac{2\pi m k_B T}{h^2}\right)^{3N/2} \left(\frac{2\pi M k_B T}{h^2 s^2}\right)^{1/2} (\beta P)^{-N-1}. \tag{20.38}$$

(4) Using the ideal gas result for the canonical partition function (20.13),

$$Q = \frac{1}{N!} \int_0^\infty dV \left(\frac{2\pi m k_B T}{h^2}\right)^{3N/2} V^N e^{-\beta PV} = \left(\frac{2\pi m k_B T}{h^2}\right)^{3N/2} (\beta P)^{-N-1}. \tag{20.39}$$

Thus, we have

$$Z_{\text{total}} = Q \left(\frac{2\pi M k_B T}{h^2 s^2}\right)^{1/2}. \tag{20.40}$$

We know $G = -k_B T \log Q$:

$$G = -Nk_B T \log \left(\frac{2\pi m k_B T}{h^2} \right)^{3/2} + Nk_B T \log \frac{P}{k_B T}, \qquad (20.41)$$

$dG = -SdT + VdP$, so

$$V = \left(\frac{\partial G}{\partial P} \right)_T = \frac{Nk_B T}{P}. \qquad (20.42)$$

This is the equation of state as expected. The enthalpy of the gas can be obtained by the Gibbs–Helmholtz equation

$$H = \left(\frac{\partial(G/T)}{\partial(1/T)} \right)_P = \frac{5}{2} Nk_B T. \qquad (20.43)$$

Q20.2 Magnetic Phenomena are all Quantum Statistical (*Bohr–van Leeuwen Theorem*)

The existence of a magnetic field may be expressed in terms of the vector potential A by replacing the (mechanical) momentum p with $p - qA$ (the canonical momentum), where q is the charge of the particle, as

$$H = \sum_i \frac{1}{2m_i} (p_i - q_i A)^2 + V. \qquad (20.44)$$

Here, A may be dependent on the position of the charges and includes mutual magnetic interactions due to induced charge motions. Compute the classical canonical partition function to obtain the Helmholtz free energy, and show it is independent of the vector potential.

Solution

This is, mathematically, a trivial question (at least formally). The (mechanical) momentum integral can be understood as the integration over the canonical momentum $p - qA$, so the classical canonical partition function cannot depend on A. That is, the system free energy is independent of the magnetic field.

Review of 4-potential
(1) *Vector potential.* Since div $B = 0$ (nonexistence of monopoles), where B is the magnetic field, we can introduce a vector field A such that curl $A = B$. This is called the *vector potential*. If B is a constant, $A = r \times B/2$ may be chosen (not unique).
(2) If an electromagnetic field is given in terms of a 4-potential $(A, \phi/c)$, $(p - qA, (1/c)(E - q\phi))$ is again a 4-vector, where q is the charge. Its length is the same as the case without the 4-potential:

$$\left(1/c^2 \right) (E - q\phi)^2 - (p - qA)^2 = m^2 c^2. \qquad (20.45)$$

Therefore, if the speed of the particle is small, we may write $E = mc^2 + E_{\text{classic}}$, where

$$E_{\text{classic}} = \frac{1}{2m}(\boldsymbol{p} - q\boldsymbol{A})^2 + q\phi. \tag{20.46}$$

Q20.3 Quartic Internal Motion

There is a 1 mole of ideal gas consisting of molecules with one internal degree of freedom. The internal motion of an individual molecule is described by the following Hamiltonian

$$H_{\text{int}} = \frac{1}{2}\mu p^2 + \frac{1}{4}\alpha q^4, \tag{20.47}$$

where μ and α are positive constants, and p and q are canonical coordinates describing the internal motion. The total Hamiltonian of the whole gas must be the sum of the Hamiltonian governing the center of mass translational motions of individual molecules and the Hamiltonians describing their internal motions (i.e., (20.47) for each molecule). We study the system classically.

(1) Let z_I be

$$z_I = \frac{1}{h}\int_{-\infty}^{\infty} dq \int_{-\infty}^{\infty} dp\, e^{-\beta H_{\text{int}}}. \tag{20.48}$$

Write down the canonical partition function Z for this ideal gas utilizing z_I. Let us assume the temperature of the gas to be T, its volume V, and the mass of each particle m.

(2) What is the constant volume specific heat of this system? Use the equipartition of energy.

(3) Although it is possible to analytically evaluate (20.48), since we take the logarithm of z_I, we have only to obtain the exponent θ in $z_I \propto T^\theta$. Get θ dimensional analytically, and confirm the answer to (2).

(4) What is the constant pressure specific heat C_P of the system?

(5) Classically, C_V does not depend on α, but quantum mechanically that is not the case. Suppose the temperature goes very close to $T = 0$ (or α becomes extremely large), what can we expect to happen to C_V?

Solution

(1) $Z = Z_{\text{ideal}} z_I^N$, where $N = N_A$, Avogadro's constant in our case. That is,

$$Z = \frac{1}{N!}\left(\frac{2\pi m k_B T}{h^2}\right)^{3N/2} z_I^N. \tag{20.49}$$

(2) The contribution of the translational motion is $3RT/2$. The contribution of p is just another kinetic energy, so $RT/2$. The contribution of q can be obtained with the aid of the equipartition of energy:

$$\left\langle q\frac{\partial H}{\partial q}\right\rangle = \langle\alpha q^4\rangle = k_B T, \tag{20.50}$$

so $RT/4$ is its contribution. Combining all of them, we get

$$\langle H \rangle = 3RT/2 + RT/2 + RT/4 = 9RT/4. \tag{20.51}$$

Hence, $C_V = 9R/4$.

(3) The integral with respect to p has the dimension of p, and must be a function of $\beta\mu$. $[\beta\mu p^2] = 1$, so $[p] = [\beta\mu]^{-1/2}$. Analogously, $[q] = [\beta\alpha]^{-1/4}$. Therefore ($h$ is Planck's constant):

$$[h z_I] = [q][p] = [\beta\mu]^{-1/2}[\beta\alpha]^{-1/4}, \implies z_i \propto T^{3/4}. \tag{20.52}$$

Therefore, $3R/4$ is the contribution of the internal degree of freedom. Consistent.

(4) Use Mayer's relation (Chapter 15). $C_P = C_V + R = 13R/4$.

(5) The internal motion is a kind of oscillation, and obviously there is a finite energy gap. Therefore, at lower temperatures, energy quantization makes excitation harder. Eventually, C_V goes to the value without the internal degree contribution. That is, $3R/2$.

21

Information and Entropy

Summary

* The amount of information may be quantified by the Gibbs–Shannon formula for entropy/information.
* Entropy quantifies how much of knowledge (information) we need to specify a microstate of a system.
* Sanov's theorem tells us why information maximization can allow plausible inference.

Key words

(Gibbs–)Shannon formula, information, bit, surprisal, Sanov's theorem

The reader should be able to:

* Explain why the Shannon formula is plausible (perhaps in terms of surprisal).
* Estimate required information in terms of yes–no questions.
* Understand why entropy maximization can give the most plausible probability.

Let us try to understand entropy more intuitively. We have already seen, with simple examples (**15.8–15.10**), that the number of questions required to determine the microstate and entropy are related. In this chapter, the soundness of this observation is explained.

21.1 The Gibbs–Shannon Formula of Entropy

Using the canonical formalism, let us compute entropy explicitly. We start from

$$TS = E - A = \operatorname{Tr}\left(H\frac{e^{-\beta H}}{Z}\right) - (-k_B T \log Z). \tag{21.1}$$

Since $H = -k_B T \log e^{-\beta H}$, we have

$$TS = k_B T \log Z - k_B T \operatorname{Tr}\left(\frac{e^{-\beta H}}{Z} \log e^{-\beta H}\right) = -k_B T \operatorname{Tr}\left(\frac{e^{-\beta H}}{Z} \log \frac{e^{-\beta H}}{Z}\right). \tag{21.2}$$

That is,

$$S = -k_B Tr \, \rho \log \rho, \tag{21.3}$$

where $\rho = e^{-\beta H}/Z$ is the canonical density operator (Chapter 19). Classically,

$$S = -k_B \int d\Gamma \, p \log p, \tag{21.4}$$

where p is the canonical distribution function. This is the formula first given by Gibbs in his famous book on the foundation of statistical mechanics. The quantum version (21.3) was proposed by von Neumann, so it is often called the *von Neumann entropy*.

The same formula as (21.4) was proposed by Shannon to quantify information, so (21.4) is often called *Shannon's formula*. It is a convenient occasion to see why such a formula is a good measure of information. Shannon did *not* ask what information was, but tried to quantify it.[1]

21.2* How about Boltzmann's Principle?

Boltzmann's principle (17.1) tells us S in terms of the microcanonical partition function w as $S = k_B \log w$. If we subscribe to the principle of equal probability, then the microcanonical distribution is characterized by the uniform distribution with probability $1/w$ on the energy shell $(E - \Delta E, E]$ or

$$\rho = \sum_\varepsilon |\varepsilon\rangle \frac{\chi_{\Delta E}(\varepsilon)}{w(E)} \langle\varepsilon|, \tag{21.5}$$

where the summation is over all the eigenvalues of the system Hamiltonian and $\chi_{\Delta E}$ is the indicator of the interval $(E - \Delta E, E]$. Therefore, the von Neumann entropy reads

$$S = -k_B Tr\left(|\varepsilon\rangle \frac{\chi_{\Delta E}}{w(E)} \log \frac{\chi_{\Delta E}}{w(E)} \langle\varepsilon|\right) = k_B \sum_{\varepsilon \in (E-\Delta E, E]} \frac{1}{w(E)} \log w(E) = k_B \log w(E), \tag{21.6}$$

because $\sum_{\varepsilon \in (E-\Delta E, E]} 1 = w(E)$. Thus, the Gibbs–Shannon (von Neumann) formula correctly gives entropy for the microcanonical case as well: the formula (21.3) or (21.4) and the principle of equal probability implies the Boltzmann principle.

21.3 How to Quantify Information

Let $\eta(m)$ be the "information" per letter we can send with a message (letter sequence) that is composed of m distinct symbols. Here, the word "information" should be understood intuitively. Let us assume that all the symbols are used evenly. Then, $\eta(m)$ must be an

---

[1] **Textbook of information theory** The best textbook of information theory (in English) is Cover, T. M. and Thomas, J. A. (1991). *Elements of Information Theory*, New York: Wiley.

increasing function of m. For example, if we can use two symbols, then we can distinguish with one symbol, e.g., the events $\{1, 2, 3\}$ vs. $\{4, 5, 6\}$ when rolling a dice. If three symbols, then the events $\{1, 2\}$, $\{3, 4\}$, $\{5, 6\}$ can be distinguished with one symbol (or letter).

Now, let us use simultaneously the second set of n symbols. We can make compound symbols by juxtaposing them as ab (just as in many Chinese characters). The information carried by each compound symbol should be $\eta(mn)$, because there are mn symbols. We could send the same message by sending all the left half symbols first and then the right half symbols later. The amount of information sent by these methods must be equal *per compound symbol*, so we must conclude that:[2]

$$\eta(mn) = \eta(m) + \eta(n). \tag{21.7}$$

Since η is an increasing function, we conclude (see **5.2**):

$$\eta(n) = c \log n, \tag{21.8}$$

where $c > 0$ is a constant. Its choice is equivalent to the choice of unit of information per letter. If we choose $c = 1/\log 2$ (i.e., $\eta(n) = \log_2 n$), we measure information in *bits*. One bit is an amount of information one can obtain from the answer to a single yes–no question (when one cannot guess the answer at all).

21.4 The Shannon Formula

We have so far assumed that all the symbols are used evenly, but such uniformity is not usual. What is the most sensible generalization of (21.8)? We can write $\eta(n) = -\log_2(1/n)$ bits; $1/n$ is the probability for a particular symbol. $-\log_2(1/n)$ may be interpreted as the expectation value of $-\log_2(\text{probability of a symbol})$. This suggests that in the case with the not-equally-probable occurrence of n symbols with probabilities $\{p_1, \ldots, p_n\}$, the information carried by the ith symbol should be defined as $-\log_2 p_i$ bits. Then, the average information in bits carried by a single symbol should be defined by

$$H(\{p_i\}) = -\sum_{i=1}^{n} p_i \log_2 p_i. \tag{21.9}$$

This is called the *Shannon information formula*.[3]

---

[2] We must send a message explaining how to combine the transferred symbols as a part of the message, but the length of the needed message is finite and independent of the length of the actual message we wish to send, so in the long message limit we may ignore this overhead.

[3] For an uncorrelated (or Bernoulli) information source.

Claude Shannon About Shannon himself and the development of information theory, see Golomb, S. W., et al. (2002). Claude Elwood Shannon (1916–2002), *Notices AMS*, **49**, 8–16. When Shannon arrived at (21.9), he asked von Neumann what it should be called. It is told that von Neumann suggested the name "entropy," adding that it was a good name because "nobody really knows what entropy really is so he will have the advantage in winning any arguments that might erupt."

21.5 Average Surprisal

The quantity $-\log_2 p_i$ that appears in (21.9) is sometimes called the *surprisal* of symbol i, because it measures how much we are surprised by encountering this symbol (smaller p should give more surprise). It may be easier to use the axioms for *surprisal* to understand the Shannon formula (21.9). The "extent of surprise" $f(p)$ we get seeing a symbol that occurs with probability p, or knowing that an event actually happens whose expected probability is p, should be:

(1) A monotone decreasing function of p (smaller p should give us bigger surprise).
(2) Nonnegative.
(3) Additive: $f(pq) = f(p) + f(q)$.

Therefore, $f(p) = -c \log p$ $(c > 0)$ is the only choice.[4] Thus, (21.9) is the average extent of surprise we expect from actually seeing a single symbol.

21.6 Entropy vs. Information

So far we have learned two pieces as to the relation between entropy and information:

(i) The change in entropy, ΔS, due to a process in a macrosystem is related to (when $\Delta S > 0$) the number of extra yes–no questions (*bits*; introduced in **15.8**) we need to determine the microstate of a macrosystem as accurately as before the process. For an ideal gas all the molecules are independent, so we must determine all the states of particles individually. Therefore, we must multiply by the number of particles to get the number of questions needed to pinpoint (within the accuracy allowed by physics) a particular microstate of the gas.
(ii) The Gibbs–Shannon formula: the expression of entropy of a system and the expression of the information carried by a collection of letters are identical.

Combining these two, we may conclude that entropy is the amount of knowledge/ information required to pinpoint the actual (micro)state of a system.

Let us repeat our previous argument from **15.7** as to $\Delta S = R \log 2$ per mole for doubling the ideal gas volume. Suppose we can know the state of each molecule (within a specified accuracy) before doubling the volume. After doubling, if we know whether each molecule is in the left or in the right half (i.e., a 1 bit/molecule information), then we can know the state of each molecule as accurately as before volume doubling.[5] The information per

---

[4] We could invoke the Weber–Fechner law in psychology.
[5]*We can use the same device to locate the particle as accurately as before. Suppose that the apparatus has a capability to locate a particle in one of the cells of size $V/2^m$ (i.e., the apparatus can get m bits of the position information of a particle). After doubling the volume the apparatus can realize the same performance in any volume V of the volume-doubled container. Therefore, if we tell the apparatus which half to look at, the apparatus can locate the particle in a cell of size $V/2^m$ (i.e., with the same absolute accuracy as before).

molecule required for this extra knowledge is one bit. This must correspond to the entropy increase $\Delta S = R \log 2$ just mentioned. The increase of entropy describes how much more the state of the system becomes diverse. To know the state as accurately as before volume-doubling, we need extra information to compensate this entropy increase. Therefore, it is sensible to identify $R \log 2$ of entropy with the N_A bits of information: $R \log 2 = N_A$ bits as alluded in Chapter 15.

Entropy S of a macrostate may be understood as the required (expected) amount of information we need to pinpoint (within the precision allowed by physics, esp., quantum mechanics) any microstate in the macrostate. Therefore, we can interpret S as a measure of disorder (or diversity) of the macrostate described microscopically.

An important observation is that it is meaningful to discuss the entropy of a single microstate, although statistical-mechanically entropy is always discussed with respect to an ensemble. The entropy of a microstate is the needed information to single it out from the microstate space.

21.7 "Thermodynamic Unit" of Information

In chemical physics entropy is often measured in eu (entropy unit = cal/(mol·K)). It may be useful to remember that 1 eu = 0.726··· bits/molecule. Some people say that the unit of entropy (e.g., J/K) and unit of information (bit) are disparate: cf. 1 bit/molecule = 9.57×10^{-24} J/(K·molecule)). This is simply because they do not think about things microscopically. If one wishes to tell each molecule to turn "to the right," the number of required messages is comparable to the number of molecules, so it is huge (in bits), but for each molecule it is about a few bits. For example, the entropy change due to a reaction involving small molecules is usually the order of a few eu. This is a reasonable value. For a summary of units see **21.12**.

21.8 Subjective and Objective Information

Since information is quantified in terms of probability, the interpretation of information depends on the interpretation of probability. In Chapter 3 we discussed subjective and objective probabilities. Accordingly, there should be subjective and objective informations. Needless to say, the information relevant to thermodynamics and statistical mechanics is the objective version in terms of objective (experimentally measurable) probabilities.

21.9 How to Quantify the Amount of Knowledge (Gained)

How much information do we need to know the outcome of a fair dice? We guess it is $\log_2 6 = 2.58$ bits. This is the entropy of a state of a dice. Suppose we are told that the face value is larger than or equal to three. How much information does this statement carry? Information is something that can reduce our extent of ignorance. After hearing this message, we know four faces are still possible, so we need two more yes–no questions to remove the remaining uncertainty completely (in order to know the actual face). Therefore, $\log_2 6 - \log_2 4 = \log_2(3/2) = 0.58 \ (> 0)$ bits must be the information carried by the message that reduced the extent of our ignorance.

☐ Let us look at some examples of information carried by messages.

(1) (cf. **Q3.2**(2)) There are two kittens. We are told that at least one of them is a male. What is the information we get from this message? In this case the initial state diversity is mm, mf, fm, and ff. After the message, the diversity is reduced to mm, fm, and mf. Therefore, $\log_2 4 - \log_2 3 = \log_2(4/3) = 0.41$ bits is the information of the message. Notice that this is less than 1 bit.

(2) (cf. **Q3.2**(3)) There are five boxes, of which one contains a prize. A game participant is asked to choose one box. After they choose one of the five boxes, the "coordinator" of the game identifies as empty three of the four unchosen boxes. What is the information of this message? In this case initially nothing is known, so the entropy of state of the system seen by them is $\log_2 5 = 2.322$ bits. The message tells us that the already chosen box has the probability of 1/5, and the remaining one has the probability of 4/5 to contain the prize. Therefore, the remaining uncertainty (entropy) after the message is $-(1/5)\log_2(1/5) - (4/5)\log_2(4/5) = 0.722$ bits. The difference is $(4/5)\log_2 4 = 1.6$ bits. That is, the message carries 1.6 bits.

21.10 Information per Letter of English[6]

If the symbols of English alphabet (+ blank) appear equally probably, what is the information carried by a single symbol? This must be $\log_2(26 + 1) = 4.755$ bits, but for actual English sentences, it is known to be about 1.3 bits. Why? Also how can we estimate this actual value? The reason is not hard to understand. Letters in actual sentences are not equally probable and are also correlated; we can guess something about other letters from the letters we have already been given. This may tell us how to estimate the information per letter.

Probably, the easiest approach is to count the letters we need to guess the whole message. For example, if 70 letters+blanks must be given to guess the whole paragraph consisting of 250 letters+blanks, then the information per letter is estimated as $(70/250)\log_2(27) \simeq 1.33$ bit. It may be fun to do this game with one's friends. This requires not only the linguistic knowledge but general cultural knowledge as well (it is not unfair, because when we read a book we use all the knowledge we can). Generally speaking, more cultivated people give lower estimates.

---

[6] Cover T. M. and King, R. C. (1978). A convergent gambling estimate of the entropy of English, *IEEE Trans. Info. Theory*, **24**, 413–421 is an excellent reference.

21.11 What is One-Half (or Any Fraction of a) Question?

We have encountered a fraction of bit of information. The reader might wonder what 0.38 yes–no questions (or 0.38 bits of information) imply. How can we ask such a question? Suppose there is 1 red ball and 999 white balls (assume that we know this fact). How many questions do we need to determine the colors of all the balls? The total entropy is in this case 9.97 bits.[7] This seems to tell us that we need only 0.01 yes–no questions to determine the color of a single ball. First, we divide the balls into two 500 ball sets and ask if one set we choose is with all the same color or not. Irrespective of the answer, this single bit question determines the color of 500 balls at once. Thus, it should not be so difficult to understand what a fraction of a question means.

21.12 Summary of Information vs. Entropy

We may safely conclude that the amount of (the average) information required to pinpoint an elementary event in the sample set Ω is proportional to

$$\eta = -\sum_{\omega \in \Omega} p(\omega) \log p(\omega). \tag{21.10}$$

If we use $\log_2$ (i.e., $(1/\log 2) \log$), this is the information in bits (the number of average yes–no questions we must ask). If we use $k_B \log$, we measure information in energy unit (per molecule). The entropy S in bits is, according to Boltzmann's principle, given by $\log_2 w(E, X)$. This is the number of yes–no questions we must ask to single out a microstate that is consistent with the macrostate (E, X). Notice that the relevant information is always of order $N k_B$ in equilibrium statistical mechanics. Increasing entropy ($\Delta S > 0$) implies that to pinpoint a microstate we must ask extra questions corresponding to ΔS. This implies that the diversity of the system microstates has increased.

Entropy and information are usually measured with different units, and the conversion rates may be summarized as follows:

1 bit/molecule = $k_B \log 2$ J/K·molecule = 9.57×10^{-24} J/(K·molecule),
1 bit/mole = $R \log 2$ = 5.7628 J/(K·mole)= 1.3774 eu,
1 J/(K·mole) = 0.17352 bit/molecule,
1 eu = 1 cal/(K·mole) = 0.7260 bits/molecule.

Bit is dimensionless and the unit on the right-hand side has a dimension. Is there inconsistency? This is because of the tradition that the temperature is measured in K and not in the energy unit. This inconsistency is benign, because we always use the same unit K for absolute temperature.

---

[7] This is $\log_2 \binom{1000}{1}$, but Stirling's formula cannot be used for an accurate estimate, so do not try to apply the Shannon formula.

21.13 Statistical Mechanics from Information Theory?

Maximizing Shannon's entropy may be interpreted as a strategy to find the least biased distribution, so we may expect that the resultant distribution is the most objective distribution. We should be able to obtain the "true distribution of microstates" p by maximizing the Gibbs–Shannon formula under the condition that we know the expectation value of energy (internal energy). This is equivalent to maximizing the following variational functional with Lagrange's multipliers β and λ (the latter for the normalization condition):[8]

$$-\sum_i p_i \log p_i - \beta \sum_i p_i E_i - \lambda \sum_i p_i. \tag{21.11}$$

This indeed gives

$$p_i \propto e^{-\beta E_i}. \tag{21.12}$$

The Shannon formula is derived logically from almost inescapable requirements about "knowing something." Therefore, the above line of argument seems to indicate that the principle of statistical mechanics can be derived directly from this fundamental conceptual basis. Some brave people concluded that this was the true basis of statistical mechanics; forget about physics. This is the so-called information-theoretical derivation of statistical mechanics. Do not subscribe to any such thoughtless statement. A fatal flaw of the above argument is clear from Sanov's theorem.

21.14 Sanov's Theorem and Entropy Maximization

There are m events and their true probabilities are given by $\{q_i\}_{i=1}^m$. Suppose we observe the occurrence of events with N statistically independent observations. We wish to make an empirical distribution π_i of the events $i \in \{1, \ldots, m\}$

$$\pi_i = \frac{1}{N} \sum_{t=1}^N \delta_{ik(t)}, \tag{21.13}$$

where $k(t) \in \{1, \ldots, m\}$ is the event that occurs for the tth sampling (or the observation at time t). The law of large numbers tells us that $\pi_i \to q_i$ for large N. We can ask the deviation from this if N is not large enough. That is, we wish to study the large deviation of the empirical distribution: What is the probability to find $\{p_i\}$ as an empirical distribution? If event i occurs n_i times ($\sum_{i=1}^m n_i = N$), the empirical probability is $p_i = n_i/N$. Therefore, the probability to get this empirical distribution from N sampling is given by the following multinomial distribution (see Appendix 3A):

---

[8] We have already done the same calculation when we discussed Boltzmann's original paper (**Q17.2**). Then, after all, isn't the argument here all right? Notice that Boltzmann did not state his assumptions carefully.

$$P(\pi \simeq p) = \prod_{i=1}^{m} \frac{N!}{(Np_i)!} q_i^{Np_i}. \tag{21.14}$$

Taking its log and using Stirling's approximation, we obtain

$$\log P(\pi \simeq p) = \log N! + \sum_{i=1}^{m} \{Np_i \log q_i - Np_i \log(Np_i) + Np_i\}. \tag{21.15}$$

That is, we get a large deviation principle for the probability distribution.

$$P(\pi \simeq p) \approx e^{-N \sum_i p_i \log(p_i/q_i)}. \tag{21.16}$$

This relation is called *Sanov's theorem*.[9] Notice that Sanov's theorem can make the statement "the most probable" objective.

This implies that maximization of the following quantity called the *Kullback–Leibler entropy* (under certain constraints):

$$K(p\|q) = \sum p_i \log \frac{p_i}{q_i} \tag{21.17}$$

gives the most probable probability distribution. That is, the Shannon-entropy maximization presupposes $q_i = 1/n$ (equal probability). It should be obvious now that information theoretical derivation of statistical physics assumes the principle of equal probability, which cannot be derived within the information-theoretical framework.

Incidentally, since (21.17) is a large-deviation rate function(al), its positivity is expected, and can be proved easily as in (19.65).

Problems

Q21.1 Elementary Question

Suppose a student cheats in a yes–no quiz by copying the answer of an A student who is correct with probability 95%.

(1) Assuming that the student has no idea about the correct answer, what is the amount of information (its expected value) they can gain (per question) by copying the A student's solution?

(2) Suppose the test is known to give more Yes solutions than No solutions (3 to 2 ratio). What is the information (per question) gained from cheating?

Solution

(1) Initially, this student has no information about the solution at all, so their knowledge state is 0, or their ignorance as to this problem (that can be answered by one YN

9   Sanov, I. N. (1957). On the probability of large deviations of random variables, *Mat. Sbornik*, **42**, 11–44. Ivan Nikolaevich Sanov (1919–1968); obituary in Borovkov, A. A. et al. (1969). *Russ. Math. Surveys*, **24**, 159.

question) is $H_i = 1$ bit (i.e., with 1 bit of information his knowledge level becomes perfect, or the initial 1 bit uncertainty becomes 0). After copying the answer their remaining uncertainty (the extent of ignorance) is the same as that of the A student. That is, $H(0.95)$:

$$H_f = H(0.95) = -(0.95 \log_2 0.95 + 0.05 \log_2 0.05) = 0.286 \text{ bits.} \qquad (21.18)$$

Therefore, their extent of ignorance is reduced by $1 - 0.286 = 0.714$ bits. This must be the information they got from cheating.

(2) Initial entropy (i.e., the intial level of ignorance) is

$$H_i = -(3/5) \log_2(3/5) - (2/5) \log_2(2/5) = 0.971. \qquad (21.19)$$

Here H_f is the same as in (1), so 0.685 bits is the gain.

Q21.2 Information-Based Microscopic Guess

The boiling temperature of acetic acid under 1 atm is 391 K, and the evaporation heat (the latent heat of evaporation) is about 23.7 kJ/mol.

(1) What is the entropy increase due to evaporation?
(2) Roughly, how many yes–no questions do you have to ask to specify the (single) molecular state in the gas phase as accurately as in the liquid phase?
(3) The evaporation entropy of ethanol is about 110 J/(K·mol). The reader should have realized a big difference between this value and the value for acetic acid obtained in (1). This is said to be due to dimerization: acetic acid gas (around the boiling point) consists of dimers $(CH_3COOH)_2$ (due to strong hydrogen bonding, but ethanol does not make dimers in the gas phase).[10] Is the entropy difference roughly consistent with this explanation (or not)? Give your opinion with your supporting argument.

Solution

(1) The entropy change due to evaporation is $\Delta S = 23700/391 = 60.6$ J/(K·mol).
(2) This corresponds to $60.6 \times 0.17 = 10.3$ bits/molecule. That is, we need about 10 yes–no questions to determine the state of each molecule as precisely as we can in the liquid phase. The volume of the gas (under the condition we are interested in) is about 200 times as large as that of the liquid. This explains about 7 to 8 bits. Not very bad.
(3) Ethanol evaporation corresponds to almost 19 bits/molecule increase of entropy, so we may say that the number of questions required for ethanol is almost doubled. If we assume that roughly two molecules behave as one in the gas phase of acetic acid, then the knowledge about one molecule tells us about one more molecule, so this is reasonable (recall entropy is additive).

---

[10] Precisely speaking, there are also tetramers, and the average number of acetic acid molecules in a single gas particle seems to be about $105/60 \simeq 1.75$.

22

Information and Thermodynamics

Summary

* Information and entropy should be handled on an equal footing thermodynamically.
* The relation between information and thermodynamics teaches us that indeed "knowledge is power."
* Observing a system, we can acquire the information quantified by the mutual information between the system and the measurement apparatus.
* In principle, we can convert the obtained mutual information into work.
* Thus, the second law of thermodynamics expanded to handle imported information I as well reads $\Delta A \leq W + k_B T I$ isothermally.
* However, if we take the cost of handling information into account, the usual second law is recovered.

Key words

conditional information, mutual information, Szilard engine, the second law of "information thermodynamics"

The reader should be able to:

* Understand how we can convert information into work at mesoscales.
* Understand that the information we can get by measurement is bounded by the mutual information between the system and the measurement apparatus.
* Understand why the second law is intact even within information thermodynamics.

22.1 Converting Information into Work

Thermodynamic entropy times temperature is energy. It can be a source of forces (entropic force; see Chapters 7, 9, 27, etc.). Then, information could be a source of force or work. Let us confirm this fact with a simple thought experiment. In a box is one particle, jumping around. The box is maintained at temperature T (with a heat bath) (see Fig. 22.1).

First, we insert a wall bisecting the box in state A. No work is required if the wall is thin. In the resultant state B we observe which side the particle is. We obtain one bit ($= k_B \log 2$) of information from this observation. Using this information (i.e., a feedback control of the system), if the particle is on the right half, we connect a weight as in C, so that the wall as a piston can do work. Notice that this procedure reduces the system entropy by $k_B \log 2$;

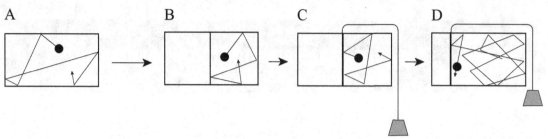

Fig. 22.1 There is a particle in a box, which is maintained at temperature T. The initial state is **A**. If we insert a wall bisecting the box, we get **B**. Next, we observe whether the particle is in the left or the right half; we get 1 bit of information. For example, if we observe that the particle is in the right half as in **B**, connect a weight to the wall as in **C** and let the wall move as a piston; we have used the information obtained by observation. The wall moves to the left due to collisions of the particle, and the weight is pulled up as in **D**, so the system has produced a mechanical work. The amount of work done is $k_BT \log 2$, because the volume has been doubled at a constant temperature. Removing the wall, we recover the initial state **A**. The "engine cycle" is complete and we can start a new cycle. Such an engine is called the *Szilard engine*.

we have used our gained knowledge to reduce the system entropy (microstate diversity). The released wall acts as a piston and moves to the left, pushed by collisions with the particle (D). The weight goes up and the system has done a work of $k_BT \log 2$, because the system volume has been doubled under a constant temperature: $W = -\int_{V/2}^{V} P dV = -k_BT \log 2$. The system entropy has returned to the original value, absorbing heat $Q = |W|$ from the environment.

Let us check that if we do not use the information acquired in B correctly, we generate heat; in the situation B in Fig. 22.1 if we connect the weight on the left side, the piston suddenly moves to the left and heat $k_BT \log 2$ is generated (and discarded to the environment). Thus, if we ignore the knowledge about the state B, with probability 1/2 we gain work $W = k_BT \log 2$ (absorbing heat $Q = k_BT \log 2$ from the heat bath) or generate heat $Q = k_BT \log 2$, wasting the potential energy. Thus, on average no work is gained, consistent with Kelvin's principle.

The cyclic process explained in Fig. 22.1 makes an engine called the *Szilard engine*. Since we have only one heat bath at temperature T, we see that information about the system's microstates apparently allows the engine to violate the second law.[1]

22.2 Second Law with Supplied Information[2]

Let us try to use the information I, however obtained, through a feedback control (as B to C in Fig. 22.1 illustrates). Without such control, the second law tells us (Chapter 16)

[1] This problem must be taken seriously and an extended thermodynamic system should be conceived that can handle observation and feedback properly. Such a system has been pursued by Sagawa in his thesis (Sagawa, T. (2013). *Thermodynamics of information processing in small systems*, Tokyo: Springer).
[2] Extremely simplified Sagawa, T. and Ueda, M. (2008). Second law of thermodynamics with discrete quantum feedback control, *Phys. Rev. Lett.*, **100**, 080403.

$$\Delta A \leq W. \tag{22.1}$$

With feedback, as we have seen at the beginning of this section, the information we acquired may be converted to work; if the process is performed ideally without any loss, we can obtain the work $k_B TI$. Let us write this work we can squeeze out with the aid of the information as $-W_I$:[3*]

$$- W_I \leq k_B TI. \tag{22.2}$$

Equation (22.1) reads $-W \leq -\Delta A$ when we extract work (see **16.2**). Therefore, the total work we get $-(W + W_I)$ written as the work $-W$ anew (since we cannot distinguish the conventional thermodynamic work and the information contribution) satisfies

$$- W \leq -\Delta A + k_B TI. \tag{22.3}$$

Therefore, the second law of information thermodynamics reads:

$$\Delta A \leq W + k_B TI. \tag{22.4}$$

That is, we can utilize information to increase the system free energy.

22.3 Clausius' Inequality with Supplied Information

Correspondingly, Clausius' inequality must be extended. We have only to reverse the discussion (see **16.3**) that connected (22.1) and Clausius' inequality. Since $\Delta E = W + Q$, from (22.4) we have

$$\Delta E - T\Delta S \leq W + k_B TI \Rightarrow Q - T\Delta S \leq k_B TI. \tag{22.5}$$

That is,

$$\Delta S \geq \frac{Q}{T} - k_B I. \tag{22.6}$$

Up to this point, we have assumed that information is supplied "for free" without any cost. We must, however, respect the "no-ghost principle": information must be carried by something. Thus, information handling cost must not be ignored. First, we must get the information by measurement.

22.4 What is Measurement?

To observe a system S is to create an interaction between the system and the measurement apparatus A in which we read the "scale." We do not call the latter process an observation (or a measurement) of the system; the actual "measurement process" is the physical

---

[3*]The work we gain is $|W_I|$ (our gain is the system's loss, so $W_I < 0$) which is bounded from above by $k_B TI$ or $-W_I \leq k_B TI$.

interaction between the system and the measurement device until the device settles down (or equilibrates) to some state. During this process we do not read the result (i.e., do not read the "scale"). Our reading occurs after the compound system S + A reaches a new equilibrium state. The subsequent reading process (e.g., using our eyeball) does not alter either the state of S or A. Thus, the usual process loosely called "measurement" or "observation" of a system consists of two steps: (i) the evolution of the compound system S + A to a final stable state called the measurement process, (ii) the reading of A. The information about the system is acquired through (ii) by us, but what would be read by us is already fixed by (i).

22.5 Conditional Entropy

What is the information we can acquire about the system by reading of the measurement apparatus? To consider this information, it is convenient to introduce conditional entropy.

After the measurement process let us assume that the probability of the occurrence of the state (s_i, a_j) of the compound system S + A is given by $P(s_i, a_j)$, where $\{s_i\}$ is the set of the system states and $\{a_j\}$ that of the measurement apparatus states. The entropy (the uncertainty or the state diversity) of the compound system is given by the joint entropy

$$H(S, A) = -\sum_{i,j} P(s_i, a_j) \log P(s_i, a_j). \tag{22.7}$$

We read the measurement apparatus (i.e., we measure the outcome of the measurement process) to get some information about the system after the measurement process, because its state s_i has statistical relationships with the reading a_j established during the measurement process. What can we learn (how much uncertainty can we reduce) about the state of the system S through this reading?

After the measurement process, if we do not perform any measurement (or ignore the result), the uncertainty about the system must be

$$H(S) = -\sum_{i} P_S(s_i) \log P_S(s_i) \tag{22.8}$$

with $P_S(s) = \sum_j P(s, a_j)$, the marginal probability for the system state.

Suppose, instead, we get the measurement (reading) result a_j. After knowing this result the remaining uncertainty about the system is the following

$$H(S \mid a_j) = -\sum_{i} P(s_i \mid a_j) \log P(s_i \mid a_j), \tag{22.9}$$

where $P(s_i \mid a_j)$ is the conditional probability of the system state when we know the apparatus reading is actually a_j. Thus, the expected value of the uncertainty about the system after measurement is obtained by averaging (22.9) over the measurement result a_j. The probability for the apparatus states a_j is given by the marginal $\sum_i P(s_i, a_j) = P_A(a_j)$. Therefore,

$$H(S \mid A) = \sum_j P_A(a_j) H(S \mid a_j) \tag{22.10}$$

$$= -\sum_j P_A(a_j) \left[\sum_i P(s_i \mid a_j) \log P(s_i \mid a_j) \right] \tag{22.11}$$

$$= -\sum_{i,j} P(s_i, a_j) \log \frac{P(s_i, a_j)}{P_A(a_j)} = H(S, A) - H(A) \tag{22.12}$$

This is called the *conditional entropy*. Here,

$$H(A) = -\sum_j P_A(a_j) \log P_A(a_j). \tag{22.13}$$

22.6 What Information Can We Get from Measurement?

If we do not know the measurement result, the uncertainty (the amount of lacking knowledge) about the system is $H(S)$. If we know the measurement result, the remaining uncertainty is, on average, $H(S \mid A)$. Therefore, knowing the measurement result reduces the uncertainty about the system by

$$I(S, A) = H(S) - H(S \mid A) = H(S) + H(A) - H(S, A) = I(A, S), \tag{22.14}$$

which is called the *mutual information* between S and A. Note the symmetry $I(A, S) = I(S, A)$. Thus, we conclude that $I(A, S)$ is the expected amount of information (knowledge) about the system we can acquire through reading the apparatus state (measurement).

We expect $I(A, S)$ is nonnegative:

$$I(A, S) = \sum_{i,j} P(s_i, a_j) \log \frac{P(s_i, a_j)}{P_S(s_i) P_A(a_j)} \geq 0. \tag{22.15}$$

This follows from the nonnegativity of the Kullback–Leibler entropy (recall **21.14**). Notice that

$$I(S, A) = H(A) - H(A \mid S) \leq H(A). \tag{22.16}$$

Here $H(A)$ is the information written in the measurement apparatus.

22.7 Cost of Handling Acquired Information[4]

Due to the no-ghost principle, the information must be associated with a certain physical entity. For example, the measurement process discussed previously increases the apparatus

---

[4] Again, this is an extremely crude version of Sagawa, T. and Ueda, M. (2009), Minimal energy cost for thermodynamic information processing: Measurement and information erasure, *Phys. Rev. Lett.*, **102**, 250602.

entropy by $H(A)$. We may understand this as the increase of entropy due to storing the acquired information. To acquire information I from measurement, we must increase the entropy of the memory by $H = H(A)$ on average.

If the measurement device imports heat Q, the entropy increase ΔS other than $H = H(A)$ obeys the usual Clausius inequality. Since $I \leq H$, Clausius' inequality reads

$$\Delta S + k_B H \geq Q/T + k_B I. \tag{22.17}$$

If W' is the needed work for this inscription of information, we get (note that $\Delta E = W' + Q$)

$$W' + k_B TH \geq \Delta A + k_B TI. \tag{22.18}$$

To use this system as a measurement device (forever), its memory must be erased after it has been used. Especially, for an engine if we do not wish to require an infinite capacity memory device, we must erase the memory. To erase memory is to remove H. That is, we add $-H$. Therefore, if simultaneously there is also other entropy change $\Delta'S$, Clausius' inequality tells us:[5]

$$\Delta'S - k_B H \geq Q'/T, \tag{22.19}$$

where Q' is the imported heat (if any) associated with the erasure process. Therefore, if we need the work W'' to erase the memory (note that $\Delta'E = W'' + Q'$)

$$W'' - k_B TH \geq \Delta'A. \tag{22.20}$$

Since the memory device returns to the original state before writing new information, this $\Delta'A$ must be equal to $-\Delta A$. Therefore, the total work W_m needed for using information I is obtained by adding (22.18) and (22.20)

$$W_m = W' + W'' \geq k_B TI. \tag{22.21}$$

22.8　Use of Information Cannot Circumvent the Second Law

Now, we have realized that the information I is not free as shown in (22.21). The work $|W|$ we can get (thus we wish $W < 0$) utilizing I is bounded as (22.3):

$$-W \leq -\Delta A + k_B TI. \tag{22.22}$$

Therefore, the total output of the engine + memory device satisfies

$$|W_m + W| = -(W_m + W) \leq -\Delta A \tag{22.23}$$

Thus, the usual second law result is recovered. That is, if we look at the whole world, the conventional second law is intact.

---

[5] **Heat required by memory erasure** If we consider only the purely informational part, ignoring $\Delta'S$, the lower bound of heat generated by the memory erasure $-Q \geq k_B TH$ is obtained (*Landauer's principle*). Note that memory erasure can be performed reversibly or irreversibly, so the generation of heat does not mean irreversibility.

22.9 Significance of Information Thermodynamics

Do not conclude, however, that since the second law is invincible, information thermodynamics is unimportant. It is possible for a system we are interested in to perform something that ordinary thermodynamics does not allow, and to pay the price somewhere else.[6] For example, it may be possible to go beyond the limit of thermodynamics for the time being by shifting the average value of fluctuations, although the price must be paid (dearly) later. There must be such cases in molecular machines. It is argued that cell senescence is due to the accumulation of defects and wastes, but cell death could have a more active implication of carrying away the debt caused by utilizing information. Thus, entropic apoptosis and entropic altruism are conceivable.

It is clear that thermodynamic entropy changes require macroscopic bits of the information change. In order to utilize information, as illustrated in Fig. 22.1, microscopic or mesoscopic degrees of freedom must be controlled individually. Thus, the possible macroscopic effect is realized through a macroscopic collection of mesoscopic entities that may be controlled individually.

Problems

Q22.1 Elementary Questions

There are two random variables X and Y that take values 0 or 1. The joint probability $P(X, Y)$ reads

$$P(0,0) = 1/2, \ P(0,1) = 0, \ P(1,0) = 1/4, \ P(1,1) = 1/4. \qquad (22.24)$$

(1) Obtain $H(X \mid Y)$ and $H(Y \mid X)$. Which is larger, and can you explain the result intuitively?
(2) Calculate the mutual information $I(X, Y)$ between X and Y.

Solution

(1) $H(X \mid Y) = H(X, Y) - H(Y)$. Therefore, if the sampling due to Y is more unbiased, less uncertainty would remain after knowing the outcome of Y. This may not be surprising, because less biased observation would give us better information.

$$H(X, Y) = (1/2) \log_2 2 + (2/4) \log_2 4 = 1.5 \text{ bits.} \qquad (22.25)$$

For Y, $P(0) = 3/4$ and $P(1) = 1/4$, so $H(Y) = (3/4) \log_2(4/3) + (1/4) \log_2 42 - (3/4) \log_2 3 = 0.811$ bits. For X, $P(0) = P(1) = 1/2$, $H(X) = 1$ bit. Therefore,

---

[6] The reader may say the crux of utilizing thermodynamics is, as noted in **16.6**, to couple free energy decreasing processes with free energy increasing processes to realize something incredible (such as making biology possible). In this sense there is nothing new.

$$H(X \mid Y) = 1.5 - 0.811 = 0.688 \text{ bits}, \quad H(Y \mid X) = 1.5 - 1 = 0.5 \text{ bits}. \tag{22.26}$$

(2) $I(X, Y) = H(X) - H(X \mid Y) = 1 - 0.688 = 0.311$ bits. Needless to say, $H(Y) - H(Y \mid X) = 0.811 - 0.5 = 0.311$ bits and is equal to $1 + 0.811 - 1.5$. They must agree.

Q22.2 Bound for Mutual Information

(1) Show

$$I(X, Y) \leq \min[H(X), H(Y)] \tag{22.27}$$

and discuss its intuitive meaning.

(2) If X and Y are independent (unrelated), $I(X, Y) = 0$.

(3) We can define a quantum analogue of mutual information using the von Neumann entropy as

$$I(X, Y) = \operatorname{Tr} \rho(X, Y) \log \rho(X, Y) - \operatorname{Tr} \rho(X) \log \rho(X) - \operatorname{Tr} \rho(Y) \log \rho(Y), \tag{22.28}$$

where $\rho(X) = \operatorname{Tr}_Y \rho(X, Y)$ and $\rho(Y) = \operatorname{Tr}_X \rho(X, Y)$. Show that the classical bound (1) can be fragrantly violated (a very clear sign of strength of the quantum correlation).

Solution

(1) Because $I(X, Y) = H(X) - H(X \mid Y) = H(Y) - H(Y \mid X)$; the conditional entropy must not be negative. Since there must be at least some correlation between X and Y, if we observe them jointly, we should obtain less uncertainty than ignoring the correlations between X and Y.

(2) Because $H(X, Y) = H(X) + H(Y)$.

(3) Suppose

$$\Psi = \sum_i \sqrt{p_i} |x_i\rangle |y_i\rangle, \tag{22.29}$$

which is a pure state, so its entropy is zero. On the other hand

$$\rho(X) = \operatorname{Tr}_Y |\Psi\rangle\langle\Psi| = \operatorname{Tr}_Y \left(\sum_i \sqrt{p_i} |x_i\rangle |y_i\rangle \right) \left(\sum_j \sqrt{p_j} \langle x_j| \langle y_j| \right) = \sum_j p_j |x_j\rangle \langle x_j|. \tag{22.30}$$

Therefore, $H(X) = -\operatorname{Tr} \rho(X) \log \rho(X) = -\sum_j p_j \log p_j > 0$. We see $H(Y)$ is identical. Therefore,

$$I(X, Y) = -2 \sum_j p_j \log p_j = 2H(X) > H(X). \tag{22.31}$$

23

Oscillators at Low Temperatures

Summary

∗ Quantized and classical harmonic oscillators are compared.
∗ In the $T \to 0$ limit, entropy ceases to change (the third law of thermodynamics).
∗ With a proper dispersion relation quantized harmonic oscillators can explain the specific heat of solids.

Key words

quantum harmonic oscillator, the third law, absolute entropy, Debye model, T^3-law

The reader should be able to:

∗ Explain intuitively (again) why quantization usually reduces specific heats.
∗ Compute the canonical partition function of the quantized harmonic oscillator.
∗ Explain the entropy behavior at low temperatures.
∗ Explain the significance of the third law.

23.1 Ideal Gas with Internal Vibration

If an ideal gas consisting of N molecules has internal degrees of freedom, the Hamiltonian of the gas consists of two parts:

$$H = H_0 + H_i, \tag{23.1}$$

where H_0 is the Hamiltonian of the translational motion (17.3) (i.e., $H_0 = \sum_{i=1}^{N} p_i^2/2m$), and H_i is the Hamiltonian governing the internal degrees of freedom, which is a sum of Hamiltonians h_i governing the internal motions of the ith molecule ($i = 1, \ldots, N$):

$$H_i = \sum_{i=1}^{N} h_i. \tag{23.2}$$

The translational degrees and internal degrees of freedom are not interacting, so they are completely independent (mechanically and statistically). Hence, the canonical partition function reads

$$Z = Z_{\text{ideal}} Z_i, \tag{23.3}$$

where Z_{ideal} is the partition function (20.13) for a monatomic classical ideal gas and Z_i is the "internal" partition function

$$Z_i = z^N \tag{23.4}$$

with

$$z = \operatorname{Tr} e^{-\beta h}. \tag{23.5}$$

Here, the subscript to denote a particular molecule has been dropped, since all the internal partition functions are identical for identical molecules.

If a molecule has a one-dimensional internal molecular vibration as diatomic molecules, h reads

$$h = \left(p^2 + \mu^2 \omega^2 q^2\right)/2\mu, \tag{23.6}$$

where μ is the (effective or reduced) mass of the oscillator and ω is its angular frequency.

23.2 Classical Statistical Mechanics of a Harmonic Oscillator

We already know that classical mechanical handling of molecular vibrations cannot explain the ideal gas specific heat (see **20.7**), but let us first calculate the classical canonical partition function:

$$z = \frac{1}{h} \int_{\mathbb{R}^2} dp dq \, e^{-\beta(p^2 + \mu^2 \omega^2 q^2)/2\mu} = \frac{1}{h} \left(\frac{2\pi\mu}{\beta} \frac{2\pi}{\mu\omega^2\beta} \right)^{1/2} = \frac{k_B T}{\hbar\omega}. \tag{23.7}$$

We know its contribution e to the internal energy from the equipartition of energy; we can confirm it as

$$e = -\frac{\partial \log z}{\partial \beta} = \beta^{-1} = k_B T. \tag{23.8}$$

Thus, as we already know well, for 1 mole of diatomic ideal gas the vibrational contribution to the constant volume specific heat C_V is $N_A k_B = R$. We also know that the actual C_V is by R smaller than the classical result at around room temperature (cf. **20.7**).

23.3 Quantum Statistical Mechanics of a Harmonic Oscillator

The energy levels (eigenvalues) of the quantized harmonic oscillator (23.6) is (**A.29**)

$$\varepsilon = \left(\frac{1}{2} + n \right) \hbar\omega, \ n = 0, 1, 2, \ldots. \tag{23.9}$$

No level is degenerate. Thus, the canonical partition function for a single oscillator reads

$$z_v \equiv \operatorname{Tr} e^{-\beta h} = \sum_{n=0}^{\infty} \exp\left[-\beta \left(\frac{1}{2} + n \right) \hbar\omega \right]. \tag{23.10}$$

Using $(1-x)^{-1} = 1 + x + x^2 + x^3 + \cdots$ ($|x| < 1$), we get the canonical partition function for a single quantized harmonic oscillator:

$$z_v = e^{-\beta\hbar\omega/2}\left(1 - e^{-\beta\hbar\omega}\right)^{-1} = \left(2\sinh\frac{\beta\hbar\omega}{2}\right)^{-1}. \tag{23.11}$$

Note that this tends to the classical result (23.7) in the $T \to \infty$ limit.

23.4 Systems Consisting of Identical Harmonic Oscillators

With regard to internal vibrations the ideal diatomic gas in the above may be understood as a collection of N identical harmonic oscillators. Let us consider, generally, a collection of M noninteracting harmonic oscillators with the individual Hamiltonian (23.6). The canonical partition function reads

$$Z_v = z_v^M, \tag{23.12}$$

where z_v is given by (23.11). This means that if we calculate thermodynamic quantities of a single oscillator, we have only to multiply M to them to obtain the thermodynamic properties of the oscillator collection.

We obtain the internal energy as:[1*]

$$E_v = -M\frac{\partial\log z_v}{\partial\beta} = \frac{M\hbar\omega}{2}\coth\left(\frac{\beta\hbar\omega}{2}\right) = M\left(\frac{1}{2}\hbar\omega + \frac{\hbar\omega}{e^{\beta\hbar\omega}-1}\right). \tag{23.13}$$

Classically, this is just Mk_BT. The contribution of the internal oscillation to the constant volume specific heat is

$$C_{Vv} = Mk_B\left(\frac{\hbar\omega}{k_BT}\right)^2\frac{e^{\beta\hbar\omega}}{(e^{\beta\hbar\omega}-1)^2}. \tag{23.14}$$

If the temperature is sufficiently high, this goes to Mk_B, the same as the classical result. If the temperature is sufficiently low ($\beta\hbar\omega \gg 1$), it is exponentially small. It goes to zero as $e^{-\beta\hbar\omega}$ in the $T \to 0$ limit due to the energy gap; recall the Schottky type specific heat (cf. **18.4**). If we compute $\hbar\omega/k_B$ and convert the vibrational energy quantum for a diatomic molecule into temperature, usually it is easily beyond 1000 K. Thus, quantization correctly explains why the molecular oscillation cannot contribute to the specific heat of the gas around room temperature.

[1*]

$$\coth\frac{x}{2} = \frac{e^{x/2}+e^{-x/2}}{e^{x/2}-e^{-x/2}} = \frac{e^{x/2}-e^{-x/2}+2e^{-x/2}}{e^{x/2}-e^{-x/2}} = 1 + \frac{2}{e^x-1}$$

The first term of the right-most expression of (23.13) is the contribution of the zero point energy, and the second term is equal to the expectation value of the number of excited phonons times the phonon energy $\hbar\omega$ (Chapter 30).

23.5 Einstein's Explanation of Small Specific Heat of Solid at Low Temperatures

Let us consider a solid with N unit cells. Then, there are $M = 3N - 6 \simeq 3N \, (\gg 1)$ vibrational modes in the solid due to the relative motions of the unit cells (see the acoustic mode in Fig. 23.1).[2] If we consider this system classically, the specific heat of this solid is $3Nk_B$. This is actually the value of the specific heat at high temperatures, and is called the *Dulong–Petit law*. However, it was long known that the specific heats of many solids around room temperature are much less than this asymptotic value.

Einstein ascribed this to the quantum effect. As a simple model, he approximated a solid as a collection of identical $M = 3N$ 1D harmonic oscillators (the *Einstein model*). The resultant specific heat is (23.14) with $M = 3N$. This becomes very small for sufficiently low temperatures, agreeing with the empirical result qualitatively.

23.6 Classically Evaluated Entropy Must Have a Serious Problem

The classical difficulty in low temperature specific heats of gases and solids implies that the classically evaluated entropy must have a serious problem at low temperatures. If we treat the harmonic oscillator system classically, its entropy reads

$$S = (E - A)/T = Mk_B + Mk_B \log(k_B T/\hbar\omega) = Mk_B \log T + \text{const.} \quad (23.15)$$

The result implies that in the $T \to 0$ limit $S \to -\infty$. Therefore, to describe a harmonic oscillator at T knowing only the state information at much lower temperature T_0 a large amount of additional information $\sim \log(T/T_0)$ is required, especially if T_0 is close to zero. This must violate a certain fundamental principle: *the finiteness principle* that a finite object with finite energy should require only a finite amount of information for its complete description (if totally isolated).[3]

23.7 The Third Law of Thermodynamics

Nernst (1864–1941) empirically found that all the quasistatic changes of entropy vanish asymptotically in the $T \to 0$ limit. In particular, all the derivatives of entropy vanish as $T \to 0$ (*Nernst's law*).

---

[2] For "mode" see **20.7**.

[3] **Consequence of the finiteness principle** If we combine Boltzmann's principle and this finiteness principle, then Boltzmann's principle must not be applicable to an extremely small phase volumes. That is, there must be "information quantum" that has the dimension of action^{3N}. Thus, notice that the existence of a fundamental quantity with the dimension of action (something like h) is required by the finiteness principle and thermodynamics.

Nernst's motivation was to get rid of indeterminacy in the chemical free energy change. The origin of internal energy E_0 and origin of entropy S_0 may be harmless as long as we discuss E or S separately. However, free energy A contains $E_0 - TS_0$ as a consequence. If we wish to use the Gibbs–Helmholtz equation to determine the free energy change from the heat associated with a reaction, we cannot get rid of an indeterminate term proportional to T. To get rid of this indeterminacy, Nernst demanded that (H is enthalpy in the following formula):

$$\Delta S = (\Delta H - \Delta G)/T \to 0 \tag{23.16}$$

is true in the $T \to 0$ limit for chemical reactions.[4] We adopt the following version generalized by Nernst as *the third law of thermodynamics*:

[III] Reversible change of entropy ΔS vanishes in the $T \to 0$ limit.

This implies that, if S is finite in the $T \to 0$ limit, then we may set the constant value to be zero (the concept of *absolute entropy* due to Planck).

We can readily show that, in contradistinction to classical harmonic oscillators, the entropy of a collection of quantized harmonic oscillators vanishes in the $T \to 0$ limit, using (23.11) and (23.13).

The absolute entropy implies that to describe a microstate of a macroscopic system at $T = 0$, the required information is subextensive, or the number of yes–no questions needed to know a microstate is zero per particle. Some people claim that the third law is a direct consequence of quantum mechanics (e.g., due to the general properties of the ground state). However, the third law is a thermodynamic principle. We must take the thermodynamic limit first, so conclusions based on mechanics of few-body systems are simple-minded.

23.8 Some Consequences of the Third Law

Here the Jacobian technique that will be explained in the next chapter will be used. Thus, the reader may simply skim the assertions and come back after Chapter 24 for the demonstrations.

Specific heats vanish in the $T \to 0$ limit as can easily be seen from the vanishing of any reversible change of S in the limit.

The temperature derivatives of any extensive variables under constant conjugate variables vanish in the $T \to 0$ limit:

$$\left(\frac{\partial X}{\partial T}\right)_x = \frac{\partial(X,x)}{\partial(T,x)} = \frac{\partial(X,x)}{\partial(T,S)}\frac{\partial(T,S)}{\partial(T,x)} = \left(\frac{\partial S}{\partial x}\right)_T. \tag{23.17}$$

Hence, no change of S in the $T \to 0$ limit implies

$$\left(\frac{\partial X}{\partial T}\right)_{x,\,\text{at }T \to 0} = 0. \tag{23.18}$$

---

[4] When Nernst proposed his assertion(s) in 1905, there was no experimental proof. Only after the publication of his assertion, did he collect old and new empirical supports.

We may exchange X and x in this calculation. Furthermore, from (23.17) the temperature change of any isothermal susceptibility vanishes in the $T \to 0$ limit; since $\delta X = \chi_T \delta x$

$$\left(\frac{\partial \delta X}{\partial T}\right)_x = \left(\frac{\partial \chi_T(x)}{\partial T}\right)_x \delta x = 0 \implies \left(\frac{\partial \chi_T}{\partial T}\right)_x = 0. \tag{23.19}$$

Let us conclude the chapter, discussing the real low temperature behavior of insulating solids.

23.9 Empirical Low Temperature Heat Capacity of Solids

The Einstein model of a solid mentioned previously is consistent with the third law. However, the C_V of the Einstein model goes to zero exponentially and is at variance with the empirical result for (insulator) solids:[5]

$$C_V \sim T^3. \tag{23.20}$$

As we already know from the Schottky type specific heat, if there is a gap between the ground state and the first excited state, the low temperature specific heat must behave as $\exp(-\beta\varepsilon)$ for some positive ε. The empirical law (23.20) implies that the actual crystals have no energy gap above the ground state. (In Chapter 35, we will learn that this is an example of a general law related to symmetry breaking associated with phase transitions.)

23.10 A Real 3D Crystal: The Debye Model

In a real crystal there is a distribution in vibrational frequencies as can be inferred from Fig. 23.1. Not all the vibrations contribute significantly to the low temperature heat capacity of solids. Elastic vibrations in a crystal can be classified into two major categories, optical and acoustic (Fig. 23.2). Only the acoustic modes are relevant. We must study the number of acoustic modes with a given angular frequency about ω.

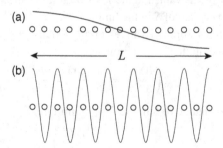

Fig. 23.1 (a) The lowest frequency mode; (b) The highest frequency mode (for a small 1D lattice). L is the lattice size.

[5] As we will learn in Chapter 29 if conducting or freely moving electrons are in the solid their contribution $C_V \sim T$ dominates at low temperatures.

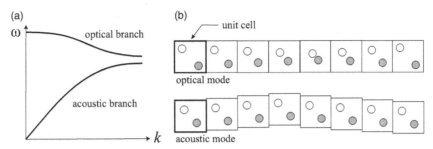

Fig. 23.2 (a) The optical modes do not displace the crystal unit cells, but the acoustic modes (here a transversal mode is depicted) displace unit cells relatively. (b) The dispersion relation (i.e., energy ($\hbar\omega$)–momentum ($\hbar k$) relation). This shows only the acoustic modes can have low energy excitations excitable at lower temperatures.

As can be seen from Fig. 23.1 in 1D direction the possible wavelengths are $\lambda = 2L, 2L/2, \ldots, 2L/N$ or in wavenumbers $k = \pi/L, 2\pi/L, \ldots, N\pi/L$, where L is the linear dimension of the crystal. We have already seen such a sequence before: the de Broglie wavelength of a free particle confined in a box which we used to study the classical ideal gas (see **17.8**; note, however, the boundary conditions are different). This implies we can compute the number of modes in the volume V just as we counted the number of eigenstates as follows.

Let the number of modes with angular frequency between ω and $\omega + d\omega$ be $\mathcal{D}(\omega)$. Then, we use the quantum-classical correspondence of states:

$$\int_0^\omega \mathcal{D}(\omega)d\omega = \frac{1}{h^3}\int d\boldsymbol{r}\int_{|\boldsymbol{p}|\leq p(\omega)} d\boldsymbol{p} = \frac{V}{h^3}\int_0^{p(\omega)} 4\pi p^2 dp. \tag{23.21}$$

If we may assume $p(\omega) = \hbar k = \hbar\omega/c$ (dispersion relation) with c being the speed of sound, the number of modes between ω and $\omega + d\omega$ is, according to (23.21),

$$\mathcal{D}(\omega) = \frac{V}{h^3}4\pi p(\omega)^2\frac{dp(\omega)}{d\omega} = 4\pi V\frac{\hbar^3}{h^3 c^3}\omega^2 = \frac{1}{2\pi^2}V\frac{\omega^2}{c^3}. \tag{23.22}$$

In reality, there are one longitudinal and two transversal modes for each ω, so the actual number of modes between ω and $\omega + d\omega$ is (23.22) times three.[6]

In contrast to the classical gas there are two important differences. $\mathcal{D}(\omega) \propto \omega^2$ holds only for low frequency modes where a crystal may be regarded as an elastic continuum body (for higher frequencies the dispersion relation is much more complicated than $\omega = ck$). Furthermore, the wavelength cannot be indefinitely small (i.e., ω cannot be too large) as seen from Fig. 23.1. Debye introduced the following approximation (the *Debye model*). Here, $\mathcal{D}(w) \propto \omega^2$ is assumed, but $\mathcal{D}(w) = 0$ beyond some cut off ω_D:

$$\mathcal{D}(\omega) = \frac{3V}{2\pi^2 c^3}\omega^2\Theta(\omega_D - \omega), \tag{23.23}$$

where ω_D is the Debye cutoff frequency (which is a materials "constant" to a good approximation) with the overall multiplicative factor three, as noted already, to take care

---

[6] The longitudinal sound speed c_l and the transversal sound speed c_t are generally different, so c is often replaced with the one defined as $3/c^3 = 1/c_l^3 + 2/c_t^3$.

of one longitudinal and two transversal modes for each ω. ω_D is fixed to have the total number of modes (= the total number of lattice cells $N \times 3$) correctly

$$\int_0^{\omega_D} \mathcal{D}(\omega)d\omega = 3N. \tag{23.24}$$

This implies $V\omega_D^3/2\pi^2c^3 = 3N$ or $3V/2\pi^2c^3 = 3 \cdot 3N/\omega_D^3$, so (23.23) can be rewritten as

$$\mathcal{D}(\omega) = \frac{3 \cdot 3N}{\omega_D^3}\omega^2\Theta(\omega_D - \omega). \tag{23.25}$$

23.11 The Debye Model: Thermodynamics

Since we know the contribution of the mode with ω to the internal energy, (23.13), the total energy (the internal energy due to lattice vibration) is

$$E = \int_0^{\omega_D} d\omega\, \mathcal{D}(\omega) \left(\frac{1}{2}\hbar\omega + \frac{\hbar\omega}{e^{\beta\hbar\omega} - 1}\right), \tag{23.26}$$

and the specific heat is given by (cf. (23.14))

$$C_V = k_B \int_0^{\omega_D} d\omega\, \mathcal{D}(\omega) \left(\frac{\hbar\omega}{k_BT}\right)^2 \frac{e^{\beta\hbar\omega}}{(e^{\beta\hbar\omega} - 1)^2}. \tag{23.27}$$

Although the integration range has an upper bound ω_D, when the temperature is low, replacing this with ∞ does not change the integral appreciably. Therefore, C_V behaves just as T^3 in the low temperature limit; the dimension of the integral is $[d\omega][\mathcal{D}][\omega/T]^2 = [\omega]^5/T^2$, and we know $[\hbar\omega/k_BT] = 1$, so the integral must behave as $\propto T^3$.

□ Introducing $\hbar\omega_D/k_B = \Theta_D$ (*Debye temperature*), we can rewrite (23.27) as

$$C_V = 3Nk_B \cdot 3 \left(\frac{T}{\Theta_D}\right)^3 \int_0^{\Theta_D/T} dx\, \frac{x^4e^x}{(e^x - 1)^2}. \tag{23.28}$$

At sufficiently low temperatures we may take the upper bound of the integral to ∞, so we can obtain

$$C_V = 3Nk_B \cdot 3 \left(\frac{T}{\Theta_D}\right)^3 \int_0^{\infty} dx\, \frac{x^4e^x}{(e^x - 1)^2} = Nk_B\frac{12\pi^4}{5}\left(\frac{T}{\Theta_D}\right)^3. \tag{23.29}$$

The actual calculation is usually done through evaluating E:

$$E = 9Nk_B\frac{T^4}{\Theta_D^3}\int_0^{\infty} dx\, \frac{x^3e^{-x}}{1 - e^{-x}} = Nk_B\frac{3\pi^4}{5}\frac{T^4}{\Theta_D^3}, \tag{23.30}$$

where

$$\int_0^{\infty} dx\, x^3\frac{e^{-x}}{1 - e^{-x}} = \int_0^{\infty} dx\, x^3 \sum_{n=1}^{\infty} e^{-nx} = 3! \sum_{n=1}^{\infty} \frac{1}{n^4} = 3!\,\zeta(4) = \frac{\pi^4}{15}. \tag{23.31}$$

Problems

Q23.1 Sackur–Tetrode Equation[7]

The entropy of the classical ideal gas is given by (20.15). It does not satisfy the third law. The microcanonical–canonical ensemble equivalence can be checked easily, and we can understand this unsatisfactory result is due to the crude approximation for lower temperatures. The microcanonical result (17.82) suggests that if T is sufficiently high, the obtained S should give a good approximation of absolute entropy.

(1) Demonstrate that canonical and microcanonical entropy formulas are identical. Since $w(E, V)$ is the phase volume/h^{3N}, it correctly counts the number of the distinguishable microstates with energy and volume close to (E, V), so Boltzmann's principle should give the absolute entropy (for higher temperatures).

Let us use (20.15) for fairly high temperatures. If we know a gas phase whose entropy S_G is known and if the ideal gas equation of state holds, we get from (20.15), using $V/N = k_B T/P$,

$$S_G = Nk_B \left\{ \frac{5}{2} \log T - \log P + \frac{5}{2} + \log \left[k_B^{5/2} \left(\frac{2\pi m}{h^2} \right)^{3/2} \right] \right\}. \tag{23.32}$$

If we use the saturated vapor pressure just above the boiling point, then $S_G = S_L + \Delta H/T$ may be used to estimate S_G, where S_L is the entropy of the liquid phase in equilibrium with the gas at pressure P and temperature T and ΔH the evaporation heat. The following expression of the pressure

$$\log P = -\frac{S_L}{Nk_B} - \frac{\Delta H}{Nk_B T} + \frac{5}{2} \log T + \frac{5}{2} + \log \left[k_B^{5/2} \left(\frac{2\pi m}{h^2} \right)^{3/2} \right] \tag{23.33}$$

is called the *Sackur–Tetrode equation* for the vapor pressure. Using various experimental results, we can check the consistency of the concept of the absolute entropy.

An example follows, where the data are all under 1 atm. The melting point of mercury is 234.2 K and the heat of fusion is 2.33 kJ/mol. The absolute entropy of solid mercury just at the melting temperature is 59.9 J/(K·mol). The entropy increase of liquid between the melting point and the boiling point is 26.2 J/(K·mol). The boiling point is 630 K and the evaporation heat is 59.3 kJ/mol.

(2) Calculate the absolute entropy of mercury gas just above the boiling point.[8]
(3) Assuming that mercury vapor is a monatomic ideal gas, obtain Planck's constant. The reader may use the value of $k_B = 1.38065 \times 10^{-23}$ J/K.

---

[7] Read: Williams, R. (2009). September, 1911, the Sackur–Tetrode equation: how entropy met quantum mechanics, *APS News*: *this month in physics history*, September.
[8] This is taken from Kubo's problem book, *Thermodynamics*, quoted in the self-study guide.

Solution

(1) Using $E = 3Nk_BT/2$ and Stirling's formula for $\Gamma(3N/2)$, it is very easy to see that the two expressions of the (monatomic) ideal gas entropy are identical.

(2) We obtain the absolute entropy of mercury gas at 1 atm just above the boiling point simply by adding required entropy increases:

$$S_G = 59.9 + 2330/234.2 + 26.2 + 59300/630 = 190.1 \ \text{J/(mol·K)}. \tag{23.34}$$

(3) From (23.33) (before S_G is replaced):

$$\log\left\{ k_B^{5/2} \left(\frac{2\pi m}{h^2} \right)^{3/2} \right\} = \frac{S_G}{Nk_B} + \log P - \frac{5}{2}\log T - \frac{5}{2} = 22.84 + 11.52 - 18.61 = 15.74. \tag{23.35}$$

If we use the known parameter values, we get 15.82, an excellent agreement. $m = 0.20059/N_A$ and this gives $h = 6.81 \times 10^{-34}\ \text{J s}$ ($6.623 \times 10^{-34}\ \text{J s}$ is the accepted value).

Q23.2 Specific Heats Near $T = 0$

Show in the $T \to 0$ limit

$$(C_P - C_V)/C_P = o[T], \tag{23.36}$$

if $V\alpha/C_P$ is bounded, where α is the (isobaric) thermal expansivity.

Solution[9]

From $TdS = C_P dT + T(\partial S/\partial P)_T dP$, we get

$$C_V = C_P + T \frac{\partial(S,T)}{\partial(P,T)} \frac{\partial(P,V)}{\partial(T,V)} = C_P + T \frac{\partial(S,T)}{\partial(P,V)} \frac{\partial(P,V)}{\partial(P,T)} \frac{\partial(P,V)}{\partial(P,T)} \frac{\partial(P,T)}{\partial(T,V)} \tag{23.37}$$

$$= C_P - T \left(\frac{\partial V}{\partial T} \right)_P^2 \bigg/ \left(\frac{\partial V}{\partial P} \right)_T = C_P - TV\alpha^2/\kappa_T, \tag{23.38}$$

or

$$(C_P - C_V)/C_P = (T\alpha/\kappa_T)(V\alpha/C_P). \tag{23.39}$$

α vanishes in the $T \to 0$ limit (see (23.18)), and κ_T is the isothermal compressibility, which is nonzero even in the $T \to 0$ limit. This concludes the solution.

---

[9] The Jacobian technique is freely used. See Chapter 24. Also see **24.11** for the relation between entropy and heat capacities.

Summary

∗ The elasticity of a rubber band is understandable in terms of entropic force.
∗ The Jacobian technique allows us to handle derivatives algebraically, helping our understanding of qualitative features (of, e.g., rubber bands).
∗ The Jacobian technique may be fully utilized, if the reader remembers

$$\frac{\partial(X,Y)}{\partial(A,B)} = -\frac{\partial(X,Y)}{\partial(B,A)} = \frac{\partial(Y,X)}{\partial(B,A)} = -\frac{\partial(Y,X)}{\partial(A,B)},$$

$$\frac{\partial(X,Y)}{\partial(Z,W)} = \frac{\partial(X,Y)}{\partial(A,B)}\frac{\partial(A,B)}{\partial(Z,W)},$$

and Maxwell's relation for conjugate pairs (X, x) and (Y, y):

$$\frac{\partial(X,x)}{\partial(y,Y)} = 1.$$

∗ Understanding a rubber band clarifies the principle of adiabatic cooling.

Key words

entropic elasticity, Jacobian, Maxwell's relation, adiabatic cooling, adiabatic demagnetization

The reader should be able to:

∗ Become familiar with the Jacobian technique.
∗ Intuitively explain rubber elasticity; get various signs of partial derivatives; and explain them intuitively.
∗ Explain adiabatic demagnetization.

24.1 Rubber Band Experiment

Let us perform simple experiments on quasistatic adiabatic processes in rubber bands. Prepare a thick rubber band (that is used to bundle, e.g., broccoli heads). We use our lips as a temperature sensor (clean the rubber band). Initially, putting the rubber band to our lips, let us confirm it is around room temperature (slightly cool). Now, hold both ends of a small portion of the band tightly between thumbs and fingers and stretch it quickly (Fig. 24.1).

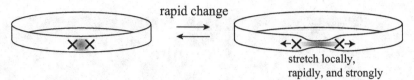

rapid change

stretch locally,
rapidly, and strongly

Fig. 24.1 Holding the band at the ×s firmly between our thumbs and fingers, let us stretch the rubber band locally, rapidly, and strongly to realize (approximately) an adiabatic and quasistatic process.

Then, let us feel the temperature of the stretched portion with our lips. It must be warm. We have just experimentally demonstrated:[1]

$$\left(\frac{\partial T}{\partial L}\right)_S > 0, \tag{24.1}$$

where L is the length of the stretched portion of the rubber band. That is, increasing L (stretching the band) while keeping entropy S (i.e., adiabatically) increases the temperature of the band. The thermodynamic coordinates for the rubber band are E and L, and its Gibbs relation reads

$$dE = TdS + FdL, \tag{24.2}$$

or

$$dS = \frac{1}{T}dE - \frac{F}{T}dL, \tag{24.3}$$

where F is the force (the component of the force parallel to the stretching direction of the rubber band) stretching the band. Since the process is adiabatic and quasistatic, S is constant. Even if we rapidly pull the band, the stretching rate we can realize is very small from the polymer point of view, so the process is (almost) quasistatic. Since the heat conduction is not a very rapid process, during quick stretch the system is virtually thermally isolated (adiabatic). Thus, S is approximately constant.

24.2 Polymer Chain is Just Like Kids Playing Hand in Hand

To begin with, let us try to understand "microscopically" what we have experimentally observed macroscopically. A rubber band is made of a bunch of polymer chains. Take a single chain that is wiggling due to thermal motion (Fig. 24.2(a)).

Stretching the chain corresponds to increasing the distance between the flags of Fig. 24.2(b) of the playing kids analogy. If the chain does not break, the spatial room for dancing is decreased, but since the kids must keep their entropy, the restricted dancing degrees of freedom (conformational entropy) must be compensated by some other degrees of freedom, shaking bodies (momentum space or motional entropy). That is, the

---

[1] If the reader needs a partial derivative review, read **24.6**: Partial Derivative Review and **24.8**: Remarks on the Notation of Partial Derivatives.

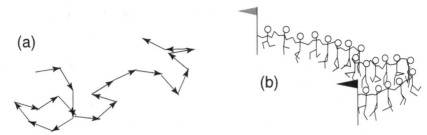

Fig. 24.2   (a): A schematic picture of a single polymer chain. Each arrow is called a monomer. (b): The polymer kid analogy. The temperature represents how vigorously kids are moving around. This also includes "vibration" of individual bodies. The figure is after Saito, N. (1967). *Polymer Physics*, Tokyo: Shokabo (the original picture was due to T. Sakai's lecture according to Saito). The entropy of a chain under constant temperature is monotonically related to the width of the range kids can play around easily, which becomes smaller if the distance between the flags is increased.

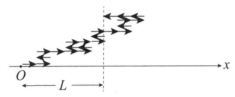

Fig. 24.3   The chain is expanded in the vertical direction to avoid cluttering. Monomers can take the positive or negative x-direction. The right and left directed monomer numbers $N_{\pm}$ ($N_+ + N_- = N$) can be used to compute the number of conformations.

temperature of the system should go up. This suggests that if the chain is stretched under constant T, the entropy of the rubber band should go down:

$$\left(\frac{\partial S}{\partial L}\right)_T < 0. \tag{24.4}$$

If we understand entropy as a measure of microscopic state diversity, this is not hard to guess, since stretching restricts the conformational diversity of the chain.

Can we conclude this from what we have already observed: (24.1)? Yes, we can, but we should first learn a technique to manipulate partial derivatives efficiently.

24.3 Freely Jointed Polymer Chain

Before explaining powerful thermodynamic techniques, let us try to understand the polymer chain system statistical-mechanically. For simplicity, let us consider a polymer chain along the x-axis (Fig. 24.3). We assume that the chain is freely jointed. That is, there is no energy cost to change its conformation at all (just as an ideal gas can change its configuration without any energy cost). The Hamiltonian of this freely jointed polymer consists of the chain kinetic energy K only, which is independent of the conformation of the chain. Thus, E depends only on T, since it is (just the averaged) kinetic energy.

24.4 Freely Jointed Polymer Entropy

We can easily compute the (conformational) entropy of the ideal rubber band using Boltzmann's principle.[2] Following the caption of Fig. 24.3, let us introduce $N_\pm$ so that $N_+ + N_- = N$, and $N_+ - N_- = X \equiv L/\ell$, where ℓ is the monomer length. We have

$$N_\pm = \frac{1}{2}(N \pm X). \tag{24.5}$$

Therefore, the (conformational portion of the) microcanonical partition function reads (note that this is exactly the same problem as the Schottky defect problem to get the entropy seen in **18.2**):

$$w(X) = \binom{N}{N_+}. \tag{24.6}$$

From this, immediately we obtain

$$S = -Nk_B \left[\frac{N_+}{N} \log \frac{N_+}{N} + \frac{N_-}{N} \log \frac{N_-}{N} \right], \tag{24.7}$$

or

$$S = -Nk_B \left[\frac{N+X}{2N} \log \frac{N+X}{2N} + \frac{N-X}{2N} \log \frac{N-X}{2N} \right]. \tag{24.8}$$

With the aid of the Gibbs relation, we find the force (note that $\ell X = L$):

$$F = -\frac{T}{\ell} \left(\frac{\partial S}{\partial X} \right) = \frac{k_B T}{2\ell} \log \frac{N+X}{N-X}. \tag{24.9}$$

This implies

$$L = N\ell \tanh(\beta \ell F). \tag{24.10}$$

Since $\tanh x \simeq x$ for small x, this implies a Hookean spring for small stretches:

$$F = (k_B T/N\ell^2)L. \tag{24.11}$$

That is, $k_B T/\langle R^2 \rangle$ is the spring constant, where $\langle R^2 \rangle = N\ell^2$ is the mean square end-to-end distance of a polymer chain.

24.5 Ideal Rubber Band

Can we explain what we have experienced at the beginning of this section using this entropy? Never. Equation (24.8) implies that if $L = X\ell$ is fixed, S is fixed. This is physically obvious, because the set of allowed conformations is completely determined by L. It is clear that we need the contribution of thermal motion. Then, the entropy should have an extra term S_e dependent on E. Let us assume S_e is not dependent on L (the *ideal rubber* model[3]):

---

[2] First, we consider only the conformation part; there is certainly a momentum contribution, which is not discussed until unit **24.5**.

[3] Recall that the ideal gas entropy (15.16) has the same structure.

$$S = Nk_B \left[\frac{N+X}{2N} \log \frac{N+X}{2N} + \frac{N-X}{2N} \log \frac{N-X}{2N} \right] + S_e(E). \tag{24.12}$$

Notice that S_e is solely determined by thermal motion (just as the corresponding term for the ideal gas; cf. (15.16)).

We can make a more detailed model of a rubber band to compute more realistic entropy, but even without such microscopic details thermodynamics can tell us many qualitative features.

To this end, we must be able to manipulate thermodynamic quantities and derivatives efficiently. We begin with a review of partial differentiation.

24.6 Partial Derivative Review[4]

Consider a two-variable function $f = f(x, y)$.[5] Partial derivatives are defined as:[6]

$$\frac{\partial f}{\partial x} \equiv f_x(x, y) = \lim_{\delta x \to 0} \frac{f(x + \delta x, y) - f(x, y)}{\delta x}, \tag{24.13}$$

$$\frac{\partial f}{\partial y} \equiv f_y(x, y) = \lim_{\delta y \to 0} \frac{f(x, y + \delta y) - f(x, y)}{\delta y}. \tag{24.14}$$

Partial differentiation is extremely tricky in general, however. For example, even if $\partial f / \partial x$ and $\partial f / \partial y$ exist at a point, f can be discontinuous at the same point (e.g., f can jump along a certain line going through the point). Also, E is once continuously differentiable with respect to S and work coordinates, so it is convenient to introduce a stronger concept of differentiability.

Let f be a function of several variables $x = (x_1, \ldots, x_n)$. We could understand f as a function of vector x. We wish to study its "linear response" to the change $x \to x + \delta x$:

$$\delta f(x) = f(x + \delta x) - f(x) = Df(x)\delta x + o[\delta x], \tag{24.15}$$

where o denotes higher order terms that vanish faster than $\|\delta x\|$,[7] when the limit $\delta x \to 0$ is taken. Here, $Df(x)$ is a linear operator: if applied on a vector $a - (a_1, \ldots, a_n)$,

$$Df(x)a = \sum_{i=1}^{n} \frac{\partial f}{\partial x_i} a_i. \tag{24.16}$$

If such a linear map $Df(x)$ is well defined, we say that f is (totally) differentiable (or strongly differentiable). If there are only two variables, we may write

$$Df(x, y)(dx, dy) = \frac{\partial f}{\partial x} dx + \frac{\partial f}{\partial y} dy. \tag{24.17}$$

When we say a function is differentiable in this chapter, it is always in this strong sense.

---

[4] Except at phase transitions thermodynamic functions are differentiable practically as many times as we wish.

[5] Whenever a function is mentioned, its domain should be stated explicitly, but usually we will not do so in this book.

[6] In case we wish to save space, we use the notation $\partial f / \partial x = \partial_x f = f_x$.

[7] Here $\| \ \|$ denotes the norm of a vector; usually, $\|(a_1, \ldots, a_n)\| = \sqrt{a_1^2 + \cdots + a_n^2}$.

24.7 Maxwell's Relations

Let us closely look at $f(x + \delta x, y + \delta y) - f(x, y)$. There are two ways to go from (x, y) to $(x + \delta x, y + \delta y)$, δx first or δy first:

$$f(x + \delta x, y + \delta y) - f(x, y) = f(x + \delta x, y + \delta y) - f(x + \delta x, y) + f(x + \delta x, y) - f(x, y)$$
$$= f_y(x + \delta x, y)\delta y + f_x(x, y)\delta x, \qquad (24.18)$$
$$f(x + \delta x, y + \delta y) - f(x, y) = f(x + \delta x, y + \delta y) - f(x, y + \delta y) + f(x, y + \delta y) - f(x, y)$$
$$= f_x(x, y + \delta y)\delta x + f_y(x, y)\delta y. \qquad (24.19)$$

The difference between these two formulas is:[8]

$$[f_y(x + \delta x, y) - f_y(x, y)]\delta y - [f_x(x, y + \delta y) - f_x(x, y)]\delta x = [f_{yx}(x, y) - f_{xy}(x, y)]\delta x \delta y. \qquad (24.20)$$

This must vanish if the surface defined by f (i.e., the surface $z = f(x, y)$ in 3-space) is at least twice differentiable:[9]

$$f_{xy} = f_{yx}. \qquad (24.21)$$

For example, for a rubber band $dE = TdS + FdL$, so

$$\left(\frac{\partial T}{\partial L}\right)_S = \left(\frac{\partial F}{\partial S}\right)_L. \qquad (24.22)$$

Such relations are called *Maxwell's relations* in thermodynamics.

24.8* Remarks on the Notation of Partial Derivatives

When a multivariate function is written in mathematics, its independent variables are specified, and is written as $f = f(x, y)$. Therefore, if we write a partial derivative $\partial f/\partial x$, it implies that the independent variable(s) other than x is fixed, so usually there is no specification of what variable(s) to fix as

$$\frac{\partial f}{\partial x} = \left(\frac{\partial f}{\partial x}\right)_y. \qquad (24.23)$$

The function f may depend on y more directly through a function $\alpha = \alpha(x, y)$, so it may be more convenient to understand the quantity f as a function of x and α as $f = g(x, \alpha(x, y))$. Accordingly, the partial derivative of f with respect to x is better considered under the condition that fixes α. In mathematics, in such a case a new function symbol $g = g(x, \alpha)$ is introduced and then we compute $\partial g/\partial x$.

In thermodynamics, this $\partial g/\partial x$ is almost always denoted as $(\partial f/\partial x)_\alpha$. That is, f is understood as the symbol for a physical quantity, and not as a function symbol (as discussed further soon). Usually, there is no accompanying specification of independent variables.

---

[8] Notation: $\partial_x(\partial_y f) = \partial_x(f_y) = f_{yx}$.
[9] **Young's theorem** If f_{xy} and f_{yx} are continuous, or if f is twice differentiable, then $f_{xy} = f_{yx}$. (If f_{xy} is not continuous, we have a counterexample.)

Thus, for example, the relevant independent variables for T depend on the situations. On the left-hand side of (24.22) T is a function of L and S, but the temperature change due to the increase in force under constant length $(\partial T/\partial F)_L$ is conceivable. In this case T is understood as a function of L and F. For example, F may be a function of L and S, as α is here. In these two cases, Ts are different functions mathematically, because the set of independent variables are distinct, but in the context of thermodynamics, we respect the identity of T as the same physical observable and use the same symbol consistently. Needless to say, the partial derivative depends on what to fix (or what is to be the set of independent variables, mathematically speaking), so the set of fixed variables must be explicitly indicated. Good examples are $C_V = T(\partial S/\partial T)_V$ and $C_P = T(\partial S/\partial T)_P$ (see **24.11**).

24.9 Jacobian Expression of Partial Derivatives[10]

To manipulate many partial derivatives, it is very convenient to use the so-called *Jacobian technique*. This technique can greatly reduce the insight and skill required in thermodynamics, especially with the Jacobian version of Maxwell's relation (24.39).

The Jacobian for two functions X and Y of two independent variables x, y is defined by the following determinant:

$$\frac{\partial(X,Y)}{\partial(x,y)} \equiv \begin{vmatrix} \left(\frac{\partial X}{\partial x}\right)_y & \left(\frac{\partial X}{\partial y}\right)_x \\ \left(\frac{\partial Y}{\partial x}\right)_y & \left(\frac{\partial Y}{\partial y}\right)_x \end{vmatrix} = \left(\frac{\partial X}{\partial x}\right)_y \left(\frac{\partial Y}{\partial y}\right)_x - \left(\frac{\partial X}{\partial y}\right)_x \left(\frac{\partial Y}{\partial x}\right)_y. \qquad (24.24)$$

In particular, we observe

$$\frac{\partial(X,y)}{\partial(x,y)} = \left(\frac{\partial X}{\partial x}\right)_y, \qquad (24.25)$$

which is the key observation of this technique. Obviously,[11]

$$\frac{\partial(X,Y)}{\partial(X,Y)} = 1. \qquad (24.26)$$

There are only two or three formulas the reader must learn by heart (they are very easy to memorize). One is straightforwardly obtained from the definition of determinants. Exchanging rows or columns switches the sign:

$$\frac{\partial(X,Y)}{\partial(x,y)} = -\frac{\partial(X,Y)}{\partial(y,x)} = \frac{\partial(Y,X)}{\partial(y,x)} = -\frac{\partial(Y,X)}{\partial(x,y)}. \qquad (24.27)$$

---

[10] In this and in **24.10**, upper case (resp., lower case) letters do not necessarily mean extensive (resp., intensive) quantities, but any variables or functions sufficiently differentiable.

[11] In this case, we regard X and Y are independent variables.

What is independent, what is dependent In the Jacobian expression, all the letters appearing upstairs are regarded as dependent variables of the variables appearing downstairs. As noted in **24.8**, in thermodynamics independent variables are often implicit, and we must explicitly indicate them when we compute partial derivatives. As seen here, if the Jacobian scheme is used, the independent variables of the context are always explicit. This is why the Jacobian technique can be mechanical.

24.10 Chain Rule in Terms of Jacobians

If we assume that X and Y are differentiable functions of a and b, and that a and b are differentiable functions of x and y, we have the following multiplicative relation:

$$\frac{\partial(X, Y)}{\partial(a, b)} \frac{\partial(a, b)}{\partial(x, y)} = \frac{\partial(X, Y)}{\partial(x, y)}. \tag{24.28}$$

This is a disguised chain rule:

$$\left(\frac{\partial X}{\partial x}\right)_y = \left(\frac{\partial X}{\partial a}\right)_b \left(\frac{\partial a}{\partial x}\right)_y + \left(\frac{\partial X}{\partial b}\right)_a \left(\frac{\partial b}{\partial x}\right)_y, \tag{24.29}$$

etc. Confirm (24.28) (use $\det(AB) = (\det A)(\det B)$).

The technical significance of (24.28) must be obvious: calculus becomes algebra. We may regard $\partial(A, B)$ just as an ordinary number: formally we can do as follows.[12,13] First, split the "fraction" and then throw in the identical factors we wish to introduce as

$$\frac{\partial(X, Y)}{\partial(x, y)} = \frac{\partial(X, Y)}{\partial(x, y)} = \frac{\partial(X, Y)}{\partial(A, B)} \frac{\partial(A, B)}{\partial(x, y)}. \tag{24.30}$$

From (24.28) we get at once

$$\frac{\partial(X, Y)}{\partial(A, B)} = 1 \left/ \frac{\partial(A, B)}{\partial(X, Y)} \right. . \tag{24.31}$$

In particular, we have:[14*]

$$\left(\frac{\partial X}{\partial x}\right)_Y = 1 \left/ \left(\frac{\partial x}{\partial X}\right)_Y \right. . \tag{24.32}$$

Using these relations, we can easily demonstrate

$$\left(\frac{\partial X}{\partial y}\right)_x = - \left(\frac{\partial x}{\partial y}\right)_X \left/ \left(\frac{\partial x}{\partial X}\right)_y \right. \tag{24.33}$$

as follows:

$$\frac{\partial(X, x)}{\partial(y, x)} \overset{(24.28)}{=} \frac{\partial(X, x)}{\partial(y, X)} \frac{\partial(y, X)}{\partial(y, x)} \overset{(24.27)}{=} - \frac{\partial(x, X)}{\partial(y, X)} \frac{\partial(X, y)}{\partial(x, y)}. \tag{24.34}$$

Then, use (24.31).

A concrete example of (24.33) is

$$\left(\frac{\partial P}{\partial T}\right)_V = - \left(\frac{\partial V}{\partial T}\right)_P \left/ \left(\frac{\partial V}{\partial P}\right)_T \right. , \tag{24.35}$$

which relates thermal expansivity and isothermal compressibility.

---

[12] **"Formal"** Here, "formally" (or "formal derivation") means "only demonstrable as mechanical transformation of formulas without any mathematical justification of the procedures." This is the standard usage of "formal" in mathematics (and throughout this book).

[13] In pragmatic thermodynamics, we can be maximally formal and seldom make any mistake.

[14*] Mathematically properly speaking, on the left-hand side we regard X and Y as functions of x and y, and $Y = Y(x, y)$ is fixed. In contrast, on the right-hand side x and y are understood as a function of X and Y, and Y is being kept constant.

For a rubber band

$$\left(\frac{\partial L}{\partial S}\right)_F = \frac{\partial(L, F)}{\partial(S, F)} = \frac{\partial(L, F)}{\partial(T, F)}\frac{\partial(T, F)}{\partial(S, F)} = \left(\frac{\partial L}{\partial T}\right)_F \left(\frac{\partial T}{\partial S}\right)_F, \tag{24.36}$$

which reads

$$\left(\frac{\partial L}{\partial S}\right)_F = \left(\frac{\partial L}{\partial T}\right)_F \bigg/ \left(\frac{\partial S}{\partial T}\right)_F = T \left(\frac{\partial L}{\partial T}\right)_F \bigg/ C_F. \tag{24.37}$$

Here, C_F is the heat capacity under constant force. It is explained in **24.11**.

24.11 Expression of Heat Capacities

The relation between heat and entropy (Clausius' equality) tells us $d'Q = TdS$, so if we differentiate this with respect to T under constant F, it must be the heat capacity under constant F. Generally speaking, the heat capacity under constant X (which can be extensive or intensive) always has the following expression:

$$C_X = T \left(\frac{\partial S}{\partial T}\right)_X. \tag{24.38}$$

The stability of the equilibrium state implies $C_X \geq 0$ (usually strictly positive). Imagine the contrary. If we inject heat into a system, its temperature goes down, so it sucks more heat from the surrounding world, and further reduces its temperature. That is, such a system becomes a bottomless heat sink. We will learn more consequences of the stability of equilibrium states later (in Chapter 25).

24.12 Maxwell's Relation (Unified)

All the Maxwell's relations can be unified in the following form:[15*]

$$\frac{\partial(X, x)}{\partial(Y, y)} = -1, \tag{24.39}$$

where (x, X) and (y, Y) are conjugate pairs. This is the third equality the reader should memorize. Do not forget that $(-P, V)$ (not (P, V)) is the conjugate pair.

Let us demonstrate this. From $\cdots + xdX + ydY + \cdots$ Maxwell's relation reads

$$\left(\frac{\partial x}{\partial Y}\right)_X = \left(\frac{\partial y}{\partial X}\right)_Y. \tag{24.40}$$

That is,

$$\frac{\partial(x, X)}{\partial(Y, X)} = \frac{\partial(y, Y)}{\partial(X, Y)}. \tag{24.41}$$

[15*]If the reader knows differential forms, this must be trivial: since dE is exact, $d^2E = d(\cdots + xdX + ydY + \cdots) = 0$. If we change only X and Y, $dx \wedge dX = -dy \wedge dY$. The ratio of the infinitesimal areas on both sides is the Jacobian. Thus, the relation is simply due to thermodynamic quantities being state variables (do not misunderstand that the formula is due to conservation laws).

This implies (mere $a/b = c/d \Rightarrow a/c = b/d$)

$$\frac{\partial(x, X)}{\partial(y, Y)} = \frac{\partial(Y, X)}{\partial(X, Y)} = -1. \tag{24.42}$$

Perhaps, the following formula may be better:

$$\frac{\partial(X, x)}{\partial(y, Y)} = 1. \tag{24.43}$$

For example, (24.22) can be obtained as follows:

$$\left(\frac{\partial T}{\partial L}\right)_S = \frac{\partial(T, S)}{\partial(L, S)} = \frac{\partial(T, S)}{\partial(L, F)}\frac{\partial(L, F)}{\partial(L, S)} = \frac{\partial(L, F)}{\partial(L, S)} = \left(\frac{\partial F}{\partial S}\right)_L. \tag{24.44}$$

24.13 Rubber Band Thermodynamics

Equipped with the Jacobian machinery, let us study the rubber band in more detail. The rubber band is elastic because of the thermal motion of the polymer chains. That is, resistance to reducing entropy is the cause of elastic bouncing. It is clearly an example of the entropic force (see Chapter 7): obstructing the natural behavior of molecules causes a "resistance force." Thus, such elasticity is called the *entropic elasticity*.[16] An important feature is that the elastic force increases with T under constant length (which is easily understood from the kid picture Fig. 24.2):

$$\left(\frac{\partial F}{\partial T}\right)_L > 0. \tag{24.45}$$

Is this related to what we have observed (24.1)? Follow the manipulation below (as a practice):

$$0 < \left(\frac{\partial T}{\partial L}\right)_S = \frac{\partial(T, S)}{\partial(L, S)} = \frac{\partial(T, S)}{\partial(L, F)}\frac{\partial(L, F)}{\partial(L, S)} = \frac{\partial(L, F)}{\partial(L, S)} \tag{24.46}$$

$$= \frac{\partial(L, F)}{\partial(T, L)}\frac{\partial(T, L)}{\partial(L, S)} = \frac{\partial(F, L)}{\partial(T, L)}\bigg/\frac{\partial(S, L)}{\partial(T, L)} = \left(\frac{\partial F}{\partial T}\right)_L \frac{T}{C_L}. \tag{24.47}$$

Or

$$0 < \left(\frac{\partial T}{\partial L}\right)_S = \frac{\partial(T, S)}{\partial(L, S)} = \frac{\partial(T, S)}{\partial(T, L)}\frac{\partial(T, L)}{\partial(L, S)} = -\frac{\partial(T, S)}{\partial(T, L)}\frac{T}{C_L} \tag{24.48}$$

$$= -\frac{\partial(F, L)}{\partial(T, L)}\frac{\partial(T, S)}{\partial(F, L)}\frac{T}{C_L} = \left(\frac{\partial F}{\partial T}\right)_L \frac{T}{C_L}. \tag{24.49}$$

That is, basically, we do not need any foresight if we recognize the starting point derivative and the derivative we wish to produce.

---

[16] In contrast, the usual elasticity is called *energetic elasticity*, which is caused by opposing increase in energy.

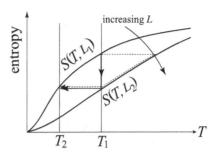

Fig. 24.4 Initially, the system is at T_1. Isothermally, L is increased as $L_1 \to L_2$ (follow the big arrows). This decreases the system entropy. Now, L is returned to the original smaller value adiabatically and reversibly. The entropy is maintained, and the temperature decreases (adiabatic cooling) to T_2. The dotted path is the one we experienced by initial rapid stretching of a rubber band with subsequent cooling (while maintaining the length).

From our microscopic imagination visualized in Fig. 24.2, we guessed

$$\left(\frac{\partial S}{\partial L}\right)_T < 0. \tag{24.50}$$

Let us derive this from (24.1):

$$\left(\frac{\partial S}{\partial L}\right)_T = \frac{\partial(S,T)}{\partial(L,T)} = \frac{\partial(S,T)}{\partial(L,S)}\frac{\partial(L,S)}{\partial(L,T)} = -\left(\frac{\partial T}{\partial L}\right)_S \frac{C_L}{T} < 0. \tag{24.51}$$

24.14 Adiabatic Cooling with a Rubber Band

If a tightly stretched rubber band is equilibrated with room temperature and then suddenly (=adiabatically) relaxed, it will cool (as we can easily feel experimentally). This is not surprising since

$$\left(\frac{\partial T}{\partial L}\right)_S > 0. \tag{24.52}$$

Here, L is reduced under constant S and so must decrease T. This is the principle of *adiabatic cooling* (see Fig. 24.4).

In Fig. 24.4 the entropy curves merge in the $T \to 0$ limit. This is the third law of thermodynamics (see **23.7**). Consequently, we cannot reach $T = 0$ with a finite number of adiabatic cooling processes.

24.15 Cooling via Adiabatic Demagnetization

Unfortunately, we cannot use a rubber band to cool a system to a very low temperature, since it becomes brittle (the chain conformational motion freezes out easily). In actual

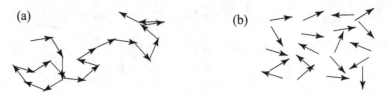

Fig. 24.5 (a), which is a copy of Fig. 24.2(a), corresponds to (b) a paramagnet, a collection of only weakly (ideally not) interacting spins. There is, however, a fundamental difference between spins and rotating monomers; while the monomers have the kinetic energy, spins do not. Thus it is possible to decouple the spin degrees of freedom from ordinary motional degrees of freedom; we can cool the spins alone.

low-temperature experiments, a collection of almost noninteracting magnetic dipoles (i.e., a paramagnetic material) is used. The system is closely related to freely jointed polymer chains as illustrated in Fig. 24.5.

The Gibbs relation of the magnetic system is

$$dE = TdS + BdM, \tag{24.53}$$

where B is the magnetic field (in the "z"-direction), and M the magnetization (the z-component). The correspondences $B \leftrightarrow F$ and $M \leftrightarrow L$ are almost perfect: M is the sum of small magnetic dipole vectors, and L is also the sum of "steps" (monomer orientation vectors) (see Fig. 24.5, and read its caption). Thus, we expect

$$\left(\frac{\partial T}{\partial M}\right)_S > 0 \tag{24.54}$$

and adiabatic cooling can be realized; first apply a strong magnetic field and align all the dipoles. We can do this slowly and isothermally. Then, turn off the magnetic field to make $M \to 0$ (demagnetization). Simply replacing L with M in Fig. 24.4, we can understand this adiabatic demagnetization strategy to cool a system.[17]

24.16 Ideal Magnetic System

We can imagine a collection of noninteracting magnetic dipoles (called spins) taking only up or down (or $s = \pm 1$) values.[18] The total magnetization of the system reads

$$M = \mu \sum_{i=1}^{N} s_i, \tag{24.55}$$

[17] Here, we discussed using the magnet to cool other systems. To do this, the spins must couple with the lattice degrees of freedom. However, if one wishes to cool the spins only, then this coupling is not needed (as discussed in **24.16**).

[18] If magnetic atoms are dilute in an insulating solid, they do not interact with each other appreciably.

where μ is the ratio of the magnetic moment and the spin (the gyromagnetic ratio). It is a good exercise to compute S as a function of M, but, as can be guessed easily from the freely jointed chain case (cf. (24.10)),

$$M = \mu N \tanh \beta \mu B, \tag{24.56}$$

where B is the magnetic field. The relation between M and B for small B corresponds to Hooke's law (24.11):

$$M = (N\mu^2/k_B T)B, \tag{24.57}$$

which is called *Curie's law*: the magnetic susceptibility $\chi = N\mu^2/k_B T$.

Just as in the freely jointed polymer model previously discussed (without the S_e term), there is no kinetic energy of spins (or of the entities carrying spins), so the entropy of this model is constant under constant M. Thus, just as in the ideal rubber model, without the term similar to S_e the model cannot explain the use of adiabatic demagnetization to cool other systems as a refrigerating mechanism. However, if we are interested in the spins themselves, then their coupling to other degrees of freedom (the so-called spin-lattice coupling) should not be large. Under this condition, the system can be described by the present model, so under the adiabatic demagnetization condition M is constant, because S is constant. Thus, since (24.57) or (24.56) implies B/T is constant, reducing B implies decreasing T.[19]

If we use the magnet as a coolant to cool other systems, we are interested in

$$\left(\frac{\partial T}{\partial B}\right)_S = \frac{\partial(T,S)}{\partial(B,S)} = \frac{\partial(T,S)}{\partial(B,M)}\frac{\partial(B,M)}{\partial(B,S)} = -\frac{\partial(B,M)}{\partial(B,S)} \tag{24.58}$$

$$= -\frac{\partial(B,T)}{\partial(B,S)}\frac{\partial(B,M)}{\partial(B,T)} = -\frac{T}{C_B}\left(\frac{\partial M}{\partial T}\right)_B. \tag{24.59}$$

Therefore, if Curie's law of the form $M = a(B/T)$ holds, then

$$\delta T = \frac{aB}{C_B T}\delta B \tag{24.60}$$

gives the cooling rate.

Problems

Q24.1 Basic Problems

(1) With $F = x\sin y$, and $y = x + z$. Express

$$\left(\frac{\partial F}{\partial x}\right)_y \quad \text{and} \quad \left(\frac{\partial F}{\partial x}\right)_z \tag{24.61}$$

in terms of x and y. ($\sin y$; $\sin y + x\cos y$)

---

[19] For a rubber band, reducing F while keeping L constant in order to change T is experimentally unthinkable.

(2) For a gas, PV and E are functions of T only. Show that actually PV/T is a constant. (Compute $(\partial E/\partial V)_T = T(\partial S/\partial V)_T - P = T(\partial P/\partial T)_V - P = 0$.)

(3) For a general gas, find the temperature change dT due to adiabatic free expansion $V \to V + dV$. [Compute $(\partial T/\partial V)_E$. $(\partial T/\partial V)_E = [P - T(\partial P/\partial T)_V]/C_V$]

(4) If M is a function of B/T, E is a function of T only. (Show $(\partial E/\partial B)_T = 0$ using $(\partial B/\partial T)_M = B/T$; quite parallel to (2).)

Q24.2 Negative Temperature is Very Hot

Let us consider a two-state-spin system containing 1 mole of spins (cf. **24.16**). Assume that under the magnetic field B, the energy gap between the up and down spin states is 600 K. Suppose the initial temperature of the magnet is -500 K.[20]

(1) What is the temperature of this system measured with an ideal gas thermometer containing 10^{-6} moles of monatomic gas particles?[21] (Assume the initial temperature of the thermometer is around room temperature, so its intial internal energy may be ignored.)

(2) If, instead, the original negative temperature system is thermally equilibrated with a 1 mole of ideal monatomic gas that is initially 200 K, what is the equilibrium temperature?

Solution

(1) The relation between the magnetization and the temperature can be solved as $(2\mu B/k_B = 600\,\mathrm{K})$

$$M = \mu N_A \tanh \frac{300}{T}. \tag{24.62}$$

Energy conservation tells us ($-\boldsymbol{M} \cdot \boldsymbol{B}$ is the energy of the magnet in the magnetic field):

$$300 N_A k_B \tanh \frac{300}{500} = -300 k_B N_A \tanh \frac{300}{T} + \frac{3}{2} n k_B T. \tag{24.63}$$

In this case we may expect that the temperature is extremely high, so

$$300 N_A \tanh \frac{300}{500} = 161 N_A \simeq \frac{3}{2} nT \implies T = 1.07 \times 10^8\,\mathrm{K}. \tag{24.64}$$

Since the temperature is outrageously high, we must pay attention to relativity. That is, the gas must be superrelativistic. Then, is the temperature higher or lower? (You can of course get the answer quantitatively.)

(2)

$$161 N_A k_B + \frac{3}{2} 200 N_A k_B = -300 N_A k_B \tanh \frac{300}{T} + \frac{3}{2} N_A k_B T. \tag{24.65}$$

---

[20] **Is $T < 0$ possible?** We have already discussed (see **Q14.1** (2)) the impossibility of $T < 0$, if the spatial degrees of freedom are involved. Here, we assume that the spins and lattice vibrations (phonons) are, for a sufficiently long time, decoupled.

[21] In this case, upon contact with the thermometer, the system temperature changes drastically, so the process cannot be a temperature measuring process. Here, we simply wish to know what happens after the drastic change.

That is, we must solve (numerically)

$$361 = -300 \tanh \frac{300}{T} + \frac{3}{2} T. \tag{24.66}$$

That is, $T = 374 \, \text{K}$.

Q24.3 Ideal Rubber Model Example

For a rubber band the tensile force F is given by

$$F = aT \left(\frac{L}{L_0} - \frac{L_0^2}{L^2} \right), \tag{24.67}$$

where a and L_0 are positive constants. Also, the constant length heat capacity C_L is independent of T.

(1) Find the entropy of the band.
(2) If the band is adiabatically and reversibly stretched from the initial length $L = L_0$ at $T = T_0$ to the final $L = 2L_0$, what is the final temperature?

Solution

(1) The Gibbs relation reads $dE = TdS + FdL$, so

$$dS = \frac{1}{T} dE - \frac{F}{T} dL \tag{24.68}$$

and

$$C_L = T \left(\frac{\partial S}{\partial T} \right)_L . \tag{24.69}$$

Thus,

$$dS = \frac{C_L}{T} dT - a \left(\frac{L}{L_0} - \frac{L_0^2}{L^2} \right) dL. \tag{24.70}$$

This is an exact form (or perfect differential), so

$$S(T, L) = S(T_0, L_0) + C_L \log \frac{T}{T_0} - a \left\{ \frac{L^2 - L_0^2}{2L_0} + L_0^2 \left(\frac{1}{L} - \frac{1}{L_0} \right) \right\}. \tag{24.71}$$

(2) Thus

$$S(T, 2L_0) = S(T_0, L_0) + C_L \log \frac{T}{T_0} - aL_0. \tag{24.72}$$

Since the change is adiabatic and reversible,

$$C_L \log \frac{T}{T_0} = aL_0 \tag{24.73}$$

or $T = T_0 e^{aL_0/C_L} > T_0$.

25

Thermodynamic Stability

Summary

* The universal stability (and evolution) criterion $\delta^2 S < 0 \, (> 0)$ is independent of the environmental constraints (say, isothermal, constant volume or not, etc.).
* Thermodynamic stability implies $\partial(X, Y)/\partial(x, y) > 0$ for conjugate pairs (X, x) and (Y, y).
* In equilibrium changes occur in the direction to discourage further changes (Le Chatelier–Braun principle). Our world is stable.

Key words

universal stability criterion, universal evolution criterion, positive definite quadratic form, Le Chatelier principle, Le Chatelier–Braun principle

The reader should be able to:

* Derive the universal stability criterion.
* Understand the significance of our world being stable.
* Give some of the crucial conclusions due to the stability criterion (say, $C_X > 0$, $C_x > C_X$, etc.).

25.1 The Need for Stability Criteria

Intuition tells us for a rubber band that

$$\left(\frac{\partial S}{\partial F}\right)_L > 0. \tag{25.1}$$

Let us check this, starting with our empirical result

$$\left(\frac{\partial T}{\partial L}\right)_S > 0. \tag{25.2}$$

What we should do first is to rewrite the partial derivatives in terms of Jacobians:

$$\left(\frac{\partial S}{\partial F}\right)_L = \frac{\partial(S, L)}{\partial(F, L)}, \quad \left(\frac{\partial T}{\partial L}\right)_S = \frac{\partial(T, S)}{\partial(L, S)}. \tag{25.3}$$

Therefore, we should keep (L, S) and introduce (T, S):

$$\left(\frac{\partial S}{\partial F}\right)_L = \frac{\partial(S, L)}{\partial(F, L)} = \frac{\partial(S, L)}{\partial(T, S)}\frac{\partial(T, S)}{\partial(F, L)} = -\frac{\partial(S, L)}{\partial(T, S)} = \frac{\partial(L, S)}{\partial(T, S)} > 0. \tag{25.4}$$

We have used a Maxwell's relation: $\partial(T, S)/\partial(F, L) = -1$ (see **24.12**).

How about the sign of

$$\left(\frac{\partial S}{\partial P}\right)_V \tag{25.5}$$

for a gas? To increase P under constant V, (usually) we have to raise the temperature, resulting in the increase of entropy, so we guess the sign must be positive. The cleverest approach may be

$$\left(\frac{\partial S}{\partial P}\right)_V = \frac{\partial(S, V)}{\partial(P, V)} = \frac{\partial(S, V)}{\partial(T, -P)}\frac{\partial(T, -P)}{\partial(P, V)} = \frac{\partial(S, V)}{\partial(T, -P)} \bigg/ \left(\frac{\partial V}{\partial T}\right)_P. \tag{25.6}$$

From this we can conclude that (25.5) and $(\partial V/\partial T)_P$ have the same sign, but to understand this statement we need the following inequality resulting from system stability:

$$\frac{\partial(S, V)}{\partial(T, -P)} > 0. \tag{25.7}$$

25.2 Two Kinds of Inequality in Thermodynamics

For a gas $(\partial V/\partial T)_P > 0$ without doubt, but this sign is not due to some reason of principle nature (in contradistinction to the inequality (25.7)). For liquid water below $4\,°C$ under the atmospheric pressure, this derivative is indeed negative.

As we will see in this chapter we encounter two different types of inequalities in thermodynamics; one class is due to some reason of thermodynamic principle, and the other is only due to materialistic accident. According to our ordinary experiences (and also due to the microscopic picture of materials) $(\partial V/\partial T)_P > 0$ looks quite natural, but thermodynamics does not say anything about this sign. That is, even if it is negative, thermodynamics would not complain, and indeed what is not forbidden by thermodynamics does happen. This seems to be the rule.

25.3 Universal Stability Criterion

Clausius told us that if a spontaneous change occurs in an isolated (or in an adiabatic)[1] system,

$$\Delta S \geq 0. \tag{25.8}$$

---

[1] As a mechanical system whether a system is isolated or only adiabatic (i.e., no net heat exchange exists between the system and the rest of the world) is totally different. In contrast, in thermodynamics, the distinction does not matter. The reader must ruminate the implication of this difference between mechanics and thermodynamics.

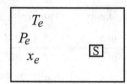

Here S is a part of a huge isolated system whose intensive parameters T_e, P_e, x_e, etc., are kept constant. This is virtually possible because the whole system is huge. Their conjugate extensive quantities S (or heat), V, X, etc., can be freely exchanged between the system S and the rest.

We use the standard trick to study a non-isolated system, S, as a small part of a huge isolated system (see Fig. 25.1) whose intensive variables are kept constant, but their conjugate extensive variables may be exchanged freely between S and its surrounding huge portion acting as a reservoir.

If something spontaneous can happen, the total entropy must increase. In the system something irreversible might have happened, so we cannot compute ΔS directly with the aid of imported quantities $\Delta E, \Delta V$, etc. However, for the reservoir, since we assume it is always in equilibrium, we can write its entropy change as

$$\Delta S_{\text{res}} = -\frac{1}{T_e}\Delta E - \frac{P_e}{T_e}\Delta V + \frac{x_e}{T_e}\Delta X. \tag{25.9}$$

Here, ΔE, etc., are the quantities seen from system S's perspective ("+ signs" for importing to S), so $-\Delta E, -\Delta V$, etc., are the quantities imported by the reservoir. That is why the signs in (25.9) are opposite to the usual Gibbs relation. Thus, the total entropy change is $\Delta S + \Delta S_{\text{res}}$, and the Clausius' inequality for the isolated system leads us to

$$\Delta S - \frac{1}{T_e}\Delta E - \frac{P_e}{T_e}\Delta V + \frac{x_e}{T_e}\Delta X + \frac{\mu_e}{T_e}\Delta N \geq 0. \tag{25.10}$$

Here, although we have not yet discussed the change of number N of particles and its conjugate variable μ (chemical potential), since it is formally quite similar to other terms, for the later convenience, the last term is added. This will be discussed in Chapter 27.

If the equilibrium state is stable, then

$$\Delta S - \frac{1}{T_e}\Delta E - \frac{P_e}{T_e}\Delta V + \frac{x_e}{T_e}\Delta X + \frac{\mu_e}{T_e}\Delta N < 0. \tag{25.11}$$

Now, let us look at ΔS more closely. If the changes are small, we can Taylor-expand ΔS into a power series of $\delta E = \Delta E$, $\delta V = \Delta V$, etc. (here Δ for the independent variables is replaced by δ to make it clear that all changes in the independent variables are small). We can separate the entropy change into the first-order small quantity δS, the second-order small quantity $\delta^2 S$, etc., as

$$\Delta S = \delta S + \delta^2 S + \cdots. \tag{25.12}$$

The first-order term reads just the Gibbs relation:

$$\delta S = \frac{1}{T_e}\delta E + \frac{P_e}{T_e}\delta V - \frac{x_e}{T_e}\delta X - \frac{\mu_e}{T_e}\delta N, \tag{25.13}$$

because the derivatives are computed around the original equilibrium state. Combining this expression, (25.11), and (25.12), we see the first-order terms cancel each other. We conclude that the (local) stability condition for the equilibrium state is

$$\delta^2 S < 0 \qquad (25.14)$$

irrespective of the constraints imposed on the system S (that is, independent of whether some extensive quantities are allowed to be exchanged or not). Thus, this is the *universal stability condition* for the equilibrium state. This is not new to us, however. Recall the concavity (equal to convexity upward) of S discussed in **14.3**. Equation (25.14) is merely its local form.

Notice that (25.14) is true for isolated systems due to the entropy maximization principle, but here it is about the general non-isolated system, so this *does not* imply maximum entropy; irrespective of S being maximum or not, $\delta^2 S < 0$ is the stability condition.[2]

25.4 Universal Stability Criterion in Terms of Internal Energy

Equation (25.11) may be rearranged as

$$\Delta E > T_e \Delta S - P_e \Delta V + x_e \Delta X + \mu_e \Delta N. \qquad (25.15)$$

Now, restricting the variations in the independent variables to be small, we can Taylor-expand ΔE just as we did for ΔS. The reader should immediately realize that a very similar logic as previously used can provide another, but equivalent universal stability criterion

$$\delta^2 E > 0. \qquad (25.16)$$

This is not new, either, since it is a local form of the convexity of the internal energy discussed in Chapter 14.

25.5 Stability and Convexity

As noted previously the universal stability conditions are essentially due to the convexity of the internal energy (or $-S$). If the thermodynamic quantities are sufficiently differentiable, the conditions reduce to the principles discussed in **25.6–25.8**. However, even if E loses twice differentiability (due to phase transitions), still the stability requires $\Delta X \Delta x \geq 0$ (see **Q25.1**(3,4); **31.2**).

---

[2] **Perturbations can be spatially nonuniform** δ here is, officially, a spatially uniform perturbation. However, as noted here and in **26.8** and the subsequent units (**26.9–26.13**), it is universal, so δ can be applied to a small portion (even to a mesoscopic portion approximately) of a system, so δ can be a spatially nonuniform perturbation, and most of them can be spontaneously realized by thermal fluctuations.

25.6 Le Chatelier's Principle

Let us study the consequences of the stability criterion (25.16): a general expression is

$$\delta^2 E = \sum_{i,j} \frac{\partial^2 E}{\partial X_i \partial X_j} \delta X_i \delta X_j > 0. \tag{25.17}$$

This is a positive definite quadratic form, and we can express it explicitly in terms of the Hessian matrix as (for example, with three variables):

$$(\delta S, \delta V, \delta N) \begin{pmatrix} \left(\frac{\partial T}{\partial S}\right)_{V,N} & \left(\frac{\partial T}{\partial V}\right)_{S,N} & \left(\frac{\partial T}{\partial N}\right)_{S,V} \\ -\left(\frac{\partial P}{\partial S}\right)_{V,N} & -\left(\frac{\partial P}{\partial V}\right)_{S,N} & -\left(\frac{\partial P}{\partial N}\right)_{S,V} \\ \left(\frac{\partial \mu}{\partial S}\right)_{V,N} & \left(\frac{\partial \mu}{\partial V}\right)_{S,N} & \left(\frac{\partial \mu}{\partial N}\right)_{S,V} \end{pmatrix} \begin{pmatrix} \delta S \\ \delta V \\ \delta N \end{pmatrix} > 0. \tag{25.18}$$

Let us assume N to be constant for simplicity. The sign of (25.18) must always be positive irrespective of the choice of the variations: δS, etc. (unless all are zero). Therefore, all the diagonal terms must be positive:

$$0 < \left(\frac{\partial T}{\partial S}\right)_V = \frac{T}{C_V}, \tag{25.19}$$

and in terms of the adiabatic compressibility $\kappa_S = -(\partial V/\partial P)_S/V$

$$0 < -\left(\frac{\partial P}{\partial V}\right)_S = \frac{1}{V\kappa_S}. \tag{25.20}$$

The reader must imagine what happens if these signs are flipped. The diagonal inequalities are called *Le Chatelier's principle*.[3] We can state the consequence in words as follows:

> In equilibrium changes occur in the direction to discourage further changes (to avoid run-away processes).

For example, if ΔS is injected (e.g., heat is injected) into the system, its temperature goes up, which makes further injection of heat more difficult. In the case of compressibility, decrease of the volume of the system increases the pressure, further squishing of the system becomes harder. Thus, Le Chatelier's principle assures us that no runaway phenomenon is realized near equilibrium.

25.7 The Le Chatelier–Braun Principle

Which is larger, the susceptibility under constant extensive quantities or that under constant intensive quantities? An example is: which is larger, C_P or C_V? A general answer is given by the Le Chatelier–Braun principle. Since $dX = \left(\frac{\partial X}{\partial x}\right)_y dx + \left(\frac{\partial X}{\partial y}\right)_x dy + \cdots$,

---

[3] Henry Louis Le Chatelier (1850–1936): Le Chatelier, H. L. (1884). Sur un énoncé général des lois des équilibres chimiques, *Compt. rend.*, **99**, 786–789 (On a general statement of the laws of the chemical equilibrium).

$$\left(\frac{\partial X}{\partial x}\right)_Y = \left(\frac{\partial X}{\partial x}\right)_y + \left(\frac{\partial X}{\partial y}\right)_x \left(\frac{\partial y}{\partial x}\right)_Y \tag{25.21}$$

$$= \left(\frac{\partial X}{\partial x}\right)_y + \left(\frac{\partial X}{\partial y}\right)_x \frac{\partial(y, Y)}{\partial(x, Y)} \tag{25.22}$$

$$= \left(\frac{\partial X}{\partial x}\right)_y + \left(\frac{\partial X}{\partial y}\right)_x \frac{\partial(y, Y)}{\partial(X, x)} \frac{\partial(X, x)}{\partial(y, x)} \frac{\partial(y, x)}{\partial(x, Y)} \tag{25.23}$$

$$= \left(\frac{\partial X}{\partial x}\right)_y - \left(\frac{\partial X}{\partial y}\right)_x^2 \left(\frac{\partial y}{\partial Y}\right)_x. \tag{25.24}$$

Here, a Maxwell's relation (see **24.12**) has been used. This implies that:[4*]

$$\left(\frac{\partial X}{\partial x}\right)_y > \left(\frac{\partial X}{\partial x}\right)_Y \text{ or } \left(\frac{\partial x}{\partial X}\right)_Y > \left(\frac{\partial x}{\partial X}\right)_y. \tag{25.25}$$

Thus, when an extensive quantity X is perturbed, the response of its conjugate x is reduced, if the indirect change in other extensive quantities (of Y in (25.25), the right-hand formula) is allowed (*Le Chatelier–Braun's principle*).[5] As already noted, a typical example is $C_P \geq C_V$: larger specific heat implies that it is harder to warm up, that is, the system becomes harder to heat up if the volume change (i.e., the indirect extensive change) is allowed.

25.8 The 2 × 2 Stability Criterion

A necessary and sufficient condition for (25.18) is the positivity of all the principal minors of the Hessian matrix in (25.18).[6] Therefore, in particular,

$$\frac{\partial(T, -P)}{\partial(S, V)} > 0. \tag{25.26}$$

[4*] It might be obvious to the reader, but we have not shown $(\partial x/\partial X)_y > 0$. We know $(\partial y/\partial Y)_X > 0$ from (25.17). Then, we have

$$\frac{\partial(x, y)}{\partial(X, y)} = \frac{\partial(x, y)}{\partial(X, Y)} \frac{\partial(X, Y)}{\partial(X, y)} = \frac{\partial(x, y)}{\partial(X, Y)} \Big/ \left(\frac{\partial y}{\partial Y}\right)_X > 0.$$

Actually, we need (25.27). Generally speaking, irrespective of the variables to keep constant diagonal elements such as $(\partial y/\partial Y)... > 0$.

[5] Braun, K. F. (1888). Über einen allgemeinen qualitativen Satz für Zustandsänderumgen nebst einigen sick anschliessenden Bemerkungen, insbesondere über nicht eindeutige Systeme, *Ann. Physik*, **269**, 337–353. The history of this principle can be found in de Heer, J. (1957). The principle of Le Chatelier and Braun, *J. Chem. Educ.*, **34**, 375–380. The form stated here is due to Ehrenfest.

K. F. Braun Karl Ferdinand Braun (1850–1918) is the inventor of the cathode-ray tube, the discoverer of principle of semiconductor diode, and shared a Nobel prize with Marconi for wireless technology.

[6] **Principal minors** When making a minor of a given determinant (= the determinant of a square submatrix), one can sample the rows and columns with the same numbers (say, 1, 3, 7, and 8 columns and 1, 3, 7, and 8 rows from the original matrix and make a determinant $\det(a_{ij})$, where $i, j \in \{1, 3, 7, 8\}$). Such minors are called *principal minors*.

Generally,

$$\frac{\partial(X, Y)}{\partial(x, y)} > 0, \tag{25.27}$$

where (x, X) and (y, Y) are conjugate pairs. This is perhaps the last formula the reader should remember when they use the Jacobian technique.

Problems

Q25.1 Elementary Problems Related to Le Chatelier–Braun

(1) Which specific heat is larger, C_B or C_M (under constant magnetic field, and under constant magnetization, respectively)?

(2) What can we say about the signs of the following partial derivatives for a fluid? Which signs are fixed by the stability of the equilibrium state?

$$\text{(i)}\ \left(\frac{\partial P}{\partial V}\right)_T, \quad \text{(ii)}\ \left(\frac{\partial T}{\partial V}\right)_S. \tag{25.28}$$

(3) Suppose a phase transition from phase I to phase II occurs upon increasing the magnetic field in the z-direction. What can we say about the relation between the magnetizations of the phases?

(4) Suppose there is a liquid that crystallizes upon heating. Discuss the latent heat for this transition.[7]

Solution

(1) This can be answered with the aid of Le Chatelier–Braun's principle:

$$\left(\frac{\partial X}{\partial x}\right)_y > \left(\frac{\partial X}{\partial x}\right)_Y. \tag{25.29}$$

Thus,

$$\left(\frac{\partial S}{\partial T}\right)_B = \frac{C_B}{T} > \left(\frac{\partial S}{\partial T}\right)_M = \frac{C_M}{T}. \tag{25.30}$$

That is, $C_B > C_M$: under constant B, M is reorganized to absorb more heat.

(2) (i) This is a diagonal element, so the sign is definite (< 0).

(ii)

$$\frac{\partial(T, S)}{\partial(V, S)} = \frac{\partial(V, T)}{\partial(V, S)}\frac{\partial(T, S)}{\partial(V, T)} = \frac{\partial(V, T)}{\partial(V, S)}\frac{\partial(T, S)}{\partial(P, V)}\frac{\partial(P, V)}{\partial(V, T)} \tag{25.31}$$

---

[7] Tombari, E., et al. (2005). Endothermic freezing on heating and exothermic melting on cooling, *J. Chem. Phys.*, **123**, 051104: α-cyclodextrin + water + 4-methylpyridine (molar ratio of 1:6:100). For this system a liquid's endothermic freezing on heating and the resulting crystal's exothermic melting on cooling occur. C_P decreases on freezing and increases on melting. Melting on cooling takes longer than freezing on heating.

Therefore, the sign is the same as $-(\partial P/\partial T)_V$, but this has no definite sign (consider water around 1 °C under 1 atm).

(3) Recall **25.5**. The internal energy must be convex, so the susceptibility must be non-negative, if M is differentiable with respect to B. At the phase transition this is not usually the case, but still the convexity must hold, so M_z must increase in the second phase.

(4) The original paper contains the answer.

Q25.2 More Stability-Related Questions

(1) What is the sign of

$$\left(\frac{\partial S}{\partial V}\right)_T \tag{25.32}$$

for liquid water below 4 °C (around the atmospheric pressure)?

(2) Using the experimental result that the reader can confirm for a rubber band (i.e., $(\partial T/\partial L)_S > 0$; see Chapter 24), find the sign of

$$\left(\frac{\partial L}{\partial S}\right)_F. \tag{25.33}$$

Then, give an intuitive explanation of its sign.

(3) There is an elastic body for which $(\partial S/\partial L)_T > 0$. Find the sign of

$$\left(\frac{\partial L}{\partial T}\right)_F, \tag{25.34}$$

where F is the tensile force, and L the length.

Solution

(1) First, write the partial derivative in terms of a Jacobian and then split the denominator and the numerator:

$$\left(\frac{\partial S}{\partial V}\right)_T = \frac{\partial(S,T)}{\partial(V,T)} = \frac{\partial(S,T)}{\partial(V,T)}\frac{---}{---}. \tag{25.35}$$

Now, look at what is given or what we wish: it is $(\partial V/\partial T)_P < 0$:

$$\frac{\partial(V,P)}{\partial(T,P)} < 0. \tag{25.36}$$

Since (25.35) already contains (V,T), we should introduce (V,P) to utilize a Maxwell's relation:

$$\frac{\partial(S,T)}{---}\frac{---}{\partial(V,T)} = \frac{\partial(S,T)}{\partial(V,P)}\frac{\partial(V,P)}{\partial(V,T)} = \frac{\partial(V,P)}{\partial(V,T)}. \tag{25.37}$$

We must introduce (T,P):

$$\frac{\partial(V,P)}{\partial(V,T)} = \frac{\partial(V,P)}{\partial(T,P)}\frac{\partial(T,P)}{\partial(V,T)} = -\left(\frac{\partial P}{\partial V}\right)_T\left(\frac{\partial V}{\partial T}\right)_P < 0, \tag{25.38}$$

because the isothermal compressibility must be positive. Thus, we have shown that isothermally shrinking the volume increases the entropy. An intuitive explanation for this behavior of water is in terms of the hydrogen-bond network that is spatially not very dense. Squishing breaks the network, increasing the system entropy.

(2)

$$\left(\frac{\partial L}{\partial S}\right)_F = \frac{\partial(L, F)}{\partial(S, F)} = \frac{\partial(L, F)}{\partial(T, S)}\frac{\partial(T, S)}{\partial(S, F)} \tag{25.39}$$

$$= \frac{\partial(T, S)}{\partial(L, S)}\frac{\partial(L, S)}{\partial(S, F)} = -\left(\frac{\partial T}{\partial L}\right)_S\left(\frac{\partial L}{\partial F}\right)_S < 0. \tag{25.40}$$

Here, a Maxwell's relation

$$\frac{\partial(L, F)}{\partial(T, S)} = 1, \tag{25.41}$$

has been used. Increase in entropy means enhancing polymer wiggling. Thus polymer ends tend to be pulled in.

(3)

$$\left(\frac{\partial L}{\partial T}\right)_F = \frac{\partial(L, F)}{\partial(T, F)} = \frac{\partial(L, F)}{\partial(T, S)}\frac{\partial(T, S)}{\partial(T, F)} = \frac{\partial(T, S)}{\partial(T, F)} \tag{25.42}$$

$$= \frac{\partial(T, S)}{\partial(T, L)}\frac{\partial(T, L)}{\partial(T, F)} > 0. \tag{25.43}$$

This is a typical situation for energetic elastic bodies.

26

Fluctuation and Response

Summary

∗ A generalized Gibbs free energy: $\tilde{G}(T, x) = -k_B T \log \operatorname{Tr} e^{-\beta(H - x\hat{X})}$ is used to compute susceptibilities.
∗ Susceptibilities are directly related to the second moments of spontaneous fluctuations (fluctuation–response relation).
∗ A universal probability distribution for mesoscopic fluctuations in equilibrium is governed by the second variation of entropy.
∗ Equilibrium work needed to create fluctuations is closely related to this universal distribution.

Key words

generalized Gibbs free energy, generalized canonical partition function, fluctuation–response relation, thermodynamic fluctuations (Einstein's theory)

The reader should be able to:

∗ Construct a convenient partition function to obtain a convenient thermodynamic potential directly.
∗ Recognize important conclusions of the fluctuation–response relation.
∗ Explain why fluctuation studies are important.
∗ Compute second moments of fluctuations, or use Einstein's theory.

26.1 The Importance of Fluctuations

We already know that the mesoscopic world is dominated by fluctuations. This allows us to glimpse the atomic world underlying the world we experience daily. We also know that the equilibrium state of a macroscopic system is always tested by thermal fluctuations; the so-called virtual change denoted by δ in the stability criterion is actually spontaneously realized by thermal fluctuations.

Thus, there is no doubt about the importance of fluctuations qualitatively. We will see that the system response to perturbation is quantitatively related to spontaneous fluctuations. Observing fluctuations is a major non-invasive means to study system responses against (semi)macroscopic perturbations.

26.2 The Generalized Gibbs Free Energy

Take a finite system and observe a work coordinate X. We assume that the system is maintained at temperature T. Let us look at the response of X to the modification of its conjugate variable x (with respect to energy). We wish to study the susceptibility

$$\chi = \left(\frac{\partial X}{\partial x}\right)_{T,\dots}. \tag{26.1}$$

Here, ... depends on the system we study. Since we wish to use $T, x, \dots$ as independent variables, we wish to have a system that can freely exchange their conjugate extensive quantities with its environment. Then, it is convenient to use the thermodynamic potential $\tilde{G}$ defined by the following Legendre transformation (see Chapter 16):

$$E \to \tilde{G} = E - TS - xX = A - xX. \tag{26.2}$$

or (convex-analytically) more conveniently

$$-\tilde{G} = \max_{S,X}[TS + xX - E]. \tag{26.3}$$

where $\tilde{G}$ is a *generalized Gibbs free energy*. Recall the original Gibbs free energy $G = A - (-P)V$ (that is, $-G = \max_{S,V}[ST + (-P)V - E]$).

26.3 Generalized Canonical Formalism

Analogy with the canonical formalism (see **18.14**, esp., (18.62)) tells us

$$-\tilde{G} = k_B T \log e^{\beta \max[TS + xX - E]} = k_B T \log \operatorname{Tr} e^{-\beta(H - x\hat{X})}, \tag{26.4}$$

where $\hat{X}$ is the observable corresponding to X (or classically, it is simply a microscopic expression of X). Thus, the generalized canonical formalism for $\tilde{G}$ requires the following *generalized canonical partition function*:

$$\tilde{Z}(T, x) = \operatorname{Tr} e^{-\beta(H - x\hat{X})} \tag{26.5}$$

and we can write

$$\tilde{G}(T, x) = -k_B T \log \tilde{Z}(T, x). \tag{26.6}$$

These relations look just like the ones we are very familiar with. The Gibbs relation now reads (including as usual μ and N; recall $\tilde{G} = A - xX$):

$$d\tilde{G} = -SdT - PdV - Xdx + \mu dN \tag{26.7}$$

or (recall $dS/k_B = \beta dE + \beta PdV - \beta xdX - \beta\mu dN$):

$$d\log \tilde{Z} = -Ed\beta + \beta PdV + Xd(\beta x) - \beta\mu dN. \tag{26.8}$$

26.4 Fluctuation–Response Relation

Equation (26.8) tells us that the susceptibility $\chi = (\partial X/\partial x)_{T,V}$ of the response X to the change of x reads (N is omitted from the subsequent formulas):

$$\chi = \beta \left(\frac{\partial^2 \log \tilde{Z}}{\partial(\beta x)^2} \right)_{T,V}. \tag{26.9}$$

Let us compute this, using the expression for the generalized canonical partition function. First, we obtain (in the following we use the classical approximation, so Tr is replaced by $\sum$ over microstates):

$$X = \left(\frac{\partial \log \tilde{Z}}{\partial \beta x} \right)_{T,V} = \frac{1}{\tilde{Z}} \sum \hat{X} e^{-\beta H + \beta x \hat{X}}. \tag{26.10}$$

Let us differentiate (26.10) once more:

$$\chi = \beta \left(\frac{\partial X}{\partial \beta x} \right)_{T,V} = -\frac{\beta}{\tilde{Z}^2} \left(\sum \hat{X} e^{-\beta H + \beta x \hat{X}} \right)^2 + \frac{\beta}{\tilde{Z}} \sum \hat{X}^2 e^{-\beta H + \beta x \hat{X}}. \tag{26.11}$$

That is,

$$= \beta \left(\langle \hat{X}^2 \rangle - \langle \hat{X} \rangle^2 \right) = \beta \langle \delta \hat{X}^2 \rangle \geq 0, \tag{26.12}$$

where $\delta \hat{X} = \hat{X} - \langle \hat{X} \rangle$.

$$\chi = \beta \langle \delta \hat{X}^2 \rangle \tag{26.13}$$

is called the *fluctuation–response relation*. As seen from its derivation the both sides of (26.13) are subjected to the identical constraints (e.g., T and V are fixed as in the present case).

26.5 Three Key Observations about Fluctuation–Response Relations

We can make three important observations from the fluctuation–response relation (26.13):

(i) The "ease" of response results from "large" "natural" fluctuations. Notice that χ describes the response to an external perturbation, but the variance of $\hat{X}$ is due to spontaneous thermal fluctuations. Gentle nudging of the system (reversible change) to measure χ just probes the spontaneity of the system.

(ii) Since X is extensive and x intensive, χ must be extensive (proportional to the number of particles in the system, N). Therefore, $\delta X = O[\sqrt{N}]$; the magnitude of fluctuations are proportional to the square root of the system size.

(iii) χ cannot be negative. This is the manifestation of the stability of the equilibrium state as we have already discussed (in Chapter 25).

We have realized that studying fluctuations is quite important; it could realize a non-invasive method to study the system response without any perturbation.

Can the reader guess the general formula when there are many variables?

26.6 Extension to Multivariable Cases

It is natural to guess the following form:

$$\chi_{ij} \left(= \frac{\partial X_i}{\partial x_j} \right) = \beta \langle \delta \hat{X}_i \delta \hat{X}_j \rangle. \tag{26.14}$$

Classically, this is correct. We introduce the following generalized canonical partition function

$$\tilde{Z}(T, \{x_i\}) = \operatorname{Tr} e^{-\beta H + \beta x_i \hat{X}_i + \beta x_j \hat{X}_j}, \tag{26.15}$$

which gives the following generalized Gibbs free energy:

$$\tilde{G} = -k_B T \log \tilde{Z}(T, x_i, x_j) \tag{26.16}$$

with the Gibbs relation: $d\tilde{G} = -S dT - X_i dx_i - X_j dx_j + \cdots$.

The susceptibility $\chi_{ij} = (\partial X_i / \partial x_j)_{T, x_j}$ (here, extensive quantities other than S, X_i and X_j are fixed but not written) may be computed as follows: since

$$X_i = \left(\frac{\partial \log \tilde{Z}}{\partial \beta x_i} \right)_{T, x_j} = \frac{1}{\tilde{Z}} \sum \hat{X}_i e^{-\beta H + \beta x_i \hat{X}_i + \beta x_j \hat{X}_j}, \tag{26.17}$$

differentiating this once more gives us (the calculation here is a classical approximation)

$$\chi_{ij} = \beta \frac{\partial X_i}{\partial \beta x_j} = -\frac{\beta}{\tilde{Z}^2} \left(\sum \hat{X}_i e^{-\beta H + \beta x_i \hat{X}_i + \beta x_j \hat{X}_j} \right) \left(\sum \hat{X}_j e^{-\beta H + \beta x_i \hat{X}_i + \beta x_j \hat{X}_j} \right)$$
$$+ \frac{\beta}{\tilde{Z}} \sum \hat{X}_i \hat{X}_j e^{-\beta H + \beta x_i \hat{X}_i + \beta x_j \hat{X}_j}. \tag{26.18}$$

That is,

$$\chi_{ij} = \beta \left(\langle \hat{X}_i \hat{X}_j \rangle - \langle \hat{X}_i \rangle \langle \hat{X}_j \rangle \right) = \beta \langle \delta \hat{X}_i \delta \hat{X}_j \rangle \geq 0, \tag{26.19}$$

where $\delta \hat{X}_i = \hat{X}_i - \langle \hat{X}_i \rangle$. This is the most general fluctuation–response relation.[1]

26.7 Onsager's Regression Hypothesis and the Green–Kubo Formula

We have seen the relation between the susceptibility and correlation functions. Onsager realized a nonequilibrium analog, a relation between transport coefficients and

---

[1] **Note on the quantum version** The reader should have realized that the derivation of (26.13) and (26.14) does not work generally for quantum mechanical cases, because noncommutative Taylor expansion is not simple, but still an analogous formula holds.

time-correlation functions, through his regression hypothesis. Its large deviation theoretical formulation is outlined in Appendix 26A.

26.8 Mesoscopic Fluctuations: Introduction to Einstein's Theory[2]

The story up to this point is about a whole finite system. If we wish to know the fluctuations for a mesoscopic portion of a system (see Fig. 26.1 later), we may wish to avoid the detailed setup given above that requires detailed specification of the ensemble (e.g., what extensive variables to fix). Recall the universal stability criterion for an equilibrium state $\delta^2 S < 0$ in Chapter 25. This is independent of the environmental constraints, and is a condition that prevents fluctuations from going wild. Then, there should be a theory that can determine the distribution of mesoscopic fluctuations independent of the environmental constraints imposed on the system. As we will see, Einstein constructed just such a theoretical framework.

The study of fluctuations is a mesoscopic scale study of a system, so it is the study of large deviation (recall **1.11** and **10.8**). Since we study a small (mesoscale) volume V in the system, the following type of large deviation principle must be natural:

$$P\left(\frac{1}{V}X(V) \sim v(y)\right) \approx e^{-VI(y)}, \tag{26.20}$$

where $X(V)/V$ is the average over volume V,[3] $v(y)$ is the volume element around y (which we may write dy) and I is the large deviation function (or rate function). If we know I, we know everything we wish to know about fluctuations. In practice, the volume is sufficiently large from the microscopic point of view. Therefore, fluctuations should not be very large, and we have only to consider the second moments to quantify fluctuations. That is, we need the quadratic approximation to I.

26.9 Einstein's Fundamental Formula for Small Fluctuations

Einstein, in 1910,[4] studied the deviation of thermodynamic observables in a small portion of a system from their equilibrium values in order to understand critical fluctuations, which we will discuss in Chapter 33.[5] Here, a heuristic approach is outlined first. To obtain the probability of fluctuations, Einstein inverted Boltzmann's principle (see **17.4**) as

---

[2] The following exposition heavily relies on Landau, L. D. and Lifshitz, E. M. (2013). *Statistical Physics*. Part 1, 3rd edition, Oxford, Butterworth-Heinemann, but perhaps is more user friendly.

[3] If X is extensive, $X(V)$ is the total amount in V (i.e., $X(V)/V$ is its density). If X is intensive, then $X(V)/V$ should be interpreted as the average value in the volume V.

[4] [In 1910 Russell and Whitehead, published *Principia Mathematica* (a further two volumes appeared in 1912 and 1913); Stravinsky's *The firebird*.]

[5] Einstein, A. (1910). Theorie der Opaleszenz von homogenen Flüssigkeitsgemischen in der Nähe des kritischen Zustandes, *Ann. Phys.*, **33**, 1275–1298.

$$w(\{X\}) = e^{S(\{X\})/k_B},\tag{26.21}$$

where $\{X\}$ collectively denotes extensive variables. Then, he postulated that the statistical weight for the value of X deviating from its equilibrium value may also be obtained by (26.21). Since we know the statistical weights, we can compute the probability of observing $\{X\}$ as

$$P(\{X\}) = \frac{w(\{X\})}{\sum_{\{X\}} w(\{X\})}.\tag{26.22}$$

The denominator may be replaced with the largest term among the summands (an analogous argument in **18.12**), so we may rewrite the formula as

$$P(\{X\}) \simeq w(\{X\})/w(\{X_{eq}\}) = e^{[S(\{X\})-S(\{X_{eq}\})]/k_B},\tag{26.23}$$

where $\simeq$ implies equality up to a certain unimportant numerical coefficient, and $\{X_{eq}\}$ is the value of $\{X\}$ that gives the largest w (maximizes the entropy), i.e., the equilibrium value. To the second order this reads

$$P(\{\delta X\}) \propto e^{-|\delta^2 S|/k_B}.\tag{26.24}$$

Here, the variation is around entropy max, so $\delta^2 S < 0$. Its sign is emphasized by writing $\delta^2 S = -|\delta^2 S|$. Einstein proposed this as the *fundamental formula for small fluctuations* in a small portion of *any* equilibrium system.

The above heuristic derivation of (26.24) assumes that a mesoscopic portion of the system is isolated. However, as noted before (**25.3** from (25.11) to (25.14)), the second-order deviation of any thermodynamic potential around its equilibrium value is given by $-T\delta^2 S$, so (26.24) is valid under *any* condition; Einstein is always correct.

26.10 Practical Form of Fluctuation Probability

To study fluctuations we need the second-order variation $\delta^2 S$ of the system entropy. This can be computed from the Gibbs relation (here δ means the so-called "virtual variation," but such variations are actually spontaneously realized by thermal fluctuations):

$$\delta S = \frac{1}{T}(\delta E + P\delta V - \mu\delta N - x\delta X)\tag{26.25}$$

as follows (the following formula is the second-order term of the Taylor expansion, so do not forget the overall factor 1/2):[6*]

[6*] If the reader has some trouble in understanding the derivation, look at a simple example: $f = f(x, y)$, where x and y are regarded as independent variables. If we can write $\delta f = X\delta x + Y\delta y$, then

$$\delta X = \frac{\partial X}{\partial x}\delta x + \frac{\partial X}{\partial y}\delta y, \quad \delta Y = \frac{\partial Y}{\partial x}\delta x + \frac{\partial Y}{\partial y}\delta y.\tag{26.26}$$

Therefore, the second-order Taylor expansion term reads

$$\delta^2 f = \frac{1}{2}\left(\frac{\partial X}{\partial x}\delta x^2 + \frac{\partial X}{\partial y}\delta y\delta x + \frac{\partial Y}{\partial x}\delta x\delta y + \frac{\partial Y}{\partial y}\delta y^2\right) = \frac{1}{2}(\delta X\delta x + \delta Y\delta y).\tag{26.27}$$

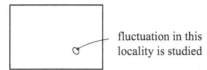

fluctuation in this
locality is studied

Fig. 26.1 The formula (26.31) applies to a small portion of a big system. For example, we can spectroscopically measure the temperature fluctuation in a small volume with appropriate probe molecules. If the observation volume is fixed, obviously we cannot choose V as an independent variable, but virtually any independent variables may be chosen to study fluctuations in the small portion of a macroscopic system.

$$\delta^2 S = \frac{1}{2}\left[\delta\left(\frac{1}{T}\right)(\delta E + P\delta V - \mu\delta N - x\delta X) + \frac{1}{T}(\delta P\delta V - \delta\mu\delta N - \delta x\delta X)\right], \quad (26.28)$$

$$\overset{(26.25)}{=} -\frac{\delta T}{2T^2}T\delta S + \frac{1}{2T}(\delta P\delta V - \delta\mu\delta N - \delta x\delta X). \quad (26.29)$$

Thus, we have arrived at the following useful expression worth remembering (actually, almost nothing to remember anew; cf. $-\delta E = -T\delta S + P\delta V - \mu\delta N - x\delta X$):[7]

$$\delta^2 S = \frac{-\delta T\delta S + \delta P\delta V - \delta\mu\delta N - \delta x\delta X}{2T} = -\frac{1}{T}\delta^2 E. \quad (26.30)$$

Do not forget the 2 in the denominator.

Consequently, the probability density of fluctuation in a mesoscopic portion (Fig. 26.1) can have the following form, which is the starting point of practical calculation of fluctuations (second moments[8]):

$$P(\text{fluctuation}) \propto \exp\left\{-\frac{1}{2k_B T}(\delta T\delta S - \delta P\delta V + \delta\mu\delta N + \delta x\delta X)\right\}. \quad (26.31)$$

26.11 Fluctuation and Reversible Work Needed to Create It

A similar calculation gives (or as already noted in (26.30)):

$$\delta^2 E = \frac{1}{2}(\delta T\delta S - \delta P\delta V + \delta\mu\delta N + \delta x\delta X). \quad (26.32)$$

Here $\delta^2 E$ can be understood as the (free) energy we must supply as the reversible work, if we wish to create the fluctuation. Therefore, we may rewrite (26.31) as

$$P(\text{fluctuation}) \propto e^{-\beta W_f}, \quad (26.33)$$

where W_f is the reversible work required to create the fluctuation. This is a practically useful formula.

---

[7] In short, the second variations of independent variables are zero (i.e., $\delta^2 x = \delta^2 y = 0$, everybody must know this): $\delta[X\delta x + Y\delta y] = \delta X\delta x + X\delta^2 x + \delta Y\delta y + Y\delta^2 y = \delta X\delta x + \delta Y\delta y.$

[7] As has already been stated when we discussed the general stability criterion, δS need not be zero.

[8] **Warning**. Note that this cannot be used to compute the higher-order moments.

26.12 Too Rapid Fluctuations

This thermodynamic formalism can be applied if $\delta^2 S$ is well defined irrespective of quantum or classical nature of the fluctuations. If the fluctuation is too rapid (i.e., $T\Delta S \sim h/\tau$ for the fluctuation time scale τ)[9], or the temperature is too low, we cannot rely on the theory. That is, for (26.24) to be applicable, we need the following condition:

$$\Delta S \gg h/T\tau. \tag{26.34}$$

26.13 How to Use the Practical Formula

To use the practical formula (26.31) we must choose the independent variables. Suppose we study a system that requires n thermodynamic coordinates (i.e., its thermodynamic space is n-dimensional). Such a system requires n conjugate pairs (T, S), $(-P, V)$, (x_i, X_i), etc. Equation (26.31) tells us that the formula is totally symmetric with respect to the intensive and extensive variables. Thus, we may choose n independent variables, arbitrarily selecting one (i.e., X or x) from each conjugate pair (x, X). Any choice will do, but a clever choice may (sometimes drastically) simplify the calculation.

After choosing the independent variables, the formula in the round parentheses of (26.31) becomes a quadratic form in independent variations (say, $\{\delta T, \delta V, \delta x\}$). Thus, the fluctuations are Gaussian. For an example, let us calculate the temperature fluctuation. We must first choose independent variables (variations). Obviously, δT must be included. We need one more independent variable (if $n = 2$). Let us choose δV as the other variation. We could choose δP, but we wish to exploit the following general fact:

$$\langle \delta X_i \delta x_j \rangle = k_B T \delta_{ij}, \tag{26.35}$$

where X_i is an extensive variable, and x_j an intensive variable; we understand that (x_i, X_i) is a conjugate pair (see **Q26.3** (1)). Since the variables are Gaussian, no correlation implies statistical independence.

Thus, δV is a convenient partner for δT, because they are statistically independent (see **Q26.5** for a not "clever" choice of a "partner"). Therefore, (26.30) reads

$$\delta^2 S = -\frac{1}{2T}\delta S \delta T + \cdots = -\frac{1}{2T}\left(\frac{\partial S}{\partial T}\right)_V \delta T^2 + \cdots = -\frac{C_V}{2T^2}\delta T^2 + \cdots. \tag{26.36}$$

Therefore, our fundamental formula (26.31) reads

$$P(\text{fluctuation}) \propto \exp\left\{-\frac{C_V}{2k_B T^2}\delta T^2 + \cdots\right\}. \tag{26.37}$$

---

[9] Due to the energy–time uncertainty principle. This is the relation between the uncertainty of the energy of a state and its relaxation time scale.

We can immediately see from this Gaussian distribution:

$$\langle \delta T^2 \rangle = k_B T^2 / C_V. \tag{26.38}$$

This is the fluctuation of the average temperature in a small volume whose constant volume heat capacity is C_V. If the observation volume is reduced, then C_V is reduced as well (recall that C_V is an extensive quantity), so the fluctuation increases. Similarly, to obtain the pressure fluctuation, we should choose P and S as independent variables, since we know $\langle \delta P \delta S \rangle = 0$. We obtain:[10]

$$\langle \delta P^2 \rangle = k_B T / V \kappa_S. \tag{26.39}$$

Such small tricks to save our energy are useful, but obviously we must know how to handle *multivariate Gaussian distribution*.

26.14* Multivariate Gaussian Distribution

A multivariate distribution is called a *Gaussian* distribution, if all of its marginal distributions are Gaussian. Or more practically, we could say that if the negative log of the density distribution function is a positive definite quadratic form (apart from a constant term due to normalization) of the deviations from the expectation values, the distribution is Gaussian:

$$f(\boldsymbol{x}) = \frac{1}{\sqrt{\det(2\pi V)}} \exp\left(-\frac{1}{2}(\boldsymbol{x} - \boldsymbol{m})^T V^{-1}(\boldsymbol{x} - \boldsymbol{m}) \right), \tag{26.40}$$

where $\langle \boldsymbol{x} \rangle = \boldsymbol{m}$, the expectation value, and V is the *covariance matrix* defined as (here, our vectors are column vectors; ^{T} implies transposition: "column $\leftrightarrow$ row")

$$V = \langle (\boldsymbol{x} - \boldsymbol{m})(\boldsymbol{x} - \boldsymbol{m})^T \rangle. \tag{26.41}$$

The reader must not have any difficulty in demonstrating that (26.40) is correctly normalized. (Hint: choose eigendirections of V as the orthogonal coordinates and rewrite the exponent in the form: $\sum_i (-y_i^2 / 2\lambda_i)$.)

In particular, for the two-variable case:

$$f(x, y) \propto \exp\left\{ -\frac{1}{2} \left(ax^2 + 2bxy + cy^2 \right) \right\}, \tag{26.42}$$

we have

$$V = \Lambda^{-1} = \left\langle \begin{pmatrix} x \\ y \end{pmatrix}(x, y) \right\rangle = \left\langle \begin{pmatrix} x^2 & xy \\ yx & y^2 \end{pmatrix} \right\rangle = \begin{pmatrix} a & b \\ b & c \end{pmatrix}^{-1} \tag{26.43}$$

$$= \frac{1}{\det \Lambda} \begin{pmatrix} c & -b \\ -b & a \end{pmatrix}. \tag{26.44}$$

---

[10] **Fluctuation in the $T \to 0$ limit** The fluctuations of the quantities that may be interpreted as the expectation values of microscopic-mechanically expressible quantities (e.g., internal energy, volume, pressure) tend to zero in the $T \to 0$ limit as we see here. However, the fluctuations of entropy and the quantities obtained by differentiating it do not satisfy the above property.

That is,

$$\langle x^2 \rangle = c/\det \Lambda, \quad \langle xy \rangle = -b/\det \Lambda, \quad \langle y^2 \rangle = a/\det \Lambda. \tag{26.45}$$

See **Q26.5** for an example.

Problems

Q26.1 Internal Energy Fluctuation

We wish to study the fluctuation of the total energy E of a closed system (a system without any material exchange) under constant volume V and temperature T (i.e., thermostatted). To what heat capacity of the system (say, under constant pressure or constant volume) is $\langle \delta E^2 \rangle$ directly related? Notice that microscopically E is just the system Hamiltonian H, so we are interested in $\langle H^2 \rangle - \langle H \rangle^2$.

Solution

$Z = \sum e^{-\beta H}$, so

$$\langle H \rangle = \left(\frac{\partial \log Z}{\partial(-\beta)} \right)_V = \frac{1}{Z} \sum H e^{-\beta H}. \tag{26.46}$$

Therefore,

$$\left(\frac{\partial \langle H \rangle}{\partial(-\beta)} \right)_V = -\frac{1}{Z^2} \left(\sum H e^{-\beta H} \right)^2 + \frac{1}{Z} \sum H^2 e^{-\beta H} = \langle H^2 \rangle - \langle H \rangle^2. \tag{26.47}$$

We know

$$\left(\frac{\partial \langle H \rangle}{\partial(-\beta)} \right)_V = k_B T^2 \left(\frac{\partial E}{\partial T} \right)_V = k_B T^2 C_V. \tag{26.48}$$

Q26.2 Easy Questions about Fluctuations

(1) Consider a small portion of a big system in equilibrium. Let us assume that the small portion contains a constant number of particles. Find the fluctuation of entropy $\langle \delta S^2 \rangle$ in terms of an appropriate heat capacity of the small portion.

(2) (Come back after Chapter 27 for chemical potential, although formally it is not hard to understand what follows.) Take a (constant volume) small portion of a solution of some substance in a solvent. Let c be the concentration of the solute. Show that $\langle \delta c \, \delta T \rangle = 0$, that is, concentration fluctuation and temperature fluctuation are uncorrelated in equilibrium.

Solution

(1) We use (needless to say, we use Einstein's thermodynamic fluctuation theory)

$$\frac{1}{k_B} \delta^2 S = -\frac{1}{2 k_B T} \delta T \delta S + \cdots . \tag{26.49}$$

Choose S and P as independent variables for convenience. Then, since δS and δP are statistically independent (**Q26.3**(1) quoted in **26.13**), we can rewrite this as

$$-\frac{1}{2k_BT}\left(\frac{\partial T}{\partial S}\right)_P \delta S^2 + \cdots = -\frac{1}{2k_BC_P}\delta S^2 + \cdots . \qquad (26.50)$$

Therefore, $\langle \delta S^2 \rangle = k_B C_P$, where C_P is the constant pressure specific heat of the small portion.

(2) This is obvious from $\langle \delta N \delta T \rangle = 0$.

Q26.3 Thermodynamic Fluctuations: General Questions

(1) Suppose X and y are a non-conjugate pair with respect to energy, X extensive and y intensive. Show that $\langle \delta X \delta y \rangle = 0$.

(2) Let X and x be a conjugate pair (wrt energy). Show $\langle \delta X \delta x \rangle = k_B T$.

(3) (Come back after Chapter 27 for chemical potential, although formally it is not hard to understand.) Express $\langle \delta \mu^2 \rangle$ in terms of a single thermodynamic derivative. The system is assumed to be described in terms of S, V, N (or their conjugate variables).

(4) Let X be an extensive quantity. What can be concluded about $\langle \delta S \delta X \rangle$? The result is intuitively suggestive, because entropy fluctuation means spatially local heat transport: that is, local temperature change.

Solution

(1) Recall that we can choose any combination of variables as independent variables as long as one variable is chosen from each conjugate pair (x, X). We know $\langle \delta X_i \delta X_j \rangle$, so in this case, we should use the all extensive independent variable set. Using (26.14), we get

$$\langle \delta X \delta y \rangle = \left\langle \delta X \sum_Y \frac{\partial y}{\partial Y} \delta Y \right\rangle = k_B T \sum_Y \frac{\partial Y}{\partial x}\frac{\partial y}{\partial Y} = k_B T \frac{\partial y}{\partial x} = 0. \qquad (26.51)$$

In more detail for those who are skeptical:

$$dx = \sum_j \left(\frac{\partial x}{\partial X_j}\right)_{X_1 \cdots \check{X}_j \cdots X_n} dX_j, \qquad (26.52)$$

where $\check{X}_j$ implies to remove the variable under the check mark. Therefore,

$$\left(\frac{\partial x_i}{\partial x_k}\right)_{x_1 \cdots \check{x}_k \cdots x_n} = \sum_j \left(\frac{\partial x_i}{\partial X_j}\right)_{X_1 \cdots \check{X}_j \cdots X_n} \left(\frac{\partial X_j}{\partial x_k}\right)_{x_1 \cdots \check{x}_k \cdots x_n}. \qquad (26.53)$$

We put (26.52) into $\langle \delta X \delta y \rangle$ (regarding X as a representative of $\{X_j\}$ and y that of $\{x_k\}$ (the derivatives are mere constants, so we can take them out of the average symbol). Now, (26.53) tells us what we want.

(2) We use (26.14).

$$\langle \delta X \delta x \rangle = \left\langle \delta X \sum_Y \frac{\partial x}{\partial Y} \delta Y \right\rangle = \sum_Y \langle \delta X \delta Y \rangle \frac{\partial x}{\partial Y} = k_B T \sum_Y \frac{\partial Y}{\partial x}\frac{\partial x}{\partial Y} = k_B T. \qquad (26.54)$$

(3) Taking (1) above into account, we should choose μ, S, V as independent variables.

$$\delta^2 S = -\frac{1}{2T}(\delta N \delta \mu + \cdots) = -\frac{1}{2T}\left(\frac{\partial N}{\partial \mu}\right)_{S,V} \delta \mu^2 + \cdots. \tag{26.55}$$

Therefore,

$$P(\delta\mu \cdots) \propto \exp\left\{-\frac{1}{2k_B T}\left(\left(\frac{\partial N}{\partial \mu}\right)_{S,V} \delta\mu^2 + \cdots\right)\right\}. \tag{26.56}$$

That is,

$$\langle \delta\mu^2 \rangle = k_B T \left(\frac{\partial \mu}{\partial N}\right)_{S,V}. \tag{26.57}$$

(4)

$$\langle \delta S \delta X \rangle = k_B T \left(\frac{\partial X}{\partial T}\right)_x. \tag{26.58}$$

That is, the temperature derivative is the cross correlation with entropy fluctuation. This is, although trivial, worth remembering.

Q26.4 Fluctuation and Le Chatelier–Braun Principle

(1) Show

$$\langle \delta x \delta X \rangle^2 \le \langle \delta x^2 \rangle \langle \delta X^2 \rangle, \tag{26.59}$$

where x and X are conjugate pair of thermodynamic variables (with respect to energy). (This is the Cauchy–Schwarz inequality.)

(2) What is the relation between this inequality and the Le Chatelier–Braun principle?

Solution

(1) The easiest way is to use the following obvious inequality valid for any real t:

$$0 \le \langle (\delta X + t\delta x)^2 \rangle = \langle \delta X^2 \rangle + 2t\langle \delta x \delta X \rangle + t^2 \langle \delta x^2 \rangle. \tag{26.60}$$

Since $\langle \delta x^2 \rangle \ge 0$, we have its discriminant to be negative: $\langle \delta x \delta X \rangle^2 - \langle \delta X^2 \rangle \langle \delta x^2 \rangle \le 0$.

(2) We know $\langle \delta x \delta X \rangle = k_B T$ (**Q26.3**(2)), and (using a clever way of calculating fluctuations):

$$\langle \delta X^2 \rangle = k_B T \left(\frac{\partial x}{\partial X}\right)_y^{-1}, \quad \langle \delta x^2 \rangle = k_B T \left(\frac{\partial X}{\partial x}\right)_Y^{-1}. \tag{26.61}$$

Therefore,

$$\left(\frac{\partial x}{\partial X}\right)_y \le \left(\frac{\partial x}{\partial X}\right)_Y, \tag{26.62}$$

which is (25.25), the basic inequality for the Le Chatelier–Braun principle.

Q26.5 Not a "Clever" Choice of Variables

Compute $\langle \delta T^2 \rangle$ with δT and δP as independent variables instead of δT and δV (T and P are not statistically independent).

Solution

$$-\frac{1}{2T}[\delta S\delta T - \delta P \delta V] = -\frac{1}{2T}\left\{ \left(\frac{\partial S}{\partial T}\right)_P \delta T^2 - 2\left(\frac{\partial V}{\partial T}\right)_P \delta T \delta P - \left(\frac{\partial V}{\partial P}\right)_T \delta P^2 \right\}. \quad (26.63)$$

Using (26.45), we should be able to compute the desired quantity. Since

$$\det \Lambda = \frac{1}{T^2}\frac{\partial(V,S)}{\partial(T,P)}, \quad (26.64)$$

$$\langle \delta T^2 \rangle = -\frac{k_B}{\det \Lambda}\frac{1}{T}\left(\frac{\partial V}{\partial P}\right)_T = -k_B T\frac{\frac{\partial(V,T)}{\partial(P,T)}}{\frac{\partial(V,S)}{\partial(T,P)}} = k_B T\frac{\partial(V,T)}{\partial(V,S)} = \frac{k_B T^2}{C_V}. \quad (26.65)$$

Q26.6 Fluctuation and Spring Constant

Inside the F_1ATPase is a rotator γ to which a long actin filament (it is a straight stiff bar of length 30 nm) is perpendicularly attached. Thus, the filament swings back and forth when the ATPase is waiting for an ATP molecule.

(1) The root-mean-square angle fluctuation of the stiff filament was 30 degrees at 290 K. If the temperature is raised by 10%, by what percentage will the angular fluctuation change? Assume that the molecular structure is not affected by this temperature change.
(2) What is the torsional spring constant of this rotator captured by the surrounding ring?
(3) Now, by adding appropriate polymers to the ambient solution, the effective viscosity of the solution is doubled. What is the mean square angle fluctuation of the filament? Assume that the polymers do not affect the ATPase itself.

Solution

(1) Suppose θ is the angular deviation around the equilibrium direction. Then, the torsional spring constant k reads

$$\tau = k\theta, \quad (26.66)$$

where τ is the torsion. Since, k^{-1} is the "susceptibility" of θ against τ, the fluctuation–response relation tells us

$$k^{-1} = \left(\frac{\partial \theta}{\partial \tau}\right)_T = \beta\langle(\delta\theta)^2\rangle. \quad (26.67)$$

That is,

$$\langle(\delta\theta)^2\rangle = k_B T/k. \quad (26.68)$$

Since we may assume k does not depend on T, the fluctuation should change by about 5%, which must be easy to guess.

(2) We must measure the angle in radians.

$$k = 1.382 \times 10^{-23} \times 290/(\pi/6)^2 = 1.46 \times 10^{-20}. \qquad (26.69)$$

The unit is J/rad. Is it reasonable? It is about 15 (pN·nm)/rad, a reasonable value.

(3) No change. The formula does not depend on the viscosity, so the amplitude of the fluctuation never changes. This is true, regardless of how gooey the solution is. It is true that the oscillation becomes slow, but then small fluctuations can be accumulated to a size as large as when the viscosity is very low.

Appendix 26A Onsager's Theory of Irreversible Processes

We wish to discuss nonequilibrium dissipative phenomena. However, we are not reckless to study nonequilibrium phenomena in general; we study small deviations from equilibrium that may be governed by linear phenomenological laws (see Chapter 8).

Thus, this Appendix is a natural continuation of Chapter 10 to outline a large deviation approach to Onsager's linear irreversible phenomenology. The reader will see the importance of correlation functions; there is a way to write down the transport coefficients in terms of flux fluctuations (the Green–Kubo formulas).

This Appendix is certainly not for the beginners; the reader can freely skip it without encountering any problem in reading through this book.

26A.1 Small Deviations from Equilibrium

If a macroscopic system is not in equilibrium but slightly away from equilibrium, the system may still be described in terms of thermodynamic variables as we have seen in Einstein's theory of equilibrium fluctuation. In linear phenomenology the driving force must be governed by the second-order deviation of an appropriate thermodynamic potential from its equilibrium value. We have learned this is always exactly $\delta^2 S$ (recall **25.3**).

Therefore, if we write the second-order deviation of the entropy from its equilibrium value as $S(X)$, where X is the deviation of thermodynamic coordinates from the equilibrium value (i.e., $X = 0$ denotes the equilibrium state in this Appendix), the linear phenomenological macroscopic laws have the following form:

$$\frac{dX}{dt} = L\frac{\partial S}{\partial X}, \qquad (26.70)$$

where L is called an *Onsager coefficient*. First, we consider the simplest case described by a single extensive quantity X (without spatial inhomogeneity). The time derivative in this equation is a coarse-grained one over the scale of Δt that appeared in Chapter 10 (i.e., an infinitesimal time span from the macroobservers' point of view).

26A.2 Mesoscopic Description of Phenomenological Laws

Next, let us try to describe the nonequilibrium system on the mesoscopic scale. We can expect a Langevin equation (Section 10)

$$\frac{dX}{dt} = L\frac{\partial S}{\partial X} + w. \tag{26.71}$$

Since the time derivative in this equation is not the truly microscopic time derivative, but the coarse-grained one, let us denote it as $\delta/\delta t$ just as explained in Section 10:

$$\frac{\delta X}{\delta t} = L\frac{\partial S}{\partial X} + w. \tag{26.72}$$

Then the theoretical framework that is most natural must be the large deviation theoretical framework (recall **1.11** and **10.11**).

26A.3 Large Deviation Formulation of Onsager's Regression Hypothesis

We assume the large deviation principle of the following form (cf. (10.34)):

$$P\left(\frac{1}{\delta t}\int_0^{\delta t} ds\, \frac{dX(s)}{ds} - L\frac{\partial S}{\partial X_i} \in v(w)\right) \approx e^{-\delta t \Gamma w^2/4}, \tag{26.73}$$

where $v(w)$ is a volume element in the set of noises and Γ a positive constant describing the magnitude of the fluctuation. This is the large deviation formulation of *Onsager's regression hypothesis* (cf. **9.10**). This means the noise is a mean zero Gaussian with the following correlation function (recall **10.11**)

$$\langle w(t)w(t')\rangle = \frac{2}{\Gamma}\delta(t - t'). \tag{26.74}$$

26A.4 Fluctuation–Dissipation Relation of the First Kind

Equation (26.74) implies $\langle w^2\rangle = 2/\Gamma\delta t$ and (26.72) implies $w = \dot{X} - L\partial S/\partial X$. Therefore, the first term that contains an $O[\sqrt{1/\delta t}]$ quantity dominates the equation (i.e., as usual at the mesoscale, noise dominates the driving force), so $\langle w^2\rangle\delta t = \langle \dot{X}^2\rangle\delta t$ or

$$\frac{2}{\Gamma} = \delta t\langle \dot{X}^2\rangle. \tag{26.75}$$

Since $\dot{X}$ is a realization of

$$\frac{\delta X}{\delta t} = \frac{1}{\delta t}\int_0^{\delta t} ds\, \frac{dX(s)}{ds}, \tag{26.76}$$

the ensemble average may be computed as

$$\frac{1}{\Gamma} = \frac{1}{2}\delta t\langle \dot{X}^2\rangle = \frac{1}{2\delta t}\int_0^{\delta t} ds\int_0^{\delta t} ds'\left\langle \frac{dX}{dt}(s)\frac{dX}{dt}(s')\right\rangle. \tag{26.77}$$

This double integration may be decomposed as

$$\int_0^{\delta t} ds \int_0^{\delta t} ds' = \int_0^{\delta t} ds \int_0^s ds' + \int_0^{\delta t} ds' \int_0^s ds = 2 \int_0^{\delta t} ds \int_0^s ds', \qquad (26.78)$$

so

$$\frac{1}{\Gamma} = \frac{1}{\delta t} \int_0^{\delta t} ds \int_0^s ds' \left\langle \frac{dX}{dt}(s) \frac{dX}{dt}(s') \right\rangle. \qquad (26.79)$$

Within the range of δt, very short from our point of view, the system may be assumed to be stationary (i.e., from the microscopic point of view δt is large and this time span is rather uniform), so the correlation does not depend on the absolute times and

$$\frac{1}{\Gamma} = \frac{1}{\delta t} \int_0^{\delta t} ds \int_0^s ds' \left\langle \frac{dX}{dt}(s-s') \frac{dX}{dt}(0) \right\rangle = \frac{1}{\delta t} \int_0^{\delta t} ds \int_0^s d\tau \left\langle \frac{dX}{dt}(\tau) \frac{dX}{dt}(0) \right\rangle. \qquad (26.80)$$

We know the noise is uncorrelated (after mesoscopically very short time), so the upper bound of the integral with respect to τ does not depend on s (even if it is replaced by ∞ the result does not change). Hence,

$$\frac{1}{\Gamma} = \int_0^\infty d\tau \left\langle \frac{dX}{dt}(\tau) \frac{dX}{dt}(0) \right\rangle. \qquad (26.81)$$

26A.5 Fluctuation–Dissipation Relation

We know, from our experience with Brownian motion (Chapter 9), that Γ cannot be chosen freely for a given system, if it is in equilibrium. The noise amplitude Γ appearing in (26.73) and the dissipation rate L are related (fluctuation-dissipation relation; see Section 10, esp., (10.20)). In the present case, the equilibrium distribution is given by $\propto e^{S/k_B}$ (recall Einstein's fluctuation theory in **26.9**), so the reader may guess the following form (i.e., (10.20) without T):

$$L = \frac{1}{k_B \Gamma}. \qquad (26.82)$$

This is correct. Let us derive this again here. The Smoluchowski (or the Fokker–Planck) equation reads (the driving force is $L \operatorname{grad} S$ and the noise amplitude is $2/\Gamma$ (recall **10.5**), so the diffusion constant is $1/\Gamma$):

$$\frac{\partial P}{\partial t} = -\operatorname{div}(PL \operatorname{grad} S - \frac{1}{\Gamma} \operatorname{grad} P). \qquad (26.83)$$

Since the equilibrium distribution must be e^{S/k_B}, indeed we get (26.82).

26A.6 Green–Kubo Relation

The fluctuation–dissipation relation (26.82) implies that the Onsager coefficient L can be described in terms of the correlation function:

$$L = \frac{1}{k_B} \int_0^\infty dt \left\langle \frac{dX}{dt}(t) \frac{dX}{dt}(0) \right\rangle. \qquad (26.84)$$

This is called the *Green–Kubo relation*. Basically, the transport coefficient L is described in terms of a time correlation function of the "changes."

26A.7 Large Deviation Formulation of Onsager's Theory: Summary

Let us summarize what we have learned, while generalizing the formulas to the cases with many variables.

The phenomenological law is, corresponding to (26.70),

$$\dot{X}_i = \sum_j L_{ij} \frac{\partial S}{\partial X_j}, \tag{26.85}$$

where L_{ij} are Onsager coefficients. Here X_i denotes deviations from equilibrium values, and S is actually $\delta^2 S$, so it is a negative definite quadratic form of X_i. The large deviation principle (Onsager's regression hypothesis) reads, corresponding to (26.73),

$$P\left(\frac{1}{\delta t}\int_0^{\delta t} ds\, \frac{dX_i(s)}{ds} - \sum_j L_{ij}\frac{\partial S}{\partial X_j} \in v(w_i)\right) \approx e^{-\delta t \sum_{ij} \Gamma_{ij} w_i w_j / 4}, \tag{26.86}$$

where

$$k_B \Gamma_{ij} = (L^{-1})_{ij} \tag{26.87}$$

is the fluctuation–dissipation relation with an explicit form corresponding to (26.84):

$$L_{ij} = \frac{1}{k_B}\int_0^\infty ds\, \left\langle \frac{dX_i}{dt}(s)\frac{dX_j}{dt}(0)\right\rangle. \tag{26.88}$$

The Langevin equation is, corresponding to (26.72),

$$\dot{X}_i = \sum_j L_{ij}\frac{\partial S}{\partial X_j} + w_i \tag{26.89}$$

with the noise satisfying $\langle w_i \rangle = 0$ and, corresponding to (26.74),

$$\langle w_i(t) w_j(s)\rangle = 2 L_{ij} k_B \delta(t-s). \tag{26.90}$$

Here, a mathematically proper interpretation of $1/\delta t$ has been used.

26A.8 Onsager Reciprocity

The average appearing in the formula for L_{ij} is computed in the equilibrium state, so it does not depend on the absolute time. Therefore,

$$\left\langle \frac{dX_i}{dt}(s)\frac{dX_j}{dt}(0)\right\rangle = \left\langle \frac{dX_i}{dt}(0)\frac{dX_j}{dt}(-s)\right\rangle = \left\langle \frac{dX_j}{dt}(-s)\frac{dX_i}{dt}(0)\right\rangle. \tag{26.91}$$

Microscopic mechanics is time-reversal symmetric (recall **11.7**), so this implies

$$\left\langle \frac{dX_i}{dt}(s)\frac{dX_j}{dt}(0)\right\rangle = \left\langle \frac{dX_j}{dt}(s)\frac{dX_i}{dt}(0)\right\rangle. \tag{26.92}$$

Hence, we conclude

$$L_{ij} = L_{ji}. \qquad (26.93)$$

This is called *Onsager's reciprocity relation*.

26A.9 Spatially Nonuniform Case: Transport Phenomena

Let us close this Appendix with a sketch of transport phenomena.

We have given a general expression for the Onsager coefficients L in (26.88), but for spatially uniform systems. The general nonequilibrium states are spatially nonuniform, so we need phenomenology for densities $\tilde{x}_i$ of extensive quantities X_i. It generally has the following form (see Chapter 8):

$$\frac{\partial \tilde{x}(r)}{\partial t} = -\operatorname{div} j(r) + \sigma, \qquad (26.94)$$

where σ is a source term we do not discuss, and j is the flux of $\tilde{x}$. Heat conduction, hydrodynamics, electric conduction, etc., all have this form. The linear phenomenological law assumes the following form for the flux:

$$j = L\nabla F, \qquad (26.95)$$

where F is the driving force (corresponding to $\partial S/\partial X_i$ in (26.85)).

Probably, the easiest approach to spatially nonuniform systems is to divide the space into cells and the variables in each cell $\{\tilde{x}\}$ are denoted with the representative cell position r as $\{\tilde{x}(r)\}$. Then, we interpret (26.85) as (we write $\partial S/\partial \tilde{x}(r) = F(r)$)

$$\frac{\partial \tilde{x}(r)}{\partial t} = \sum_{r'} L(r, r') F(r'). \qquad (26.96)$$

Extending (26.88), let us proceed quite formally:

$$\sum_{r'} L(r, r') F(r') = \frac{1}{k_B} \int_0^\infty dt \sum_{r'} \left\langle \frac{\partial \tilde{x}(r, t)}{\partial t} \frac{\partial \tilde{x}(r')^T}{\partial t} \right\rangle F(r'), \qquad (26.97)$$

$$= \frac{1}{k_B} \int_0^\infty dt \sum_{r'} \langle \operatorname{div} j(r, t) \operatorname{div} j(r')^T \rangle F(r'), \qquad (26.98)$$

$$= -\operatorname{div}_r \frac{1}{k_B} \int_0^\infty dt \sum_{r'} \langle j(r, t) j(r')^T \rangle \nabla_{r'} F(r') \qquad (26.99)$$

Here, j is the fluctuating part of the flux (nonsystematic part). We have used Gauss' theorem, interpreting the spatial sum as a spatial integral: $\sum_{r'} \operatorname{div} j(r')^T F(r') = \sum_{r'} \operatorname{div} [j(r')^T F(r')] - \sum_{r'} j(r')^T \nabla_{r'} F(r')$, and ignored the surface integral coming from the first term.

26A.10 Green–Kubo Formula for Transport Coefficients

If F changes slowly over the correlation length of the fluctuating flux, we may set $r' = r$ in F, and rewrite (26.96) with (26.99) as

$$\frac{\partial \tilde{x}(r)}{\partial t} = -\mathrm{div}_r \frac{1}{k_B} \left[\int_0^\infty dt \sum_{r'} \langle j(r,t)j(r')^T \rangle \nabla_r F(r) \right]. \tag{26.100}$$

Comparing this with (26.94) and (26.95), we conclude, at least formally, that

$$L = \frac{1}{k_B} \int_0^\infty dt \int dr \, \langle j(r,t)j(0)^T \rangle, \tag{26.101}$$

where we have assumed that the equilibrium state is spatially uniform. If we introduce

$$J(t) = \int dr j(r,t), \tag{26.102}$$

then we get the *Green–Kubo formula* for transport coefficients:

$$L = \frac{1}{Vk_B} \int_0^\infty dt \, \langle J(t)J(0)^T \rangle. \tag{26.103}$$

Here, V is the volume of the system (or the domain being paid attention to).[11]

---

[11] Do not forget that j is the fluctuating part of the flux. If we use the whole flux, then the time integral in (26.103) may not exist, because $\lim_{t\to\infty} J$ may not be zero. To avoid this, we can simply subtract this value from J, and define $J^- = J - \lim_{t\to\infty} J$ to obtain the final result:

$$L = \frac{1}{Vk_B} \int_0^\infty dt \, \langle J^-(t)J^-(0)^T \rangle. \tag{26.104}$$

Notice that j appears in div, so we may choose the constant subtraction term so that L is well defined.

Summary

* Open systems require the mass action $dZ' = \sum \mu dN$ to extend the first law.
* If a chemical is exchanged among various parts of a system, the equilibrium condition is that its chemical potential is everywhere identical.
* The chemical potential generally has the form $\mu = \mu^{\ominus} + k_B T \log a$, where activity a is related to the concentration of the chemical.
* Colligative properties are universal properties of dilute systems solely due to the entropic effect of concentrations.
* Algebraic expression of a chemical reaction gives the equilibrium condition $\sum_i \nu_i \mu_i = 0$, which leads to the concept of equilibrium constants.

Key words

mass action, chemical potential, Gibbs–Duhem relation, phase equilibrium, Clapeyron–Clausius equation, osmotic pressure, van 't Hoff's law, Raoult's law, colligative properties, chemical reaction, signed stoichiometric coefficients, law of mass action, equilibrium constant, van 't Hoff's equation

The reader should be able to:

* Explain what the chemical potential is, and understand various equilibrium conditions in terms of chemical potentials.
* Understand colligative properties in a unified fashion.
* Understand the shifting direction of the reaction when T or P is altered.

27.1 Open Systems

So far we have discussed systems not allowed to exchange substances with their surrounding world (although at various places our Gibbs relation already included the μdN term, only formally). They are (materially) *closed systems*. In this chapter we will discuss

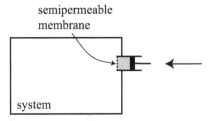

Fig. 27.1 How to measure a chemical potential.

open systems for which exchange of their component chemicals with their environments is allowed. We must extend the first law (from **13.1**) as follows:

$$\Delta E = W + Q + Z, \qquad (27.1)$$

where Z is called the *mass action*, which describes the energetic contribution of materials exchange.

27.2 Mass Action and Chemical Potential

To begin with, for simplicity, we assume only one chemical may be exchanged. We need one variable N to specify its amount in the system (in moles or in numbers[1]). We prepare a semipermeable rigid (and diathermal) membrane that allows the passage of this chemical only. Setting up a device as illustrated in Fig. 27.1 and injecting (only) this chemical into the system (or sucking it up from the system), we can (in principle) measure the necessary work (necessary mass action) $d'Z$ to inject dN molecules into the system. This is written as

$$d'Z = \mu dN. \qquad (27.2)$$

Here, it is explicitly noted that Z is not a state function (recall **12.7**), and μ is called the *chemical potential* of this chemical. If we have more than one chemical that we can exchange independently,[2] we can appropriately generalize this setup and write

$$d'Z = \sum_i \mu_i dN_i, \qquad (27.3)$$

where N_i denotes the amount of the ith chemical and μ_i the corresponding chemical potential.

---

[1] In this book N always implies the number of particles, and we will not use moles unless clearly stated otherwise. Thermodynamics only handles the situation that the atomic nature of the material is not discernible, but here we adopt an eclectic and pragmatic attitude.

[2] **Independent chemicals** What is independently exchanged and what not may not always be a simple question, but usually, common sense tells us the right answer. For example, if we want to inject water into a system, inevitably, we inject OH^-, H_3O^+, etc. as well, but their amounts are "slaved" to the total amount of water when T and P (and other thermodynamic variables) are fixed (chemical equilibrium). Therefore, we may conclude that only one component matters, and this answer is in agreement with our common-sense conclusion.

Intuitively, the chemical potential of a chemical of a system is the measure of the system's capability to export the chemical (so it is also the measure of our "effort" required to inject the same chemical). Therefore, in a stable equilibrium increasing N increases μ, a consequence of the stability criterion (cf. **Q26.3** (3)).

27.3 The Gibbs Relation with Mass Action

Including the mass action, the full form of the Gibbs relation reads, as we have already written,

$$dE = TdS - PdV + \mu dN + xdX + \cdots, \tag{27.4}$$

or

$$dS = \frac{1}{T}dE + \frac{P}{T}dV - \frac{\mu}{T}dN - \frac{x}{T}dX + \cdots. \tag{27.5}$$

Be careful about the signs. Needless to say, if we have several independently modifiable chemicals, we must replace μdN with a sum over these chemicals $\sum_i \mu_i dN_i$.

Other thermodynamic potentials read

$$dA = -SdT - PdV + \mu dN + xdX + \cdots, \tag{27.6}$$

$$dG = -SdT + VdP + \mu dN + xdX + \cdots. \tag{27.7}$$

27.4 The Gibbs–Duhem Relation

If the system size is increased by a fraction $\delta\lambda$ by joining a small fraction of the identical system in the identical equilibrium state (Fig. 27.2), all the extensive quantities are multiplied by $1 + \delta\lambda$. However, all the conjugate intensive quantities are intact. Therefore, (27.4) applied to this situation reads

$$E\delta\lambda = (TS - PV + \mu N + xX + \cdots)\delta\lambda, \tag{27.8}$$

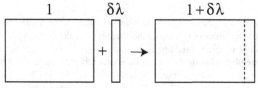

Fig. 27.2   A fraction $\delta\lambda$ of the system identical to the original system is added. Then, the increase of the extensive quantity X is $\delta X = X\delta\lambda$ (i.e., $X \rightarrow X + X\delta\lambda$).

or (see **Q27.1**):[3]

$$E = TS - PV + \mu N + xX + \cdots . \tag{27.9}$$

This implies

$$dE = (TdS - PdV + \mu dN + xdX + \cdots) + (SdT - VdP + Nd\mu + Xdx + \cdots), \tag{27.10}$$

but (27.4) is true, so we must conclude that

$$SdT - VdP + Nd\mu + Xdx + \cdots = 0. \tag{27.11}$$

This relation is called the *Gibbs–Duhem relation*. This tells us how the chemical potential changes as a function of T, P, etc., if there is only one chemical species:

$$d\mu = -\frac{S}{N}dT + \frac{V}{N}dP - \frac{X}{N}dx + \cdots . \tag{27.12}$$

If we combine (27.9) with the definition of the Gibbs free energy, we obtain

$$G = \mu N + xX + \cdots . \tag{27.13}$$

In many practical situations, there are various chemicals but no other work coordinates than V, so this becomes

$$G = \sum_i \mu_i N_i. \tag{27.14}$$

Notice that (27.13) can be directly obtained by the same approach as is used to derive (27.9).

27.5* Gibbs Relation for Densities

As an application of the Gibbs–Duhem relation let us derive the Gibbs relation for densities (extensive quantities per volume): $e = E/V$ (internal energy density), $s = S/V$ (entropy density), $n = N/V$ (number density), $\bar{x} = X/V$, etc. From (27.9) we obtain

$$e = Ts - P + \mu n + x\bar{x}. \tag{27.15}$$

On the other hand, dividing the Gibbs–Duhem relation (27.11) with V, we get

$$sdT - dP + nd\mu + \bar{x}dx = 0 \tag{27.16}$$

Therefore, differentiating (27.15) and using (27.16), we get

$$de = Tds + \mu dn + xd\bar{x}. \tag{27.17}$$

---

[3] **The energy zero** Since the origin of energy may be shifted freely, isn't an equation with an absolute value of energy meaningless? This is a good question. In fact, if we wish to use (27.9), we must stick to a convention as to the origin of the energy (or how to measure the energy). With this caution, the equation is very useful to discuss, e.g., quantum ideal gases (see Chapter 29).

27.6 Equilibrium Condition

Suppose two systems I and II are joined to exchange heat, volume, and chemicals while the whole system is in isolation (or thermally isolated and no work nor mass action is provided from outside). Then, the equilibrium condition must be the maximization of the total entropy (recall **14.2**). Let S_X be the entropy of system X. Then, $S = S_I + S_{II}$. If we assume thermal and pressure equilibration have already been attained (see Chapter 14), the remaining equilibrium condition is

$$\delta S = -\sum_i \left(\frac{\mu_{iI}}{T} \delta N_{iI} + \frac{\mu_{iII}}{T} \delta N_{iII} \right) = 0. \qquad (27.18)$$

Assuming that there is no chemical reaction changing the N_is, we must conclude $\delta N_{iI} + \delta N_{iII} = 0$. Therefore, the equilibrium condition is

$$\mu_{iI} = \mu_{iII} \qquad (27.19)$$

for each chemical i that can be (independently) exchanged between the two subsystems. This is quite natural, because a chemical potential is a "strength" of a system to export the corresponding chemical.

27.7 Phase Equilibria

We will discuss phase transitions in depth later (Chapter 31 and beyond),[4] but let us discuss some elementary and important aspects of phase coexistence associated with first-order phase transitions (discontinuous phase transitions).[5]

If two phases of a pure substance coexist (just as liquid water with floating ice), we may regard different phases as different compartments I and II in contact with each other through the interface. Let us study the condition for the equilibrium of these two phases (*phase equilibrium*) under constant T and P. The equilibrium condition is the minimum of the Gibbs free energy of the whole system (Chapter 15). Since there is only one chemical component,

$$0 = \delta G = \mu_I \delta N_I + \mu_{II} \delta N_{II}, \qquad (27.20)$$

Therefore, if the system is materially closed, $\delta N = \delta N_I + \delta N_{II} = 0$, so we must conclude

$$\mu_I = \mu_{II}. \qquad (27.21)$$

---

[4] What is the phase transition? Here, the reader has only to consider familiar examples such as melting of ice or boiling of water.

[5] **Continuous and discontinuous phase transitions** There are two classes of phase transitions, continuous and discontinuous (see **31.7**). In the former, there is no jump in any extensive quantities across the transition, but something strange can happen (say, their derivatives = susceptibilities diverge). In contrast, for discontinuous phase transitions at least one extensive quantity changes discontinuously at the phase transition point. For example, when ice melts, the volume jumps.

This is the phase equilibrium condition, which may be rewritten as

$$\Delta\mu = 0, \tag{27.22}$$

where Δ implies the change due to the phase transition.

27.8 The Clapeyron–Clausius Equation

It is often interesting to know what happens, e.g., to the boiling point if the pressure is reduced (cf. vacuum distillation). To this end we must understand how the chemical potential changes. The Gibbs–Duhem relation (27.12) gives us

$$d\mu = v\,dP - s\,dT, \tag{27.23}$$

where $v = V/N$ and $s = S/N$. These densities depend on the phases, but T and P are common to the coexisting phases, so (27.22) implies

$$\Delta v\,dP = \Delta s\,dT, \tag{27.24}$$

where dT and dP are changes along the phase coexistence line (the white arrow in Fig. 27.3), and $\Delta v = v_{\mathrm{I}} - v_{\mathrm{II}}$ and $\Delta s = s_{\mathrm{I}} - s_{\mathrm{II}}$. (27.24) implies

$$\left.\frac{dP}{dT}\right|_{\text{coexist.}} = \frac{\Delta s}{\Delta v} = \frac{s_{\mathrm{I}} - s_{\mathrm{II}}}{v_{\mathrm{I}} - v_{\mathrm{II}}}, \tag{27.25}$$

which is called the *Clapeyron–Clausius equation*. Here, v may be interpreted as the molar volume (volume/one mole) and s as the molar entropy. Since the entropy change Δs is related to the latent heat (in terms of enthalpy) $\Delta h = h_{\mathrm{I}} - h_{\mathrm{II}}$ as $\Delta s = \Delta h/T$ with the phase transition temperature T (recall **15.10**), we can also write

$$\left.\frac{dP}{dT}\right|_{\text{coexist}} = \frac{\Delta h}{T\Delta v}. \tag{27.26}$$

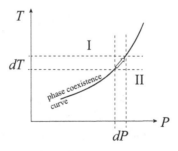

Fig. 27.3 What happens to the phase coexistence temperature if the phase coexistence pressure is changed by *dP*? Here, the thick curve describes the phase transition line between phase I and phase II.

27.9 Chemical Potential of Ideal Gas

What can thermodynamics say about the chemical potential? From (27.23), if the temperature is fixed

$$d\mu = \frac{V}{N}dP. \tag{27.27}$$

Therefore, if we know the equation of state (the PVT relation), we can say something about the chemical potential. For example, for an ideal gas we have

$$d\mu = \frac{k_B T}{P}dP. \tag{27.28}$$

That is,

$$\mu(T, P) = \mu\left(T, P^{\ominus}\right) + k_B T \log\left(P/P^{\ominus}\right). \tag{27.29}$$

Here $\ominus$ denotes a certain standard state. In practice, we often write

$$\mu(T, P) = \mu^{\ominus}(T) + k_B T \log P, \tag{27.30}$$

where $\mu^{\ominus}(T)$ is called the chemical potential of a standard state.

Let us compute the chemical potential of an ideal gas statistical-mechanically. We have computed the Helmholtz free energy, so we may obtain

$$\mu = \left(\frac{\partial A}{\partial N}\right)_{T,V}. \tag{27.31}$$

We know (from Chapter 20):

$$Z = \frac{1}{N!}\left[\left(\frac{2\pi m k_B T}{h^2}\right)^{3/2} V\right]^N = \frac{1}{N!}\left[\frac{\sqrt{2\pi}}{\lambda_T}L\right]^{3N} = \left[\frac{e n_Q V}{N}\right]^N, \tag{27.32}$$

where we have used Stirling's formula and $n_Q = (\sqrt{2\pi}/\lambda_T)^3$ (*quantum number density*) with $\lambda_T = (h^2/mk_B T)^{1/2}$, the thermal de Broglie wavelength. Numerically, n_Q may be computed as

$$n_Q = 9.88 \times 10^{29}\hat{m}^{3/2}\left(\frac{T}{300}\right)^{3/2} \simeq \hat{m}^{3/2}\left(\frac{T}{300}\right)^{3/2} \times 10^{30}, \tag{27.33}$$

where $\hat{m}$ is the mass of the particle in atomic mass unit (e.g., for water $\hat{m} = 18$). We obtain

$$A = Nk_B T \log(N/Vn_Q) - Nk_B T, \tag{27.34}$$

so

$$\mu = k_B T \log\frac{N}{Vn_Q} = k_B T \log\frac{n}{n_Q} = k_B T \log\frac{P}{k_B T n_Q}, \tag{27.35}$$

where n is the number density. This indeed has the form (27.30).

The chemical potential of the ith component of the ideal gas mixture can be obtained with the aid of Dalton's law of partial pressures (Chapter 2):

$$\mu_i(T, P) = \mu_i^{\ominus}(T) + RT \log P_i, \tag{27.36}$$

where P_i is the partial pressure of the ith gas.

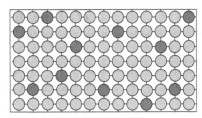

Lattice solution: N solvent molecules (pale gray) and n solute molecules (dark gray).

27.10 Chemical Potential of Ideal Solutions

Let us model the solution as a lattice model: each lattice point of the system is occupied either by a solvent molecule or by a solute molecule (Fig. 27.4). We ignore the interactions among molecules (except for the volume exclusion due to occupying the lattice points); we call such a solution an *ideal solution*. The N solvent molecules are mixed with n solute molecules to make a solution containing mole fraction $x = n/(N + n)$ of the solute. Let us discuss a dilute solution, $0 < x \ll 1$.

Let the chemical potential of the pure solvent be $\mu_0^{\ominus}(T, P)$ and that of the pure solute $\mu_s^{\ominus}(T, P)$. Then, the initial Gibbs free energy (before mixing) is $G = N\mu_0^{\ominus} + n\mu_s^{\ominus}$. After mixing, the Gibbs free energy (recall $G = E - TS + PV$; here T and P are constant) of the total system will change:

$$\Delta G = \Delta E - T\Delta S + P\Delta V, \tag{27.37}$$

but if we assume that the solution is ideal, its volume and energy do not depend on the concentration of the solute, so we may assume ΔE and ΔV are both zero. Therefore,

$$\Delta G = -T\Delta S. \tag{27.38}$$

That is, G changes only due to the mixing entropy. We model the solution as a lattice gas mixture as illustrated above, so we have (recall **18.2**):

$$\Delta S = k_B \log \binom{N + n}{n} = -Nk_B \log(1 - x) - nk_B \log x \tag{27.39}$$

immediately from Boltzmann's formula. Therefore, the Gibbs free energy after mixing is,

$$G = N\mu_0^{\ominus} + n\mu_s^{\ominus} + Nk_BT \log(1 - x) + nk_BT \log x. \tag{27.40}$$

Although we can readily read off the chemical potentials after mixing, let us honestly differentiate G to get the chemical potentials:

$$\mu_{\text{solv}} = \left(\frac{\partial G}{\partial N}\right)_{T,P,n} = \mu_0^{\ominus} + k_BT \log(1 - x), \tag{27.41}$$

$$\mu_{\text{solute}} = \mu_s^{\ominus} + k_BT \log x. \tag{27.42}$$

27.11 Raoult's Law

We could interpret the mixture as the mixed liquid consisting of two liquids I and II. The chemical potentials of the components are given by

$$\mu_I = \mu_I^{\ominus} + k_B T \log x_I, \tag{27.43}$$

$$\mu_{II} = \mu_{II}^{\ominus} + k_B T \log x_{II}, \tag{27.44}$$

where x_I (x_{II}) is the mole fraction of component I (II), the temperature is T, and the chemical potential of pure substances are marked with $\ominus$. We measure these chemical potentials using the coexisting vapor phase above the liquid mixture.

(i) Find pure phase chemical potentials $\mu_I^{\ominus}$ and $\mu_{II}^{\ominus}$: We assume that the vapors are ideal gases and their chemical potentials have the form (27.36). Denote the chemical potentials of individual pure gases at T and under the atmospheric pressure (in atm[6]) as $\mu_{IG}^{\ominus}$, $\mu_{IIG}^{\ominus}$. If we write the vapor pressures of pure liquid as P_{I0}, P_{II0} (in atm) at T, we have

$$\mu_I^{\ominus} = \mu_{IG}^{\ominus} + k_B T \log P_{I0}, \tag{27.45}$$

$$\mu_{II}^{\ominus} = \mu_{IIG}^{\ominus} + k_B T \log P_{II0}. \tag{27.46}$$

Therefore, combining these results with (27.43) and (27.44), we get the chemical potentials of the components in the liquid phase as

$$\mu_I = \mu_{IG}^{\ominus} + k_B T \log P_{I0} + k_B T \log x_I, \tag{27.47}$$

$$\mu_{II} = \mu_{IIG}^{\ominus} + k_B T \log P_{II0} + k_B T \log x_{II}. \tag{27.48}$$

(ii) Relate the actual gas phase chemical potentials to partial pressures: If the partial pressures in the vapor equilibrating with the mixture are P_I and P_{II}, then their chemical potentials read μ_{IG} and μ_{IIG} that are identical to μ_I and μ_{II}, respectively:

$$\mu_{IG} = \mu_{IG}^{\ominus} + k_B T \log P_I = \mu_I, \tag{27.49}$$

$$\mu_{IIG} = \mu_{IIG}^{\ominus} + k_B T \log P_{II} = \mu_{II}. \tag{27.50}$$

(iii) Using the results (i) and (ii). We have

$$\mu_{IG}^{\ominus} + k_B T \log P_{I0} + k_B T \log x_I = \mu_{IG}^{\ominus} + k_B T \log P_I, \tag{27.51}$$

$$\mu_{IIG}^{\ominus} + k_B T \log P_{II0} + k_B T \log x_{II} = \mu_{IIG}^{\ominus} + k_B T \log P_{II}. \tag{27.52}$$

Consequently, we have arrived at *Raoult's law*:

$$P_I = x_I P_{I0}, \quad P_{II} = x_{II} P_{II0}. \tag{27.53}$$

---

[6] Here, pressures measured in a particular unit appear in the logarithm, so we must stick to the chosen unit when we use the formulas.

That is, the partial vapor pressure of a component is identical to its pure vapor pressure times its mole fraction in the solution.[7] Equation (27.53) implies that the amount of a gas dissolved in a solvent is proportional to the gas pressure (*Henry's law*).[8]

27.12 Osmotic Pressure

The osmotic pressure π is the required "extra" pressure to prevent the solvent molecules from flowing into the solution from the pure solvent phase through the semipermeable membrane that blocks solute molecules (see Fig. 27.5). That is, the pressure of the solution side must be increased as $P \rightarrow P + \pi$. This result is, as we have already seen, an important ingredient of Einstein's theory of Brownian motion as entropic force (see Chapter 9). The both sides of the membrane must have the same chemical potential for the solvent molecules, because they can go through the membrane. Let μ_{solv} be the chemical potential of the solvent molecules in the solution and $\mu^{\ominus}$ be the chemical potential of the pure solvent. Then,

$$\mu_{\mathrm{solv}}(P + \pi, T) = \mu^{\ominus}(P + \pi, T) + k_B T \log(1 - x), \qquad (27.54)$$

which must be identical to the pure solvent chemical potential $\mu^{\ominus}(P, T)$. Therefore,

$$- \mu^{\ominus}(P + \pi, T) + \mu^{\ominus}(P, T) = k_B T \log(1 - x) \simeq -k_B T x. \qquad (27.55)$$

As we have already seen $d\mu = (V/N)dP$ under constant temperature, so (27.55) reads (Taylor expansion)

$$- (V/N)\pi = -k_B T(n/N) \implies \pi V = n k_B T, \qquad (27.56)$$

where we have used $x \simeq n/N$ (see Fig. 27.4). This is called *van 't Hoff's law*.

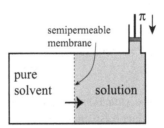

Fig. 27.5 The horizontal arrow indicates the tendency of solvent molecules to invade. The solute molecules are blocked by the semipermeable membrane, but solvent molecules are not hindered.

---

[7] François-Marie Raoult (1830–1901). He also pointed out the melting point depression (see **Q27.5** (1)) for the first time (1878). This was a key to demonstrate that electrolytes indeed dissociate (Arrhenius 1884).

[8] William Henry (1774–1836). The law was published in 1803, the year of the publication of Dalton's atomic theory.

27.13 Colligative Properties

Raoult's law, van 't Hoff's law, Henry's law, boiling-point elevation, melting-point depression (see **Q27.5**), etc., are all independent of the chemical properties of the solute and are due to the $\log(1 - x)$ or $\log x$ term (the entropic terms, recall **27.10**) in the chemical potential, so we may understand them in a unified fashion. These phenomena are traditionally said to exhibit the *colligative properties*.[9]

27.14 Chemical Reactions

Here, an elementary exposition of equilibrium chemical reactions is given. Without (highly irreversible) chemical reactions unambiguously demonstrating atomism was difficult at least.

Since the general formulas may look cumbersome, here, we use the following reaction to illustrate the general formulas:

$$N_2 + 3H_2 \leftrightarrow 2NH_3. \tag{27.57}$$

This formula implies that one molecule (or 1 mole[10]) of nitrogen reacts with three molecules (or 3 moles) of hydrogen to produce two molecules (or 2 moles) of ammonia. This does not mean that four molecules react at once; it is a summary of an appropriate set of elementary reactions.[11]

The left-hand side of (27.57) is called the *original system* (or *reactant system*) and the right-hand side the *product system*. The coefficients 2, 3 and (not explicitly written) 1 (for nitrogen) are called *stoichiometric coefficients*.

If we use the *sign convention* that the stoichiometric coefficients for the product system are all positive and those for the original system all negative, we may write the reaction in an algebraic form

$$- N_2 - 3H_2 + 2NH_3 = 0. \tag{27.58}$$

Thus, generally any reaction may be written as

$$\sum v_i C_i = 0, \tag{27.59}$$

where v_i are *signed stoichiometric coefficients* for chemical C_i; $v_i > 0$ (resp., $v_i < 0$) implies C_i is a product (resp., a reactant).

---

[9] Etymology of colligative: ligated together

[10] **Mole** The quantity of one mole (mole number) is, as already noted, defined by the same number (Avogadro's number) of molecules as contained in 12 g of ^{12}C.

[11] **Elementary reactions** As an actual *elementary reaction* in the gas phase, a reaction such as (27.57) is very unusual, because most elementary reactions are unimolecular decay (decomposition) or binary collision type reactions.

For a gas mixture with the partial pressure P_i of chemical i, as noted before, we may write its chemical potential per mole as:[12]

$$\mu_i = \mu_i^\ominus + RT \log P_i. \tag{27.60}$$

Here, $\mu_i^\ominus$ is the chemical potential for, e.g., $P_i = 1$ (in, say, MPa, atm, etc.[13]). In solutions, the chemical potential of a solute i is written as

$$\mu_i = \mu_i^\ominus(T, P) + RT \log a_i, \tag{27.61}$$

where a_i is called the *activity* of chemical i, which is close to the mole fraction x_i when the solution is dilute.[14]

27.15 Equilibrium Condition for Reactions: The Law of Mass Action

The equilibrium condition $\Delta G = 0$ for the reaction (27.59) under constant T and P reads

$$0 = \sum_i v_i \mu_i = \sum_i v_i[\mu_i^\ominus(T, P) + RT \log a_i]. \tag{27.62}$$

Or,

$$-\Delta G^\ominus \equiv -\sum_i v_i \mu_i^\ominus(T, P) = RT \log \left(\prod_i a_i^{v_i} \right). \tag{27.63}$$

The left-hand side does not depend on the chemical composition of the system, so we introduce the *chemical equilibrium constant* $K(T, P)$ according to

$$K(T, P) = e^{-\Delta G^\ominus/RT} = \frac{\cdots a_p^{v_p} \cdots}{\cdots a_r^{-v_r} \cdots}, \tag{27.64}$$

where, in the right-most expression, the numerator have all the products, and the denominator all the reactants. Equation (27.64) is called the *law of mass action*. Note that all the exponents in the above formula are positive.

Large K implies that the reaction favors the product system in equilibrium (the reaction shifts to the right). The equilibrium constant for the reaction (27.57) is given by

$$K(T, P) = \frac{[NH_3]^2}{[N_2][H_2]^3}. \tag{27.65}$$

Here, $[X]$ generally describes the partial pressure (or fugacity) of chemical X in the gas phase reaction or the mole fraction (or activity) in the solution.

---

[12] Up to this point we studied everything per molecule, so k_B appeared in the expression of chemical potentials. However, in the chemical reaction part of this chapter, the chemical potentials per mole will be used, so k_B is everywhere replaced with R.

[13] Since a dimensional quantity appears in the logarithm, when we use such formulas, we must stick to the unit being used.

[14] **Fugacity** If the gases are not ideal, then the partial pressure P_i cannot be used in (27.60). Generally, the fugacity f_i for gas i replaces P_i. In the dilute limit we have $f_i \to P_i$.

In principle, the chemical equilibrium constant may be computed statistical-mechanically. However, except for ideal gasses, it is prohibitively hard to compute the needed chemical potentials. For almost all interesting examples of chemical reactions, statistical-mechanical calculations are useless.

27.16 Shift of Chemical Equilibrium

If we differentiate the equilibrium constant with respect to T, we can obtain the *heat of reaction*, that is, ΔH (enthalpy change) due to reaction. The Gibbs–Helmholtz equation (**18.7** or its analog for the Gibbs free energy) tells us

$$\left(\frac{\partial \log K}{\partial T}\right)_P = \left(\frac{\partial(-\Delta G^{\ominus}/RT)}{\partial T}\right)_P = \frac{1}{RT^2}\left(\frac{\partial(\Delta G^{\ominus}/T)}{\partial(1/T)}\right)_P = \frac{\Delta H^{\ominus}}{RT^2}, \qquad (27.66)$$

where $\Delta H^{\ominus}$ is the enthalpy change for the "standard state." This is called *van 't Hoff's equation*. Note that in reactions the change Δ always implies (product system) − (original system).

Equation (27.66) tells us that if the reaction is exothermic (exoergic, i.e., $\Delta H^{\ominus} < 0$), then increasing the temperature shifts the reaction to reduce heat generation. That is, K decreases and the reaction tends to shift back from the product side to the original reactant side. This is an example of Le Chatelier's principle (se Chapter 25) asserting that "the response to a perturbation is in the direction to reduce its effect."

Similarly,

$$\left(\frac{\partial \log K}{\partial P}\right)_T = -\frac{\Delta V^{\ominus}}{RT}, \qquad (27.67)$$

where $\Delta V^{\ominus}$ is the volume change due to reaction for the "standard state." Equation (27.67) also illustrates Le Chatelier's principle. Needless to say, it is a manifestation of the stability of our world.

Problems

Q27.1 Euler's Theorem about Homogeneous Functions

If a function of $x_1, x_2, \ldots, x_n$ defined on an appropriate region[15] of $\mathbb{R}^n$ (n-dimensional real space) satisfies for any positive real number λ

$$f(\lambda x_1, \lambda x_2, \ldots, \lambda x_n) = \lambda^{\alpha} f(x_1, x_2, \ldots, x_n), \qquad (27.68)$$

f is called a *homogeneous function of degree α*, where α is a real number.

(1) If f is differentiable, then $\partial f/\partial x_k$ for any $k \in \{1, 2, \ldots, n\}$ is a homogeneous function of degree $\alpha - 1$. (This can be shown easily by partially differentiating (27.68).)

---

[15] This can be a cone whose apex is at the origin.

(2) If f is differentiable, a necessary and sufficient condition for f to be a homogeneous function of degree α is (*Euler's theorem* for homogeneous functions):

$$\sum_k x_k \frac{\partial f}{\partial x_k} = \alpha f. \tag{27.69}$$

Read the end of the solution for thermodynamic implications.

Solution

Differentiating (27.68) with respect to λ and setting $\lambda = 1$, we obtain (27.69). To show the converse we need a method to obtain the general solution of a quasilinear partially differential equation (27.69).[16]

Extensive quantities are homogeneous functions of degree 1 of the amounts of chemical species in the system; extensive quantities are homogeneous functions of degree 1 of extensive quantities. (1) with $\alpha = 1$ implies that intensive quantities are zeroth-order homogeneous functions of extensive quantities (amount of chemical species). (27.9) is (2) with $\alpha = 1$ applied to E as a function of extensive quantities.

Q27.2 Dilution Limit

Suppose a chemical reaction changes a chemical A whose concentration is n_A into B whose concentration is 0. Then, this chemical reaction produces "infinite" free energy that can be converted into work. Thermodynamically nothing is wrong, but this sounds too good. What happens, actually?

It is said that the "grade" of chemical energy is lower than the "grade" of mechanical energy, and the "grade" of heat energy is further lower than the "grade" of chemical energy. What does this refer to?

Solution

Since $\log 0 = -\infty$, discarding a chemical into an empty space can provide "infinite" (Gibbs) free energy. Therefore, we can "in principle" extract as much work as we wish from the reaction. We must devise a machine to utilize this. Notice that chemical reactions cannot change the long-range interactions (in condensed phases due to charge shielding).

To convert chemical energy into mechanical (or potential) energy, the former must be converted into a potential energy stored in a particular conformation. There are basically two categories of mechanism according to the relation between the conformational change and the energy change: (a) to capture the conformation with high energy or (b) to capture the right conformation that enables the efficient conversion of chemical energy to the potential energy.

A typical category (a) mechanism is the "jump (or diffusion) and catch" mechanism. Waiting for a convenient mechanical displacement (jump) to happen, the system captures the right conformation by paying chemical free energy. If we have a nice chemical to

---

[16] See, for example, Note 3.4 of Oono, Y. (2013). *The Nonlinear World*, Tokyo: Springer.

discard to prevent the reversal of the catch process, we can secure a conformation with a large potential energy. However, the captured conformation must already have a large potential energy before the chemical reaction ("catch"), and must be a product of a rare fluctuation that chemical reactions cannot enhance. Thus, even if we have a power to wield, we must be extremely patient to employ it. The mechanism (a) requires a trade-off. If we wish to utilize the full free energy, usually we must be extremely patient.

A typical category (b) mechanism is to select a right conformation in which the charge distribution rearrangement due to the chemical reaction (say, the hydrolysis of ATP) can actually exert considerable force in the predetermined direction to convert the chemical energy to the mechanical (potential) energy. In this case the waiting time may not be very long compared with (a), but utilization of the concentration difference as a source of large free energy is not allowed.

For electric motors the efficiency of converting electromagnetic energy into mechanical energy could be 98% or more without slowing it down; we need not be patient. If we are patient enough, the total chemical energy could be converted into work in contrast to heat. This is the reason why the grade of chemical energy is said to be between the mechanical and thermal energies.

Q27.3 Vapor Pressure of Silicon: An Elementary Question

The chemical potential μ_s of the atom in a solid is essentially given by the binding energy Δ of the atom in the solid: $\mu_s = -\Delta$. Obtain the formula for the equilibrium vapor pressure of the solid, and estimate the vapor pressure at room temperature of silicon for which $\Delta = 3\,\text{eV}$.

Solution

We may assume that the gas is ideal, so its chemical potential is given by $\mu = k_B T \log(n/n_Q)$. The chemical potential of the atom in the solid is $-\Delta$. Therefore the equilibrium condition (the identity of chemical potentials in two phases) gives

$$n = n_Q e^{-\beta \Delta} \implies P = k_B T n. \tag{27.70}$$

We know $\hat{m} = 28$, so (27.33) gives $n_Q = 28^{3/2} \times 10^{30} \simeq 1.5 \times 10^{32}$. Therefore, ($k_B T \sim 0.026\,\text{eV}$):

$$P = k_B T n_Q e^{-\beta \Delta} = 1.38 \times 10^{-23} \times 300 \times 1.5 \times 10^{32} e^{-3/0.026} = 4.8 \times 10^{-39} (\text{Pa}). \tag{27.71}$$

Q27.4 Making Diamond: An Elementary Question

The Gibbs free energy change ΔG at 300 K under 1 atm (almost 0 in the present context) for graphite $\longrightarrow$ diamond per 1 mole carbon is about 2.9 kJ. The molar volume of graphite is $5.3 \times 10^{-6}\,\text{m}^3$ and that of diamond is $3.4 \times 10^{-6}\,\text{m}^3$. Beyond what pressure can this reaction (at 300 K) be spontaneous (with an appropriate catalyst)? $((\partial \Delta G/\partial P)_T = \Delta V$ gives about 15 kbar.)

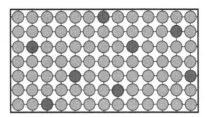

Fig. 27.6 Lattice solution model we already discussed in **27.10**.

Q27.5 Colligative Properties: Continued

Let us continue the lattice solution example to understand the effect of impurities on the phase transition temperatures.

We know the chemical potential of the solvent molecules (majority) reads

$$\mu_L = \mu_L^{\ominus} + k_B T \log(1 - x). \tag{27.72}$$

Here, x is the mole fraction of the solute molecules (dark gray particles in Fig. 27.6) in the solution. Here $\mu_L^{\ominus}$ is the chemical potential of the pure solvent liquid. Let us write the chemical potential of the pure solid phase of the solvent as $\mu_S^{\ominus}$.

(1) If we cool the solution, a solid phase of the solvent molecules emerges. When solidification occurs, impurity molecules (i.e., solute molecules) are largely excluded from the emerging solid. Let us idealize (not a bad approximation) the solid phase to be pure. Let T_m be the melting point of the pure substance. This implies

$$\mu_L^{\ominus}(T_m, P) = \mu_S^{\ominus}(T_m, P). \tag{27.73}$$

What is the equilibrium coexistence temperature of the pure solid and the solution with a mole fraction x of impurity molecules under the same pressure? Compute the melting temperature shift ΔT to order x. Assume that the latent heat of melting is L (per molecule).

(2) We may assume that the vapor of the solvent at T may be approximated as an ideal gas. At T its pressure is P. Now, we add the solute molecules as impurity to this solvent. Assume that the solute molecules cannot escape the liquid phase, so the vapor phase still consists of pure solvent molecules. What is the vapor pressure change ΔP due to the addition of the mole fraction x of the impurity molecules?

(3) The reader must have obtained $\Delta P/P = -x$ from (2), assuming the gas volume is much larger than the liquid volume so the latter may be ignored. The pressure $|\Delta P|$ is 2850 Pa, when 23.3 g of a substance is solved in 100 g of water at 100 °C. What is the molecular weight of this substance?

Solution

(1) Let us assume that at temperature $T_m + \Delta T$ the equilibrium between the solvent crystal and the solution holds:

$$\mu_S^{\ominus}(T_m + \Delta T, P) = \mu_L^{\ominus}(T_m + \Delta T, P) + k_B(T_m + \Delta T) \log(1 - x). \tag{27.74}$$

That is, to order x, we may Taylor-expand this condition as

$$\frac{\partial[\mu_S^\ominus(T_m, P) - \mu_L^\ominus(T_m, P)]}{\partial T_m}\Delta T = -k_B T_m x. \tag{27.75}$$

Denoting extensive quantities per molecule with lower case letters corresponding to the standard notation, the Gibbs relation gives $d\mu = -sdT + vdP$, so (27.75) reads (the "pure sign" $\ominus$ is omitted)

$$\left[s_L(T_m, P) - s_S(T_m, P)\right]\Delta T = -k_B T_m x. \tag{27.76}$$

From the latent heat L

$$s_L(T_m, P) - s_S(T_m, P) = L/T_m. \tag{27.77}$$

Thus, we have reached

$$\Delta T = -\frac{k_B T_m^2}{L}x < 0. \tag{27.78}$$

That is, the melting point is lowered by the amount proportional to the impurity concentration. This is called the *melting-point depression*.

(2) Let us denote the ideal gas chemical potential as $\mu_G(T, P)$. From the general form, we may write

$$\mu_G(T, P) = \mu_G^\ominus(T) + k_B T \log P. \tag{27.79}$$

Therefore, the equilibrium between the pure solvent and its vapor requires at pressure P

$$\mu_L^\ominus(T, P) = \mu_G^\ominus(T) + k_B T \log P, \tag{27.80}$$

and the equilibrium between the solution and its vapor requires at pressure $P + \Delta P$

$$\mu_L^\ominus(T, P + \Delta P) + k_B T \log(1 - x) = \mu_G^\ominus(T) + k_B T \log(P + \Delta P). \tag{27.81}$$

Subtracting (27.80) from (27.81), we obtain

$$\mu_L^\ominus(T, P + \Delta P) - \mu_L^\ominus(T, P) + k_B T \log(1 - x) = k_B T \log(1 + \Delta P/P). \tag{27.82}$$

Taylor-expanding this to order x, we get

$$\left(\frac{\partial \mu_L}{\partial P}\right)_T \Delta P - k_B Tx = \frac{k_B T}{P}\Delta P. \tag{27.83}$$

The partial derivative here gives the volume (per molecule) of the solvent liquid, which may be neglected relative to the gas volume $k_B T/P$. After this approximation we get the famous equation

$$\Delta P = -xP. \tag{27.84}$$

We can discuss the boiling point elevation similarly.

(3) Let M be the molecular weight of the solute. Then, since the ambient pressure is 1 atm (as seen from the boiling point of water):

$$x = \frac{23.3/M}{100/18 + 23.3/M} = \frac{|\Delta P|}{P} \tag{27.85}$$

or

$$\frac{23.3}{M} = \frac{(100/18)|\Delta P|/P}{1 - |\Delta P|/P} \simeq \frac{100}{18}\frac{|\Delta P|}{P} = \frac{100}{18}\frac{2850}{1.013 \times 10^5}. \tag{27.86}$$

Hence, $M \simeq 149$.

Q27.6 Ammonia Synthesis

Let us consider the reaction to synthesize ammonia:

$$N_2 + 3H_2 \longrightarrow 2NH_3. \tag{27.87}$$

Its equilibrium constant may be written in terms of partial pressures as

$$K = \frac{P^2_{NH_3}}{P_{N_2}P^3_{H_2}} = 1.5 \times 10^{-5}, \tag{27.88}$$

if the (partial) pressures are in atm (at 500 °C).

(1) If we wish to synthesize ammonia, is it more or less advantageous to increase the total pressure of the reaction vessel?
(2) Suppose the atomic ratio of N and H is 1:3 (i.e., the stoichiometric ratio). If 90% of reactants are converted into ammonia in equilibrium, what is the total pressure P of the mixture? Treat the gases as ideal gases.

Solution

(1) This is le Chatelier. Since the volume decreases if the reaction proceeds, increasing pressure should shift the reaction to the ammonia side. Or more precisely, we use

$$\left(\frac{\partial \log K}{\partial P}\right)_T = -\frac{\Delta V}{RT}, \tag{27.89}$$

where ΔV is the volume change due to the reaction. In this case it is negative, so increasing P increases K, that is, the reaction shifts to the right.

(2) We need the final mole fractions of the chemicals. If $100x\%$ of N_2 has been converted into ammonia and there was 1 mole of nitrogen gas, the total number of molecules in moles is

$$1 - x + 3(1 - x) + 2x = 4 - 2x. \tag{27.90}$$

Therefore, the partial pressures after equilibration read

$$P_{N_2} = P\frac{1 - x}{4 - 2x}, P^3_{H_2} = P\frac{3 - 3x}{4 - 2x}, P^2_{NH_3} = P\frac{2x}{4 - 2x}. \tag{27.91}$$

Hence,

$$K = \frac{[2x/(4 - 2x)]^2}{P^2[(1 - x)/(4 - 2x)][(3 - 3x)/(4 - 2x)]^3} = \frac{16x^2(2 - x)^2}{27P^2(1 - x)^4} \tag{27.92}$$

For $x = 0.9$

$$P^2 = 5808/K = 3.87 \times 10^8, \tag{27.93}$$

$P = 1.97 \times 10^4$ atm.

Q27.7 Saha Equation for Ionization

Consider the ionization reaction of hydrogen atoms: H $\longleftrightarrow$ p + e. Find the equilibrium constant $K = P_p P_e / P_H$ for this reaction. All the components may be treated as ideal gases. The ionization potential of hydrogen atom is I.

Solution

The general formula (27.36) for the chemical potential reads

$$\mu = k_B T \log \frac{P}{k_B T n_Q(m)}, \tag{27.94}$$

with

$$n_Q(m) = \left(\frac{2\pi m k_B T}{h^2} \right)^{3/2}. \tag{27.95}$$

We may set $n_Q(m_H) = n_Q(m_p)$. We must take the ionization potential into account. μ_H is stabilized by I, so

$$\mu_H = k_B T \log \frac{P_H}{k_B T n_Q(m_H)} - I. \tag{27.96}$$

Therefore, the equilibrium condition

$$\Delta \mu = \mu_p + \mu_e - \mu_H = 0 \tag{27.97}$$

implies

$$\log \frac{P_p P_e}{P_H} - \log \frac{k_B T n_Q(m_p) n_Q(m_e)}{n_Q(m_H)} + \beta I = 0 \tag{27.98}$$

Therefore,

$$\log \frac{P_p P_e}{P_H} = \log \left[k_B T n_Q(m_e) \right] - \beta I \tag{27.99}$$

Thus, we have arrived at

$$\frac{P_p}{P_H} = \frac{k_B T}{P_e} \left(\frac{2\pi m_e k_B T}{h^2} \right)^{3/2} e^{-\beta I}. \tag{27.100}$$

This is called the *Saha equation*. When this equation is used, the electron density $P_e / k_B T$ is an input.

28 Grand Canonical Ensemble

Summary

* The grand canonical formalism can handle chemostatted open systems.
* The grand canonical formalism is convenient to discuss adsorption of molecules on surfaces.
* Lifting the constraints on N, the grand canonical formalism facilitates quantum statistics.
* Thanks to ensemble equivalence, the grand canonical formalism can be used to study even isolated systems (if large enough).
* Pressures of fermions and bosons are compared.
* $PV = 2E/3$ for "any" ideal gas.

Key words

grand (canonical) partition function, grand (canonical) ensemble, fermion, boson, Bose–Einstein distribution, Fermi–Dirac distribution, Langmuir isotherm

The reader should be able to:

* Understand why PV is directly obtained from the grand canonical partition function.
* Recognize salient features of Bose–Einstein and Fermi–Dirac distributions.
* Understand the pressures of boson and fermion systems intuitively.
* Derive the one-particle density of states.

28.1 Grand Canonical Formalism

We wish to describe an open system that is chemostatted and thermostatted with a chemical potential μ and temperature T. Then, we should Legendre-transform E as follows:

$$E \to E - ST - \mu N \, (= -PV), \tag{28.1}$$

that is (if there are no other thermodynamic coordinates),

$$PV = \max_{S,N}[TS + \mu N - E] = \max_{N}[\mu N - A]. \tag{28.2}$$

Comparing this with (26.4) we see

$$PV = k_B T \log \operatorname{Tr} e^{-\beta(H-\mu N)}. \tag{28.3}$$

Thus, we see

$$\frac{PV}{T} = k_B \log \Xi(T, V, \mu) \tag{28.4}$$

with

$$\Xi(T, V, \mu) = \operatorname{Tr} e^{-\beta(H-\mu N)}. \tag{28.5}$$

The function Ξ is called the *grand (canonical) partition function*, which describes the system thermostatted and chemostatted with a reservoir at temperature T and chemical potential μ. Recall that μ is the work needed to push one molecule into the system, so by adjusting μ we can regulate the average number of particles in the system. If a system is macroscopic, fluctuation of the total number of particles is negligible compared with the total number, so we may use μ to control N. The equivalence of this formalism with the canonical formalism is discussed in **Q28.4**. Be sure to recognize that if the system is large enough (as we will see if $\log N/N \ll 1$) the ensemble equivalence tells us that even if the system is totally isolated, we may use a chemostatted system to study its thermodynamics. That is, we may use the grand canonical formalism to study isolated systems.

The Gibbs relation reads

$$d\left(\frac{PV}{T}\right) = -Ed\frac{1}{T} + \frac{P}{T}dV + Nd\frac{\mu}{T}. \tag{28.6}$$

This has a (statistical-mechanically) more convenient form:

$$d(\beta PV) = d \log \Xi = -Ed\beta + \beta PdV + Nd(\beta\mu). \tag{28.7}$$

28.2 Example: Adsorption

Let us solve an example problem (which is made deliberately slightly complicated) to become familiar with the use of chemical potentials and the grand canonical formalism. Suppose there is a gas mixture consisting of two distinct molecular species A and B. The mixture is an ideal gas and the partial pressure of X is P_X ($X = A$ or B). The gas is in equilibrium with an adsorbing metal surface on which there are N adsorption sites. Molecule X adsorbed at a site has ε_X (which is often negative) more energy than the one in the gas phase, where $X = A$ or B. Each surface site can accommodate at most one A molecule, and at most two B molecules. One adsorbed A atom has two different (internal) states (with the same energy), and one adsorbed B molecule has one state, but if two B molecules are adsorbed to the same site, then they can together have five states with the same energy (the caption of Fig. 28.1 summarizes the system). We wish to know the surface concentration of the atoms when the surface is in equilibrium with the gas mixture. The reader may assume that the gas phase is huge, so its composition does not change due to adsorption. That is, the gas phase is a chemical reservoir with given chemical potentials for the chemicals.

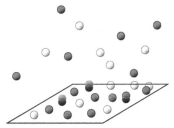

Fig. 28.1 Adsorption of gas particles on a metal surface: *A*: white (two internal states when adsorbed), *B*: gray (if singly adsorbed, with single internal state; if doubly adsorbed, with five internal states).

Since we do not know how many particles are on the surface, it must be convenient to use the grand canonical ensemble. Assuming the chemical potentials of *A* and *B* are μ_A and μ_B, respectively, write down the grand canonical partition function of the metal surface. We have only to study "single adsorption site grand canonical ensemble" Ξ_1 (recall a similar situation **18.9**). The answer is $\Xi = \Xi_1^N$:

$$\Xi = \left[1 + 2e^{-\beta(\varepsilon_A - \mu_A)} + e^{-\beta(\varepsilon_B - \mu_B)} + 5e^{-2\beta(\varepsilon_B - \mu_B)} \right]^N. \tag{28.8}$$

The needed chemical potentials can be computed as (see (27.36)):

$$\mu_A = k_B T \log(\beta P_A / n_{QA}), \quad \mu_B = k_B T \log(\beta P_B / n_{QB}). \tag{28.9}$$

Here, n_{QX} is the "quantum number density" depending on *T* and the mass (see (27.33)), and $P_{QX} = n_{QX} k_B T$ may be called the "quantum pressure." We know (we ignore the volume (or the area) change):

$$d \log \Xi = -E d\beta + N_A d(\beta \mu_A) + N_B d(\beta \mu_B), \tag{28.10}$$

so we obtain

$$N_A = N \frac{2e^{-\beta(\varepsilon_A - \mu_A)}}{1 + 2e^{-\beta(\varepsilon_A - \mu_A)} + e^{-\beta(\varepsilon_B - \mu_B)} + 5e^{-2\beta(\varepsilon_B - \mu_B)}} \tag{28.11}$$

and

$$N_B = N \frac{e^{-\beta(\varepsilon_B - \mu_B)} + 10e^{-2\beta(\varepsilon_B - \mu_B)}}{1 + 2e^{-\beta(\varepsilon_A - \mu_A)} + e^{-\beta(\varepsilon_B - \mu_B)} + 5e^{-2\beta(\varepsilon_B - \mu_B)}}. \tag{28.12}$$

28.3 Microstates for Noninteracting Indistinguishable Particle Systems

Let us consider a system consisting of noninteracting particles. Suppose the states of a single particle are numbered as $i = 1, 2, \ldots$ If all the particles are indistinguishable,[1] then in order to specify a microstate of a system consisting of such particles, we have only to

---

[1] Recall that this is the most natural situation, since molecules of the same chemical species are not distinguishable (recall **20.3**).

count the number n_i of particles in the ith one-particle state. The number n_i is often called the *occupation number* of the ith one-particle state (do NOT confuse the microstates of the whole system and the one-particle states). Or, we have only to make a table of the occupation numbers $\{n_1, n_2, \ldots\}$; we may identify this table and the microstate.

28.4 Grand Partition Function of Indistinguishable Particle System

Let ε_i be the one-particle energy of the ith one-particle state. The total energy E and the total number of particles N of the microstate $\{n_1, n_2, \ldots\}$ can be written as

$$E = \sum_{i=1} \varepsilon_i n_i \tag{28.13}$$

and

$$N = \sum_{i=1} n_i. \tag{28.14}$$

These summations are over all the one-particle states $i = 1, 2, \ldots$ Then, the grand canonical partition function must be

$$\Xi(\beta, \mu) = \sum_{\{n_1, n_2, \cdots\}} e^{-\beta E + \beta \mu N}, \tag{28.15}$$

where the summation is over all the possible occupation tables (microstates) $\{n_1, n_2, \ldots\}$. Using the microscopic descriptions of E and N, (28.13) and (28.14), we can rearrange the summation as (using the same logic as we used in the canonical formalism; recall **18.9**):

$$\Xi = \prod_i \Xi_i, \tag{28.16}$$

where

$$\Xi_i \equiv \sum_{n_i} \exp[-\beta(\varepsilon_i - \mu)n_i]. \tag{28.17}$$

Here, the summation is over all the possible occupation numbers allowed to the ith state: $n_i = 0, 1, \ldots$ The sum Ξ_i may be called the grand canonical partition function for the ith one-particle state.

28.5 Bosons and Fermions

In our world it seems that there are only two kinds of particles:

bosons: Any one-particle state has no upper bound for its occupation number.
fermions: The occupation number for any one-particle state is at most one (the Pauli exclusion principle).

There is the so-called *spin-statistics theorem* that the particles with half odd integer spins are fermions, and those with integer spins are bosons.[2] The rule applies also to composite particles such as hydrogen atoms. Thus, H and T atoms are bosons, but their nuclei are fermions. The D and ^{3}He atoms are fermions. The ^{4}He atom is a boson, and so is its nucleus.

For a neutral system consisting of positive and negative charged particles (e.g., the usual electron–nucleus system) it is proved that at least positive or negative species must be all fermions for the system to be stable. Here, "stable" means that there is a positive number B such that the system energy E satisfies $E > -NB$, where N is the number of particles in the system. That is, for the world to be stable, we need fermions.[3]

28.6 Ideal Boson Systems

For bosons, any number of particles can occupy the same one-particle state, so

$$\Xi_i = \sum_{n=0}^{\infty} e^{-\beta(\varepsilon_i - \mu)n} = \left(1 - e^{-\beta(\varepsilon_i - \mu)}\right)^{-1}. \tag{28.18}$$

The mean occupation number of the ith state is given by

$$\langle n_i \rangle = \sum_{n=0}^{\infty} n e^{-\beta(\varepsilon_i - \mu)n} / \Xi_i, \tag{28.19}$$

so we conclude

$$\langle n_i \rangle = \left(\frac{\partial \log \Xi_i}{\partial \beta \mu}\right)_\beta = k_B T \left(\frac{\partial \log \Xi_i}{\partial \mu}\right)_T = \frac{1}{e^{\beta(\varepsilon_i - \mu)} - 1}. \tag{28.20}$$

This distribution is called the *Bose–Einstein distribution*. (See Fig. 28.3 later.)

If the (one-particle) ground-state energy is zero, then the ground-state occupancy is

$$\langle n_{\text{ground}} \rangle = \frac{1}{e^{-\beta\mu} - 1}, \tag{28.21}$$

but this should not be negative, so μ cannot be positive. That is, the chemical potential must be *smaller* than the (one-particle) ground-state energy (which is usually set to be 0) to maintain the positivity of the average occupation number.

28.7 Ideal Fermion Systems

For fermions, the occupation number of a particular one-particle state is 0 or 1 (the *Pauli exclusion principle*), so

[2] Haag, R. (1996). *Local Quantum Physics: Fields, Particles, Algebras*, Second revised and enlarged edition, Berlin: Springer, Section II.5.

[3] Lieb, E. H. and Seiringer, R. (2010). *The Stability of Matter in Quantum Mechanics*, Cambridge: Cambridge University Press; a good introduction (a review of this book) is found in Slovej, J. P. (2011). *Bull. Am. Math. Soc.*, **50**, 169–174.

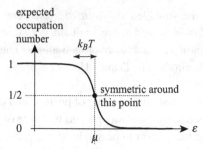

Fig. 28.2 The Fermi–Dirac distribution function. The cliff has a width of order $k_B T$, and μ is called the *Fermi level*. The symmetry noted in the figure is the so-called *particle–hole symmetry*.

$$\Xi_i = \sum_{n=0}^{1} e^{-\beta(\varepsilon_i - \mu)n} = 1 + e^{-\beta(\varepsilon_i - \mu)}. \qquad (28.22)$$

The mean occupation number of the ith state is given by

$$\langle n_i \rangle = \sum_{n=0}^{1} n e^{-\beta(\varepsilon_i - \mu)n} / \Xi_i, \qquad (28.23)$$

so we conclude

$$\langle n_i \rangle = \left(\frac{\partial \log \Xi_i}{\partial \beta \mu} \right)_\beta = k_B T \left(\frac{\partial \log \Xi_i}{\partial \mu} \right)_T = \frac{1}{e^{\beta(\varepsilon_i - \mu)} + 1}. \qquad (28.24)$$

This distribution is called the *Fermi–Dirac distribution*. See Fig. 28.3 (later) to compare with the boson case.

It is important to recognize the qualitative features of this Fermi–Dirac distribution function (see Fig. 28.2). The distribution has a cliff of width of order $k_B T$. In the $T \to 0$ limit, it has a vertical cliff at $\varepsilon = \mu$, which is called the *Fermi level*.[4] Notice that $\mu > 0$ is required, if the temperature is low enough.

28.8 Classical Limit

The distribution functions of the occupation numbers are quite different from the classical distribution function obtained by Maxwell and Boltzmann (see Fig. 28.3). The difference should be due to the quantum interference among particles when the number density is not low. Therefore, in order to obtain the classical limit, we must take the occupation number 0 limit to avoid quantum interference among particles. The chemical potential μ is a measure of the "strength" of the chemostat to push particles into the system. Thus, we must make the chemical potential sufficiently small: $\mu \to -\infty$.

---

[4] **Fermi energy and Fermi level** Do not forget that μ for a fixed N is temperature dependent. The energy of the highest occupied state at $T = 0$ is called the *Fermi energy*. The *Fermi level* denotes the chemical potential μ. However, in solid state physics sometimes this is called the Fermi energy. In this book the latter usage is strictly avoided.

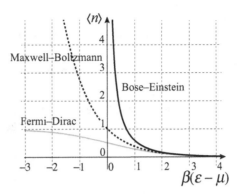

Fig. 28.3 Comparison of $\langle n \rangle$ for the Bose–Einstein (black), the Maxwell–Boltzmann (dotted) and the Fermi–Dirac (gray) cases. Needless to say, the Maxwell–Boltzmann case is meaningful only when both the quantum cases almost agree. The expression of $\langle n \rangle$ should be memorized for these cases.

In this limit both Bose–Einstein (28.21) and Fermi–Dirac distributions (28.23) reduce to the Maxwell–Boltzmann distribution as expected:

$$\langle n_i \rangle \to \mathcal{N} e^{-\beta \varepsilon_i}, \tag{28.25}$$

where $\mathcal{N} = e^{\beta \mu}$ is the normalization constant determined by the total number of particles in the system. See Fig. 28.3 to compare the classical and quantum cases.

28.9* Elementary Illustrations

Before going to the equations of state, let us try to build our intuition.

(i) Suppose there are only three one-particle states with energies 0, ε, and 3ε ($\varepsilon > 0$), and there are three indistinguishable particles. Make a table of all the microstates of the three-particle system for bosons and for fermions. The answer is Fig. 28.4.

(ii) Suppose there are 100 indistinguishable spinless bosons or fermions whose sth one-particle state has an energy $\varepsilon_s = (s - 1)\varepsilon$ ($s = 1, 2, \ldots$; $\varepsilon > 0$).[5] These particles do not interact. For the boson case, at $T = 0$ all the particles are in the lowest energy one-particle state (see Fig. 28.5). For fermions, all the low-lying one-particle states are completely filled up to some energy level that is determined by the Fermi level μ. Notice that the ground-state energy of the fermion and boson systems are quite different (if the energy origin is the one-particle ground state).

An example of low-lying excited microstates is also in Fig. 28.5(b). For the boson case all the particles have equal chance to be excited, but in the case of fermions, only the particles near the Fermi level (μ) can be excited (and excited particles leave *holes*). This should tell us something about the specific heat of these systems (cf. Fig. 29.1 later).

---

[5] "Spinless" implies that these particles do not have any internal degrees of freedom.

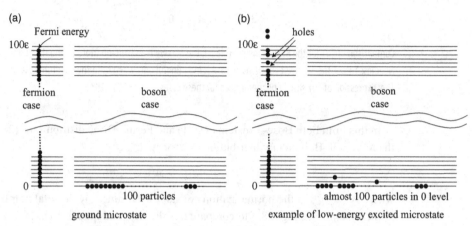

one particle level	fermion case	boson case								
3ε	-•--	-------	-------	•	• ••	-------	•	••	•••
ε	-•--	-------	•	••		•••	••	•	
0	-•--	•••	••	•		•			
total energy	4ε	0	ε	3ε	2ε	4ε	6ε	3ε	5ε	7ε	9ε

Fig. 28.4 The table of microstates: three indistinguishable particles in three one-particle states (energies 0, ε, and 3ε); fermion (left most) and boson cases. Vertical columns correspond to individual microstates.

Fig. 28.5 Ground states and low-lying excited microstates for fermions (left in each panel) and for bosons (right). Do not confuse one-particle states and microstates. Here, the lowest energy (ground state) microstate is described in (a), and one example of low-energy excited microstates is illustrated in (b).

28.10 Pressure of Ideal Systems (Same N)

The distinction between fermions and bosons shows up clearly in pressure (in what follows, the upper signs are for bosons, and the lower ones for fermions):

$$\frac{PV}{k_B T} = \log \Xi = \mp \sum_i \log \left(1 \mp e^{-\beta(\varepsilon_i - \mu)} \right). \tag{28.26}$$

If T, V, N are the same, then the pressures of the system consisting of the particles with the same single-particle energy states (i.e., with the same one-particle density of states) have the following ordering (BE = Bose–Einstein, MB = Maxwell–Boltzmann, FD = Fermi–Dirac):

$$P_{FD} > P_{MB} > P_{BE}. \tag{28.27}$$

To show this requires some trick, so a demonstration follows, but intuitively this can be understood by the extent of effective particle–particle attraction as is illustrated in Fig. 28.6. The figure not only suggests the pressures, but also suggests the extent

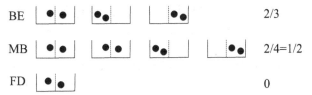

Fig. 28.6 Two-particle two-box illustration of statistics. The fractions in the right denote the relative weights of the states for which effective attraction can be seen (i.e., the states with two particles in the same box). (BE = Bose–Einstein, MB = Maxwell–Boltzmann, FD = Fermi–Dirac)

of particle density fluctuations (see **Q28.9**). The particle density fluctuations in a boson system are larger than those in a fermion system. As seen from the 'effective attraction weights' in Fig. 28.6, the fermion system exhibits the largest pressure; fermions avoid each other (Pauli's exclusion principle), so they hit the wall more often than bosons.

☐ Equation (28.27) may be demonstrated as follows: Classically, $PV = Nk_BT$, so we wish to demonstrate (N and $\langle N \rangle$ need not be distinguished, since we consider macrosystems):

$$\log \Xi_{FD} > \langle N \rangle > \log \Xi_{BE}. \tag{28.28}$$

Let us see the first inequality.[6]* Writing $x_j = e^{-\beta(\varepsilon_j - \mu)}$, we have

$$\log \Xi_{FD} - \langle N \rangle = \sum_j \left[\log(1 + x_j) - \frac{x_j}{1 + x_j} \right]. \tag{28.29}$$

We are done, because for $x > 0$:[7]*

$$\log(1 + x) - \frac{x}{1 + x} > 0. \tag{28.30}$$

Similarly, we can prove the second inequality in (28.28).

28.11* Pressure of Ideal Systems (Same μ)

If T, V, μ are the same, then the ordering is reversed:

$$P_{BE} > P_{MB} > P_{FD}. \tag{28.31}$$

To see $P_{BE} > P_{FD}$ let $x = e^{-\beta(\varepsilon_i - \mu)}$ and we compare $-\log(1 - x)$ and $\log(1 + x)$ (that is, each summand in (28.26)): $1/(1 - x) = 1 + x + x^2 + \cdots > 1 + x$ for $x \in (0, 1)$.

---

[6]*The reader might wonder why we cannot use "Ξ_{MB}" to demonstrate the formula; the reason is that μ in this grand partition function and that in Ξ_{FD} or Ξ_{BE} are distinct. Remember that we keep N fixed; inevitably μ depends on statistics, so we cannot easily compare the Boltzmann factor $e^{\beta(\varepsilon - \mu)}$ in each term.
[7]*Consider the derivatives.

28.12 Universal *P–E* Relation: Introduction

Let $\mathcal{D}(\varepsilon)d\varepsilon$ denote the number of single-particle states whose energy is between ε and $\varepsilon + d\varepsilon$. $\mathcal{D}(\varepsilon)$ is called the one-particle state density (or *density of states* of the one-particle system). If we know this, the pressure (28.26) can be rewritten as

$$PV = \mp k_B T \int d\varepsilon\, \mathcal{D}(\varepsilon) \log\left(1 \mp e^{-\beta(\varepsilon-\mu)}\right). \tag{28.32}$$

In the present case $\mathcal{D}(\varepsilon)$ is the density of states for a particle confined in a 3D box of volume V, which we will denote by $\mathcal{D}_t(\varepsilon)$, where the subscript t implies the translational degrees of freedom.

We know for a classical ideal gas (Chapter 2)

$$PV = \frac{2}{3}E, \tag{28.33}$$

where E is the internal energy. "Miraculously" (but see **28.15**), this is true for noninteracting fermions and bosons (and their mixtures) as well. The formulas for PV and E are quite different from the classical case. Here,

$$E = \int d\varepsilon\, \mathcal{D}_t(\varepsilon)\varepsilon\langle n(\varepsilon)\rangle = \int d\varepsilon\, \mathcal{D}_t(\varepsilon)\frac{\varepsilon}{e^{\beta(\varepsilon-\mu)} \mp 1}. \tag{28.34}$$

28.13 Density of States

To demonstrate (28.33) for quantum ideal gases, we need the density of states $\mathcal{D}_t(\varepsilon)$. A quick way to obtain $\mathcal{D}_t(\varepsilon)$, which may be appropriate for statistical mechanics books, is as follows. We count the number of energy eigenstates for a single particle up to some energy ε. To do this we use the classical–quantum mechanics correspondence: the number of quantum states in the phase volume element $d^3q d^3p$ is given by $d^3q d^3p/h^3$ (Chapter 20). Then, the total number of quantum states for a single particle whose energy is less than or equal to ε must be given by

$$\frac{1}{h^3}\int_{\boldsymbol{q}\in V} d^3q \int_{|\boldsymbol{p}|<\sqrt{2m\varepsilon}} d^3p = \int_0^\varepsilon d\varepsilon'\, \mathcal{D}_t(\varepsilon'), \tag{28.35}$$

that is,

$$\int_0^\varepsilon d\varepsilon'\, \mathcal{D}_t(\varepsilon') = \frac{4\pi}{h^3}V\int_0^{\sqrt{2m\varepsilon}} p^2 dp. \tag{28.36}$$

Differentiating this with ε, we get

$$\mathcal{D}_t(\varepsilon) = \frac{4\pi}{h^3}V\frac{\sqrt{2m}}{2\sqrt{\varepsilon}}\left(\sqrt{2m\varepsilon}\right)^2 = 2\pi V\left(\frac{2m}{h^2}\right)^{3/2}\varepsilon^{1/2}. \tag{28.37}$$

The reader can easily extend this approach to higher or lower dimensional spaces, and to the cases with other p–ε relations (dispersion relations) (see **Q29.1**). That $\mathcal{D}_t(\varepsilon) \propto \varepsilon^{1/2}$ is important (worth memorizing) for the case with $\varepsilon \propto p^2$ in 3-space.

That is, $\mathcal{D}_t(\varepsilon) = \gamma V \varepsilon^{1/2}$, where $\gamma = 2\pi(2m/h^2)^{3/2}$ is a constant. This can be obtained by a dimensional analytical idea as well. Here $\mathcal{D}_t(\varepsilon)h^3 d\varepsilon$ must have the dimension of the phase volume whose dimension is L^3 times [momentum]^3, so $\mathcal{D}_t(\varepsilon)h^3 \propto V\sqrt{\varepsilon}^3/\varepsilon = V\varepsilon^{1/2}$.

Note the following relation that can easily be seen from $\int \varepsilon^{1/2} d\varepsilon = (2/3)\varepsilon^{3/2} = (2\varepsilon/3)\varepsilon^{1/2}$:

$$\int_0^\varepsilon d\varepsilon \mathcal{D}_t(\varepsilon) = \frac{2}{3}\varepsilon \mathcal{D}_t(\varepsilon). \tag{28.38}$$

28.14 Universal *P–E* Relation: Demonstration[8]

The pressure can be rewritten as (the fundamental theorem of calculus):

$$PV = \mp k_B T \int d\varepsilon \left[\int_0^\varepsilon d\varepsilon' \mathcal{D}_t(\varepsilon') \right]' \log\left(1 \mp e^{-\beta(\varepsilon - \mu)}\right). \tag{28.39}$$

Performing an integration by parts, we get

$$PV = \mp k_B T \left[\left(\int_0^\varepsilon d\varepsilon' \mathcal{D}_t(\varepsilon') \right) \log\left(1 \mp e^{-\beta(\varepsilon - \mu)}\right) \right]_0^\infty$$
$$\pm k_B T \int d\varepsilon \left[\int_0^\varepsilon d\varepsilon' \mathcal{D}_t(\varepsilon') \right] \frac{d}{d\varepsilon} \log\left(1 \mp e^{-\beta(\varepsilon - \mu)}\right). \tag{28.40}$$

The first term vanishes (check this[9*]), so

$$PV - \pm k_B T \int d\varepsilon \left[\int_0^\varepsilon d\varepsilon' \mathcal{D}_t(\varepsilon') \right] \frac{\pm \beta e^{-\beta(\varepsilon - \mu)}}{1 \mp e^{-\beta(\varepsilon - \mu)}} \tag{28.41}$$

$$\overset{(28.38)}{=} \pm \frac{2}{3} k_B T \int d\varepsilon \, \varepsilon \mathcal{D}_t(\varepsilon) \frac{\pm \beta e^{-\beta(\varepsilon - \mu)}}{1 \mp e^{-\beta(\varepsilon - \mu)}} \tag{28.42}$$

$$= \frac{2}{3} \int d\varepsilon \, \mathcal{D}_t(\varepsilon) \frac{\varepsilon}{e^{\beta(\varepsilon - \mu)} \mp 1} = \frac{2}{3} E. \tag{28.43}$$

---

[8] **Dimensional analytic approach** According to the dimensional analysis E, V, m (the particle mass), and h can produce only one dimensionless quantity: $VE^{3/2}m^{3/2}/h^3$. This is because $[V] = L^3$, $[h] = [E]T$, and $[E] = ML^2/T^2$ imply $[VE^{3/2}M^{3/2}/h^3] = L^3M^{3/2}/[E]^{3/2}T^3 = 1$. This implies that $dE/E = -(2/3)dV/V$, since m is constant for a given gas. Under any mechanical change entropy S is constant, so $dE + PdV = 0$. Combining this with $dE/E = -(2/3)dV/V$, we get $-(2E/3V)dV + PdV = 0$ or $PV = (2/3)E$.

[9*] That $\varepsilon \to 0$ for the boson case is only slightly tricky, but notice that any positive power of ε times $\log \varepsilon \to 0$ in the $\varepsilon \to 0$ limit.

28.15 Virial Equation of State

We derived the universal relation $PV = 2E/3$ for any ideal gas with a detailed calculation. Although it was said that all the ideal gas results agree "miraculously" in the above, P, V, and E are purely mechanically meaningful quantities, so the relation should not be dependent on the statistics. Indeed, this is the case. We use the *virial theorem*:[10]

$$\langle K \rangle = -\frac{1}{2}\left\langle \sum q_i \cdot F_i \right\rangle, \tag{28.44}$$

where K is the total kinetic energy, F_i is the total force acting on the ith particle whose position is q_i, and $\langle \, \rangle$ is the ensemble average.

☐   Consider the canonical ensemble average[11] $\langle \sum_i q_i(t)p_i(t) \rangle$, which must be time-independent:

$$\left\langle \sum_i q_i(t)p_i(t) \right\rangle = \mathrm{Tr}\left[\sum_i (e^{-iHt/\hbar}q_i e^{iHt/\hbar})(e^{-iHt/\hbar}p_i e^{iHt/\hbar})\rho \right]$$

$$= \mathrm{Tr}\left[\sum_i q_i p_i e^{iHt/\hbar}\rho e^{-iHt/\hbar} \right] = \left\langle \sum_i q_i(0)p_i(0) \right\rangle, \tag{28.45}$$

where ρ is the canonical density operator, which is invariant (Chapter 19). Therefore,

$$\left\langle \sum_i \left(\frac{dq_i}{dt}p_i + q_i\frac{dp_i}{dt} \right) \right\rangle = 0. \tag{28.46}$$

Since $\sum \dot{q}_i p_i = 2K$, this leads to (28.44) with the aid of Heisenberg's (or Newton's) equation of motion (**A.18, A.36**).

Let us consider a gas confined in a box made of a potential well U_W. Then, the Hamiltonian of the ideal gas may be written as $K + U_W$. Equation (28.44) or (28.46) gives us

$$2\langle K \rangle = \sum \left\langle q_i \frac{\partial U_W}{\partial q_i} \right\rangle. \tag{28.47}$$

Since U_W is the effect of the box walls, its derivative is nonzero only very close to the walls. Here, $-\partial U_W/\partial q_i$ is the force the walls exert on the particle. Therefore, if we choose the surface element dS (outward normal) at q on the wall, the sum of $-q_i\partial U_W/\partial q_i$ near the surface element must be $-Pq \cdot dS$, where P is the pressure. Notice that dS points outward (outward normal), so the negative sign is needed.

$$\sum \left\langle q_i \frac{\partial U_W}{\partial q_i} \right\rangle = \int_{\partial V} q \cdot P dS = P \int_{\partial V} q \cdot dS = 3PV, \tag{28.48}$$

---

[10]   This is originally due to Clausius; the ensemble average here is replaced by the time average in the original version. Shizhong Zhang urged the author to avoid any use of time averaging in equilibrium statistical thermodynamics.

[11]   Actually, any ensemble that is time-independent would do.

where Gauss's theorem ($\operatorname{div} \boldsymbol{q} = 3$) has been used. Consequently, (28.47) gives us the ideal gas equation of state in general:

$$PV = \frac{2}{3}\langle K \rangle. \tag{28.49}$$

Problems

Q28.1 Classical Ideal Gas with the Aid of Grand Canonical Ensemble

Let us study the classical ideal gas with the aid of the grand canonical ensemble. Let μ be the chemical potential of the gas particles.

(1) Compute the grand canonical partition function for a monatomic ideal gas. Assume that the mass of the atom is m.
(2) Find the internal energy and the pressure as a function of chemical potential μ.
(3) Suppose the expectation value of the number of particles is N. How is the chemical potential determined?
(4) Are the results obtained here (especially the results of (2)) consistent with what we already know?

Solution

(1) If we are asked the question before discussing quantum ideal gases, we would proceed as

$$\Xi = \sum_{N=0}^{\infty} \frac{1}{N!} \left(\frac{2\pi m k_B T}{h^2} \right)^{3N/2} V^N e^{\beta \mu N} = \exp\left[\left(\frac{2\pi m k_B T}{h^2} \right)^{3/2} V e^{\beta \mu} \right]. \tag{28.50}$$

However, this approach uses the classical distinguishable particle picture (so $1/N!$ appears). Let us use the classical limit of quantum ideal gases as discussed in **28.8** (notice that $\mu \ll 0$):

$$\Xi_i = 1 + e^{\beta \mu} e^{-\beta \varepsilon_i}, \quad \langle n_i \rangle = e^{\beta \mu} e^{-\beta \varepsilon_i}. \tag{28.51}$$

This means

$$\Xi = \prod_i (1 + e^{\beta \mu} e^{-\beta \varepsilon_i}) = 1 + e^{\beta \mu} \sum_i e^{-\beta \varepsilon_i} = 1 + e^{\beta \mu} Z_1, \tag{28.52}$$

$$N = e^{\beta \mu} \sum_i e^{-\beta \varepsilon_i} = e^{\beta \mu} Z_1, \tag{28.53}$$

where Z_1 is the one-particle canonical partition function given by (20.2):

$$Z_1 = V \left(\frac{2\pi m k_B T}{h^2} \right)^{3/2}. \tag{28.54}$$

Equation (28.52) agrees with (28.50), since $\mu \ll 0$.

(2) From this we get

$$P = \frac{k_B T}{V} \log \Xi = \frac{k_B T}{V} e^{\beta \mu} Z_1 = \frac{N k_B T}{V}. \tag{28.55}$$

Since

$$d \log \Xi = -E d\beta + \beta P dV + N d(\beta \mu), \tag{28.56}$$

$$E = -\left(\frac{\partial \log \Xi}{\partial \beta} \right)_{\mu/T, V} = T^2 \left(\frac{\partial k_B \log \Xi}{\partial T} \right)_{\mu/T, V}$$

$$= \frac{3}{2} k_B T \left(\frac{2\pi m k_B T}{h^2} \right)^{3/2} V e^{\beta \mu} = \frac{3}{2} N k_B T, \tag{28.57}$$

consistent with $3PV/2$.

(3) Since (or as we already know from (28.53)):

$$N = \frac{\partial}{\partial \beta \mu} \log \Xi = \left(\frac{2\pi m k_B T}{h^2} \right)^{3/2} V e^{\beta \mu}, \tag{28.58}$$

we obtain

$$\mu = k_B T \log \left[\frac{N}{V} \left(\frac{2\pi m k_B T}{h^2} \right)^{-3/2} \right]. \tag{28.59}$$

The result agrees with (27.36).

(4) Thus, the results of the grand canonical ensemble completely reproduce the properties of the classical ideal gas. Notice that we did not introduce the extra factor $1/N!$ in this approach. This confirms that the particle distinguishability is the source of $N!$ (recall **20.3**). [Taste a bit of surprise, removing $N!$ from (28.5) (or (28.50)).]

Q28.2 Poisson Distribution

To study the exchange of particles between a "small volume" v and its surroundings the grand canonical partition function with the chemical potential μ is convenient. Let $Z_n = (1/n!) f^n$ be the canonical partition function for the small volume v with n ideal gas particles.

(1) Show $\langle n \rangle = f e^{\beta \mu}$ and the probability to have n particles in v is given by

$$P(n) = \frac{1}{n!} \langle n \rangle^n e^{-\langle n \rangle}, \tag{28.60}$$

which is the *Poisson distribution* with average $\langle n \rangle$ (its variance is also $\langle n \rangle$).

(2) The same problem may be seen as follows. Let N be the number of particles in the "world." Then, the binomial distribution tells us

$$P(n) = \binom{N}{n} p^n (1-p)^{N-n}, \tag{28.61}$$

where p is the probability to find a particle to be in v. Setting $p = \langle n \rangle / N$ and taking $N \to \infty$ limit, reproduce (28.60).

Solution

(1) The probability to find n particles in the volume is

$$P(n) = \frac{1}{\Xi} Z_n e^{\beta n \mu} = \frac{1}{n! \, \Xi} (f e^{\beta \mu})^n \qquad (28.62)$$

with

$$\Xi = \sum_n Z_n e^{\beta n \mu} = \exp\left(f e^{\beta \mu}\right). \qquad (28.63)$$

Since

$$\langle n \rangle = \frac{\partial \log \Xi}{\partial \beta \mu} = f e^{\beta \mu}, \qquad (28.64)$$

we can rewrite $P(n)$ just as (28.60).

(2) For $N \gg n$ $\binom{N}{n} \simeq N^n / n!$, so in the $N \to \infty$ limit

$$P(n) = \frac{N^n}{n!} \left(\frac{\langle n \rangle}{N}\right)^n \left(1 - \frac{\langle n \rangle}{N}\right)^N \to \frac{1}{n!} \langle n \rangle^n e^{-\langle n \rangle}. \qquad (28.65)$$

Q28.3 Shannon Formula Works

If we use the grand canonical distribution function in the Gibbs–Shannon information formula (21.4), we get the thermodynamically correct entropy. An analogous assertion for the von Neumann entropy formula (21.3) also holds for grand canonical density operator.

Solution

The density distribution function reads

$$\rho = \frac{1}{\Xi} e^{-\beta(H - \mu N)}. \qquad (28.66)$$

Therefore,

$$H(\rho) = -\sum_N \int d\Gamma_N \frac{1}{\Xi} e^{-\beta(H - \mu N)} [-\beta H + \beta \mu N - \log \Xi] \qquad (28.67)$$

$$= \beta E - \beta \mu N + \beta P V = S/k_B. \qquad (28.68)$$

Here, $E = TS - PV + \mu N$ has been used.

Q28.4 Adsorption

In a big box is an ideal gas A whose mass is m per molecule. Inside the box is a surface with N adsorption sites that can accommodate at most one A molecule per site. When one A is adsorbed, its energy is $-\varepsilon$ ($\varepsilon > 0$, i.e., ε lower than the free state) and has z internal states with the same energy. When the pressure of the box is P, and temperature T, what is the fraction θ of the surface occupied by A?

Solution

We regard the gas phase as a chemical reservoir, whose chemical potential μ may be calculated later (or see (27.36)). Under this chemical potential the grand canonical partition function of the adsorbing surface can be written as

$$\Xi = \prod_{i=1}^{N} \Xi_i, \tag{28.69}$$

where Ξ_i is the "grand canonical partition function for the ith adsorption site." We may write

$$\Xi_i = 1 + z e^{-\beta(-\varepsilon-\mu)} = 1 + z e^{\beta(\varepsilon+\mu)}. \tag{28.70}$$

The expectation value of the total number M of the adsorbed A molecules is N times expected number of A at a single adsorption site:

$$M = \frac{\partial \log \Xi}{\partial \beta \mu} = N \frac{z e^{\beta(\varepsilon+\mu)}}{1 + z e^{\beta(\varepsilon+\mu)}}. \tag{28.71}$$

Now, let us obtain (or copy) μ: $e^{\beta\mu} = P/P_Q$, where $P_Q = n_Q k_B T$ with $n_Q = (2\pi m k_B T/h^2)^{3/2}$. Therefore, $\theta = M/N$ reads

$$\theta = \frac{z\alpha(P/P_Q)}{1 + z\alpha(P/P_Q)} = \frac{P}{P + P_Q/z\alpha}, \tag{28.72}$$

where $\alpha = e^{\beta\varepsilon}$. This formula is called the *Langmuir isotherm*.

Q28.5 Equivalence of Canonical and Grand Canonical Ensembles

Let us check the equivalence of grand canonical and canonical ensembles. That is, if we compute thermodynamic quantities in the thermodynamic limit, both give the same answers.

The grand partition function $\Xi(T, \mu)$ and canonical partition function $Z(T, N)$ (the ground-state energy is taken to be the origin of energy) are related as

$$\Xi(T, \mu) = \sum_{N=0}^{\infty} Z(T, N) e^{\beta\mu N}. \tag{28.73}$$

Here, to be consistent with the contribution of the all zero microstate $\{0, 0, \ldots, 0\}$ to the grand canonical partition function (see (28.15)–(28.17)), we must set the $N = 0$ canonical partition function to be $Z(T, 0) = 1$.

Let us assume that the system consists of N (which is variable) particles in a box of volume V and the total interaction potential Φ among particles is bounded from below by a number proportional to the number of particles N in the system: $\Phi \geq -NB$, where B is a (positive) constant. The system Hamiltonian generally has the form of $H = K + \Phi$, where K is the kinetic energy.

Through answering the following almost trivial questions, we can demonstrate the ensemble equivalence.

(1) Show that there is a constant a such that

$$Z(T,N) \leq \left(\frac{aV}{N}\right)^N. \tag{28.74}$$

Here, for simplicity, show classically:

$$Z(T,N) \leq Z_0(T,N)e^{\beta NB}, \tag{28.75}$$

where Z_0 is the canonical partition function for the ideal gas. This is just (28.74).

(2) Show that the infinite sum defining the grand partition function actually converges. The reader may use (28.74) and $N! \sim (N/e)^N$ freely.

(3) Choose N_0 so that

$$\sum_{N=N_0}^{\infty} Z(T,N)e^{\beta\mu N} < 1. \tag{28.76}$$

Show that this N_0 may be chosen to be proportional to V (that is, N_0 is at most extensive).

(4) Show the following almost trivial bounds:

$$\max_N Z(T,N)e^{\beta\mu N} \leq \Xi(T,\mu) \leq (N_0+1)\max_N Z(T,N)e^{\beta\mu N}. \tag{28.77}$$

This implies

$$\max_N[\mu N - A] \leq PV \leq (N_0+1)\max_N[\mu N - A]. \tag{28.78}$$

Thus, PV is obtained from A via the canonical method plus thermodynamics (see (28.2)) and PV obtained directly from the grand canonical method agree as long as $\log N/N \ll 1$.

Solution

(1) The canonical partition function reads

$$Z(T,N) = \frac{1}{N!}\int d\Gamma\, e^{-\beta(K+\Phi)} \leq \frac{1}{N!}\int d\Gamma\, e^{-\beta K}e^{\beta BN} = Z_0(T,N)e^{\beta NB}, \tag{28.79}$$

where Z_0 is the canonical partition function of the ideal gas. We know the kinetic part may be factorized into the individual particle contributions, and $N! \sim (N/e)^N$, so there must be a satisfying the inequality.

Remark. Quantum mechanically, generally we demonstrate $Z(T,N) \geq e^{-\beta\mu_1 N + c}$ for $\mu_1 > \mu$ with c being a constant. The following steps required in the equivalence proof may be used with small modifications.

(2) The grand partition function is a positive term series and each term is bounded from above by the estimate in (1). Therefore,

$$\Xi(T,\mu) = \sum_{N=0}^{\infty} Z(T,N)e^{\beta\mu N} \leq \sum_{N=0}^{\infty}\left(\frac{aV}{N}\right)^N e^{\beta\mu N} = \sum_{N=0}^{\infty}\left(\frac{aVe^{\beta\mu}}{N}\right)^N. \tag{28.80}$$

That is, with the aid of Stirling's formula: $(1/N)^N \sim 1/N! \, e^N$,

$$\Xi(T, \mu) \leq \sum_{N=0}^{\infty} \frac{1}{N!} (aVe^{\beta\mu-1})^N = \exp\left(aVe^{\beta\mu-1}\right). \qquad (28.81)$$

Here $\Xi(T, \mu)$ is a positive term series bounded from above, so it must converge to a positive number.

(3) This is the tail estimation to majorize it. Any crude choice will do, so we first "overestimate" the sum beyond N_0 as

$$\sum_{N=N_0}^{\infty} Z(T, N)e^{\beta\mu N} \leq \sum_{N=N_0}^{\infty} \frac{1}{N!} (aVe^{\beta\mu-1})^N \simeq \sum_{N=N_0}^{\infty} \left(\frac{aVe^{\beta\mu}}{N}\right)^N \qquad (28.82)$$

For example, if we assume

$$\frac{aVe^{\beta\mu}}{N_0} < 0.1, \qquad (28.83)$$

then

$$\sum_{N=N_0}^{\infty} Z(T, N)e^{\beta\mu N} < \sum_{N=N_0}^{\infty} 0.1^N. \qquad (28.84)$$

The sum on the right-hand side is obviously bounded by 0.2 (by 1/9, at worst $N_0 = 1$). Thus, the choice (28.83) is enough. Such N_0 can clearly be chosen proportional to V.

(4) The grand partition function is a sum of positive terms, so it must be larger than any one term, especially larger than the largest term, in it:

$$\max_N Z(T, N)e^{\beta\mu N} \leq \Xi(T, \mu). \qquad (28.85)$$

Notice that the largest term cannot be less than 1, because the $N = 0$ term is 1. To obtain the upper bound Ξ is divided into the sum up to $N_0 - 1$ and that beyond $N_0 - 1$:

$$\Xi(T, \mu) = \sum_{N=0}^{N_0-1} Z(T, N)e^{\beta\mu N} + \sum_{N=N_0}^{\infty} Z(T, N)e^{\beta\mu N}. \qquad (28.86)$$

The second term on the right-hand side is bounded by 1, which is not larger than the maximum term in the sum, so it is bounded by $\max_N Z(T, N)e^{\beta\mu N}$. Therefore,

$$\Xi(T, \mu) \leq \sum_{N=0}^{N_0-1} Z(T, N)e^{\beta\mu N} + \max_N Z(T, N)e^{\beta\mu N}. \qquad (28.87)$$

The sum in this inequality must be less than the number of terms times the largest term:

$$\sum_{N=0}^{N_0-1} Z(T, N)e^{\beta\mu N} \leq N_0 \max_N Z(T, N)e^{\beta\mu N}. \qquad (28.88)$$

Therefore, we have

$$\Xi(T, \mu) \leq (N_0 + 1) \max_N Z(T, N)e^{\beta\mu N}. \qquad (28.89)$$

Combining this with (28.85) we get the desired result.

Q28.6 Very Elementary Problem about Ideal Particles

There is a system in which each particle can assume only three states with energies 0, ε, and ε ($\varepsilon > 0$, i.e., excited states are doubly degenerate). There are two indistinguishable particles without spin.

- (1F) When the particles are noninteracting fermions, write down the canonical partition function of this 2-particle system.
- (2F) Give the probability of finding N $(= 0, 1, 2)$ particles in the ground state.
- (3F) Compute the average occupation number N_0 of the ground state. Are the limits $T \to \infty$ and $T \to 0$ reasonable?
- (1–3B) Repeat the same problems assuming that the particles are noninteracting bosons.
- (4) In the high temperature limit what is the most important observation?

Solution

Here "degenerate" means that the energies happen to be identical but the states are clearly distinguishable like the three $2p$ orbits in the hydrogen atom.

(1F) To compute the canonical partition function, the reader must itemize all the microstates.

microstate	0	ε	ε	total energy
1	1	1	0	ε
2	1	0	1	ε
3	0	1	1	2ε

Hence,

$$Z = 2e^{-\beta\varepsilon} + e^{-2\beta\varepsilon}.$$

(2F) Let us write the desired probabilities as $P(N)$.

$$P(1) = \frac{2e^{-\beta\varepsilon}}{2e^{-\beta\varepsilon} + e^{-2\beta\varepsilon}} = \frac{2}{2 + e^{-\beta\varepsilon}}, \quad P(0) = \frac{e^{-2\beta\varepsilon}}{2e^{-\beta\varepsilon} + e^{-2\beta\varepsilon}} = \frac{1}{e^{\beta\varepsilon} + 2}, \quad P(2) = 0.$$

(3F)

$$\langle N_0 \rangle = P(1) = \frac{2}{2 + e^{-\beta\varepsilon}}$$

$T \to \infty$: $\langle N_0 \rangle = 2/3$ (yes, all the microstates are equally probable).

$T \to 0$: $\langle N_0 \rangle = 1$ (yes, the lowest level must surely be occupied).

(1B) To compute the canonical partition function, the reader must itemize all the microstates.

microstate	0	ε	ε	total energy
1	1	1	0	ε
2	1	0	1	ε
3	0	1	1	2ε
4	2	0	0	0
5	0	2	0	2ε
6	0	0	2	2ε

Hence,

$$Z = 1 + 2e^{-\beta\varepsilon} + 3e^{-2\beta\varepsilon}.$$

(2B)

$$P(1) = (2/Z)e^{-\beta\varepsilon}, \quad P(0) = (3/Z)e^{-2\beta\varepsilon}, \quad P(2) = 1/Z.$$

(3B)

$$N_0 = \frac{2 + 2e^{-\beta\varepsilon}}{1 + 2e^{-\beta\varepsilon} + 3e^{-2\beta\varepsilon}}.$$

$T \to \infty$: $N_0 = 2/3$ (yes, all the microstates are equally probable, and must be the same as the fermion case).

$T \to 0$: $N_0 = 2$ (yes, all the particles must be there).

(4) Both agree as noted above.

Q28.7 Elementary Problem on Ideal Particles

There are 120 indistinguishable spinless bosons whose nth one-particle state has an energy $\varepsilon_n = n\varepsilon$ ($n \in \mathbb{N}$). These particles do not interact. When the system is in equilibrium with the particle reservoir (chemostat) of temperature T and chemical potential μ, on average 119 particles occupy the one-particle ground state ($n = 0$), and one particle occupies the one-particle first excited state ($n = 1$). The other one-particle states are negligibly occupied.

(1) How many microstates are there with energy 4ε?
(2) Find the chemical potential μ in terms of ε (or compute μ/ε).
(3) Is the second excited state occupied only negligibly? Compute $\langle n_2 \rangle$. $\langle n_2 \rangle / \langle n_1 \rangle$ is not terribly small, so the reader might think that the problem is not self-consistent. Comment on this observation.

Solution

(1) This is equivalent to the question of expressing 4 as a sum of positive integers, ignoring the summation order (the partition problem of 4). $4 = 1 + 1 + 1 + 1 = 1 + 1 + 2 = 2 + 2 = 1 + 3 = 4$, so there are 5 ways. (The number $p(n)$ of partitions of a positive integer n is called the partition function in number theory: $p(4) = 5$.)
(2) Since

$$\langle n_0 \rangle = \frac{1}{e^{-\beta\mu} - 1} = 119, \tag{28.90}$$

$$\langle n_1 \rangle = \frac{1}{e^{\beta(\varepsilon - \mu)} - 1} = 1, \tag{28.91}$$

we have

$$-\beta\mu = \log(120/119) = 0.00836825, \tag{28.92}$$

$$\beta(\varepsilon - \mu) = \log 2 = 0.693147. \tag{28.93}$$

Hence, $\beta = 0.684779/\varepsilon$ and $\mu = -0.0122204\varepsilon$. Clearly recognize that μ is negative (does not exceed the ground-state energy).

(3) $\beta(2\varepsilon - \mu) = 2\beta(\varepsilon - \mu) - (-\beta\mu) = 2 \times 0.693147 - 0.00836825 = 1.37793$, so $\langle n_2 \rangle = 0.337$. Actually, $\langle N \rangle$ is about 0.5 particles more than 120. This is inevitable because $\log 120/120 = 0.04$, so a few percent error must be tolerated. Our result is much better than this crude error estimate.

Q28.8 T required for Classical Behavior

If we have 10^{20} neutrons confined on a surface of $1\,\mathrm{m}^2$, how high should the temperature be for the system to be classical (as a 2D system)?

Solution

As is explained in **28.8**, $e^{-\beta\mu} \gg 1$ is required. Thus, the question boils down to the estimation of μ. The chemical potential is determined by the number density $n = 10^{20}/\mathrm{m}^2$. If the behavior is classical, then the chemical potential can be calculated as in Chapter 20 (but in 2-space) and (for a neutron $m = 1.675 \times 10^{-28}$ kg):

$$e^{\beta\mu} = \left(\frac{nh^2}{2\pi m k_B T} \right) = \frac{(6.626 \times 10^{-34})^2 \times 10^{20}}{2\pi \times 1.675 \times 10^{-28} \times 1.38 \times 10^{-23}} \frac{1}{T} = \frac{3020}{T} \ll 1 \quad (28.94)$$

This suggests that we need a temperature much higher than 10^4 K.

Q28.9 Occupation Number Fluctuation

From the fluctuation response relation we know

$$\left(\frac{\partial N}{\partial \mu} \right)_T = \beta \langle \delta n^2 \rangle, \quad (28.95)$$

but for noninteracting particle cases we can separate each one-particle state, so

$$\left(\frac{\partial \langle n_i \rangle}{\partial \mu} \right)_T = \beta \langle \delta n_i^2 \rangle. \quad (28.96)$$

Using this, compute the occupation number fluctuation for bosons, fermions, and classical particles.

Solution

For the classical case, the number distribution in a small volume must obey the Poisson distribution (see **Q28.2**), so $\langle \delta n^2 \rangle = \langle n \rangle$. For fermions if n is the actual occupation number of a state, it is 0 or 1, so $n^2 = n$. Hence, $\langle \delta n^2 \rangle = \langle n^2 \rangle - \langle n \rangle^2 = \langle n \rangle - \langle n \rangle^2$. Then, we guess "by symmetry" for bosons $\langle \delta n^2 \rangle = \langle n \rangle + \langle n \rangle^2$. Let us confirm our guesses:

$$\langle \delta n_i^2 \rangle = k_B T \frac{\partial}{\partial \mu} \frac{1}{e^{\beta(\varepsilon - \mu)} \mp 1} = e^{\beta(\varepsilon - \mu)} \frac{1}{(e^{\beta(\varepsilon - \mu)} \mp 1)^2}. \quad (28.97)$$

Notice that $e^{\beta(\varepsilon-\mu)} = 1/\langle n \rangle \pm 1$, so

$$\langle \delta n_i^2 \rangle = \langle n \rangle \pm \langle n \rangle^2. \qquad (28.98)$$

Here, the upper (resp., lower) sign corresponds to bosons (resp., fermions). Thus, we have confirmed our guess; for the classical case $\langle n \rangle \ll 1$ tells us the answer from this result. Notice that the result agrees with our intuition illustrated in Fig. 28.6.

Q28.10 Fermi Level of Intrinsic Semiconductor

In the solid the one-particle energy levels make a band structure due to the periodic crystal structure (see Fig. 28.7).

If the band gap E_G between the valence band (i.e., the filled band) and the conduction band (i.e., unoccupied band) is not too large, an electron in the valence band can be thermally excited to an electron in the conduction band, leaving a hole behind. Such a solid is called an *intrinsic semiconductor*. Let us assume that electrons (resp., holes) may be understood as a free particle of effective mass m_e (resp., m_h). If we choose the energy origin to be the top of the valence band, the energy of the conducting electron with momentum $\boldsymbol{p}$ is given by $\varepsilon_c = E_G + p^2/2m_e$, and that of the hole by $\varepsilon_v = -p^2/2m_h$.

(1) Suppose there are N electrons in the system. Then, the total number of microstates in the valence band must be $N/2$ (taking the two spin states into account). Write down the equation describing the equality of the number of holes and the excited electrons. Let ε_c (resp., ε_v) denote the energy level in the conducting (resp., valence) band.

(2) Find the Fermi level μ. Assume E_G sufficiently large.

Solution

(1) The total number of electrons can be expressed as

$$N = \sum_c \frac{2}{e^{\beta(\varepsilon_c-\mu)} + 1} + \sum_v \frac{2}{e^{\beta(\varepsilon_v-\mu)} + 1}, \qquad (28.99)$$

where $\sum_c$ is the summation over all the levels in the conduction band, and $\sum_v$ that over all the levels in the valence band. Since the total number of the levels in the valence band must be $N/2$ (i.e., $\sum_v 1 = N/2$), this equation may be rewritten as

$$\sum_c \frac{1}{e^{\beta(\varepsilon_c-\mu)} + 1} = \sum_v \left[1 - \frac{1}{e^{\beta(\varepsilon_v-\mu)} + 1} \right] = \sum_v \frac{1}{e^{\beta(-\varepsilon_v+\mu)} + 1}. \qquad (28.100)$$

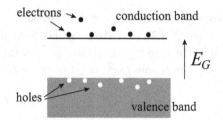

electrons   conduction band

E_G

holes

valence band

Fig. 28.7   The distribution of conducting electrons and holes in an intrinsic semiconductor.

The left-hand side of 28.100 is (1/2 of) the number of electrons in the conduction band, so the right-hand side must be (1/2 of) the number of holes in the valence band.

(2) If E_G is sufficiently large, then the number of excited electrons, and, consequently, the number of holes left, must not be large, so we may handle excited electrons and holes as classical ideal gases. Then, we can calculate their chemical potentials in terms of their effective masses and the number densities n for electrons and p for holes. For excited electrons in the conduction band (under the classical approximation):

$$n = 2 \sum_c \frac{1}{e^{\beta(\varepsilon_c - \mu)} + 1} = 2 \int_0^\infty d\varepsilon\, \mathcal{D}_t(\varepsilon) e^{-\beta(E_G + \varepsilon - \mu)} = 2 \left(\frac{2\pi m_e k_B T}{h^2} \right)^{3/2} e^{-\beta E_G} e^{\beta \mu},$$

$$(28.101)$$

where $\mathcal{D}_t$ is (28.37) with $m = m_e$ (and $V = 1$ for a unit volume). Therefore, for excited electrons

$$\mu = E_G + k_B T \log \frac{n}{2} \left(\frac{h^2}{2\pi m_e k_B T} \right)^{3/2}. \qquad (28.102)$$

For holes this reads

$$p = 2 \sum_v \frac{1}{e^{\beta(-\varepsilon_v + \mu)} + 1} = 2 \int_0^\infty d\varepsilon\, \mathcal{D}_t(\varepsilon) e^{-\beta(\varepsilon + \mu)} = 2 \left(\frac{2\pi m_h k_B T}{h^2} \right)^{3/2} e^{-\beta \mu},$$

$$(28.103)$$

where $\mathcal{D}_t$ is (28.37) with $m = m_h$. Therefore,

$$\mu = -k_B T \log \frac{p}{2} \left(\frac{h^2}{2\pi m_h k_B T} \right)^{3/2}. \qquad (28.104)$$

Since $p = n$, averaging the above two equations, we get

$$\mu = \frac{1}{2} E_G + \frac{3}{4} k_B T \log \frac{m_h}{m_e}. \qquad (28.105)$$

Q28.11 Conducting Electrons in n-Type Semiconductor

Most practical semiconductors are extrinsic semiconductors: the *n-type semiconductors* have electron donors with the electron energy level E_D below the conduction band; the *p-type semiconductors* have electron acceptors with the energy level E_A above the valence band (see Fig. 28.8). Determine the number density of the conducting particles (for electrons n_c and for holes p_c) for these two types of extrinsic semiconductors. Assume that N_D (resp., N_A) is the donor (resp., acceptor) number density. Assume that each donor or acceptor can accommodate only one electron.

Solution

The *n*-type case. The easiest approach is to use the grand canonical formalism applied to the donors. Let μ be the chemical potential of the electrons in the conduction band. Then, we may regard the band as an electron reservoir and the grand partition function for the donors reads

$$\Xi = (1 + 2e^{-\beta(-E_D - \mu)})^{N_D}, \qquad (28.106)$$

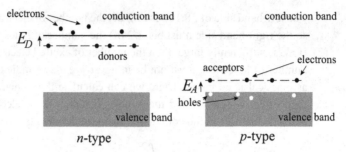

n-type p-type

Fig. 28.8 For the n-type extrinsic semiconductor donor impurities donate electrons to the conduction band. For the p-type extrinsic semiconductor acceptor impurities extract electrons from the valence band to create conducting holes.

because the donor-accommodated electrons can have two different spins and their energy is $-E_D$ relative to the reservoir (the conduction band). Therefore, the occupation number of the donors is

$$n_D = \frac{\partial \log \Xi}{\partial \beta \mu} = N_D \frac{2e^{\beta(E_D+\mu)}}{1 + 2e^{\beta(E_D+\mu)}}. \tag{28.107}$$

The chemical potential is the chemical potential of the electrons in the conduction band. These electrons are not degenerate, so we may handle them as a classical gas. Let us denote the number of conducting electrons as n_c. Then, just as (28.101) but without E_G, we have

$$n_c = 2 \left(\frac{2\pi m_e k_B T}{h^2} \right)^{3/2} e^{\beta \mu}. \tag{28.108}$$

Combining this and (28.107), we obtain

$$\frac{n_c(N_D - n_D)}{n_D} = \left(\frac{2\pi m_e k_B T}{h^2} \right)^{3/2} e^{-\beta E_D}. \tag{28.109}$$

The p-type case. Analogously to the n-type case, the grand partition function for the acceptors reads

$$\Xi = (1 + 2e^{-\beta(E_A-\mu)})^{N_A}, \tag{28.110}$$

because the excited electron from the valence band can have two spin states. Therefore, the occupation number of the acceptors is

$$n_A = N_A \frac{2e^{-\beta(E_A-\mu)}}{1 + 2e^{-\beta(E_A-\mu)}}. \tag{28.111}$$

However, in the present case, we are interested in the holes, so we introduce the hole density

$$p_A = N_A - n_A = N_A \frac{1}{1 + 2e^{-\beta(E_A+\mu)}} \tag{28.112}$$

The chemical potential is now flipped to denote the chemical potential of the holes in the valence band. Holes are not degenerate, so we may handle them as a classical gas. Let us denote the number of conducting holes as p_c. Then, just as (28.108) but the sign of μ flipped

$$p_c = 2 \left(\frac{2\pi m_h k_B T}{h^2} \right)^{3/2} e^{-\beta \mu}.$$

(28.113)

Combining this and (28.112), we obtain

$$\frac{p_c(N_A - p_A)}{p_A} = 4 \left(\frac{2\pi m_h k_B T}{h^2} \right)^{3/2} e^{-\beta E_A}.$$

(28.114)

29

Ideal Quantum Systems at Low Temperatures

Summary

* The low temperature behaviors of free fermions may be understood intuitively from the excitations of the one-particle states near the Fermi level.
* At sufficiently low temperatures a macroscopic number of particles occupy the one-particle ground state (the Einstein condensation) for free boson systems.

Key words

Fermi energy, Einstein condensation, condensate

The reader should be able to:

* Calculate various quantities for $T = 0$ fermion systems.
* Explain why the deviations from the $T = 0$ limit of E and μ start with the order T^2 terms.
* Explain intuitively why $C_V \propto T$ for fermions close to $T = 0$.
* Understand why the Einstein condensation occurs.
* Compute various thermodynamic quantities for free bosons below the condensation temperature.

Some representative values of the Fermi energy ε_F (the highest one-particle energy at $T = 0$) for metals are given in Table 29.1. Here n is the number density of electrons, v_F is the representative speed of the electron with the Fermi energy (notice that it is about 1% of the speed of light, so no relativistic correction is important); T_F is Fermi energy in K.

29.1 Noninteracting Fermion Pressure at $T = 0$

The equation of state reads $PV = 2E/3$,[1] so let us compute the internal energy at $T = 0$. The one-particle states are completely filled up to the chemical potential $\mu(0)$ at $T = 0$, which is determined by

---

[1] The origin (zero point) of the energy is the one-particle ground state. We must stick to this convention throughout the calculations in this chapter.

	n in m^{-3}	v_F in m/s	ε_F in eV	T_F in K
Table 29.1 Some representative values of the Fermi energy, ε_F				
Li	4.6×10^{16}	1.3×10^6	4.7	5.5×10^4
Na	2.5	1.1	3.1	3.7
Cu	8.50	1.56	7.0	8.2
Ag	5.76	1.38	5.5	6.4

$$N = \int_0^{\mu(0)} d\varepsilon \, \mathcal{D}_t(\varepsilon) = \gamma V \int_0^{\mu(0)} \varepsilon^{1/2} d\varepsilon = \frac{2}{3}\gamma V \mu(0)^{3/2}, \qquad (29.1)$$

where $\mathcal{D}_t(\varepsilon) = \gamma V \varepsilon^{1/2}$ with $\gamma = 2\pi(2m/h^2)^{3/2}$ (see (28.37)). We do not need details, but

$$\mu(0) \propto n^{2/3} \qquad (29.2)$$

is worth remembering, where n is the number density.

The internal energy at $T = 0$ is given by

$$E(0) = \int_0^{\mu(0)} d\varepsilon \, \mathcal{D}_t(\varepsilon)\varepsilon = \gamma V \int_0^{\mu(0)} \varepsilon^{3/2} d\varepsilon = \frac{2}{5}\gamma V \mu(0)^{5/2} = \frac{3}{5}N\mu(0). \qquad (29.3)$$

That $E(0) \propto N\mu(0)$ should be obtainable by dimensional analysis alone. Equation (29.3) implies

$$PV = \frac{2}{5}N\mu(0). \qquad (29.4)$$

Notice that this is usually very large. The reader might have realized that at $T = 0$, $E = -PV + \mu(0)N$ (recall **27.4**), so our familiar $E = 2PV/3$ gives (29.4) immediately.

29.2 Low Temperature Specific Heat of Fermions (Intuitively)

Let us intuitively discuss the electronic heat capacity of metals at low temperatures. We may assume that the Fermi–Dirac distribution is (almost) a step function. We can infer from the width of the "cliff region" of the Fermi–Dirac distribution that the number of excitable electrons is $\propto Nk_BT$ at T (recall Fig. 28.2). We know generally that the specific heat is proportional to the number of the excitable degrees of freedom, so $C_V \propto T$ at lower temperatures (see Fig. 29.1). Thus, at sufficiently low temperatures this dominates the heat capacity of metals (where T^3, coming from the lattice vibration (**23.9**), is much less than the electron contribution).

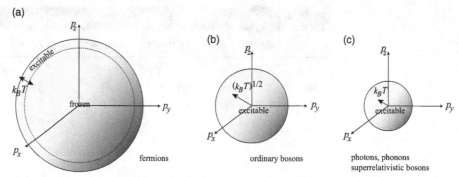

Fig. 29.1 Intuitive understanding of low temperature C_V. The specific heat C_V is proportional to the number of degrees of freedom that may be excited at temperature T. The idea may be used in any dimensional space. For fermion systems, among the occupied one-particle states, only the particles within the width $\sim k_B T$ near the top of the occupied states can be excited, so $C_V \propto T$ in any ∂-space (a). For bosons the momentum space up to $|\boldsymbol{p}| \sim \sqrt{T}$ (i.e., the square root of the energy) can be excited (b), so the specific heat should be proportional to $T^{3/2}$ in 3-space as we will see in (29.30). For superrelativistic bosons (or photons and phonons) all the particles occupying the one-particle states up to $|\boldsymbol{p}| \sim T$ (recall the energy is proportional to $|\boldsymbol{p}|$) can be excited (c), so in 3-space the specific heat is proportional to T^3 as Debye showed (Chapter 23; see also Chapter 30).

29.3 Low Temperature Approximation for Fermions

To get the result in **29.2** by calculation, we need a way to estimate the contribution of the cliff width. Let us look at the formula for N:

$$N = \int_0^\infty \mathcal{D}_t(\varepsilon) \frac{1}{e^{\beta(\varepsilon - \mu)} + 1}. \tag{29.5}$$

Let $f(\varepsilon) = 1/(e^{\beta(\varepsilon - \mu)} + 1)$. Since we know $-f'(\varepsilon)$ is concentrated sharply around $\varepsilon = \mu$ (Fig. 29.2) as a δ-function $\delta(\varepsilon - \mu)$, if T is small, we wish to exploit this fact:

$$\int_0^\infty \left[\int_0^\varepsilon \mathcal{D}_t(\varepsilon') d\varepsilon' \right]' f(\varepsilon) d\varepsilon = \int_0^\varepsilon \mathcal{D}_t(\varepsilon') d\varepsilon' f(\varepsilon) \Big|_{\varepsilon=0}^\infty - \int_0^\infty d\varepsilon \left[\int_0^\varepsilon \mathcal{D}_t(\varepsilon') d\varepsilon' \right] f'(\varepsilon)$$

$$= - \int_0^\infty d\varepsilon \left[\int_0^\varepsilon \mathcal{D}_t(\varepsilon') d\varepsilon' \right] f'(\varepsilon). \tag{29.6}$$

Now, f' is localized around μ, so we need the quantity in [ ] only near $\varepsilon = \mu$. Therefore, let us Taylor-expand it as follows:

$$\int_0^\varepsilon \mathcal{D}_t(\varepsilon') d\varepsilon' = \int_0^\mu \mathcal{D}_t(\varepsilon') d\varepsilon' + \mathcal{D}_t(\mu)(\varepsilon - \mu) + \frac{1}{2} \mathcal{D}_t'(\mu)(\varepsilon - \mu)^2 + \cdots. \tag{29.7}$$

Introducing this into (29.6), we obtain

$$\int_0^\infty \mathcal{D}_t(\varepsilon') f(\varepsilon') d\varepsilon' = \int_0^\mu \mathcal{D}_t(\varepsilon') d\varepsilon' - \frac{1}{2} \mathcal{D}_t'(\mu) \int_0^\infty (\varepsilon - \mu)^2 f'(\varepsilon) d\varepsilon + \cdots. \tag{29.8}$$

Detailed calculation is not given here, but it is clear that the correction to the $T = 0$ value is of order T^2; the integral in the second term has the dimension of energy squared, so it

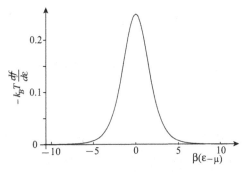

Fig. 29.2   The derivative of the Fermi distribution. Its half width is about $2k_B T$ and the height is $\beta/4$.

must be proportional to $(k_B T)^2$. If we wish to compute E in the above calculation, $\mathcal{D}_t$ is replaced by $\varepsilon \mathcal{D}_t$, but the expansion method is exactly the same, so the correction term is proportional to T^2. That is, although we do not go through any detailed calculation, we can conclude with a certain positive number α (because E must increase with T) that:[2]

$$E(T) = E(0) + \frac{1}{2}\alpha T^2 + \cdots . \tag{29.9}$$

Therefore, as we have expected above, $C_V = \alpha T$ for sufficiently small T.

29.4 Low Temperature Entropy of Fermion Systems

We know

$$\left(\frac{\partial S}{\partial T}\right)_V = \frac{C_V}{T} = \alpha. \tag{29.10}$$

This implies that under V, N constant condition $S(T) = S(0) + \alpha T$, but $S(0) = 0$ is assumed usually (Planck's absolute entropy; Chapter 23), so we conclude that for sufficiently small T

$$S(T) = C_V. \tag{29.11}$$

29.5 Low Temperature Behavior of Chemical Potential

Now, let us study the T dependence of μ. The reader may have probably guessed that $\mu(T) = \mu(0) - O[T^2]$ from the previous calculation. We know $\mu = G/N = (E - ST + PV)/N = (5E/3 - ST)/N$ for noninteracting systems. Therefore, we confirm our guess with the aid of (29.9) and (29.11): in 3-space

[2] A more detailed calculation gives $\alpha = \pi^2 k_B^2 N / 2\mu(0)$.

$$N\mu = \frac{5}{3}\left(E(0) + \frac{1}{2}\alpha T^2\right) - \alpha T^2 = \frac{5}{3}E(0) - \frac{1}{6}\alpha T^2. \tag{29.12}$$

What happens if the spatial dimension is 1? It is an increasing function of T for sufficiently low T.

29.6 Einstein Condensation for Noninteracting Bosons

Now, let us study low temperature noninteracting bosons.

Let us take the ground state energy of the system to be the origin of energy as usual. Then, the chemical potential cannot be positive. The total number of particles in the system of free bosons is given by

$$N = \sum_i \frac{1}{e^{\beta(\varepsilon_i - \mu)} - 1}, \tag{29.13}$$

where the summation is over all the one-particle states. If T is sufficiently small, the first term corresponding to the single-particle ground state can become very large (see Fig. 29.3; also recall Fig. 28.5), so in general it is dangerous to approximate (29.13) by an integral with the aid of a smooth density of state as in the fermion case (in case of fermions, each term cannot be larger than 1, so there is no problem at all with this approximation).

Let us look at the difficulty in approximating (29.13) by an integration in 3-space. In this case the density of states has the form $\mathcal{D}_t(\varepsilon) = \gamma V \varepsilon^{1/2}$ with some positive constant γ $(= 2\pi(2m/h^2)^{3/2})$. Let us write the continuous approximation to (29.13) as N_1:

$$N_1(T, \mu) \equiv \gamma V \int_0^\infty d\varepsilon \, \frac{\varepsilon^{1/2}}{e^{\beta(\varepsilon - \mu)} - 1}. \tag{29.14}$$

Here N_1 is a function of T and μ. It is an increasing function of T, and also an increasing function of μ. For a given T, if we can choose μ (which must be negative) satisfying

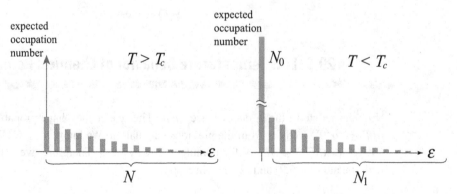

Fig. 29.3   If $T < T_c$, where T_c is the Einstein condensation temperature, the ground state is occupied by $N_0 = O[N]$ particles, so the approximation of (29.13) by an integral becomes grossly incorrect.

$N = N_1$, then we can describe the system with a continuum approximation of the grand canonical ensemble with this μ.[3]

Now, let us decrease the temperature. Then, N_1 decreases, so to keep $N_1 = N$ we must increase μ. However, we cannot indefinitely increase μ; $\mu = 0$ is the upper limit. Since $N_1(T, \mu) \leq N_1(T, 0)$,

$$\gamma V \int_0^\infty d\varepsilon \, \frac{\varepsilon^{1/2}}{e^{\beta(\varepsilon - \mu)} - 1} \leq \gamma V \int_0^\infty d\varepsilon \, \frac{\varepsilon^{1/2}}{e^{\beta\varepsilon} - 1} = \gamma V (k_B T)^{3/2} \int_0^\infty dz \frac{z^{1/2}}{e^z - 1}. \quad (29.15)$$

The integral on the right-hand side is finite. That is, with a positive constant A we may write

$$N_1(T, \mu) \leq A V T^{3/2}. \quad (29.16)$$

The equality holds when $\mu = 0$. Thus N_1 can be made indefinitely close to 0 by reducing T. However, the system should have N bosons independent of T, so there must be a temperature T_c at which

$$N = N_1(T_c, 0) = A V T_c^{3/2} \quad (29.17)$$

and for $T < T_c$

$$N > N_1(T, 0). \quad (29.18)$$

The temperature T_c is called the *Einstein condensation temperature*,[4] below which the continuum approximation breaks down. Notice that T_c is determined by (29.17):

$$T_c \propto n^{2/3}. \quad (29.19)$$

Since the system must have N particles, the remaining $N_0 = N - N_1$ must occupy certain one-particle states. The only possibility is the ground state which is not properly taken into account by the continuum approximation. A macroscopic number $N_0 (= N - N_1)$ of particles fall into the lowest energy one-particle state (see Fig. 29.3). Here, "macroscopic" implies "extensive," that is, N_0/N is a positive number in the large system size limit ($N \to \infty$ limit, the thermodynamic limit). This phenomenon is called the *Einstein condensation* (or the Bose–Einstein condensation).[5] Notice that only the one-particle ground state can be occupied by a macroscopic number of particles below T_c (see **Q29.2**).

29.7 Non-condensate Population

From (29.16) for $\mu = 0$ we know $N_1(T) = A V T^{3/2}$. $N = N_1(T_c)$, so for $T \leq T_c$ (see Fig. 29.4):

$$N_1 = N \left(\frac{T}{T_c} \right)^{3/2}. \quad (29.20)$$

---

[3] Notice that if $\mu < 0$ no finite number of one-particle states can contain a macroscopic number of particles.

[4] Usually, this is called the Bose–Einstein condensation temperature, because this is due to the peculiarity of the Bose–Einstein distribution. The condensation phenomenon was discovered by Einstein in 1924 (this was the year that de Broglie proposed his wave; Hubble recognized Andromeda is another galaxy).

[5] Perhaps it is more common to call this phenomenon "Bose–Einstein condensation," but, as already noted, it is solely due to Einstein's argument.

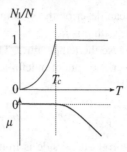

The ratio N_1/N of non-condensate atoms has a singularity at the Einstein condensation point T_c. This is a telltale sign of the existence of a phase transition. The lower panel describes the chemical potential.

Thus the condensate population reads for $T \leq T_c$.

$$N_0 = N \left[1 - \left(\frac{T}{T_c} \right)^{3/2} \right]. \tag{29.21}$$

Notice that there is a singularity at $T = T_c$.

We can estimate the chemical potential when the Einstein condensation occurs by applying the Bose–Einstein distribution to the ground state: $N_0 = 1/(e^{-\beta\mu} - 1)$. μ must be very close to 0, so

$$\frac{1}{N_0} = e^{-\beta\mu} - 1 \implies \frac{1}{N_0} = -\beta\mu \implies \mu = -\frac{k_B T}{N_0}. \tag{29.22}$$

Indeed, it is very small, since N_0 is macroscopic.

29.8 Einstein Condensation Does Not Occur in 2- and 1-Space

No Einstein condensation occurs in one- and two-dimensional free spaces, because N_1 is not bounded from above. For example, in 2-space

$$N_1 = \int_0^\infty \mathcal{D}_2(\varepsilon) \frac{1}{e^{\beta(\varepsilon-\mu)} - 1} d\varepsilon, \tag{29.23}$$

where $\mathcal{D}_2(\varepsilon)$ is the density of states of a single particle in 2-space. Let us repeat our quick derivation:

$$\int_0^\varepsilon d\varepsilon \, \mathcal{D}_2(\varepsilon) = \frac{V}{h^2} \int_{p^2/2m \leq \varepsilon} d\boldsymbol{p} = \frac{2\pi V}{h^2} \int_0^{\sqrt{2m\varepsilon}} p \, dp, \tag{29.24}$$

so

$$\mathcal{D}_2(\varepsilon) = \frac{2\pi V}{h^2} \sqrt{2m\varepsilon} \frac{d\sqrt{2m\varepsilon}}{d\varepsilon} = cV, \tag{29.25}$$

where c is a constant. Therefore,

$$N_1 \propto V \int_0^\infty \frac{1}{e^{\beta(\varepsilon-\mu)} - 1} d\varepsilon. \tag{29.26}$$

We know N_1 must be an increasing function of μ and the largest possible μ is zero for bosons, so

$$\int_0^\infty \frac{1}{e^{\beta(\varepsilon-\mu)} - 1} d\varepsilon \leq \lim_{\mu \to 0} \int_0^\infty \frac{1}{e^{\beta(\varepsilon-\mu)} - 1} d\varepsilon. \qquad (29.27)$$

The limit on the right-hand side blows up to infinity from the contribution close to $\varepsilon = 0$. Therefore, for any N and T, we can find $\mu < 0$ such that $N = N_1$. Thus, there is no Einstein condensation.

The 1D case is left for the reader.

29.9 Continuum Approximation is Always Valid for E and P

Notice that the integral expression for E or PV is still legitimate, because for these quantities the ground state does not contribute at all.[6] The Einstein condensate (i.e., N_0) does not contribute to internal energy.

29.10 Low Temperature Heat Capacity of Boson Systems

Below T_c we may set $\mu = 0$, so

$$E = \int_0^\infty d\varepsilon\, \mathcal{D}_t(\varepsilon) \frac{\varepsilon}{e^{\beta\varepsilon} - 1}. \qquad (29.28)$$

In 3-space, we know $\mathcal{D}_t(\varepsilon) = \gamma V \varepsilon^{1/2}$ with γ being a positive constant. Therefore, for $T < T_c$

$$E = \gamma V \int_0^\infty d\varepsilon\, \varepsilon^{1/2} \frac{\varepsilon}{e^{\beta\varepsilon} - 1} = \gamma V \beta^{-5/2} \int_0^\infty d(\beta\varepsilon) \frac{(\beta\varepsilon)^{3/2}}{e^{\beta\varepsilon} - 1} \propto V T^{5/2}. \qquad (29.29)$$

From this,[7] the low temperature heat capacity is, as we guessed in Fig. 29.1,

$$C_V \propto \left(\frac{T}{T_c}\right)^{3/2}. \qquad (29.30)$$

Here C_V is proportional to the number of degrees of freedom excitable at around T.

---

[6] A careful calculation shows that the ground-state contributions to E and P (with the aid of the grand canonical formalism) are of order $\log N$, which we may ignore. If not, the grand canonical formalism cannot be applied in any case to closed systems.

[7] More easily, we can simply count the power of ε. Here, we have $d\varepsilon$, $\varepsilon^{1/2}$ and ε, so $\varepsilon^{5/2}$ is the "dimension of the integral." The only relevant quantity with the dimension of energy is $k_B T$, so this integral must be proportional to $T^{5/2}$.

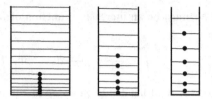

Fig. 29.5 If the system is compressed, the spacings between energy levels widen; if expanded, they shrink.

Let Us Build Our Intuition

The reader might think ideal quantized systems are artificial, but, e.g., electrons in the metal behave, with good approximation, as if they are in a vacuum without any mutual interaction. Also the superfluidity and the Einstein condensation may be closely related.

Before going to examples, let us summarize what we should not forget:

(i) How the (one-particle) energy levels of a single particle confined in a box changes if the volume is changed: If squished, the level spacings widen, and if expanded, they shrink (Fig. 29.5; see (17.17)).

(ii) If particles' energies shift with the energy levels (as if the particles just perch on the shifting energy levels), then the system changes adiabatically and reversibly, and the system entropy stays constant.[8]

(iii) Do not forget a thermodynamic relation $E = TS - PV + \mu N$.[9]

(iv) $PV = 2E/3$ is universal.[10] Also for bosons below T_c, $\mu = 0$.

29.11 Isothermal Compression

Let us halve the volume isothermally: $V \to V/2$. The isothermal condition implies that T of the initial and the final equilibrium states are the same, but anything can happen during the change. What happens to the internal energy and the pressure, if T is sufficiently low?[11]

Fermion Case

Let us consider that $T \simeq 0$. Before performing any calculation let us guess the result using elementary quantum mechanics, that is, Fig. 29.5. This clearly tells us that the internal energy must increase. Since $P \propto E/V$, obviously. P more than doubles.

To be quantitative, if T is very close to zero, we can use the quantitative results for the $T = 0$ case. We could go back to our earlier calculation, but here, let us proceed

---

[8] Such a process certainly preserves entropy, but it is far from necessary.

[9] As long as we keep the energy origin to be the one-particle ground state.

[10] As long as the system is three-dimensional and the kinetic energy is given by $mv^2/2$.

[11] When it is stated that temperature is "sufficiently low," this implies for fermions that the system is almost degenerate (the cliff of the Fermi–Dirac distribution is very sharp) and for bosons it is below T_c.

from scratch, although we already know $E \propto N\mu(0)$. From the thermodynamic relation $E = TS - PV + \mu N$, we see at $T = 0$, $E = -2E/3 + \mu(0)N$, so we get $E = (3/5)\mu(0)N$ as (29.3). We need $\mu(0)$, which is determined by (let us repeat):

$$N = \int_0^{\mu(0)} \mathcal{D}(\varepsilon)d\varepsilon. \tag{29.31}$$

Power counting: $\mathcal{D} \propto V\varepsilon^{1/2}$ and ε coming from integration implies

$$N \propto V\mu(0)^{3/2}. \tag{29.32}$$

Since N is constant, $\mu(0) \propto V^{-2/3}$, so $E \propto V^{-2/3}$ as well. Therefore, $E \to 2^{2/3}E$ by halving the volume isothermally. The pressure is obtained from $PV = 2E/3$, so $P \to 2^{5/3}P$.

Now, how does the situation change if $T > 0$? In this case, the states below the Fermi energy are not completely filled, but have holes. Therefore, although the levels are pushed up just as in Fig. 29.5, some of the particles can fall into the holes, so we may expect that the extent of increasing of E would be reduced. This is consistent with the high temperature classical case in which we know E does not change.

Boson Case

This question is not very easy if $T > T_c$, but if $T < T_c$, we can express the T-dependence of many quantities easily, because $\mu = 0$. But before any calculation let us guess the result. Look at the levels in Fig. 29.5. This implies that the condensate becomes harder to be excited because the energy gap widens by compression. Also already excited particles pushed up with the energy levels tend to tumble down into the condensate. Therefore, we expect that N_0 increases by compression, so E should go down.

Let us check this. Since $\mu = 0$, we can compute everything.

$$E = \int_0^\infty \mathcal{D}_t(\varepsilon)\varepsilon^{1/2}\frac{\varepsilon}{e^{\beta\varepsilon} - 1}d\varepsilon, \tag{29.33}$$

but without any calculation this tells us that $E \propto V$ isothermally. Thus, simply E is halved: $E \to E/2$. Since $PV = 2E/3$, P is V-independent (as we know well).

That is, the condensate is a pressure buffer. By compression we have expected that N_0 increases. How much? We can calculate by "power counting" just as we did above,

$$N_1 = \int_0^\infty \mathcal{D}_t(\varepsilon)\frac{1}{e^{\beta\varepsilon} - 1}d\varepsilon \propto VT^{3/2}, \tag{29.34}$$

N_1 is simply halved, so we can know how much N_0 increases.

29.12 Adiabatic Free Expansion

Let us suddenly double the volume $V \to 2V$ while thermally isolating the system. Irrespective of the particle nature, the system does not do any work, and does not get

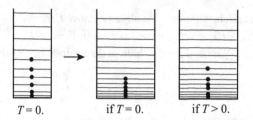

$T = 0.$ if $T = 0.$ if $T > 0.$

Fig. 29.6 Let us consider the fermion case. Suppose initially $T = 0$. If the temperature were $T = 0$ even after expansion, then there is no way to keep E constant, so $T > 0$. That is, T should increase. This argument should be effective even when the initial temperature is positive.

any heat, so E is kept constant. We may be interested in T and P of the system. Whenever we are interested in P, recall $PV = 2E/3$. Since E is constant, P is halved irrespective of the statistics.

How about T? For fermions, let us guess what would happen. See Fig. 29.6. The temperature T must be increased. To be quantitative is not easy in this case.[12]

For the classical case, we know T does not change, because E is a function of T only.

Probably, the reader guesses that for bosons ($T < T_c$ case), the temperature goes down. An intuitive argument may go as follows: since the level spacings narrow, particles in the condensate (numerous) tend to evaporate to have more energy easily, so if we wish to keep E, we must cool the system. Thus, T must go down. In this case, the reader might say the already excited particles come down with the levels, so we must heat the system, but at least at sufficiently low temperatures, N_0 is much larger than N_1 and dominates the scene. Simply perform easy calculations: we already know for $T < T_c$, $E \propto VT^{5/2}$, so clearly T goes down. We could even quantitatively predict the temperature.

29.13 Adiabatic Reversible Expansion

As the last exercise, let us study the adiabatic reversible expansion. Thermodynamically, S is constant, so (since S and N are constant) the Gibbs relation reads

$$dE = -PdV. \tag{29.35}$$

We should use $PV = 2E/3$. Combining these two, we obtain

$$dE/E = -(2/3)dV/V \tag{29.36}$$

or $EV^{2/3}$ is constant. This is true for any noninteracting particle system. Thus, from this, we can easily obtain how P changes.

---

[12] We can study the sign of

$$\left(\frac{\partial T}{\partial V}\right)_E = \frac{3}{2C_V}\left[\left(\frac{\partial P}{\partial V}\right)_T V + P\right], \tag{29.37}$$

where we have used $PV = 2E/3$, so its sign is not thermodynamically dictated.

How about the temperature? For fermions, for $T > 0$ we need quantitative $E(T)$ that we avoided. What if $T = 0$ initially? Since the entropy cannot change, we should expect $T = 0$ after the expansion. Or, according to the elementary quantum mechanical consideration, "very gentle change" means that the particles move with the energy levels (they keep sitting on the levels). Thus, the $T = 0$ case in Fig. 29.6 indeed happens. Therefore, $T = 0$, unchanged. Is this consistent with our quantitative expression for E: $E \propto N\mu(0)$? We know $\mu(0) \propto n^{2/3} \propto V^{-2/3}$, so from this we indeed obtain $EV^{2/3}$ is constant. Therefore, everything is consistent.

How about bosons with $T < T_c$? Since the entropy should not change, we expect that the distribution does not change (scales as the energy level spacings shrink). Therefore, we expect N_0 does not change. Of course, the elementary quantum argument tells us the same. Is this correct? Since $E \propto VT^{5/2}$ and since $EV^{2/3}$ must be constant, $V^{5/3}T^{5/2}$ or $VT^{3/2}$ must be constant. We know $N_1 \propto VT^{3/2}$, so indeed we have confirmed that N_1 is constant.

Problems

Q29.1 Density of One-Particle States

Find the density of states $\mathcal{D}(\varepsilon)$ (for the translational degrees of freedom) of a single particle in the volume V in 2-space with the superrelativistic dispersion relation $\varepsilon = c|\boldsymbol{p}|$.

Solution

Here, the most general solution is given with a (fairly detailed) explanation: in ∂-space with the dispersion relation $\varepsilon = \alpha|\boldsymbol{p}|^\gamma$, where α is a positive constant. Our strategy is always the same. If we can study the classical phase volume, dividing it with h^∂ in ∂-space, we can obtain the number of states for a single particle. The single-particle states with energy not exceeding ε corresponds to those with the momenta satisfying $|\boldsymbol{p}| \leq (\varepsilon/\alpha)^{1/\gamma}$. Therefore, the number of the single-particle states with energy less than ε may be written in two ways:

$$\int_0^\varepsilon d\varepsilon' \, \mathcal{D}(\varepsilon') = \frac{1}{h^\partial} \int_{\boldsymbol{q} \in V} d^\partial q \int_{|\boldsymbol{p}| \leq (\varepsilon/\alpha)^{1/\gamma}} d^\partial p. \tag{29.38}$$

The right-hand side may be rewritten by computing the position-coordinate integral (that gives V) and by introducing the polar coordinate system

$$\int_0^\varepsilon d\varepsilon' \, \mathcal{D}(\varepsilon') = \frac{1}{h^\partial} V \int_0^{(\varepsilon/\alpha)^{1/\gamma}} S_{\partial-1} p^{\partial-1} dp, \tag{29.39}$$

where $S_{\partial-1}$ is the volume of $\partial - 1$-unit sphere (see **Q17.5**, but no detail needed here). To obtain $\mathcal{D}(\varepsilon)$, we simply differentiate the above identity:[13*]

$$\mathcal{D}(\varepsilon) = \frac{V}{h^\partial} S_{\partial-1} (\varepsilon/\alpha)^{(\partial-1)/\gamma} \frac{d(\varepsilon/\alpha)^{1/\gamma}}{d\varepsilon} = \frac{V}{h^\partial} S_{\partial-1} \frac{\varepsilon^{(\partial-1)/\gamma}}{\alpha^{\partial/\gamma}} \frac{d\varepsilon^{1/\gamma}}{d\varepsilon} = \frac{S_{\partial-1} V}{\gamma h^\partial} \frac{\varepsilon^{\partial/\gamma-1}}{\alpha^{\partial/\gamma}}. \tag{29.40}$$

[13*]$f(x) = g(x)$ means $f'(x) = g'(x)$ if the first equality is an identity (and differentiable) in x.

For $\eth = 3$ and $\gamma = 2$ (the usual particle in 3-space), $S_2 = 4\pi$, $\alpha = 1/2m$ and we recover the formula (28.37). For the problem case $\eth = 2$, $\gamma = 1$, $S_1 = 2\pi$, and $\alpha = c$, so

$$\mathcal{D}(\varepsilon) = 2\pi \frac{V}{h^2} \frac{\varepsilon}{c^2}. \tag{29.41}$$

Q29.2 Do We Have Only to Treat the Ground State as Special Below T_c?

For an ideal Bose gas in 3-space we know the following integral expression is not always correct:

$$N = \sum_{i=0}^{\infty} \langle \hat{n}_i \rangle = \int_0^{\infty} d\varepsilon \, \mathcal{D}_t(\varepsilon). \tag{29.42}$$

It is because the expression ignores a large number of particles in the one-particle ground state. Thus, we are taught that if we count the number N_0 of the particles occupying the one-particle ground state and if we add this to N_1, then the number of particles in the system may be expressed correctly. However, still some of the readers may not have been convinced yet: why only ground state? Don't we have to consider the first excited state as well and have to perform the following calculation, for example:

$$\frac{N}{V} = \frac{1}{V}\langle \hat{n}_0 \rangle + \frac{1}{V}\langle \hat{n}_1 \rangle + \frac{1}{V}\int_0^{\infty} d\varepsilon \, \mathcal{D}_t(\varepsilon) \, ? \tag{29.43}$$

Let us perform a slightly more honest calculation (to recognize clearly that Einstein is always correct):

(1) Our energy coordinate convention is that the ground state is always 0: $\varepsilon_0 = 0$. Let us assume that the system is a cube of edge length L: $V = L^3$. Show that the lowest excited one-particle state energy ε_1 is a function of V.

(2) Compare the occupation numbers of the one-particle ground state and the one-particle first excited states (triply degenerate). That is, compute the ratio $(\langle \hat{n}_0 \rangle / (\langle \hat{n}_1 \rangle + \langle \hat{n}_2 \rangle + \langle \hat{n}_3 \rangle)) = \langle \hat{n}_0 \rangle / 3\langle \hat{n}_1 \rangle$ for a very small negative chemical potential, μ,[14] required by the Einstein condensation. How big is it as a function of V? Assume that the particle number density is given (say, n).

(3) We just saw in (2) except for $\langle \hat{n}_0 \rangle$ other expectation values are not extensive. That is, the ground state is really special. Excited states cannot contribute an extensive quantity unless infinitely many of them are collected. Explain that the contribution of all the excited states may be obtained accurately by replacing the summation with integration (as usual).

Solution

(1) This calculation is just as we did in Chapter 17 (see (17.17)):

$$\varepsilon_n = \frac{h^2}{8mV^{2/3}} \left(n_x^2 + n_y^2 + n_z^2 \right), \tag{29.44}$$

---

[14] Which is not zero, because the system is finite. Cf. (29.22).

where ns are positive integer quantum numbers. Therefore, the energy difference between the ground state and the first excited state is (compare $(1,1,1)$ and $(1,1,2)$ states):

$$\Delta\varepsilon = 3\frac{h^2}{8mV^{2/3}}. \tag{29.45}$$

This is the energy of the first excited state ε_1 according to our convention (to set $\varepsilon_0 = 0$).

(2) We have

$$\frac{\langle\hat{n}_0\rangle}{3\langle\hat{n}_1\rangle} = \frac{e^{\beta(\varepsilon_1-\mu)} - 1}{3(e^{-\beta\mu} - 1)}. \tag{29.46}$$

We know below T_c $\beta\mu = O[N^{-1}]$ $(< 0$; cf (29.22)). Furthermore, we know $\varepsilon_1 = O[V^{-2/3}] = O[N^{-2/3}]$, because the number density is fixed. Since $T > 0$ is a fixed temperature, however small it is (or however large β is), if we take a sufficiently large V, we may regard $\beta\varepsilon_1$ to be sufficiently small ($\beta\mu$ is much smaller than this), so we may expand as

$$\frac{\langle\hat{n}_0\rangle}{3\langle\hat{n}_1\rangle} = \frac{\varepsilon_1 - \mu}{-3\mu} = -\frac{1}{3}\varepsilon_1/\mu = O[N^{1/3}] \gg 1, \tag{29.47}$$

since we assume that the particle number density is fixed. Thus, we see that only the one-particle ground state is occupied by an extensive number of particles; any finite sum of the occupation numbers of one-particle excited states is far less than N_0 for large systems.

(3) Let $\{f(i)\}$ be a monotone decreasing sequence of positive integers and assume $\sum_{i=1}^{\infty}f(i)$ converges. Define monotone decreasing (piecewise linear continuous) functions $f_L(x)$ as $f_L(i-1) = f(i)$ for $i = 1,2,\ldots$ and $f_U(x)$ as $f_U(i) = f(i)$ for $i = 1,2,\ldots$ and $f_U(0) = f_U(1)$ (see Fig. 29.7). Then,

$$\int_0^{\infty} f_L(x)dx \le \sum_{i=1}^{\infty}f(\varepsilon) \le \int_0^{\infty} f_U(x)dx. \tag{29.48}$$

As can easily be seen from Fig. 29.7

$$\int_0^{\infty} f_U(x)dx - \int_0^{\infty} f_L(x)dx \le f(1). \tag{29.49}$$

Therefore, the difference divided by V is extremely small.

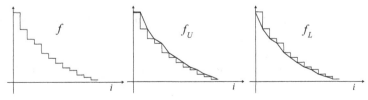

Fig. 29.7   The thick curve in the center is f_U; that in the right is f_L.

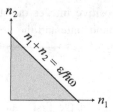

Fig. 29.8 The relation between n_1, n_2, and $\varepsilon/\hbar\omega$. Here, the zero-point energy ($\hbar\omega$) has been subtracted from ε.

Q29.3 The 2D Harmonic Trap

There is a 2-harmonic trap $U = (1/2)\alpha x^2$,[15] where x is the distance from the origin, and α is a positive constant. We know the single-particle energy levels in this trap are denoted as

$$\varepsilon = \hbar\omega(1 + n_1 + n_2), \tag{29.50}$$

where $n_1, n_2 \in \mathbb{N} = \{0, 1, 2. \ldots\}$, and ω is a positive constant.

(1) The density of states $\mathcal{D}(\varepsilon)$ is the number of states with energy between ε and $\varepsilon + d\varepsilon$. Here, however, to make the one-particle ground state to be with zero energy, let us subtract the zero-point energy $\hbar\omega$ in the following. Therefore, we know

$$\int_0^\varepsilon d\varepsilon' \, \mathcal{D}(\varepsilon') = \sum_{n_1+n_2 \in [0, \varepsilon/\hbar\omega]} 1. \tag{29.51}$$

Noting that the sum on the right-hand side is essentially the area of the shaded triangle in Fig. 29.8, obtain $\mathcal{D}(\varepsilon)$.
(2) Is there an Einstein condensation in this 2D trap?
(3) Find the number of particles $N_0(T)$ occupying the one-particle ground state.
(4) Find T_c as a function of N. For $N = 3000$, and $\omega = 10^9$ rad/s estimate T_c. (Use $\zeta(2) = \pi^2/6$.)

Solution

(1) As explained, the right-hand side must be the area of the triangle, so

$$\int_0^\varepsilon d\varepsilon' \, \mathcal{D}(\varepsilon') = \frac{1}{2} \left(\frac{\varepsilon}{\hbar\omega} \right)^2. \tag{29.52}$$

Differentiating this with ε, we obtain

$$\mathcal{D}(\varepsilon) = \frac{\varepsilon}{(\hbar\omega)^2} \propto \varepsilon. \tag{29.53}$$

(2) Let us compute (the expectation value of) the total number N_1 of particles in the excited states for $\mu = 0$:

$$N_1 = \int_0^\infty d\varepsilon \, \mathcal{D}(\varepsilon) \frac{1}{e^{\beta\varepsilon} - 1} \propto \int_0^\infty d\varepsilon \, \frac{\varepsilon}{e^{\beta\varepsilon} - 1} = (k_B T)^2 \int_0^\infty dz \, \frac{z}{e^z - 1}. \tag{29.54}$$

---

[15] This can be realized on graphene.

This integral is finite: the potentially dangerous contribution comes from small z, because for large z the integrand is exponentially small. For small z the integrand tends to a constant, so the integral is finite, and N_1 can be indefinitely small for sufficiently small T. Therefore, $N_0 = N - N_1$, the condensate population, must be macroscopic. That is, we can expect an Einstein condensation.

(3) If $\mu = 0$, we know from (2) that $N_1(T) \propto T^2$. Therefore, $N_0(T) = N[1 - (T/T_c)^2]$.

(4) To estimate T_c we need the value of A in $N_1 = AT_c^2$: at T_c

$$N = N_1 = \int_0^\infty d\varepsilon' \frac{\varepsilon'}{(\hbar\omega)^2} \frac{1}{e^{\beta\varepsilon'} - 1} = \left(\frac{k_B T}{\hbar\omega}\right)^2 \int_0^\infty \frac{x}{e^x - 1} dx = \zeta(2) \left(\frac{k_B T}{\hbar\omega}\right)^2.$$
$$(29.55)$$

Therefore, we conclude

$$T_c = \frac{\hbar\omega}{\pi k_B}(6N)^{1/2} = \frac{1.0546 \times 10^{-34} \times 10^9}{3.14 \times 1.381 \times 10^{-23}} \times \sqrt{18000} = 32.61 \times 10^{-25+23}$$
$$= 3.26 \times 10^{-1} \text{ K}. \qquad (29.56)$$

Q29.4 Peculiar Features of Condensate

Demonstrate the following peculiar features of ideal boson gas below T_c.

(1) The pressure is independent of the particle number density (we already showed this, but for completeness' sake it is stated again). It is reasonable to conclude that the volume of the condensate is subextensive.

(2) The Einstein condensation can occur at any high temperature, if sufficiently compressed.

Solution

(1) We know below T_c $\mu = 0$, so $E \propto V$ (cf. (29.29)). Hence, $PV = 2E/3$ implies that P is independent of the volume. This implies that however small V is, the pressure does not change. Thus, we must conclude that the volume of the condensate must be subextensive. Needless to say, this happens because the ideal bosons do not have any hard core.

(2) This is seen from (29.16). However high the temperature is, if V is sufficiently small, $N_1 < N$ occurs. That is, we can make the Einstein condensation at any temperature by compression. This is quite different from the usual gas–liquid phase transition which requires sufficiently low temperatures (see **31.1**).

Q29.5 Volume Increase of Ideal Quantum Gas

Assume that the particles in the gas do not interact, and answer the following questions for **both** ideal bosons and ideal fermions (both without any internal degree of freedom).

(1) The volume V of the gas is increased under constant internal energy. Does the temperature decrease? Assume that the initial temperature is sufficiently low (below T_c for bosons).

(2) The volume V is increased under constant entropy. Does the temperature decrease?

(3) The volume V is increased under constant temperature. Does the pressure decrease?

Solution

(1) If the volume is increased, the level spacings decrease.

Fermions: For fermions, if this happens at a very low temperature, then particles must be excited to go beyond the Fermi energy. Thus, T increases. More quantitatively, let us squarely consider

$$\left(\frac{\partial T}{\partial V}\right)_E = \frac{\partial(T, E)}{\partial(V, E)} = \frac{\partial(T, E)}{\partial(V, T)}\frac{\partial(V, T)}{\partial(V, E)} = -\frac{1}{C_V}\left(\frac{\partial E}{\partial V}\right)_T. \tag{29.57}$$

Using the Gibbs relation, we get

$$\left(\frac{\partial E}{\partial V}\right)_T = T\left(\frac{\partial S}{\partial V}\right)_T - P, \tag{29.58}$$

but

$$\left(\frac{\partial S}{\partial V}\right)_T = \frac{\partial(S, T)}{\partial(V, T)} = \frac{\partial(S, T)}{\partial(V, P)}\frac{\partial(V, P)}{\partial(V, T)} = \left(\frac{\partial P}{\partial T}\right)_V \tag{29.59}$$

with the aid of a Maxwell's relation. Using $P = 2E/3V$,

$$\left(\frac{\partial S}{\partial V}\right)_T = \frac{2C_V}{3V}. \tag{29.60}$$

Thus,

$$\left(\frac{\partial E}{\partial V}\right)_T = \frac{2}{3V}(TC_V - E). \tag{29.61}$$

For free fermions, we know E at low temperatures has a big T-independent chunk E_0, so for sufficiently low temperatures, this derivative must be negative. Hence, (29.57) must be positive.

Bosons: We know classically T is constant, so we could guess that for bosons T must decrease. Since the level spacings shrink generally, the gap between the ground state and the first excited state also shrinks. This destabilizes the condensate (makes the particles in it easier to "evaporate" into the non-condensate state). Thus, the total energy tends to increase. Therefore, we must cool the system to keep E constant. That is, T goes down as expected. Needless to say, under this condition the contribution from the non-condensate is opposite, but even its low-lying energy states are much less populated than the ground state below T_c, so we must cool the system. Quantitatively, the boson case is easy: Since we may assume $\mu = 0$, we know

$$E = TS - PV = TS - \frac{2}{3}E \implies E = \frac{3}{5}TS. \tag{29.62}$$

Expanding V reduces the condensate, so S increases. Hence, T must be decreased under constant E. In this case we can do better: we can explicitly obtain E from integration

$$E = \int d\varepsilon \, \mathcal{D}_t(\varepsilon)\frac{\varepsilon}{e^{\beta\varepsilon} - 1} \propto VT^{5/2}, \tag{29.63}$$

so T must be decreased as $V^{-5/2}$. We could use (29.61) which is also correct for noninteracting bosons. We know $C_V \propto T^{\theta}$ for some $\theta > 0$ (of course, we know $\theta = 3/2$, but we do not need the exact value), so

$$TC_V - E = \frac{\theta}{1+\theta} TC_V > 0, \tag{29.64}$$

which is the opposite of fermions.

(2) Since the entropy is constant, we may imagine a situation in which the particles move with the energy levels. However, the level spacings decrease, so excitation would be easier with the initial temperature. To keep S constant we must maintain the shape of the occupation number distribution. In particular, the condensate population in case of bosons and the cliff shape in case of fermions must be maintained. Hence, T must be decreased. To be quantitative, we need

$$\left(\frac{\partial T}{\partial V}\right)_S = \frac{\partial(T,S)}{\partial(V,S)} = \frac{\partial(T,S)}{\partial(V,P)}\frac{\partial(V,P)}{\partial(V,S)} = -\frac{\partial(V,P)}{\partial(V,S)} \tag{29.65}$$

$$= -\frac{\partial(V,T)}{\partial(V,S)}\frac{\partial(V,P)}{\partial(V,T)} = -\frac{T}{C_V}\left(\frac{\partial P}{\partial T}\right)_V = -\frac{2T}{3V} < 0. \tag{29.66}$$

Here, $P = 2E/3V$ has been used.

(3) $(\partial P/\partial V)_T < 0$ thermodynamically.

30

Photons and Internal Motions

Summary

* Radiation field is understood as a collection of quantized harmonic oscillators.
* The resultant Planck's radiation formula gives a finite energy density of radiation field.
* The internal degrees of freedom have (often well-separated) characteristic energy scales.

Key words

photon gas, Planck's radiation formula, ultraviolet catastrophe, Stefan–Boltzmann law, internal degrees of freedom, rotational partition functions

The reader should be able to:

* Derive Planck's formula.
* Recognize clearly the main features of Planck's formula.
* Itemize internal degrees of freedom of a molecule and telling their energy scales (in K).
* Sketch the molecular ideal gas specific heat as a function of T.

30.1 Quantization of Harmonic Degrees of Freedom

Photons and phonons are obtained through quantization of the systems that can be described as a collection of harmonic oscillators.[1] Possible energy levels for the ith mode whose angular frequency is ω_i[2] are $(n + 1/2)\hbar\omega_i$, where $n = 0, 1, 2, \ldots$ (**A.29**). The canonical partition function of a system with modes $\{\omega_i\}$ is given by (cf. Chapter 23)

$$Z(\beta) = \prod_i \left(\sum_{n_i=0}^{\infty} e^{-\beta(n_i+1/2)\hbar\omega_i} \right), \tag{30.1}$$

since no modes interact with each other. Here, the product is over all the modes. The sum in the parentheses gives the canonical partition function for a single harmonic oscillator,

---

[1] That is, the system whose Hamiltonian is quadratic in canonical coordinates (quantum mechanically in the corresponding operators).

[2] What "mode" means was explained in **20.7** footnote 6.

which we have already computed (Chapter 23). The canonical partition function (30.1) may be rewritten as:

$$Z(\beta) = \left[\prod_i \left(e^{-\beta \hbar \omega_i / 2} \right) \right] \prod_i \left(\sum_{n_i} e^{-\beta n_i \hbar \omega_i} \right) = \left[\prod_i \left(e^{-\beta \hbar \omega_i / 2} \right) \right] \Xi(\beta, 0), \quad (30.2)$$

where

$$\Xi(\beta, 0) = \prod_i \left(\sum_{n=0}^{\infty} e^{-\beta n \hbar \omega_i} \right) = \prod_i (1 - e^{-\beta \hbar \omega_i})^{-1}, \quad (30.3)$$

which may be obtained from the definition of the grand partition function (see (28.16) and (28.17)) by setting $\varepsilon_i = \hbar \omega_i$, and $\mu = 0$. As long as we consider a single system, the total zero-point energy of the system $\sum_i \hbar \omega_i / 2$ is constant and may be ignored by shifting the energy origin.[3]

Therefore, the canonical partition function of the system consisting of harmonic modes (or equivalently, consisting of photons or phonons) may be written as the bosonic grand partition function with a zero chemical potential $\Xi_{BE}(\beta, 0)$ (recall **28.6**), regarding each mode, $\hbar \omega_i$, as a one-particle state energy. From this observation, the reader should immediately recognize that T dependence of various thermodynamic quantities can be computed easily (or dimensional analytic approaches allow us to guess many T-dependent behaviors).

30.2 Warning: Grand Partition Function with $\mu = 0$ is Only a Gimmick[4]

The thermodynamic potential for the system consisting of photons or phonons is the Helmholtz free energy A whose independent variables are T and V. The expected number $\langle n_i \rangle$ of phonons (photons) of mode i is completely determined by the temperature T and the volume V. Notice that we do not have any more "handle" like μ to modify the expectation value.

If we set formally $\mu = 0$, then $dA = -SdT - PdV$, so we have $A = -PV$ (see Chapter 27 around the Gibbs–Duhem equation in **27.4**). That is, our observation $\log Z(\beta) = \log \Xi(\beta, 0)$ is consistent with thermodynamics. Thus, we may consistently describe systems consisting of phonons and photons in terms of the grand partition function

---

[3] **Warning: zero-point energies can shift** However, if the system is deformed or chemical reactions occur, the system zero-point energy can change, so we must go back to the original formula with the total zero-point energy and take into account its contribution. For the electromagnetic field, the change of the total zero-point energy may be observed as force. This is the *Casimir effect*.

[4] This is emphasized by Akira Shimizu. There is a report on the Einstein condensation of photons (Klaers, J. et al. (2011). Bose–Einstein condensation of photons in an optical microcavity, *Nature* **468**, 545–548). In this case, the experiment is performed in an artificial environment in which the number of photons mimics a conserved quantity.

for noninteracting bosons with zero chemical potential (as long as the zero-point energy is constant).

However, do not understand this relation to indicate that the chemical potentials of photons and phonons are indeed zero; actually they cannot be defined. The relation is only a mathematical formal relation that is useful sometimes.[5]

30.3 Expectation Number of Photons

The $\mu = 0$ boson analogy tells us that the average number of photons of a harmonic mode with angular frequency ω is given by (as we know, cf. (23.13) and footnote 1*):

$$\langle n \rangle = \frac{1}{e^{\beta \hbar \omega} - 1}. \tag{30.4}$$

30.4 Internal Energy of Photon Systems

The photon contribution to the internal energy of a system may be computed just as we did for the Debye model (see Chapter 23). We need the density of states (i.e., photon spectrum is equal to the distribution of the frequencies of the modes) $\mathcal{D}(\omega)$. The internal energy of all the photons is given by

$$E = \sum_{\text{modes}} \langle n(\omega) \rangle \hbar \omega = \int_0^\infty d\omega \, \mathcal{D}(\omega) \frac{\hbar \omega}{e^{\beta \hbar \omega} - 1}. \tag{30.5}$$

This is the internal energy without the contribution of zero-point energy.

A standard way to obtain the density of states $\mathcal{D}(\omega)$ is to study the wave equation governing the electromagnetic waves, but here we use our usual shortcut (28.13). The dispersion relation for photons is $\varepsilon = c|\boldsymbol{p}| = \hbar \omega$, so

$$\int_0^\omega \mathcal{D}(\omega') d\omega' = \frac{V}{h^3} \int_{|\boldsymbol{p}| \leq \hbar \omega / c} d^3 \boldsymbol{p}. \tag{30.6}$$

Differentiating the above equality, we obtain exactly as in **23.10**:

$$\mathcal{D}(\omega) = \frac{4\pi V}{h^3} \left(\frac{\hbar \omega}{c} \right)^2 \frac{\hbar}{c} = \frac{V \omega^2}{2\pi^2 c^3}. \tag{30.7}$$

Photons have two polarization directions, so the actual density of the modes is this formula times 2.

---

[5] **When chemical potential is definable** Intuitively speaking, chemical potential may be defined only for particles we can "pick up." More precisely speaking, if no (conserved) charge of some kind (say, electric charge, baryon number) is associated with the particle, its chemical potential is a dubious concept.

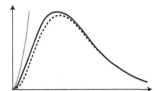

Fig. 30.1 Classical electrodynamics gives the Rayleigh–Jeans formula (30.12) (gray); this is the result of equipartition of energy. The density is not integrable due to the contributions from the UV modes (the total energy diverges). Wien reached (30.13) empirically (dotted curve). Planck's formula gives the black curve.

30.5 Planck's Distribution, or Radiation Formula

The internal energy dE_ω and the number dN_ω of photons in $[\omega, \omega + d\omega)$ in a box of volume V are given by

$$dE_\omega = 2\mathcal{D}(\omega)\frac{\hbar\omega}{e^{\beta\hbar\omega} - 1}d\omega, \qquad (30.8)$$

$$dN_\omega = 2\mathcal{D}(\omega)\frac{1}{e^{\beta\hbar\omega} - 1}d\omega. \qquad (30.9)$$

The factor 2 comes from the polarization states (i.e., $\mathcal{D}$ here is given by (30.7)).

Therefore, the energy density $u(T, \omega)$ at temperature T due to the photons with the angular frequencies in $[\omega, \omega + d\omega)$ reads

$$u(T, \omega) = \frac{\omega^2}{\pi^2 c^3}\frac{\hbar\omega}{e^{\beta\hbar\omega} - 1}. \qquad (30.10)$$

This is *Planck's radiation formula* (1900[6]) (see Fig. 30.1).

30.6 Salient Features of Planck's Law

It is important to know some qualitative features of Planck's formula (Fig. 30.1):

(i) Planck's law can explain why the spectrum blue-shifts as temperature increases: The peak position is $2.82 k_B T$ corresponding to the wavelength $\lambda = 2.9 \times 10^6/T$ nm.[7]
(ii) The total energy density $u(T) = E/V$ of a radiation field at temperature T is finite. The total energy density $u(T)$ is obtained by integration:

$$u(T) = \int_0^\infty d\omega\, u(T, \omega). \qquad (30.11)$$

With Planck's law (30.10) this is always finite (we will study this in **30.7**).

[6] [In 1900: The Paris World Exhibition; the Boxer Rebellion in Qing dynasty China; de Vries rediscovers Mendel's laws; Sibelius' *Finlandia* premiered; Evans began to unearth Knossos.]
[7] **The Sun** The surface temperature of the Sun is about 5800 K, and its radiation roughly follows Planck's law of $T = 5500$ K with the peak wavelength about 525 nm (green).

(iii) In the classical limit $\hbar \to 0$, we get

$$u(T, \omega) = \frac{k_B T \omega^2}{\pi^2 c^3} \left(= 2\mathcal{D}(\omega) k_B T \right), \tag{30.12}$$

which is the formula obtained by classical physics (i.e., the equipartition of energy; recall the last portion of **20.6**) called the *Rayleigh–Jeans formula*. Upon integration, the classical limit gives an infinite $u(T)$. This divergence is obviously due to the contribution from the high frequency modes. Thus, this difficulty is called the *ultraviolet catastrophe*, which destroyed classical physics.

(iv) In the high frequency limit $\hbar\omega \gg k_B T$ Planck's law (30.10) goes to

$$u(T, \omega) \simeq \frac{k_B T}{\pi^2 c^3} \omega^2 e^{-\beta\hbar\omega}, \tag{30.13}$$

which was empirically proposed by Wien (1896) (indicating clearly the existence of the energy gap corresponding to the quantum $\hbar\omega$).

30.7 Statistical Thermodynamics of Black-body Radiation

The total energy density $u(T)$ of the radiation field at T is given by

$$u(T) = \int_0^\infty \frac{\omega^2}{\pi^2 c^3} \frac{\hbar\omega}{e^{\beta\hbar\omega} - 1} d\omega = \beta^{-4} \int_0^\infty \frac{(\beta\omega)^2}{\pi^2 c^3} \frac{\hbar\beta\omega}{e^{\beta\hbar\omega} - 1} d(\beta\omega). \tag{30.14}$$

We see (as seen above)

$$u(T) \propto T^4, \tag{30.15}$$

which is called the *Stefan-Boltzmann law*.[8]

Since we know the T^3-law of the phonon low temperature specific heat (the Debye theory; **23.10**), this should be expected. This is understandable by counting the number of degrees of freedom (as explained in Fig. 29.1). Clearly recognize that the radiation field problem and the low temperature specific heat of crystals are the same problem.

30.8 Black-body Equation of State

The proportionality (30.15) was obtained purely thermodynamically by Boltzmann before the advent of quantum mechanics (as shown below). The proportionality constant contains $\hbar$, so it was impossible to obtain the constant theoretically before Planck.

Photons may be treated as ideal bosons with $\mu = 0$, so the equation of state is immediately obtained as (the "2" in front of the following integral is due to the polarization states):

$$\frac{PV}{k_B T} = \log \Xi = -2 \int d\varepsilon \, \mathcal{D}(\varepsilon) \log(1 - e^{-\beta\varepsilon}). \tag{30.16}$$

---

[8] The proportionality constant can be computed as $k_B^4 \pi^2 / 15\hbar^3 c^3$. cf. 23.11.

For 3D superrelativistic particles, $\mathcal{D}(\varepsilon) \propto \varepsilon^2$, so

$$\int_0^\varepsilon d\varepsilon\, \mathcal{D}(\varepsilon) = \frac{1}{3}\varepsilon \mathcal{D}(\varepsilon). \tag{30.17}$$

This gives us (review what we did to derive $PV = 2E/3$ for the ordinary particles; recall **28.13** and **28.14**):

$$PV = \frac{1}{3}E. \tag{30.18}$$

Just as $PV = 2E/3$ is a result of pure mechanics, (30.18) is a result of pure electrodynamics, so this was known before quantum mechanics. Boltzmann started with (30.18) to obtain the Stefan–Boltzmann law as follows.[9]

☐ Since we know generally

$$E = TS - PV = TS - \frac{1}{3}E, \tag{30.19}$$

$$ST = \frac{4}{3}E \text{ or } S = \frac{4}{3}\frac{E}{T}. \tag{30.20}$$

Differentiating S wrt E under constant V and noting $(\partial S/\partial E)_V = 1/T$, we obtain

$$\frac{1}{T} = -\frac{4}{3T^2}\left(\frac{\partial T}{\partial E}\right)_V E + \frac{4}{3T} \tag{30.21}$$

or

$$\frac{1}{3T} = \frac{4}{3T^2}\left(\frac{\partial T}{\partial E}\right)_V E, \tag{30.22}$$

that is, under constant V

$$\frac{dE}{E} = 4\frac{dT}{T}. \tag{30.23}$$

This implies the Stefan–Boltzmann law $E \propto T^4$. Compare this with (23.30) in **23.11**.

30.9 Internal Degrees of Freedom of Classical Ideal Gas

If noninteracting particles are sufficiently dilute ($\mu \ll 0$), we know classical ideal gas approximation is admissible. However, the internal degrees of freedom may not be handled classically, because energy gaps may be huge. We have already glimpsed at this when we discussed the gas specific heat (Chapter 23).

Let us itemize internal degrees of freedom of a molecule:

i) Each atom has a nucleus, and its ground state could have a nonzero nuclear spin. This interacts with electronic angular momentum in the atom to produce the *ultrafine structure*. The splitting due to this effect is very small, so for the temperature range

---

[9] Stefan deduced the law from Tindal's data in 1879. [In 1879: Edison patents the first practical electric light bulb; Ibsen's *A Doll's House* first performed; Frege's *Begriffsschrift* was published; Einstein was born.]

relevant to the gas phase we may assume all the levels are energetically equal. Thus, (usually) we can simply assume that the partition function is multiplied by a constant g = degeneracy of the nuclear ground state.[10]

ii) Electronic degrees of freedom has a large excitation energy (of order of ionization potential $\sim$ a few electronvolts), so unless the ground state of the orbital electrons is degenerate, we may ignore it.[11]

iii) If a molecule contains more than one atom, it can exhibit rotational motion. The quantum of rotational energy (Θ_R in **30.10**) is usually of order 10 K.[12]

iv) Also such a molecule can vibrate. The vibrational quantum (Θ_V in **30.10**) is of order 1000 K.[13]

30.10 Rotation and Vibration

Notice that there is a wide temperature range, including room temperature, where we can ignore vibrational excitations and can treat rotation classically (Fig. 30.2). Thus, equipartition of energy applied to translational and rotational degrees of freedom can explain the specific heat of many gases.

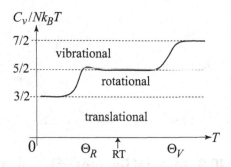

Fig. 30.2 The constant volume specific heat. RT means room temperature. Θ_R is the rotational energy quantum (in K) and Θ_V the vibrational energy quantum (in K). The hump in the rotational contribution can only be found by actual computation (no simple physical reason for it).

***

[10] **Nuclear spin-rotation interference in light homonuclear diatomic molecules** In the case of light homonuclear diatomic molecules (e.g., H_2, D_2, T_2), nuclear spins could interfere with rotational degrees of freedom through quantum statistics, so we cannot take the nuclear effect simply by g. This effect is historically important in showing protons are fermions, but is not discussed in this book (see Kubo's problem book). For not-so-light homonuclear diatomic molecules, usually we may ignore this effect.

[11] **What if the ground state is degenerate?** If the ground state is degenerate, then it could have a fine structure with an energy splitting of order a few hundred kelvin due to the spin-orbit coupling. For ground state oxygen ($^3P_2$) the splitting energy is about 200 K, so we cannot simply assume that all the states are equally probable nor that only the ground slate is relevant.

[12] However, for H_2 it is 85.4 K. For other molecules, the rotational quantum is rather small: N_2: 2.9 K; HCl: 15.1 K.

[13] N_2 3340 K; O_2: 2260 K; H_2: 6100 K.

The Hamiltonian for the internal degrees of freedom for a diatomic molecule reads

$$H = \frac{1}{2I}\hat{J}^2 + \hbar\omega\left(\hat{n} + \frac{1}{2}\right), \tag{30.24}$$

where I is the moment of inertia, $\hat{J}$ the total angular momentum operator and $\hat{n}$ the phonon number operator. Therefore, the partition function for the internal degrees of freedom reads $z_i = z_r z_v$: the rotational contribution is:[14]

$$z_r = \sum_{J=0}^{\infty}(2J + 1)e^{-(\Theta_R/T)J(J+1)}, \tag{30.25}$$

with $\Theta_R = \hbar^2/2k_B I$ and the vibrational contribution is

$$z_v = \sum_{n=0}^{\infty} e^{-(\Theta_V/T)(n+1/2)}, \tag{30.26}$$

with $\Theta_V = \hbar\omega/k_B$.

If the temperature is sufficiently low, then

$$z_r \simeq 1 + 3e^{-2\Theta_R/T}. \tag{30.27}$$

The contribution of rotation to specific heat is

$$C_{\text{rot}} \simeq 3Nk_B\left(\frac{\Theta_R}{T}\right)^2 e^{-2\Theta_R/T}. \tag{30.28}$$

For $T \gg \Theta_R$, we may approximate the summation by integration (large Js contribute, so we may approximate $J \simeq J + 1$):

$$z_r \simeq 2\int_0^{\infty} dJ J e^{-J^2(\Theta_R/T)} = \frac{T}{\Theta_R}. \tag{30.29}$$

This gives the rotational specific heat, but it is more easily obtained by the equipartition of energy, because the rotational energy is a quadratic form. Thus, $C_{\text{rot}} = k_B$ (per molecule) in the high temperature limit for a diatomic molecule (because two orthogonal rotations are allowed; each degree of freedom classically contributes $k_B/2$; recall (20.25)).

The vibrational contribution has already been studied in Chapter 23.

Problem

Q30.1 Einstein's A and B

Let us consider a system with two energy levels with energy 0 (ground state) and ε (> 0) (excited state). If many such systems are maintained at temperature T, we know the

---

[14] A rigid rotor is a symmetric top that cannot rotate around its symmetry axis. This restricted eigenspace is characterized by the total angular momentum quantum number J with the rotational energy $\hbar^2 J(J + 1)/2I$ which is $2J + 1$-tuple degenerate.

occupation ratio of these states must be given by a Boltzmann factor $e^{-\beta\varepsilon}$. Now, let us assume that the system is also interacting with the electromagnetic field (radiation field) of frequency $\nu = \varepsilon/h$ with radiation energy density ρ (equal to the number density of the photons of energy $h\nu$; no other field interacts with the system because the energy exchange is through photons of energy $h\nu$).[15]

Interacting with the radiation field, the system may be excited at the rate $\rho B^\uparrow n_0$, where n_0 is the number of the systems in their ground state and $B^\uparrow$ a positive constant (imagine a collisional reaction between the atom and the photon). It is also de-excited at the rate $\rho B^\downarrow n_\varepsilon$, where n_ε is the number of the systems in their excited state and $B^\downarrow$ a positive constant. It is also known that there is the so-called "spontaneous radiation": even if there exists no radiation field, a photon is emitted spontaneously from the excited state with the rate An_ε, where A is a positive constant.

If T is very large and the intensity of the radiation field is very strong, then $n_\varepsilon \simeq n_0$, and the spontaneous radiation may be ignored relative to the induced transitions by radiation, so we must assume $B^\uparrow = B^\downarrow = B$, a common constant. In equilibrium there must be a transition balance between going up and going down. From this balance derive Planck's radiation formula $\rho \propto 1/(e^{\beta h\nu} - 1)$.

Solution

The equilibrium condition must be (up rate = down rate):

$$B\rho n_0 = (A + B\rho)n_e. \tag{30.30}$$

This implies

$$\rho = \frac{A/B}{e^{\beta h\nu} - 1}. \tag{30.31}$$

---

[15] We follow the argument by Einstein in Einstein, A. (1917). Zur Quantentheorie der Strahlung, *Phys. Z.* **18**, 121–128 (To the quantum theory of radiation). [In 1917: The February and October Revolutions occurred in Russia; the Balfour Declaration expressed British support for a Jewish homeland in Palestine.]

PART III

ELEMENTS OF PHASE TRANSITION

31

Phases and Phase Transitions

Summary

∗ Qualitative change of phases is the phase transition, which corresponds to a certain singularity of thermo-dynamic potentials.
∗ Phase coexistence conditions (under given T and P) are a set of equalities among chemical potentials. Gibbs' phase rule, $f = c + 2 - \phi$, follows from the condition.
∗ Thermodynamic limit is absolutely needed to rationalize phase transitions statistical-mechanically.

Key words

phase, phase transition, phase diagram, coexistence curve, triple point, kelvin scale, Gibbs' phase rule, first-order phase transition, second-order phase transition, Ising model, thermodynamic limit

The reader should be able to:

∗ Understand thermodynamically what phase transitions are.
∗ Draw the phase diagram of an ordinary one-component fluid on the PT-plane.
∗ Sketch $G(T, P)$ (with justification) for an ordinary fluid.
∗ Reproduce the logic behind Gibbs' phase rule.

If there are interactions among particles, there are various phases as exemplified by ice, liquid, and vapor of water. First, we discuss how to describe thermodynamically what we experience. Then, we discuss whether statistical mechanics can understand phase transitions.

31.1 What is a Phase?

Under different conditions (say, at various (T, P)) a system can exhibit qualitatively different properties. When this happens, we say the system (or the material) is in different phases. We may say an equilibrium state is in a single phase, if it is macroscopically homogeneous. To understand the equilibrium macrofeatures of a substance is to understand its various phases and their characteristic features. Therefore, we wish to map out what

Table 31.1 Possible characterization of three phases of matter		
	long-range order	coherence
solid	Y	Y
liquid	N	Y
gas	N	N

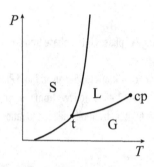

Fig. 31.1 A representative phase diagram of a system consisting of particles interacting with the Lennard-Jones interaction (11.2) (see Fig. 11.1) in the (*P*, *T*)-space. S: solid; L: liquid; G: gas; t: triple point; cp: critical point. The curves denote the phase boundaries where phase transitions occur. Since the "boundary" between L and G ends at cp, globally we cannot have a well-defined liquid or gas phase.

happens at various points in the thermodynamic space or at least in the space spanned by certain convenient thermodynamic parameters (e.g., *T*, *P*), i.e., we wish to construct the *phase diagram* (see, e.g., Fig. 31.1). To understand the world we wish to understand where the state boundaries are and what the features of the territories are. To understand the boundaries corresponds to the understanding of phase transitions, and to understand the features of the territories corresponds to characterizing individual phases.

In the previous paragraph, "qualitative differences" do not imply quantitative difference such as soft–hard, hot–cold, hue changes, etc., but existence–non existence of some properties such as symmetry, long-range correlation, etc. For example, solid, liquid, and gas phases may be characterized in Table 31.1.[1]

Here, "long-range" correlation implies that if we know a position of a particle, we can tell the position of another particle far away (a macroscopic distance away) from the first one. Crystalline spatial regularity implies long-range spatial ordering of the particles. For fluid phases we cannot have this property. This property can either exist or not. This is one of the qualitative differences between solid and fluid phases. To distinguish fluid phases is not easy. One possibility may be "coherence." We know gases can be compressed easily but

---

[1] This is a rather microscopic attempt to characterize the usual three phases of matter due to J. D. Bernal (1901–1971). A more macroscopic characterization such as fluid dynamical properties are not thermodynamic. Thermodynamic characterization is not so trivial. Usually, we discuss compressibility, thermal expansion coefficient, and elastic responses. The reader should try their own.

liquids cannot; they can be as incompressible as solids. This must be due to the interactions ("touching") among molecular hard cores. "Coherence" implies that each particle has at least four repulsive interactions with its surrounding particles simultaneously.

Since some qualitative properties appear or disappear upon crossing a phase boundary, something "singular" can happen thermodynamically (e.g., loss of differentiability or continuity of some thermodynamic quantities). We say there is a *phase transition*, if such singularities can be observed.

However, a precise definition of "phase" is actually rather difficult. Near the phase boundaries we may clearly distinguish the phases, but the "territory which a phase occupies" may not be well defined as in the case of gases and liquids (see Fig. 31.1). Therefore, here, the concept of "phase" is used "locally" when precise statements are needed. We say the states are distinct phases, if they cannot be changed into each other without a phase transition (i.e., thermodynamic singularity) in the domain of our interest in the phase diagram.

31.2 Statistical Thermodynamic Key Points Relevant to Phase Transitions

For a given system any equilibrium state is described (uniquely) as a point in its thermodynamic space spanned by its thermodynamic coordinates (internal energy and work coordinates). For example, a point in the thermodynamic space can describe phase coexistence unambiguously as noted in Chapter 12. This can be clearly seen from the solid–liquid–gas phase diagram in the thermodynamic space (see Fig. 31.2). In this diagram, for example, the gray S + G region describes the coexistence of solid and gas phases. A point inside the big dark gray triangle tells us the relative amount of the three phases.[2]

The convexity of the internal energy implying $\Delta X \Delta x > 0$ (see Chapter 25) unambiguously tells us how various quantities change upon phase transition between the phases. For example, for a phase transition from phase I to phase II, if II is a high temperature phase, then $S_I < S_{II}$ (i.e., $\Delta T \Delta S > 0$). Or, if II is a higher pressure phase than I, then $-V_{II} > -V_I$, i.e., $V_I > V_{II}$ (i.e., $\Delta(-P)\Delta V > 0$, since the conjugate variable of V is not P but $-P$).

Near the critical point fluctuations become very large. Then, the fluctuation–response relation (Chapter 26):

$$\chi = \left(\frac{\partial X}{\partial x}\right)_y = \beta \langle \delta X^2 \rangle \tag{31.1}$$

explains why susceptibilities blow up at criticality (Chapter 33).

---

[2] A good exposition of the phase diagram in the thermodynamic space is in the Introduction by A. S. Wightman to Israel, R. B. (1979). *Convexity in the Theory of Lattice Gases*, Princeton; Princeton University Press.

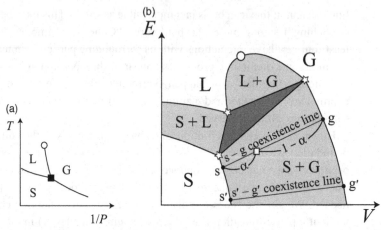

Fig. 31.2 The (schematic) phase diagram in the thermodynamic space (larger diagram; b). The open circle indicates the critical point, and the black square (in the small diagram (a)) and the dark gray big triangle (in (b)) denote the triple point. The pale gray zones are two-phase coexisting states. The aim of this schematic diagram is to exhibit that the phase coexistence lines and the triple point in the ordinary phase diagram (a) are resolved so that we can even tell the relative amount of coexisting phases at a given point in the thermodynamic space. In the S + G region a line connecting the black dots "s" on the solid phase boundary and "g" on the gas phase boundary illustrates an example of a series of coexistence states of a particular solid phase s and a particular gas phase g (another such line connecting s′ and g′ is also illustrated). The location on the line can indicate the ratio of these two particular phases. For example, the coexistence state indicated by a white square on the coexistence line with the segment length ratio $\alpha : (1 - \alpha)$ consists of solid "s" and gas "g" phase ratio $(1 - \alpha) : \alpha$ (this is called the *lever rule*). Analogous lines may be drawn in the S + L and L + G coexistence regions. A point inside the big dark gray triangle tells us the relative amount of the three phases indicated by white stars.

31.3 Phase Coexistence Condition: Two Phases

If there is a phase transition from one phase to another, there may or may not be a coexistence of these phases. Let us assume the phase coexistence is possible (as in the case of ice and liquid water) for a pure substance and discuss what is required for phases to coexist and what the requirements entail.[3] If phases I and II coexist in an isolated box, the two phases must exchange heat (energy), volume, and molecules. We know $S = S_I + S_{II}$ must be maximized, so the equilibrium condition is the identity of T, P/T, and μ/T between the two phases as we discussed previously (in Chapters 14 and 26):

$$T_I = T_{II}, \quad P_I = P_{II}, \quad \mu_I = \mu_{II}. \tag{31.2}$$

---

[3] **Can we tell if two phases coexist?** It is impossible to know whether the given two phases coexist or not from the knowledge that there is a first-order phase transition between them, although, usually, we may say the phases satisfying the thermodynamic coexistence conditions do coexist. Thus, precisely speaking, the so-called coexistence condition is the condition required if there is indeed a phase coexistence.

The last equality in (31.2) is

$$\mu_{\mathrm{I}}(T,P) = \mu_{\mathrm{II}}(T,P). \tag{31.3}$$

This functional relation determines a curve called the *coexistence curve* in the T–P diagram. From (31.3) we already obtained the Clapeyron–Clausius relation (27.26).

Along this line the Gibbs free energy G of the whole system may be written as

$$G = N_{\mathrm{I}}\mu_{\mathrm{I}} + N_{\mathrm{II}}\mu_{\mathrm{II}}. \tag{31.4}$$

Therefore, without changing the value of G, any mass ratio of the two phases can coexist. Thus, in the phase diagram in the thermodynamic space the phase coexistence relation is described as a boundary "disputed land strip" instead of a curve (see the pale gray zones in Fig. 31.2(b) and read the caption as to the "lever rule").

31.4 Phase Coexistence Condition

How many phases can coexist at a given T and P for a pure substance? Suppose we have ϕ coexisting phases. The following $\phi - 1$ conditions must be satisfied:

$$\mu_{\mathrm{I}}(T,P) = \mu_{\mathrm{II}}(T,P) = \cdots = \mu_{\phi}(T,P). \tag{31.5}$$

We believe that for the generic case, μs are sufficiently functionally independent. To be able to solve for T and P, we can allow at most two independent relations. That is, at most three phases can coexist at a given T and P for a pure substance.

For a pure substance, if three phases coexist, T and P are uniquely fixed. This point on the T–P diagram is called the *triple point*. The kelvin scale of temperature is *defined* so that the triple point of water is at $T = 273.16\,\mathrm{K}$ (since 1954[4]), and $t = T - 273.15$ defines the temperature in Celsius scale. A triple point water cell is used to calibrate precision thermometers.

31.5 Gibbs' Phase Rule

Consider a more general case of a system consisting of c chemically independent components (i.e., with c components we can change independently[5]).

Suppose there are ϕ coexisting phases. The equilibrium conditions are:

(1) T and P must be common to all the phases.
(2) The chemical potentials of the c chemical species must be common to all the phases.

---

[4] [In 1954: The hydrogen bomb was tested on Bikini Atoll; The Battle of Dien Bien Phu ended the first Indo-China war.]

[5] For example, as we have already discussed in Chapter 27, H_3O^+ in pure water should not be counted, if we count H_2O among the independent chemical components.

To specify the composition of a phase we need only $c - 1$ mole fractions. Thus, the chemical potential for a chemical species depends on T, P, and $c - 1$ mole fractions $(x^1, x^2, \ldots, x^{c-1})$. The mole fractions are generally different from phase to phase, so we must introduce $\phi(c - 1)$ different mole fractions, adding subscripts to distinguish phases as $x_X^1, x_X^2, \ldots, x_X^{c-1}$ ($X = 1, \ldots, \phi$). That is, we have $2 + \phi(c - 1)$ unknown variables, T, P, and x_X^j ($j = 1, \ldots, c - 1, X = 1, \ldots, \phi$). We have $\phi - 1$ equalities among the chemical potentials in different phases for all the c chemical species as ($j = 1, \ldots, c$):

$$\mu_I^j(T, P, x_I^1, x_I^2, \ldots, x_I^{c-1}) = \mu_{II}^j(T, P, x_{II}^1, x_{II}^2, \ldots, x_{II}^{c-1}) = \cdots = \mu_\phi^j(T, P, x_\phi^1, x_\phi^2, \ldots, x_\phi^{c-1}),$$
$$(31.6)$$

so the number of equalities we have is $(\phi - 1) \times c$. Consequently, for the generic case we can choose $f = 2 + \phi(c - 1) - c(\phi - 1) = c + 2 - \phi$ variables freely. We have thus arrived at the *Gibbs phase rule*:[6]

$$f = c + 2 - \phi. \qquad (31.7)$$

This number f is called the number of *thermodynamic degrees of freedom*. For example, for a pure substance $f = 3 - \phi$, so we can have phase coexistence curves (1-manifolds, $f = 1$) between two phases ($\phi = 2$) and triple points (0-manifolds, $f = 0$) among three phases ($\phi = 3$).

31.6　How *G* Behaves at Phase Boundaries

What happens to the Gibbs free energy at the phase transition point under constant T or P? The reader must be able to sketch it in the ordinary fluid case. Note the usual Gibbs relation:

$$dG = -SdT + VdP. \qquad (31.8)$$

Under constant P, the Gibbs free energy G may be sketched as Fig. 31.3. If P is the critical pressure, the liquid–gas (LG) transition "break" disappears: G becomes differentiable.

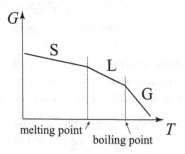

Fig. 31.3　Typical behavior of the Gibbs free energy G for a pure substance (S: solid, L: liquid, G: gas). The free energy loses differentiability at first-order phase transition points. Note that the slope reflects the entropies of the phases. The Gibbs free energy G must be concave as a function of T (and P), because $-G = \max_{S,V}[TS + (-P)V - E]$ (recall **16.12**).

---

[6] As astute readers have probably sensed already, the derivation excludes all the possible accidental functional relations among chemical potentials.

However, the specific heat has a singular behavior at the critical temperature. That is, the LG transition becomes second order. We will discuss this later in more detail (in Chapter 33).

Try to sketch G under constant T as a function of P.

31.7 Classification of Phase Transitions

Usually, phase transitions are classified into *first-order phase transitions* and the rest called continuous phase transitions or *second-order phase transitions*. For the first-order phase transition at least one thermodynamic density (i.e., one extensive quantity per volume) changes discontinuously, but for the second-order phase transition there is no discontinuity in thermodynamic densities. The liquid–gas transition at the critical pressure is a second-order phase transition as noted just previously.

Phase transitions in many interesting cases occur between more ordered and less ordered phases;[7] it is between the low entropy state and the high energy (high enthalpy) state. For example, melting is (in most cases) the transition from a low entropy solid to a high energy liquid. Protein folding is the transition from a higher energy random coil state to a lower entropy folded state.[8]

In practice, a first-order phase transition occurs if the ordered phase loses its stability "catastrophically." In other words, the first-order phase transition occurs when a slight loss of order favors reduction of "interactions" maintaining the order. Thus, there is no equilibrium state with significantly reduced order (see Chapter 36). In contrast, for second-order phase transitions, the interactions that maintain order do not decrease and withstand the order-destroying thermal fluctuations to the point that the order really vanishes.

31.8 A Typical Example of Second-Order Phase Transition

A typical second-order phase transition is the one between the paramagnetic disordered phase and the ferromagnetic ordered phase we can observe in magnets.

A magnet can be understood as a lattice of spins interacting with each other locally in space. The interaction between two spins has a tendency to align them in parallel. At higher temperatures, due to vigorous thermal motions, this interaction cannot quite bring order among spins, but at lower temperatures the entropic effect becomes less significant, so spins order globally. There is a special temperature T_c below which this ordering occurs. We say an *order–disorder (phase) transition* occurs at this temperature.

---

[7] Some authors criticize that the distinction between "order" and "disorder" is often subjective, but in many cases the distinction is mathematically clearly definable (e.g., with the aid of group theory; see Chapter 35).

[8] However, do not have a prejudice that biologically natural states of proteins are equilibrium states. Many large proteins may well be in metastable states when they function biologically normally. Think how we can prove experimentally that a particular protein is in equilibrium.

31.9 The Ising Model

The *Ising model* is the simplest model exhibiting the order–disorder transition. At each lattice point is a (classical) spin s which takes only $+1$ (up) or -1 (down). A nearest-neighbor spin pair (i, j) has the following interaction energy:

$$- Js_i s_j, \qquad (31.9)$$

where J is called the *coupling constant*, which is positive in our example (ferromagnetic case; if spins are parallel, interaction energy is lowered). We assume all the spin–spin interaction energies are superposable. Thus, the system Hamiltonian of an Ising model is given by

$$H = - \sum_{\langle i, j \rangle} Js_i s_j, \qquad (31.10)$$

where $\langle \, , \, \rangle$ implies the nearest-neighbor pair. The summation is over all the nearest-neighbor pairs on the lattice.

The (generalized canonical) partition function for this system reads

$$Z = \sum_{\{s_i = \pm 1\}} e^{-\beta H + \beta h M}, \qquad (31.11)$$

where h is the magnetic field, and $M = \sum s_i$ (the magnetization). The summation is over all the spin configurations specified by all the s_i values ($+1$ or -1). In other words, the summation is over all the ± 1 arrangement patterns on the lattice.

31.10 Fundamental Questions about Phase Transitions

From the statistical mechanics point of view, the most important question is why such qualitative changes can occur at all. Actually, a more fundamental question is: can statistical mechanics ever describe phase transitions? Up to the early 1930s such doubts existed. Now, we are fairly sure that statistical mechanics correctly describes various phase transitions. *However*, do not forget that we cannot yet explain statistical-mechanically why ordinary molecules can make crystals below some finite temperature.

31.11 The Necessity of Thermodynamic Limit: Densities and Fields

If the system size is finite, the sum in (31.11) is a finite sum of positive terms. Each term in this sum is analytic in T and h, so the sum itself becomes analytic in T and h

(i.e., very smooth).[9] Furthermore, Z cannot be zero, because each term in the sum is strictly positive. Therefore, its logarithm is analytic in T and h; the free energy of the finite lattice system cannot exhibit any singularity. That is, there is no thermodynamic singularity, and consequently, there is no phase transition for this finite system.

In the actual system we study experimentally, there are only a finite number of atoms. However, this number is huge, so we believe the situation is correctly and practically captured by the large system limit (the thermodynamic limit). Consequently, the question of phase transitions from the statistical mechanics point of view is this: is there any singularity in $A = -k_B T \log Z$ in the large system limit?

If we take such a limit, extensive quantities including A are not finite generally, so they become meaningless. Therefore, we introduce the *densities* (extensive quantities per unit volume or per particle). The thermodynamic coordinates are replaced by the energy and work coordinate densities. They are intensive quantities, but must be distinguished from the conjugate intensive quantities such as T, P, so the usual intensive quantities are often called thermodynamic *fields*. We take a system size infinity limit while keeping the densities constant. This limit is called the *thermodynamic limit*.[10] We ask whether the free energy density exhibits certain singularities or not in this limit. The reader might not wish to go into mathematics, but at least clearly recognize that qualitative changes require loss of analyticity.

31.12* Statistical Mechanics in the Thermodynamic Limit

How can we study the thermodynamic limit of a system? The Hamiltonian in this limit is meaningless; it is generally a nonconvergent infinite sum. Consequently, $e^{-\beta H}$ is almost always 0 or ∞ even if definite.

A natural approach to study an infinitely big system is to take all its possible finite volume subsets V,[11] with various boundary conditions B on ∂V (more precisely stated below) and to prepare, for each volume-boundary condition pair (V, B), the canonical distribution:

$$\mu_{V,B} = \frac{1}{Z_{V,B}} e^{-\beta H_V(B)}, \qquad (31.12)$$

where $H_V(B)$ is the Hamiltonian for the subsystem on V including the surface interactions B across ∂V.[12] Make the totality of $\mu_{V,B}$ and find the set $\mathcal{M}$ of all the accumulation

---

[9] A more accurate mathematical term here than analytic is "holomorphic."
Introductory complex analysis Two introductory textbooks for complex function theory are mentioned here. Priestley, H. A. (1990). *Introduction to Complex Analysis*, Oxford: Oxford, is a friendly introductory book; a slightly more advanced modern text book may be: Rao, M. and Stetker, H. (1991). *Complex Analysis: An Invitation*, Singapore: World Scientific.

[10] Does such a limit exist? This is also a fundamental question never considered till the 1950s. This is a far easier question than the existence/nonexistence question of phase transitions. We asked such a question in Chapter 12.

[11] But its boundary ∂V is not fractal. That is, the volumes are van Hove volumes (see Chapter 12).

[12] More precisely, B is specified by the fixed state on V^c (i.e., the outside of V) and the interactions between V and V^c across ∂V.

points.[13] The elements of $\mathcal{M}$ are called the *Gibbs measures* (or Gibbs states).[14] At least one Gibbs state exists. When $\mathcal{M}$ changes qualitatively (e.g., if $\mathcal{M}$ bifurcates to contain more than one state) with the change of, e.g., T, we say that a phase transition occurs. It is also natural to expect that at such a point thermodynamic functions should have certain qualitative changes (have singularities).

Problems

Q31.1 Phase Transition: Basic Questions

Are the following statements correct or incorrect? If incorrect, the reader must explain why or give a counterexample for the statement. If correct, then provide a brief supporting argument. (However, most questions will be discussed in the subsequent chapters. This is a sort of preview.)

(1) One of the thermodynamic densities must exhibit discontinuity for a phase transition to occur.
(2) When a solid phase melts to a liquid phase, the entropy always increases.
(3) No 1D system can exhibit phase transition, if the interaction range is finite.
(4) When a first-order phase transition occurs between phase I and II, these two phases can coexist.

Solution

(1) No. This is required only for first-order phase transitions.
(2) No. We have already seen a counterexample (**Q25.1**(4)). Another famous counterexample is ^{3}He. Crystallization localizes atoms, so the spin–spin coupling due to exchange of particles is reduced, and the spin order that exists in liquid is lost (look up the Pomeranchuk effect).
(3) Yes. This is according to Peierls' argument or from the Perron–Frobenius theorem (see Chapters 32 and 34).
(4) No. A counterexample is the 2-Ising model (see Chapter 32).

Q31.2 Thermodynamic Space of 3-Ising Model

Consider an Ising magnetic system in 3-space. There is a second-order order–disorder phase transition at $T = T_c$, if the magnetic field $B = 0$ (i.e., $B_c = 0$).

---

[13] To this end we need a topology on $\mathcal{M}$. See, e.g., Ruelle, D. (2004). *Thermodynamic Formalism: The Mathematical Structure of Equilibrium Statistical Mechanics*, Cambridge: Cambridge University Press (Cambridge Mathematical Library) (probably too terse for physicists).

[14] Precisely speaking, $\mathcal{M}$ is often a subset of the totality of the Gibbs states for a system, but all the pure states (equal to the states observable in an actual single sample) are in $\mathcal{M}$.

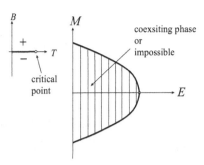

Fig. 31.4 Below T_c $M \neq 0$ is possible with $B = 0$. If a nonzero B is applied, one of the up or down phases remains as an equilibrium state. Thus, the "lined" region (on the right) is with $B = 0$ (recall Fig. 31.2). Outside of this "coexistence" region, for higher T of the temperature axis, magnetic field must be applied to have nonzero M. Needless to say, too large $|M|$ is not realizable, so the thermodynamic space is bounded vertically.

(1) What is the thermodynamic space for this system (or what are the thermodynamic coordinates for the system)?
(2) Sketch the phase diagram of this magnet in its thermodynamic space.

Solution

(1) E and M must be the thermodynamic coordinates. B is intensive, and corresponds to P in fluids.
(2) We must draw a phase diagram on (E, M). The usual diagram is on (B, T). The rough sketch is shown in Fig. 31.4.

Q31.3 Melting Heat for Tetrachlorocarbon

The melting temperature of tetrachlorocarbon (CCl_4) depends on the pressure as follows:

$$T_m = 250.56 + 3.952 \times 10^{-7} P - 2.10 \times 10^{-16} P^2, \qquad (31.13)$$

where T is measured in K and P in Pa. At $P = 1000\,\text{atm}$, the melting causes the volume increase of $\Delta V = 3.06 \times 10^{-6}\,\text{m}^3$ per mole.

(1) Find the latent heat of melting per mole of tetrachlorocarbon at 1000 atm. Notice that 1 atm = 101 325 Pa.
(2) Is the melting entropy change reasonable?

Solution

(1) We use the Clapeyron–Clausius equation (27.26) (subscript CC in the following means along the coexistence curve as noted by the white arrow in Fig. 27.3.):

$$\left(\frac{\partial T_m}{\partial P} \right)_{CC} = \frac{T_m \Delta V}{\Delta H} \qquad (31.14)$$

or

$$\Delta H = T_m \Delta V \Big/ \left(\frac{\partial T_m}{\partial P}\right)_{CC} \tag{31.15}$$

$$\left(\frac{\partial T_m}{\partial P}\right)_{CC} = 3.952 \times 10^{-7} - 4.20 \times 10^{-16} P = 3.43 \times 10^{-7} \tag{31.16}$$

in K/Pa. Therefore, ($T_m = 288.5$ K at 1000 atm):

$$\Delta H = 288.5 \times (3.06 \times 10^{-6}/3.43 \times 10^{-7}) = 2573\,\mathrm{m}^3 \cdot \mathrm{Pa}, \tag{31.17}$$

or 2.5 kJ/mol.

(2) $\Delta S = 8.65$ J/K·mol is the increase of entropy by melting. This corresponds to 8.65/5.76 = 1.5 bits/molecule. This is a bit too small for ordinary melting, suggesting that some disordering process is already going on in the crystal.

Indeed, there are two crystalline phases I (mp = 250 K at 1 atm) and II (mp = 225 K at 1 atm). The melting entropy of I is about 10 J/K·mol, and that of II is about 20 J/K·mol, which is 3.6 bits/mol. This is a reasonable magnitude. Thus, the crystal discussed earlier that melts at 288.5 K at 1000 atm must be phase I, and some disordering phase transition (perhaps with molecular rotation) at a lower temperature must have occurred.

Phase Transition in ∂-Space

Summary

* Statistical mechanics (apparently) can explain various phases and phase transitions.
* Spatial dimensionality is crucial to the phase ordering and to the existence of the order–disorder phase transition.
* If the interaction is long-ranged, phase transitions can occur even in 1-space.
* The second-order phase transition for magnets, fluids, and binary fluid mixtures may be understood in a unified fashion (universality).

Key words

Peierls' argument, Kac potential, van der Waals gas, Maxwell's rule, Tonks gas

The reader should be able to:

* Intuitively understand why spatial dimensionality matters.
* Explain Peierls' basic idea.
* Derive Tonk's equation of state or understand why it is "obvious."

32.1 Order Parameter

To characterize the order in the system we define an *order parameter* which is nonzero only in the ordered phase: magnetization per particle $m = \langle s \rangle = M/N = (1/N)\sum s_i$ for the Ising model is a good example. Thus, the fundamental question about the Ising model is whether $M/N = m = \langle s \rangle$ converges to zero or not in the thermodynamic limit. Notice that the Hamiltonian is invariant under global spin flip $s_i \leftrightarrow -s_i$ (for all the spins simultaneously), so $m = 0$ is very natural. Appreciate the nontriviality of the fact that the $m > 0$ or $m < 0$ equilibrium state exists at lower temperatures despite the spin-flip symmetry of the system Hamiltonian. That is, the equilibrium state may not obey the symmetry of the system.[1]

---

[1] If this, called *spontaneous symmetry breaking* (Chapter 35), occurs, certainly there is a thermodynamic singularity, so there is a phase transition; any qualitative change implies singularity. However, the converse is not always obvious.

$$+ +$$
$$+ + + - - - - - - - - - + + + + + + + +$$
$$\longleftarrow L \longrightarrow$$

Fig. 32.1   Top: completely ordered state of 1-Ising model; Bottom: an Ising chain with a spin-flipped island of size L (+ implies up spins and − down spins).

32.2 Spatial Dimensionality is Crucial

For the existence of a phase transition the spatial-dimensionality of the system is often crucial.

Let us consider a 1-Ising model (Ising chain), whose total energy reads:[2]

$$H = -J \sum_{-\infty < i < +\infty} s_i s_{i+1}, \tag{32.1}$$

where s_i can take ± 1. Compare the energies of the following two spin configurations (Fig. 32.1; + denotes the up spins and − down spins): The bottom one has a larger energy than the top by $2J \times 2$ due to the existence of the two mismatching edges. However, this energy difference is independent of the size L of the island. Therefore, as long as $T > 0$ there is a finite chance of making big (macroscopic) down spin islands amidst the ocean of up spins. If a down spin island becomes large, there is a finite probability for a large lake of up spins on it. This implies that no ordering is possible for $T > 0$.[3]

As can be easily guessed there is no ordered phase in any one-dimensional lattice system with local interactions for $T > 0$.

32.3 There is an Ordered Phase in 2-Space

Consider the 2-Ising model. Imagine there is an ocean of up spins (see Fig. 32.2). To make a circular down spin island of radius L, we need $4\pi JL$ more energy than the completely ordered phase. This energy depends on L, making the formation of a larger island harder.[4] Actually, we can show that the probability to make such a contour cannot be larger than $e^{-\beta(4\pi JL)}$ according to Peierls' inequality (see **Q32.4**(3)), so let us call this type of

---

[2] Here, the sum is over an infinite lattice. Needless to say, this is only a formal expression. For a more respectable approach see **31.12**.

[3] **Entropic effect: defect locations** There is also an entropic factor disfavoring ordering; we can locate the island boundary at any place we like, so if the island size is L, then the entropy of order $\log L$ reduces the free energy of the system with an island by $\sim T \log L$. This factor totally eclipses the energetic effect that is system-size independent.

[4] **Entropic effect: boundary fluctuations** In this case as well we must take the entropic effect into account. The phase boundary would wiggle, and shore locations could wander. The latter effect contributes $\sim \log$ (island size), so in higher dimensions, it is not important. The boundary wiggling modifies the interface energy into the interface free energy, reducing the energetic penalty to make phase boundaries at higher temperatures. However, at low temperatures, we may (qualitatively) ignore the effect.

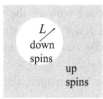

Fig. 32.2 An intuitive argument for the stability of ordered phase.

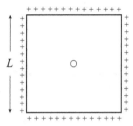

Fig. 32.3 All the boundary spins are fixed to be up. What happens to the central spin in the circle in the $L \to \infty$ limit?

argument relying on the cost to flip spins a *Peierls' argument*. To destroy a global order in 2-space we need a system-size dependent amount of energy, so for sufficiently low temperatures, the ordered phase is unlikely to be destroyed spontaneously. Of course, small local islands could form, but they never become very large. Hence, we may conclude that an ordered phase could exist at sufficiently low temperatures.[5]

Prepare a $L \times L$ square lattice, and fix all the edge spins upward (see Fig. 32.3). If T is not low, eventually the probability $P(s_0 = +1)$ of the center spin s_0 to be up converges to $1/2$ for large L. If this is always true for any T, it implies no spin ordering occurs. What Peierls (1907–1995) actually stated, in 1936,[6] was the following theorem: $P(s_0 = +1) > 1/2$ at sufficiently low temperatures (see **Q32.4**).

Thus we have learned that spatial dimensionality is crucial for the ordering of the phase for the system with short-range interactions.[7]

32.4 Interaction "Range" is also Crucial

What happens if the range of interactions is not finite and the intensity of interaction decays sufficiently slowly? A Peierls' type argument is still applicable (see **Q32.3**). Obviously,

[5] **Spin dimension** There is at least one more crucial factor governing the existence of phase transition. It is the *spin dimension*: the degree of freedom of each spin. Ising spins cannot point in different directions, only up or down (their spin dimension is 1). However, the true atomic magnets can orient in any direction (their spin dimension is 3). This freedom makes ordering harder. Actually, in 2-space ferromagnetic long-range ordering at $T > 0$ by spins with a spin dimension larger than 1 is impossible.

[6] [In 1936: Keynes published *The General Theory of Employment, Interest and Money*; the Turing machine was invented; Germany reoccupied the Rhineland.]

[7] Here, "short range" implies the interaction vanishes beyond some finite range, or the strength of the interaction decays sufficiently quickly (say, faster than $1/r^{\partial+1}$).

if each spin can interact with all the spins in the system uniformly, an ordered phase is possible even in 1-space. If the coupling constant, J, decays slower than $1/r^2$, then an order–disorder phase transition is known to be still possible at a finite temperature in 1-space.

If the interaction is long ranged, then the system may not behave thermodynamically normally (the fourth law may be violated). However, if interaction is infinitesimal, then thermodynamics may be saved even if the interaction does not decay spatially. As we see next such a system is essentially the system described by the van der Waals equation of state.

32.5 Van der Waals Model: "Derivation"

Van der Waals (1837–1923) proposed the following equation of state (*van der Waals equation of state* (1873[8]):

$$P = \frac{Nk_BT}{V - Nb} - \frac{aN^2}{V^2}, \tag{32.2}$$

where a and b are positive materials constants. Here, P, N, T, V have the usual meaning in the equation of state of gases. His key ideas are:

(1) The existence of the real excluded volume due to the molecular core should reduce the actual volume in which molecules can roam from V to $V - Nb$ (recall **8.13**); this would modify the ideal gas law to $P_{HC}(V - Nb) = Nk_BT$. Here, subscript HC implies "hard core."

(2) The attractive binary interaction reduces the actual pressure from P_{HC} to $P = P_{HC} - a(N/V)^2$, because the particles colliding with the wall are on average pulled back by the particles in the bulk.

Let us derive this equation of state from its entropy. We know that the entropy of a classical ideal monatomic gas reads (see (15.16)):

$$S = S_0 + Nk_B \log \frac{V}{N} + \frac{3N}{2}k_B \log \frac{E}{N}. \tag{32.3}$$

Notice that E in this formula is just the total kinetic energy K. For the van der Waals model:

(1) implies $V \to V - Nb$,

(2) implies $K = E + aN^2/V$: if every particle interacts uniformly with all other particles, the interaction energy should be proportional to N^2. The interaction parameter is $-a/V$, which is infinitesimal (since V is macroscopic) to make the total interaction energy extensive. Thus, the total internal energy $E = K - aN^2/V$.

Therefore, the entropy of the van der Waals gas should read

$$S = S_0 + Nk_B \log \frac{(V - Nb)}{N} + \frac{3N}{2}k_B \log \frac{(E + aN^2/V)}{N}. \tag{32.4}$$

---

[8] [In 1873: Verne published *Around the World in 80 Days*; Schliemann discovered Priam's Treasure.]

This gives

$$\frac{1}{T} = \left(\frac{\partial S}{\partial E}\right)_V = \frac{3}{2}\frac{Nk_B}{E + aN^2/V},$$ (32.5)

and

$$\frac{P}{T} = \left(\frac{\partial S}{\partial V}\right)_E = \frac{Nk_B}{V - Nb} + \frac{3}{2}\frac{Nk_B}{E + aN^2/V}\left(-\frac{aN^2}{V^2}\right).$$ (32.6)

Combining these two, we obtain the van der Waals equation of state:

$$\frac{P}{T} = \left(\frac{\partial S}{\partial V}\right)_E = \frac{Nk_B}{V - Nb} - \frac{aN^2}{TV^2}.$$ (32.7)

32.6 Liquid–Gas Phase Transition Described by the van der Waals Model

The most noteworthy feature of the equation that fascinated Maxwell is that liquid and gas phases are described by a single equation.

Let us study the general behavior of (32.7). The first term is $\propto 1/(V - Nb)$, so it blows up at $V = Nb$. It is basically the ideal gas equation shifted to the right by Nb along the V axis. The second term becomes very important if T is small and is $\propto 1/V^2$. Since $1/V^2$ becomes larger more quickly than $1/(V - Nb)$ if V is not too close to Nb, (32.7) becomes non-monotonic as a function of V for sufficiently low T. That is, as in Fig. 32.4, the PV-curve wiggles, if T is small enough. We know, however, thermodynamically, $(\partial P/\partial V)_T$ cannot be positive (cf. Le Chatelier's principle **25.6**). This "wrong behavior" must be due to the attractive interactions. Van der Waals guessed that this singular behavior implied a phase transition, a gas–liquid phase transition.

Maxwell proposed a liquid–gas coexistence condition (*Maxwell's rule*, see Fig. 32.4; for a derivation see **Q32.2**).

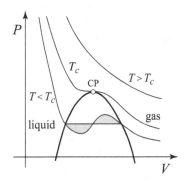

Fig. 32.4 The thick curve is the *coexistence curve* below which no single phase can stably exist. The liquid–gas coexistence pressure for a given temperature is determined by *Maxwell's rule*: the two shaded regions have the same area.

32.7 Kac Potential and van der Waals Equation of State

The van der Waals equation of state is heuristically derived, but what is really the microscopic model that gives it, if any? A proper understanding of van der Waals's idea is

$$P = P_{HC} - \frac{1}{2}\varepsilon n^2, \tag{32.8}$$

where P_{HC} is the hard sphere fluid pressure,[9] and the subtraction term is the average effect of attractive forces ($n = N/V$). As it is, this equation exhibits the non-monotonic (i.e., not thermodynamically realizable) PV curve just as the van der Waals equation of state, so there cannot be any microscopic model for (32.8).[10] However, this equation augmented with Maxwell's rule is thermodynamically legitimate, and indeed it is the equation of state of the gas interacting with the *Kac potential* which is spherically symmetric:

$$\phi(r) = \phi_{HC}(r/\sigma) + \gamma^3 \phi_0(\gamma r/\sigma), \tag{32.9}$$

where $\phi_{HC}(x)$ is the hard-core potential: 0 beyond $x = 1$ and ∞ for $x \leq 1$, σ is the hard-core diameter, and ϕ_0 is an attractive tail. The parameter γ is a scaling factor; we consider the $\gamma \to 0$ limit (long range but infinitesimal interaction).

32.8 The 1D Kac Potential System may be Computed Exactly

The hard-sphere fluid pressure P_{HC} is not exactly obtainable if the spatial dimension is $2, 3, \ldots$, but in 1-space, we can obtain it exactly.[11] Therefore, the 1D Kac model (augmented van der Waals model) is exactly solvable, and exhibits a phase transition.

Let us use the fact that the ideal gas law can be obtained purely mechanically as Bernoulli demonstrated long ago (Chapter 2). Look at the trajectories of the particles (with diameter or length σ; see Fig. 32.5). Collisions are just exchange of velocities (the momentum conservation). Therefore, if we trace the trajectories of the centers of mass of the particles "subtracting" the particle sizes, then the trajectories behave just as shown in Fig. 32.5(b): point masses going through each other ballistically (ideal gas). Therefore, the equation of state of the hard "rods" must be identical to the ideal gas law in the space not covered by the particles (the free volume), that is $L - N\sigma$. Writing L as V the volume, we get

$$P(V - N\sigma) = Nk_B T. \tag{32.10}$$

---

[9] For a collection of hard spheres there is no gas–liquid transition at any temperature.

[10] Roughly speaking, if the interaction potential is not too long ranged, if it does not allow packing infinitely many particles into a finite volume, and if the energy density is bounded from below, then the normal thermodynamics is guaranteed.

[11] Originally by Lewis Tonks (1897–1971), so 1D hard-core gas is called the *Tonks gas*.

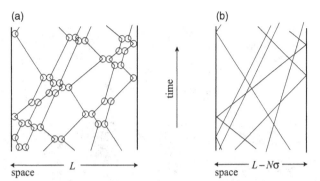

Fig. 32.5 Trajectories of hard balls of diameter σ in a 1D box of size L (a) is just the trajectories of noninteracting point masses in a 1D box of size $L - N\sigma$ (b) as Van der Waals inferred.

That is, van der Waals' idea is correct in 1-space. Therefore, we understand that the 1D Kac model gives

$$P = \left[\frac{Nk_BT}{V - N\sigma} - \frac{1}{2}\varepsilon \left(\frac{N}{V} \right)^2 \right]_{\text{Maxwell}}. \qquad (32.11)$$

This equation gives a liquid–gas phase transition (due to the infinite-range interaction even in 1-space).[12]

32.9 What are the Strategies to Study Phase Diagrams?

We have seen that statistical mechanics can describe phase transitions. Now, let us survey how to study the phase diagram for a given substance.

We wish to map out the equilibrium states in the space spanned by, say, T, P, and other intensive parameters (usually) as we have seen in Fig. 31.1. To study an ordinary geographical map we usually pay attention to the territorial boundaries first. This means we must understand phase transitions. As we will learn near the phase transitions (especially second-order phase transitions) fluctuations can be so big that any theory ignoring them does not make sense (see Chapter 33). However, far away from phase transition points, we may often ignore (at least qualitatively) fluctuations and simple theoretical methods work.

Thus, the study of the phase diagrams consists of two pillars: "renormalization-group theory" (Chapter 33) that can handle violent fluctuations and "mean-field theory" (Chapter 34) that is convenient if we may ignore fluctuations.[13]

---

[12] It is not explained here why Maxwell's rule is also derived from the model.

[13] Exactly solvable models are also important (as we have already seen), but only a very limited number of models can be studied. Numerical methods are also important to enhance our intuition as some of the quoted pictures exemplify.

fluid

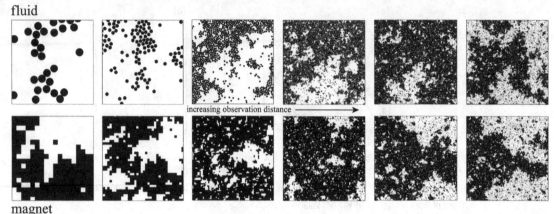

magnet

Fig. 32.6 The 2-fluid and 2-Ising models at the critical point are compared at various scales. Notice that the fluctuations of the up-spin density or number density of the particles (black portions) look more and more similar if we observe these models from increasing distances. (The figures are from Ashton's video with the kind permission of Dr. Douglas Ashton.)

32.10 Magnets, Liquids, and Binary Mixtures Share some Common Features

We can interpret the Ising model as a lattice model of a fluid (lattice gas) with a correspondence (for example): "up" (resp., "down") → lattice point "occupied" (resp., "empty"). Then, something proportional to magnetic field h may be interpreted as the chemical potential (large positive h implies more up spins corresponding to more particles). This correspondence suggests that there are common features in the phase diagrams of a magnet and of a fluid system. Miraculously, the analogy becomes especially appealing near the critical point as illustrated in Fig. 32.6.

We may also interpret the Ising model as a lattice fluid mixture of "up" molecules and "down" molecules, so the phase diagram of a binary mixture should also share some features with that of magnets (see Fig. 32.7). Thus, we can discuss the magnetic system as a representative example.[14]

32.11 The Ising Model in D-Space: A Brief Review

We already know spatial dimensionality is crucial for the existence/nonexistence of phase ordering and consequently phase transitions. Phase ordering is possible because the order can resist thermal fluctuations. To this end microscopic entities must stand "arm in arm." The number of entities with which each entity directly interacts (cooperates) crucially depends on the spatial dimensionality. Also important is the number of independent

---

[14] However, we must not forget that mathematically, we do not even know the existence of phase transitions for fluids.

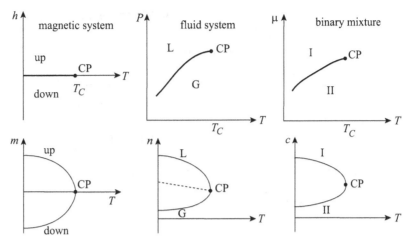

Fig. 32.7 The correspondence of the phase diagrams among the magnet, the fluid, and the binary liquid mixture systems. T: temperature, T_c: the critical temperature, h: magnetic field, P: pressure, μ: chemical potential of component I, m: magnetization per spin, n: the number density, c: the concentration of component I. For the magnetic system, the spins are assumed to be the Ising spins (only two directions are allowed, up or down), and "up" (resp., "down") in the figure means majority of the spins point upward (resp., downward) (ferromagnetically ordered). L implies the liquid phase and G the gas phase. I and II denote phases with different compositions. The following correspondences are natural: for the fields $h \leftrightarrow P \leftrightarrow \mu$; for the order parameters $m \leftrightarrow (n_L - n_G) \leftrightarrow (c_I - c_{II})$.

"communication routes" between two spatially separated points, because in order to order all the members must know what others are doing; higher dimensionality generally means harder to sever communications between two points. This is the intuitive reason why spatial dimensionality matters (higher dimensionality generally favors ordering).

Let us survey the effect of spatial dimensionality on the Ising model **31.9**.

The 1-Ising Model

We can obtain the free energy (with a magnetic field) exactly as we will see in Chapter 34 by, e.g., the transfer matrix method; the phase transition does not occur for $T > 0$ as we have intuitively seen earlier.

The 2-Ising Model

(1) The exact solution by Onsager (1944) gives the free energy without magnetic field. There is a phase transition at $T_c > 0$ as we already know (see **Q32.4**).
(2) Below the phase transition temperature T_c there are only two phases corresponding to the up spin phase and the down spin phase, but up and down phases cannot stably coexist, because the boundary curve fluctuates violently.
(3) Near T_c there are various nontrivial *critical divergences*.[15]

---

[15] Here, "nontrivial" means that fluctuations are so large and so strongly correlated that we cannot use the ordinary central limit theorem of **4.7** to describe the fluctuations correctly.

The 3-Ising Model

(1) No exact evaluation of the free energy is known, but it is easy to demonstrate that $T_c > 0$.
(2) It is known that at sufficiently low temperatures there is phase coexistence.
(3) The critical divergences are nontrivial just as in 2-space.[16]

Beyond 3-Space

Although no exact free energy is known, the existence of positive T_c is easy to demonstrate, and the critical divergences around this point are believed to be the same and not nontrivial for all $d \geq 4$.[17]

32.12 Fluctuation can be Crucial in $\eth < 4$

As we have seen previously, near the critical point, or the point where "strong order" disappears at last, fluctuations are quite important if the spatial dimensionality is less than four. Beyond 4D, however, the effect of fluctuations may not be so pathological, and perhaps we may largely ignore the correlation effects. This observation is relevant to the study of phase transitions.

Problems

Q32.1 The van der Waals Equation of State

(1) Show that the critical point is defined by

$$\left(\frac{\partial P}{\partial V}\right)_T = \left(\frac{\partial^2 P}{\partial V^2}\right)_T = 0. \tag{32.12}$$

(2) For the van der Waals equation of state, find the universal (i.e., a, b-independent) ratio $P_c V_c / k_B T_c$.
(3) Obtain the *reduced equation of state* $P_r = f(V_r, T_r)$ for the van der Waals gas. Here, $P_r = P/P_c$, $V_r = V/V_c$, and $T_r = T/T_c$ are reduced variables. (The reader can work with a 1 mole gas.)
(4) Near the critical point $\pi \equiv P_r - 1$ may be expanded in powers of $\tau \equiv T_r - 1$ and $n \equiv n_r - 1$, where $n_r = 1/V_r$ is the reduced number density. Find the coefficients A to C.

---

[16] However, this is not a proved statement, because the existence of the critical exponents is not established.
[17] This has been established for the dimensions strictly greater than four.

$$\pi = A\tau + B\tau n + Cn^3 + \cdots. \tag{32.13}$$

For the significance of this expression, read the solution.

Solution

(1) The condition for the criticality is for two extremal points in the van der Waals equation of state at lower temperatures to coalesce into a single point. Therefore, the critical point corresponds to the inflection point of the PV-curve (the point where the local maximum and minimum points coalesce). This implies the two conditions stated in the problem.

(2) The equation of state we start with is

$$P = \frac{Nk_BT}{V - Nb} - \frac{aN^2}{V^2}, \tag{32.14}$$

and the two conditions in (1) read

$$-\frac{Nk_BT_c}{(V_c - Nb)^2} + 2\frac{aN^2}{V_c^3} = 0, \tag{32.15}$$

$$2\frac{Nk_BT_c}{(V_c - Nb)^3} - 6\frac{aN^2}{V_c^4} = 0. \tag{32.16}$$

Taking the ratio of these two equations, we get $V_c - Nb = 2V_c/3$. That is, $V_c = 3Nb$. From the first equality (32.15) we get $k_BT_c = 8a/27b$. Now, with the aid of the equation of state, we get $P_c = a/27b^2$. Combining all the results, we get

$$\frac{P_cV_c}{Nk_BT_c} = \frac{(a/27b^2)(3Nb)}{N(8a/27b)} = \frac{3}{8}. \tag{32.17}$$

That is, unless the ratio is $3/8$, a gas does not obey the van der Waals equation of state.

(3) Introducing $P = (a/27b^2)P_r$, $k_BT = (8a/27b)k_BT_r$ and $V = (3b)V_r$ into the van der Waals equation of state, we get

$$\frac{a}{27b^2}P_r = \frac{(8a/27b)NT_r}{3bNV_r - bN} - \frac{aN^2}{9N^2b^2V_r^2}, \tag{32.18}$$

that is,

$$P_r = \frac{(8/3)T_r}{V_r - (1/3)} - \frac{3}{V_r^2}. \tag{32.19}$$

Thus all the different van der Waals gas equations of state can be expressed as a single master equation of state. This is called the *law of corresponding states*. The actual imperfect gases obey the corresponding law fairly well,[18] so one can guess the equation of state fairly accurately from the critical point data.

(4) If the reader is *confident* about their analytical muscle, they may leave all the following calculation to, e.g., Mathematica.

---

[18] Guggenheim, E. A. (1945). The principle of corresponding states, *J. Chem. Phys.*, **13**, 254–261.

We can rewrite (32.19) as

$$\pi + 1 = \frac{(8/3)(1+\tau)}{1/(n+1) - 1/3} - 3(1+n)^2 = \frac{8(1+\tau)(1+n)}{2-n} - 3(1+n)^2 \quad (32.20)$$

$$= 4(1+\tau)(1+n)\left(1 + \frac{n}{2} + \frac{n^2}{4} + \cdots\right) - 3(1+n)^2 \quad (32.21)$$

$$= 1 + 4\tau + 6n\tau + \frac{3}{2}n^3 + 3n^2\tau + \cdots . \quad (32.22)$$

That is,

$$\pi = 4\tau + 6n\tau + \frac{3}{2}n^3 + \cdots . \quad (32.23)$$

This implies that $A = 4$, $B = 6$, and $C = 3/2$. Just below the critical point, $\tau < 0$ and $n = 0$ should still be a solution. Thus, $\pi = 4\tau$ and $n = \pm 2\sqrt{-\tau}$. This implies that the density jumps by $4\sqrt{-\tau}$ just below the critical point. As we will see in Chapter 33, this means that the critical exponent $\beta = 1/2$.

Q32.2 Thermodynamic Justification of Maxwell's Rule

If the temperature is sufficiently low, the PV-curve given by the van der Waals equation of state implies

$$\frac{\partial P}{\partial V} = -\frac{Nk_BT}{(V - N\sigma)^2} + a\frac{N^2}{V^3} \nless 0. \quad (32.24)$$

That is, it is thermodynamically unrealizable. Actually, gas–liquid coexistence occurs when this "unphysical behavior" happens, and the coexistence temperature T is determined by Maxwell's rule. The usual derivation of Maxwell's rule uses (abuses) thermodynamics where the system is thermodynamically unrealizable. However, it is possible to avoid this abuse and still we can thermodynamically demonstrate Maxwell's rule. The coexistence condition for phase A and phase B is the agreement of P, T, and μ. The $\mu_B(T,P) - \mu_A(T,P)$ of the difference in the Gibbs free energy must be computable along a path that passes only thermodynamically stable phases (that is, the broken curve in Fig. 32.8). Since

$$G = E - ST + PV, \quad (32.25)$$

if we compute $E_B - E_A$ and $S_B - S_A$, then $G_A = G_B$ allows us to compute the difference of PV, that is, $P(V_A - V_B)$.

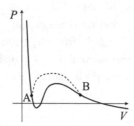

Fig. 32.8   Thermodynamically sound derivation of Maxwell's rule.

(1) Compute $E_B - E_A$.

(2) Compute $S_B - S_A$.

(3) Since $G_B - G_A = 0$, these results allow us to compute $P(V_B - V_A)$. Confirm that this and the result obtained by the naive abuse of thermodynamics:

$$\int_A^B P\,dV = \int_{V_A}^{V_B} dV \left(\frac{Nk_BT}{V - Nb} - \frac{aN^2}{V^2} \right), \tag{32.26}$$

where T is kept constant in the calculation, happen to agree.

Solution

(1) We compute the internal energy difference as

$$E_B - E_A = \int_A^B dE = \int_A^B \left[\left(\frac{\partial E}{\partial V} \right)_T dV + \left(\frac{\partial E}{\partial T} \right)_V dT \right], \tag{32.27}$$

where the temperatures at A and at B are identical. The specific heat C_V for the van der Waals gas is identical with that for a (monatomic) ideal gas. Since $T_A = T_B$, the second term is always zero. From $dE = TdS - PdV$ we get

$$\left(\frac{\partial E}{\partial V} \right)_T = T \left(\frac{\partial S}{\partial V} \right)_T - P = T \left(\frac{\partial P}{\partial T} \right)_V - P = a \left(\frac{N}{V} \right)^2. \tag{32.28}$$

Therefore,

$$E_B - E_A = \int_A^B a \left(\frac{N}{V} \right)^2 dV. \tag{32.29}$$

(2) $S_B - S_A$ can also be obtained, if we note $T_A = T_B$. The temperature derivative gives $(C_V/T)dT$, but this is a function of T only (V-independent), so if the initial and the final temperatures are the same, it cannot contribute to the integral. Therefore,

$$S_B - S_A = \int_A^B dS = \int_A^B \left(\frac{\partial S}{\partial V} \right)_T dV. \tag{32.30}$$

Thus, we get

$$T(S_B - S_A) = T \int_A^B \left(\frac{\partial P}{\partial T} \right)_V dV = T \int_A^B \frac{Nk_B}{V - N\sigma} dV. \tag{32.31}$$

(3) Since $G_B - G_A = 0$ and since the initial and the final T and P are the same,

$$P(V_B - V_A) = T(S_B - S_A) - (E_B - E_A). \tag{32.32}$$

Equations (32.31) and (32.29) imply that the right-hand side of the above equality is path-independent. Therefore, Maxwell's rule has been justified thermodynamically.

Q32.3 Phase Transition in 1D Long-Range System

Discuss the phase transition in a 1D spin system whose coupling constant behaves as r^{-q} ($q < 2$) beyond some distance r_0 (it may be assumed that the coupling constant for $r < r_0$ is J, constant). No rigorous argument is wanted. Simply follow the intuitive Peierls' type argument.

Solution

Assume initially that all the spins are up. Then, we flip L contiguous spins. The required energy is estimated as

$$\Delta E = 2 \int_{L/2+\delta}^{\infty} dx \int_{-L/2}^{L/2} dy \frac{1}{(x-y)^q}. \tag{32.33}$$

The contribution of the short-range interactions is irrelevant. It is easy to see that ΔE is not smaller than

$$\int_{L}^{\infty} dx \int_{-L/2}^{L/2} dy \frac{1}{(x-y)^q} = O[L^{2-q}]. \tag{32.34}$$

This energy increases indefinitely with L if $q < 2$. As we have seen for the nearest-neighbor interaction system in the text, the contribution of entropy that fatally cripples system ordering is of order $\log L$, so the entropy effect cannot destroy the order even at finite temperatures.[19]

Q32.4 Demonstration of Peierls' Theorem

Let us impose the all up spin boundary condition to the 2-Ising model on the finite square (see Fig. 32.3). Then, we wish to take a thermodynamic limit. If the spin at the center of the square is more likely to be up than to be down, we may conclude that there is a long-range order.

Let γ be a closed Bloch wall (i.e., the boundary between up and down spin domains; this does not mean that the domain enclosed by γ is a totally up or down domain (lakes can contain islands with ponds with islets, etc.; the wall corresponds to the shore lines). The probability, $P_V(\gamma)$, to find such a wall in the system with volume, V, has the following estimate (our intuitive contour argument utilizes something like this estimate):

$$P_V(\gamma) \le e^{-2\beta J|\gamma|}, \tag{32.35}$$

where $|\gamma|$ is the total length of the contour γ, $\beta = 1/k_B T$, and J is the usual ferromagnetic coupling constant. (This naturally looking inequality needs a proof; see (3) for a smart proof.) For (1) and (2) assume this.

(1) Let P_V^0 be the probability of the spin at the origin to be down. Show

$$P_V^0 \le \sum_{\gamma} e^{-2\beta J|\gamma|}, \tag{32.36}$$

where the summation is over all the possible contours surrounding the origin.

[19] According to a rigorous argument even for $q = 2$ phase transition occurs at a finite temperature.

(2) Estimate crudely the number of contours γ of length $|\gamma|$ surrounding the origin, and find the upper bound of the sum in (32.36) and show that if β is sufficiently large, $P_V^0 < 1/2$ for large V.

(3) To complete our demonstration show (32.35).

Solution

The argument here is, although simplified, almost rigorous.

(1) The event that the spin at the origin is down occurs only if at least one Bloch wall surrounds the origin. Let $P(\gamma)$ be the probability that there is a closed Bloch wall γ surrounding the origin. Then,

$$P_V^0 \leq P(\text{at least there is one Bloch wall surrounding the origin}) = P(\cup\{\exists\gamma\}), \tag{32.37}$$

where $\cup\{\exists\gamma\}$ is the event that there is at least one closed wall around the origin, irrespective of its shape. Since $P(A \cup B) \leq P(A) + P(B)$, $P(\cup\{\exists\gamma\}) \leq \sum_\gamma P(\gamma)$, The sum is over all the closed curves around the origin. We use (32.35) to get (32.36).

(2) Notice that the wall is a walk on the dual lattice (Fig. 32.9). To draw a closed curve on the dual lattice of length $|\gamma|$ surrounding the origin, we must start at some point. Let the starting point to be the closest point on γ to the origin. The number of candidate points for this cannot exceed $|\gamma|^2$. Let us start a random walk of length $|\gamma|$. There is no guarantee that the walk makes a closed curve, but all the curves satisfying the desired condition can be drawn as trajectories of the random walk (even ignoring the self-avoiding nature). Since the number of distinct walks (without immediate retracing of the previous step) is $4 \times 3^{|\gamma|-1}$, we have

$$P_V^0 \leq \sum_\gamma e^{-2\beta J|\gamma|} \leq \frac{4}{3} \sum_{|\gamma|=4}^\infty |\gamma|^2 3^{|\gamma|} e^{-2\beta J|\gamma|}. \tag{32.38}$$

Here, we have used the fact that the smallest closed curve on the lattice surrounding the origin has length 4 (the smallest square on the dual lattice) (see Fig. 32.9). Therefore, if β is sufficiently large, for any V we can make $P_V^0 < 1/2$.

Equation (32.38) tells us that the spin at the origin points upward more likely than downward. The required β may be unrealistically large due to the crudeness of the

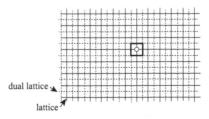

dual lattice

lattice

Fig. 32.9 Dual square lattice: the dotted lattice is the dual lattice of the full lined square lattice. The white disk is the central spin position and the smallest Bloch wall is the thick square surrounding it.

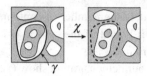

Fig. 32.10   Configurations with and without γ. This correspondence is one to one.

estimation, but still it is finite (that is $T > 0$). Thus, we have shown that the system orders at some low but positive temperature.[20]

(3) This is called *Peierls' inequality*.[21] To demonstrate this, the energy in the volume V is estimated (let us denote the number of lattice points in V by the same symbol V):

$$E(\phi) = -2JV + 2J|\partial\phi|, \tag{32.39}$$

where $\partial\phi$ denotes the totality of the Bloch walls appearing in the spin configuration ϕ (in V). The spin configurations on V is divided into the ones including γ denoted by Φ_γ and the rest denoted by Φ_γ^-. We may write

$$P_V(\gamma) = \sum_{\phi \in \Phi_\gamma} e^{-\beta E(\phi)} \Bigg/ \sum_\phi e^{-\beta E(\phi)}. \tag{32.40}$$

Here, in the numerator the sum is over all the spin configurations containing the Bloch wall γ. Next, let us define a one-to-one map $\chi : \Phi_\gamma \to \Phi_\gamma^-$ that flips all the spins inside γ (see Fig. 32.10). The map χ reduces the number of adjacent up–down spin pairs by $|\gamma|$, so the Boltzmann factor must be multiplied by $e^{2\beta J|\gamma|}$. Therefore, if we replace every element in Φ_γ with the corresponding element in Φ_γ^-, we must multiply $e^{-2\beta J|\gamma|}$ to cancel this Boltzmann factor:

$$P_V(\gamma) = \frac{\sum_{\Phi_\gamma} e^{-\beta E(\phi)}}{\sum_\phi e^{-\beta E(\phi)}} = e^{-2\beta J|\gamma|} \frac{\sum_{\Phi_\gamma^-} e^{-\beta E(\phi)}}{\sum_\phi e^{-\beta E(\phi)}} \le e^{-2\beta J|\gamma|}. \tag{32.41}$$

---

[20] Incidentally, the sum in (32.38) has the following form $\sum \exp(|\gamma|(\log 3 - 2\beta J + 2 \log |\gamma|/|\gamma|))$, so for a large island with large $|\gamma|$, if $2\beta J > \log 3$ or $k_B T < 2J/\log 3$ we expect ordering. Therefore, this tells us $k_B T_c$ is estimated as $2J/\log 3 = 1.82\,J$. The exact answer is about $2.27\,J$, so it is not a very bad estimate.

[21] Chapter 2 Section 1 of Sinai, Ya. G. (1982). *Theory of Phase Transitions: Rigorous Results*, Oxford: Pergamon Press.

Critical Phenomena and Renormalization

Summary

∗ In the second-order phase transition all the length scales couple; small-scale fluctuations couple and build up into larger scale fluctuations.

∗ Large fluctuations near the critical point imply universality.

∗ Scaling + coarse-graining is the renormalization group approach that can extract universal features.

Key words

correlation length, critical exponent, scaling relation, Kadanoff picture, renormalization group, (genuine and trivial) universality, bifurcation, central limit theorem

The reader should be able to:

∗ Illustrate what happens if a second-order phase transition is approached by changing T.

∗ Explain why universality holds near critical points.

∗ Grasp the basic and intuitive idea of renormalization groups.

This chapter explains the basic ideas of the renormalization group approach. The reader can browse through the first half (up to around **33.11**) and then skip the rest to go to the following chapters.

33.1 Typical Second-Order Phase Transition and Correlation Length

As a typical second-order phase transition, let us look at the phase transition point T_c (also called the critical point) of the 2-Ising model (see Fig. 33.1). If we approach T_c from the high temperature side, up and down spin islands grow as T is reduced. The typical size ξ of the islands is called the *correlation length*. Mathematically, it is defined as the decay rate of the following covariance called the spatial correlation function:[1]

---

[1] To be precise, $\langle s(\boldsymbol{r})s(0) \rangle - \langle s \rangle^2 \sim r^{-(\eth-1)/2}e^{-|\boldsymbol{r}|/\xi}$ (proved above T_c and for sufficiently below T_c for the Ising model), but the exponential factor dominates.

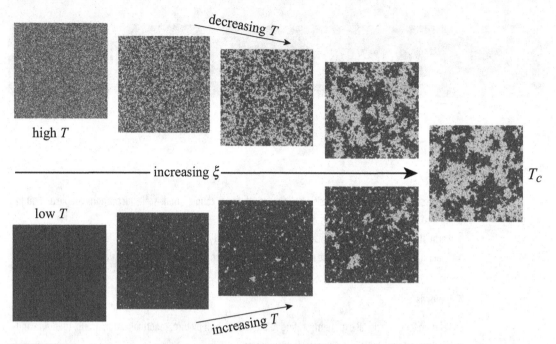

Fig. 33.1   Temperature dependence of spin fluctuations of a 2-Ising model. The right-most figure corresponds to the critical point. The upper half is the disordered high temperature phase, and the lower half is the ordered phase. The correlation length ξ increases from left to right. Black denotes down spins.

$$\langle s(\boldsymbol{r})s(0)\rangle \sim e^{-|\boldsymbol{r}|/\xi}, \tag{33.1}$$

where $s(\boldsymbol{r})$ is the spin at position $\boldsymbol{r}$. We see ξ grows and actually diverges at T_c as $\xi \sim |T - T_c|^{-\nu}$, where ν is a universal constant (one of the critical exponents we will encounter in the following). From the low temperature side, in the almost completely ordered phase appear flipped spins like blinking stars. They become spin-flipped islands, growing bigger and bigger as T increases (i.e., ξ grows as the lower series in Fig. 33.1 exhibits).

The patches (islands) of size up to $\sim \xi$ appear and disappear, so big fluctuations occur near the critical point. This is the *critical fluctuation*. Since large-scale changes cannot be completed very quickly, we see that the dynamics of mesoscopic or large-scale patterns becomes sluggish near the critical point (called *critical slowing down*). These fluctuations are not only big but also statistically highly correlated; the spins within distance ξ behave similarly. Thus, even at the scale we can observe the system is not simply governed by the law of large numbers.

What happens to the correlation function if ξ diverges? It can be shown that if the correlation does not decay exponentially, then

$$\langle s(\boldsymbol{r})s(0)\rangle \sim \frac{1}{r^{d-2+\eta}} \tag{33.2}$$

is the general expression for the order parameter correlation function. Here, $\eta \ (\in (0, 1])$ is another example of the critical exponents.

Table 33.1 Ising critical exponents			
Exponents	2-space	3-space	$d(\geq 4)$-space
α	0 (log)	0.11	0 (jump)
β	1/8	0.325	1/2
γ	1.75	1.24	1
ν	1	0.63	1/2

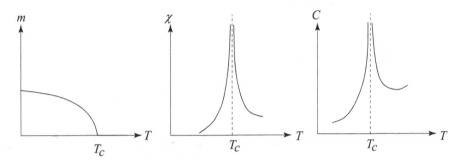

Fig. 33.2   Schematic illustrations of singular behaviors near the critical point.

33.2 Divergence of Susceptibilities and Critical Exponents

We know the fluctuation–response relation (Chapter 26). For example, the (isothermal) magnetic susceptibility χ is directly related to the variance of magnetization:

$$\chi_T = \beta \langle \delta M^2 \rangle. \tag{33.3}$$

We just learned the big critical fluctuation, so we can expect that the susceptibilities become very large near the critical point. Empirically the magnetic susceptibility diverges as (for $h = 0$, without magnetic field) (see Fig. 33.2):

$$\chi \sim |\tau|^{-\gamma} \ (h = 0), \tag{33.4}$$

where $\tau = (T - T_c)/T_c$ and $\gamma \ (> 0)$ is another critical exponent.

We cannot expect smooth increase of the magnetization m from zero below T_c:

$$m \sim (-\tau)^\beta \ (h = 0, \tau < 0). \tag{33.5}$$

The divergence of energy (or entropy) fluctuation causes the divergence of the constant magnetic field specific heat C_B (cf. **Q26.1**, **Q26.2**) as

$$C_B \sim |\tau|^{-\alpha} \ (h = 0). \tag{33.6}$$

Here α, β, γ, are positive numbers and are called *critical exponents*. Representative values can be found in Table 33.1. They are universal numbers. For example, for any fluid or binary mixture, or magnets (with an easy axis = Ising magnets) these numbers are common. They are determined by the nature of our world, not by material-scientific details.[2]

33.3 Critical Exponent (In)equalities

It was empirically noted that several relations hold among these exponents such as

$$\alpha + 2\beta + \gamma = 2. \tag{33.7}$$

Thermodynamically, we can prove (*Rushbrooke's inequality*; assuming the existence of these exponents)

$$\alpha + 2\beta + \gamma \geq 2. \tag{33.8}$$

33.4* Proof of Rushbrooke's Inequality: A Good Review of Thermodynamics

Demonstration of this inequality gives us an excellent opportunity to review elementary thermodynamics. Even around the critical point the system does not become thermodynamically unstable. Therefore, inequalities required by the thermodynamic stability conditions (e.g., **25.8**) remain valid. For example,[3] thermodynamic stability of a magnet implies (the Gibbs relation in this case is $dE = TdS + BdM$):

$$\frac{\partial(S, M)}{\partial(T, B)} \geq 0. \tag{33.9}$$

This inequality can be written explicitly, by expanding the determinant defining the Jacobian, as

$$\left(\frac{\partial S}{\partial T}\right)_B \left(\frac{\partial M}{\partial B}\right)_T \geq \left(\frac{\partial S}{\partial B}\right)_T \left(\frac{\partial M}{\partial T}\right)_B = \left(\frac{\partial M}{\partial T}\right)_B^2, \tag{33.10}$$

where a Maxwell's relation (recall **24.12**):

$$\frac{\partial(S, T)}{\partial(B, M)} = 1 \tag{33.11}$$

has been used to obtain the second equality. This implies

$$\frac{1}{T} C_B \chi \geq \left(\frac{\partial M}{\partial T}\right)_B^2, \tag{33.12}$$

---

[2] However, statistical-mechanically, we do not know whether these exponents are well defined. That is, we do not have any general mathematical demonstration of the existence of the divergence behaviors described here.

[3] This requires twice differentiability of the potential, so it does not hold exactly at the critical point, but we may use it in any neighborhood.

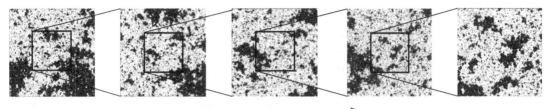

successive magnifications

Fig. 33.3 A configuration of 2-Ising model at the criticality has a statistical self-similarity. Going to the left means to observe the system from successively farther away. Since the pattern stays statistically similar, it should not be sensitive to microscopic details. The figure was made from Ashton's video (with Dr. Douglas Ashton's kind permission).

Introducing the definitions of the critical exponents given above, we obtain

$$|\tau|^{-\alpha}|\tau|^{-\gamma} \geq |\tau|^{2(\beta-1)}. \tag{33.13}$$

Here, we have ignored all the finite coefficients near the critical point (such as T^{-1}).[4] Equation (33.13) implies that

$$|\tau|^{-(\alpha+2\beta+\gamma-2)} \geq 1 \tag{33.14}$$

is required for $\tau \to 0$. Therefore, the quantity in the parentheses must be non-negative:

$$\alpha + 2\beta + \gamma \geq 2. \tag{33.15}$$

33.5 Fluctuation and Universality

Look at Fig. 33.3 which is a configuration of a 2-Ising model exactly at $T = T_c$. Statistically, the configurations look indistinguishable, although the five configurations are obtained by successive magnification of a single snapshot of a configuration at the critical point. This scale invariance implies that fluctuations are enormous and even macroscopic properties would be insensitive to microscopic details (*universality*). Indeed, the critical exponents for fluids, binary mixtures, and 3-Ising magnets are identical (cf. Fig. 32.7). Also, look at Fig. 32.6 again.

33.6 Universality: Trivial and Nontrivial

The reader might say we already know examples of universality. For example, we know $PV = 2E/3$ for any ideal gas irrespective of statistics, if the spatial dimensionality is 3, and the dispersion relation is $\varepsilon \propto p^2$ (Chapter 28). Or we know $PV = E/3$ for phonons

---

[4] We have assumed that the critical point is not zero; The 1-Ising model has $T_c = 0$, but this is a pathological case.

and photons in 3-space. This is quite universal. However, it is due to the universality (common quantitative feature) of the elementary entities making up the systems. In this sense, universality is unsurprising, and trivial.

In contradistinction, the universality near the critical point is obviously not due to some common quantitative features at the microscopic level. Of course, the system must exhibit a critical phenomenon, but we use only three features: the interaction is short-ranged, the order parameter is a scalar, and the system is 3D. Thus, the reason for the universality is not in the common nature of the system constituents.

Furthermore, the response to system changes is quite different from the trivial case above. If one adds certain interactions, the "trivial" universality is lost in infinitely different ways according to the infinitely different perturbations. In contrast, in the case of the critical phenomena, if one turns on perturbations modifying interactions, T_c changes sensitively and also the actual values of susceptibilities are altered, but the main features (e.g., critical exponents) do not change. Thus, the universality of the second-order phase transition deserves to be called the *genuine universality*.

33.7 What is Statistical Mechanics For?

The phase-transition points (e.g., T_c) and the values of susceptibilities sensitively depend on materialistic details as mentioned earlier. We also noted that there is no use of theory to study chemical equilibrium constants in **27.15**. Generally speaking, it is impossible to calculate materials constants very accurately through implementing the theoretical formalism of statistical thermodynamics. Then, what is statistical mechanics for? Statistical mechanics should try to understand (and compute) universal features of many-body systems that are insensitive to quantitative details. Needless to say, demonstrating the existence of some features (say, a phase transition) is an important target of statistical mechanics. As we will see soon in the case of critical phenomena there is a hope that statistical mechanics can obtain universal features quantitatively. Actually, it is fair to say that the true role of statistical mechanics was consciously recognized as the study of universality and not of fetish details (as the actual value of T_c) through the study of critical phenomena.

33.8 Kadanoff Construction

Without any simulation Kadanoff (1937–2015) completely understood the structure illustrated in Fig. 33.3 and succeeded in elucidating the general features of critical phenomena with an ingenious intuitive picture (Fig. 33.4). If the original system has a temperature $\tau = (T - T_c)/T_c$ and magnetic field h, then from our stepped-back point of view the system looks as if it has these parameters scaled (increased; farther away from the critical point) to $\tau \ell^{y_1}$ and $h\ell^{y_2}$; the exponents y_1 and y_2 must be positive, where ℓ is the shrinking rate (> 1). This is a *guess* or *hypothesis*, but seems to explain everything neatly as we will see below.

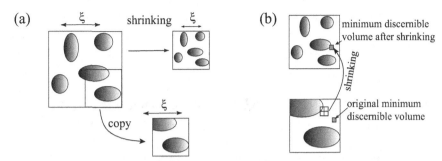

Fig. 33.4 (a) The Kadanoff construction. "Shrinking" corresponds to looking at the system from distance with fixed eyesight, that is, scaling + coarse-graining. The outcome corresponds to the system away from the critical point; the correlation length ξ becomes smaller (cf. Fig. 33.1). (b): If we step back and the distance between us and the sample becomes ℓ times as large as the original distance (in the figure $\ell = 2$), the actual *linear dimension* of the minimum discernible volume becomes ℓ-times as large as the original minimum discernible volume.

33.9 Scaling Law

Let us write $m = \mathcal{M}(\tau, h)$ (this is the equation of state for the magnetic system). After one stepping back, the volume of the region recognized as a unit cube to us would be actually the cube with edge ℓ (see Fig. 33.4(b)) before stepping back.

Let us put a prime, ′, to the quantities observed after stepping back. We look at the magnetic energy stored in the minimum discernible block $h'm'$ (after shrinking). The energy should be a much better additive quantity than the local magnetic moment (since energy is additive even microscopically), so we expect

$$h'm' = \ell^d hm. \tag{33.16}$$

Since $h' = h\ell^{y_2}$, we obtain:[5]

$$m = \ell^{-d}(h'/h)m' = \ell^{y_2-d}\mathcal{M}(\tau', h') = \ell^{y_2-d}\mathcal{M}(\tau\ell^{y_1}, h\ell^{y_2}). \tag{33.17}$$

This is the *scaling relation* for the equation of state. It should be clearly recognized that this is an *identity* that holds for any positive number ℓ. Therefore, we may set $|\tau|\ell^{y_1} = 1$. Thus, we obtain from (33.17) ($\tau < 0$ to have non-zero magnetization):

$$m(\tau, 0) = |\tau|^{(d-y_2)/y_1} m(-1, 0). \tag{33.18}$$

That is (compare with (33.5)),

$$\beta = \frac{d - y_2}{y_1}. \tag{33.19}$$

[5] B. Widom (1965) (in Equation of state in the neighborhood of the critical point, *J. Chem. Phys.*, **43**, 3898) realized that if we assume this generalized homogeneous function form of the equation of state, critical phenomena can be understood. The Kadanoff construction explains this – see Kadanoff, L. P. (1966). Scaling laws for Ising models near T_c, *Physics*, **2**, 263–272.

We can also conclude from the derivative of (33.17) with respect to h:

$$\gamma = \frac{2y_2 - d}{y_1}. \tag{33.20}$$

33.10 Critical Exponent Equality

To obtain α we must compute the specific heat, which is available as the second derivative of the free energy with respect to T (recall **24.11**). The (singular part of the) free energy $f_s = \mathcal{F}_s(\tau, h)$ per minimum discernible volume unit scales as:[6]

$$f_s = \mathcal{F}_s(\tau, h) = \ell^{-d} \mathcal{F}_s(\tau \ell^{y_1}, h \ell^{y_2}). \tag{33.21}$$

This comes from $f'_s = \ell^d f_s$ due to the extensivity of the free energy. If we differentiate (33.21) with h, we get (33.17). Differentiating (33.21) twice with respect to τ (that is, T), we obtain

$$C(\tau, h) = \ell^{2y_1 - d} C(\tau \ell^{y_1}, h \ell^{y_2}). \tag{33.22}$$

Therefore,

$$\alpha = \frac{2y_1 - d}{y_1}. \tag{33.23}$$

From (33.19), (33.20), and (33.23) we obtain Rushbrooke's equality:

$$\alpha + 2\beta + \gamma = 2. \tag{33.24}$$

33.11 Renormalization Group (RG) Transformation

Kadanoff's idea (the Kadanoff construction) consists of two parts: coarse-graining and scaling (shrinking). The crux of the idea is: if the system is at the critical point, the configuration is invariant under coarse-graining, $\mathcal{K}$, with an appropriate scaling $\mathcal{S}$. That is, if we define $\mathcal{R} = \mathcal{K}\mathcal{S}$, then thermodynamic observables (densities and fields) are invariant under the application of $\mathcal{R}$ at T_c. To apply $\mathcal{R}$ is to observe the system from distance with a fixed eyesight. Figure 33.5(a) illustrates how iterative operations of $\mathcal{R}$ drive the statistical configurations at various temperatures.

Operating $\mathcal{R}$ is called a *renormalization group* (RG) *transformation*. We can understand its iterative applications as multiplication of $\mathcal{R}$; doing nothing corresponds to the unit element. Therefore, the totality of the renormalization group transformations is informally called a *renormalization group*.[7] According to Kadanoff's original idea, the image due

---

[6] The free energy itself has a large nonsingular part that does not contribute to the singular behaviors near the critical point (cf. **Q33.2**).

[7] The inverse may not be defined, so it is usually a monoid. For the concept of "group" see Chapter 35.

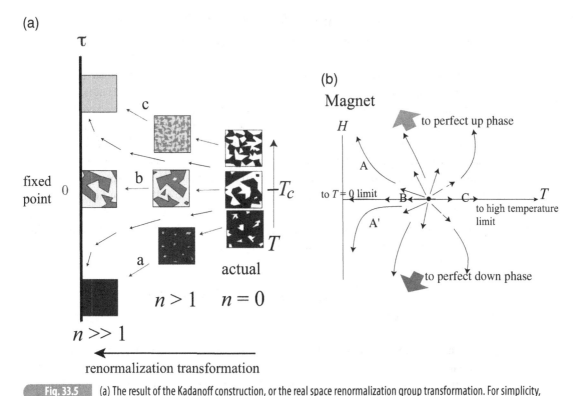

Fig. 33.5 (a) The result of the Kadanoff construction, or the real space renormalization group transformation. For simplicity, $h = 0$ (i.e., on the phase diagram in (b) we study the system only along the curve BC). Here, $\tau = (T - T_c)/T_c$ and n is the number of times we operate the renormalization group transformation $\mathcal{R}$; we start from the actual configurations ($n = 0$) at various temperatures. As $\mathcal{R}$ is applied successively, the configurations are transformed as the arrows indicate. The leftmost vertical line denotes the destination after many applications of $\mathcal{R}$. a, b, c correspond to the trajectories a, b, c in Fig. 33.6. Only when the starting point is just right, the system can stay at $\tau \sim 0$. The low temperature states are driven to one of the ordered phases at $T = 0$; in the illustration it happens to be totally "down." If the starting point is $T > T_c$, the state is driven to $T = \infty$ state. (b) RG flows "projected" (see the text) on the phase diagram of the Ising magnet. There are five ultimate destinations (high temperature limit, phase boundary, critical point, all up and all down low temperature states).

to $\mathcal{R}$ is the same system under a different condition (e.g., at a different temperature), so we may understand that $\mathcal{R}$ transforms a thermodynamic state into another (of the same system); then, we may imagine that successive applications of $\mathcal{R}$ define a flow on the phase diagram of the same materials system under study. This view is illustrated in Fig. 33.5(b).[8]

---

[8] As we will see soon, this flow does not generally flow on the phase diagram (of a given material). In terms of Fig. 33.5(a), the flow diagram exhibits what happens to the "actual" configurations. The renormalization flows move as $n = 1, 2, \ldots$ to the left, starting from the "actual slice." The flows in Fig. 33.5(b) are, intuitively, an approximate projection of these RG flow lines onto the actual system.

33.12 Renormalization Group Fixed Point

At the fixed point $\mathcal{R}\xi = \xi$ should hold for the correlation length ξ. Since $\mathcal{S}$ definitely shrinks the system, this condition is satisfied only if $\xi = 0$ or $\xi = \infty$. That is, the phases without spatial correlation at all or critical points are the only possible fixed points. Notice that if we understand these fixed points, we understand the general structure of the phase diagram. The ordinary bulk phases from our macroscopic point of view do not have any appreciable correlation distance, so they are close to the $\xi = 0$ fixed points. To understand their macroscopic properties we need not worry (qualitatively) about spatial correlations of fluctuations (see footnote 9). This is the reason why the so-called mean-field theory (see Chapter 34) is useful. Thus, to understand the phase diagram, we use mean-field theory to understand the bulk phases not too close to the critical points,[9] and use renormalization group theory to understand the features near the critical points.

33.13 Renormalization Group Flow

We may interpret the renormalization group transformation as a map from a (generalized) canonical distribution μ to another (generalized) canonical distribution $\mu' = \mathcal{R}\mu$. We can imagine effective Hamiltonians H and H' (it is customary that β is absorbed in Hs) according to

$$\mu = \frac{1}{Z}e^{-H}, \ \mu' = \frac{1}{Z'}e^{-H'}. \tag{33.25}$$

We may write $H' = \mathcal{R}H$. Therefore, we can imagine that successive applications of $\mathcal{R}$ defines a flow (RG flow) in the space of Hamiltonians (or models or systems). This idea is illustrated in Fig. 33.6 (Fig. 33.5(a) actually illustrates the pattern changes along a, b, or c in Fig. 33.6). In Fig. 33.6 H^* is a fixed point with an infinite correlation length of the RG flow. Its stable manifold (mfd)[10] is called the *critical surface*. The Hamiltonian of the actual material, say, magnet A, changes (do not forget that β is included in the definition of the Hamiltonian in (33.25)) as the temperature changes along the trajectory denoted by the curve with "magnet A." It crosses the critical surface at its critical temperature. The renormalization transformation uses the actual microscopic Hamiltonian of magnet A at various temperatures as its initial conditions. Three representative RG flows for magnet A are depicted: a is slightly above the critical temperature; b exactly at T_c of magnet A (b' is the corresponding RG trajectory for magnet B, a different material; both b and b' are on the critical surface); c is slightly below the critical temperature (these a, b, c correspond to

---

[9] This does not mean that we can use the original microscopic Hamiltonian when we utilize a mean-field approach; we must use an appropriately renormalized Hamiltonian. Therefore, a precise statement is: there is a (model) Hamiltonian (with short-range interactions) that can be used to describe the macroscopic features of a bulk phase with the aid of a mean-field approach.

[10] **Stable manifold** For a fixed point x, the totality of points y flowing into x is called the *stable manifold* of x.

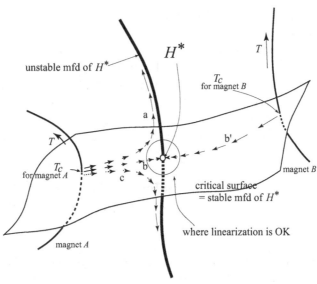

Fig. 33.6   A global picture of renormalization group flow in the Hamiltonian space $\mathcal{H}$. The explanation is in the text. "mfd" = manifold. The thick curves emanating from H^* denote the direction that the Hamiltonians are driven away from the fixed point by renormalization. This curve corresponds to the leftmost thick line in Fig. 33.5(a).

those in Fig. 33.5(a)). Do not confuse the trajectory (black curve) of the actual microscopic system as temperature changes and the trajectories (successive arrows; RG flow) produced by the RG transformation.

If we understand H^*, we understand all the universal features of the critical behaviors of all the magnets crossing its critical surface.

33.14   Central Limit Theorem and Renormalization[11]

We have realized that near the second-order phase transition/critical point, fluctuations become large, and also strongly correlated. How can we handle such a strongly correlated fluctuating system? We wish to know the distribution of fluctuations at the mesoscopic scale (because the correlation lengths are at the mesoscopic scale). If we understand the mesoscopic fluctuation statistics, we should be able to compute macroscopic observables. How can we study the mesoscopic fluctuation statistics? The usual central limit theorem is the tool to study mesoscopic fluctuations if random variables are independently and identically distributed (iid) (recall **4.7**):

> If X_i are assumed to be iid with a finite variance, V, and zero mean, the density distribution function of $S_N/\sqrt{N}$ converges to a Gaussian distribution function with mean 0 and variance V.

[11] See Chapter 3 of Oono, Y. (2013). *The Nonlinear World*, Tokyo: Springer) for other types (interpretations) of renormalization.

According to the Kadanoff construction, we make a block spin (by summing nearby spins) and then scale it appropriately as we see in (33.17). Blocking corresponds to constructing the sum S_N, and scaling corresponds to dividing with $\sqrt{N}$. At the renormalization group fixed point we expect the "blocked and scaled" spins obey a definite distribution function. Thus, we understand that renormalization group procedure is an extension of the ordinary central limit theorem to highly correlated random variables. Unfortunately, mathematically, this generalization is still in its infancy.

Let us have a taste of quantitative realization of the renormalization idea in **33.15**.

33.15 Detailed Illustration of Real Space Renormalization Group Calculation[12]

Kadanoff's idea, which may be summarized as follows, allows us to compute, e.g., critical exponents: Introduce some method $\mathcal{K}$ to coarse-grain the system. This method also dictates the spatial scale reduction rate ℓ. The coarse-graining method $\mathcal{K}$ may be understood as a map from a configuration S (this may be a field or spin configuration $\{s_i\}$) of the original system to a configuration of the reduced system. Figure 33.7 illustrates two examples. The important point of $\mathcal{K}$ is that it is a map: given a configuration S, $\mathcal{K}(S)$ is unique. However, it is not an injection (one-to-one map), since it is a kind of coarse-graining.

Triangular Lattice 2-Ising Coarse-Graining

Let us study the triangular lattice Ising model. It is generally the case that coarse-graining produces multispin interactions, even if the original model contains only binary spin interactions as in the present example. However, we wish to be as simple as possible, so

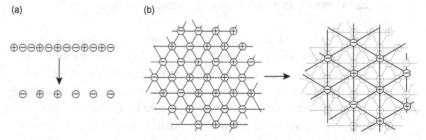

(a)   (b)

Fig. 33.7   (a): Decimation of 1-Ising model, $\ell = 2$ (see **Q33.1**). (b): Blocking of three spins of the triangular lattice 2-Ising model. $\ell = \sqrt{3}$. The value of the block spin is determined by the majority rule: the block spin is up (down) if two or more spins being blocked are up (down).

[12] To understand renormalization group approaches the best way is to follow a few examples to nurture the reader's intuition; Leo Kadanoff told the author that it was easy to invent an RG if we knew the answer (the system behavior). Chaikin, P. M. and Lubensky, T. C. (1995). *Principles of Condensed Matter Physics*, Cambridge, Cambridge University Press contains excellent explanations and examples.

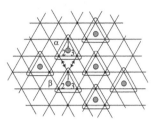

Fig. 33.8 Triangular lattice and the block spins α and β. 1, 2, 3 denote the original spins (small black dots). The rounded triangles denote block spins, and small gray disks indicate the positions of the block spins.

we use a (crude but still interesting) approximation that under $\mathcal{K}$ illustrated in Fig. 33.7(b), the Hamiltonian preserves its shape (that is, we assume that the RG flow does not leave the phase diagram of this particular system; recall Fig. 33.5):[13]

$$H = \sum K s_i s_j + h s_i \rightarrow H' = \sum K' s'_\alpha s'_\beta + h' s'_\alpha, \tag{33.26}$$

where s'_α, etc. denote the block spins defined by the majority rule: if two or more spins are up (down) in the block, the block spin is up (down).

How to Specify Block Spins

Figure 33.8 explains the block spins more explicitly. For simplicity, let us study the small h case; we ignore its effect on the coarse-grained coupling constant. Since we are interested in the macroscopic global behavior of the mode, we need not worry about the intrablock spin interactions.[14] Therefore, the "block spin α"–"block spin β" interaction energy must be equal to the sum of the interaction energies among the original spins belonging to different blocks. As can be seen from Fig. 33.8, we may demand

$$K' s'_\alpha s'_\beta = K(s_{\alpha 2} s_{\beta 1} + s_{\alpha 3} s_{\beta 1}) \tag{33.27}$$

on average (we cannot demand this exactly). That is, the block spin α–β interaction is supported by two "actual" interactions: interactions between $\beta 1$ spin and $\alpha 2$ and $\alpha 3$ spins.

The Spin–Block Spin Relation

If we wish to relate K and K', we must relate s and s'. The basic idea is that near the critical point the correlation length ξ is large, so

$$K' s'_\alpha s'_\beta = K(\langle s_{\alpha 2}\rangle_{s'_\alpha} \langle s_{\beta 1}\rangle_{s'_\beta} + \langle s_{\alpha 3}\rangle_{s'_\alpha} \langle s_{\beta 1}\rangle_{s'_\beta}), \tag{33.28}$$

---

[13] More accurate handling of this problem can be seen in Niemeijer, Th. and van Leeuwen, J. M. J. (1973). Wilson theory for spin systems on a triangular lattice, *Phys. Rev. Lett.*, **31**, 1411–1414.

[14] They shift the origin of the free energy, but it has nothing to do with the correlation length, so they correspond to the nonsingular part of the free energy. Recall that we discussed the singular part of the free energy; we are picking up the singular part only. See **33.10**.

Table 33.2 The original spin configuration

$S_{\alpha 1}$	+1	+1	+1	−1
intra-block energy	−3K	+K	+K	+K

where $\langle s \rangle_{s'}$ is the average of the original spin s in the block spin whose value is s' (a conditional average), and

$$s'_\alpha = \mathrm{sgn}(\langle s_{\alpha 1}\rangle_{s'_\alpha}). \tag{33.29}$$

Table 33.2 tells us the original spin configuration compatible with $s'_\alpha = +1$ (i.e., the majority up; $s_{\alpha 1}$ spin is circled). The last line in the table is the intra-block energy of the block spin that determines how a particular internal configuration is likely. Therefore, we obtain

$$\langle s_{\alpha 1}\rangle_+ = \frac{e^{3K} + e^{-K} + e^{-K} - e^{-K}}{e^{3K} + e^{-K} + e^{-K} + e^{-K}} = \frac{e^{3K} + e^{-K}}{e^{3K} + 3e^{-K}} \equiv \phi(K). \tag{33.30}$$

By symmetry $\langle s_{\alpha 1}\rangle_- = -\langle s_{\alpha 1}\rangle_+$, so we can write

$$\langle s_{\alpha 1}\rangle_{s'_\alpha} = \phi(K)s'_\alpha. \tag{33.31}$$

The $K \to K'$ Relation

Equation (33.28) now reads

$$K's'_\alpha s'_\beta = 2K\phi(K)^2 s'_a s'_\beta, \tag{33.32}$$

or

$$K' = 2K\phi(K)^2. \tag{33.33}$$

The $h \to h'$ Relation

Since we have assumed that h is small, we may simply ignore its effect on K', and we require

$$h's'_\alpha = h(s_{\alpha 1} + s_{\alpha 2} + s_{\alpha 3}), \tag{33.34}$$

so we immediately obtain

$$h' = 3h\phi(K). \tag{33.35}$$

$\mathcal{R}$ Has Been Constructed

This completes our construction of $\mathcal{R} : (K, h) \to (K', h')$ with $\ell = \sqrt{3}$ (from the geometry: Fig. 33.8).

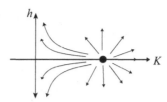

Fig. 33.9   The RG flow for the triangular lattice Ising model in the (K, h)-plane. The black dot denotes the location of the nontrivial fixed point $(K^*, 0)$. The origin is also a fixed point. This figure corresponds to Fig. 33.5 (b); larger K corresponds to lower temperature. The black dot corresponds to H^* and the K axis to the thick curve through H^* in Fig. 33.6.

Fixed Points of $\mathcal{R}$

Let us look for fixed points of $\mathcal{R}$, (K_F, h_F), determined by

$$K_F = 2K_F\phi(K_F)^2, \;\; h_F = 3h_F\phi(K_F). \tag{33.36}$$

Here $K_F = 0$ is certainly a solution, but $\phi = 1/\sqrt{2}$ gives $K^* = (1/4)\log(1 + 2\sqrt{2}) \simeq 0.3356\cdots$. For all K_F $h_F = 0$ is a solution. There is no other finite solution.[15*] That is, $(K, h) = (0, 0)$ or $(K^*, 0)$ is the fixed point.[16] From the correspondence explained in Fig. 33.9 the unstable fixed point $(K^*, 0)$ in the (K, h)-plane corresponds to the critical point seen from far away; The thick K axis corresponds to the thick curve through H^* in Fig. 33.6.

Flow Near the Fixed Point: Linear Approximation to $\mathcal{R}$

Although we studied both τ and h, let us study $h = 0$ as before and the flow along the unstable manifold of H^* (the thick curve in Fig. 33.5). As can be seen from $\mathcal{R}\tau = \tau\ell^{y_1}$ and since $K - K^*$ is essentially τ, we must study the local behavior of $\mathcal{R}$ near the critical fixed point. Let $K = K^* + \delta K$ and $K' = K^* + \delta K'$. Then the linear approximation of $\mathcal{R}$ along the K curve is

$$\mathcal{R}(K^* + \delta K) = K^* + \left(\frac{d\mathcal{R}}{dK}\right)_{\text{at } K^*} \delta K = K^* + \delta K'. \tag{33.37}$$

We may identify (K' is given by (33.33)):

$$\ell^{y_1} = \left.\frac{dK'}{dK}\right|_{K=K^*} = 1.634\cdots. \tag{33.38}$$

Therefore, since $\ell = \sqrt{3}$,

$$y_1 = \log 1.634/\log\sqrt{3} \simeq 0.8939\cdots. \tag{33.39}$$

Its exact value is 1 (related to the critical exponent α).[17] The reader may think the result is not impressive (the mean-field theory gives 2).

[15*] Don't divide the equation with zero.
[16] $K_F = \infty$ is a fixed point corresponding to the ordered phases or $T = 0$.
[17] Actually, $1/y_1 = \nu$, the exponent for the correlation length.

Problem

Q33.1 Simple 1D Renormalization

As a simple example, let us study the 1-Ising model with the aid of *decimation* illustrated in Fig. 33.7(a). This procedure thins the spins through summing over a subset of spins, keeping the rest fixed. The original partition function reads (here, $K = \beta J$)

$$Z = \sum_{s,\sigma} \cdots e^{K(s_{-1}\sigma_0 + \sigma_0 s_1)} \cdots , \qquad (33.40)$$

where spins at the even lattice positions are written as σ. Sum over all σ states and make a decimated chain whose Hamiltonian is $-K' \sum_n s_{2n-1} s_{2n+1}$. Find K' in terms of K and show that there is no phase transition for $T > 0$.

Solution

The result is a product of the terms of the following form:

$$\sum_{\sigma_0 = \pm 1} e^{K(s_{-1}\sigma_0 + \sigma_0 s_1)} = 2\cosh K(s_{-1} + s_1). \qquad (33.41)$$

Equating this with the form $\propto e^{K' s_{-1} s_1}$, we can fix K' with the aid of the fact that $s^2 = 1$:[18*]

$$K' = \frac{1}{2} \log \cosh 2K. \qquad (33.42)$$

Thus, we have constructed a map from the original Hamiltonian to the coarse-grained Hamiltonian with $\ell = 2$:

$$H = \sum_{i \in \mathbb{Z}} K s_i s_{i+1} \to H' = \sum_{i \in 2\mathbb{Z}} K' s_i s_{i+2}. \qquad (33.43)$$

Starting from some positive K and iterating (33.42), we see (e.g., graphically) clearly that $K \to K' \to \cdots \to 0$ quickly (that is, the system is driven to the high temperature disordered fixed point), consistent with the fact that there is no phase transition for $T > 0$.

[18*] If one wishes to be more formal, assume (33.41) is equal to $e^{A + K' s_{-1} s_1}$. If the spin sum is zero, $2 = e^{A - K'}$; otherwise, $2\cosh(2K) = e^{A + K'}$, so we obtain $e^{2K'} = \cosh(2K)$. The free energy of the original system cannot be determined by H' alone, but depends on A as well. In reality, the contribution of A is larger, but it does not have the singularity that is responsible for the critical singularity. The free energy f_s in Kadanoff's theory (see **33.10**) was just the singular part removing this contribution of A.

34 Mean Field and Transfer Matrix

Summary

∗ The mean-field theory that ignores the effects of fluctuations is formulated with the aid of conditional expectations.
∗ The mean-field theory should not be uncritically used, but can be a useful tool to be used first.
∗ 1D short-ranged systems may be studied exactly by the transfer matrix technique.
∗ There is no phase transition in 1D short-ranged systems.

Key words

mean-field theory, transfer matrix, Perron–Frobenius eigenvalue

The reader should be able to:

∗ Set up the mean-field equation; understand the formulation in terms of the conditional probability.
∗ Solve (or qualitatively understand) the mean-field equation graphically.
∗ Recognize the limitations of the mean-field approach.
∗ Set up transfer matrices for 1D finite-range models.
∗ Understand the assertions of the Perron–Frobenius theorem.

34.1 The Mean-Field Idea

Sufficiently away from critical points/second order phase transition points, the equilibrium average of a function of several spins $f(s_0, s_1, \ldots, s_n)$ may be computed through separately averaging all the spins. Furthermore, if we assume that fluctuations are not large, the following approximation looks appealing:

$$\langle f(s_0, s_1, \ldots, s_n) \rangle \simeq f(\langle s_0 \rangle, \langle s_1 \rangle, \ldots, \langle s_n \rangle), \tag{34.1}$$

where the average $\langle \ \rangle$ denotes the equilibrium ensemble average. This is the original basic idea of the *mean-field* approach.

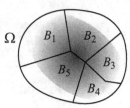

Fig. 34.1 Equation (34.3) is illustrated here. Suppose A is some quantity (shade in the figure) distributed on the sample space Ω which is partitioned into mutually exclusive events $B_1, \ldots, B_5$. In each partition we can define the average on it $E(A \mid B_i)$. If the probability for event B_i to happen is $P(B_i)$, the average of A over Ω is given by $\sum_j P(B_j) E(A \mid B_j)$, i.e., (34.3).

34.2 Quantitative Formulation of Mean-Field Theory

Let us define the *conditional expectation value* $E(A \mid B)$ of A under the occurrence of event B as

$$E(A \mid B) = \sum_{a \in A} a P(a \mid B), \tag{34.2}$$

where $P(a \mid B)$ is the conditional probability for A to have a particular value a under the condition B.

If $\cup_i B_i = \Omega$ and $B_i \cap B_j = \emptyset$ for $i \neq j$, $\{B_i\}$ is called a *partition* of the sample space Ω (Fig. 34.1). Then, we have an elementary identity of probability theory:[1*]

$$E(A) = E\left[E(A \mid B_j)\right] = \sum_j P(B_j) E(A \mid B_j). \tag{34.3}$$

That is, the average of conditional expectations over all the possible conditions is equal to the unconditional average. This is the key relation for the mean-field approach.

To illustrate how to use (34.3) let us apply it to the Ising model (from **31.9**) on a $\mathfrak{d}$-cubic lattice.[2] Its Hamiltonian (with the contribution of an external magnetic field h) reads (see (31.10) and (31.11)):

$$H = -J \sum_{\langle i, j \rangle} s_i s_j - h \sum_i s_i, \tag{34.4}$$

where s_i is the Ising spin at lattice site i taking the values ± 1, J is a positive coupling constant, and the first summation is over the nearest-neighbor site pairs denoted by $\langle i, j \rangle$.

Let us choose as B_i the totality of the spin configurations specified by a particular configuration $\{s_1, \ldots, s_{2\mathfrak{d}}\}$ of the spins that interact with the "central spin" s_0 on a $\mathfrak{d}$-cubic

---

[1*]Equation (34.3) follows from $P(a_j) = \sum_i P(a_j \mid B_i) P(B_i)$ (see Chapter 3), $E(A) = \sum_j a_j P(a_j)$, and $E(A \mid B_i) = \sum_j a_j P(a_j \mid B_i)$ as

$$E(A) = \sum_j a_j P(a_j) = \sum_j a_j \left[\sum_i P(a_j \mid B_i) P(B_i)\right] = \sum_i P(B_i) \left[\sum_j a_j P(a_j \mid B_i)\right] = \sum_i P(B_i) E(A \mid B_i).$$

[2] Remember that throughout this book spatial dimensionality is denoted by $\mathfrak{d}$ (the lower case d in Fraktur).

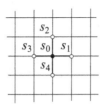

The central spin s_0 (black dot) and its nearest-neighbor surrounding spins $s_1, \ldots, s_{2\mathfrak{d}}$ (white dots; $\mathfrak{d} = 2$ in this figure). All the bonds connected to the central spin (thick bonds) are blocked by its nearest-neighbor spins; the central spin is walled by nearest-neighbor spins. Therefore, if they are fixed as $+1$ or -1, s_0 is decoupled from the rest of the world.

lattice (Fig. 34.2; $\mathfrak{d} = 2$ in the figure). Notice that if $s_1, \ldots, s_{2\mathfrak{d}}$ are fixed, the central s_0, which is interacting only with these neighboring spins, is totally decoupled from the rest of the world. Therefore, to study the distribution of $s_0 = \pm 1$, we have only to compute its energy in the given local environment. Then, we can make the Boltzmann factors for particular s_0 states ($+1$ or -1) to average s_0 as follows:[3]

$$E(s_0|s_1, \ldots, s_{2\mathfrak{d}}) = \frac{\sum_{s_0 = \pm 1} s_0 e^{\beta J s_0 (s_1 + \cdots + s_{2\mathfrak{d}}) + \beta h s_0}}{\sum_{s_0 = \pm 1} e^{\beta J s_0 (s_1 + \cdots + s_{2\mathfrak{d}}) + \beta h s_0}} = \tanh[\beta h + \beta J (s_1 + \cdots + s_{2\mathfrak{d}})].$$
(34.5)

Because $E(s_0) = E(E(s_0|s_1, \ldots, s_{2\mathfrak{d}}))$ (see (34.3)), we obtain

$$\langle s_0 \rangle = \langle \tanh[\beta h + \beta J (s_1 + \cdots + s_{2\mathfrak{d}})] \rangle.$$
(34.6)

This is an *exact* relation into which we may introduce various approximations to construct mean field approaches.

34.3 The Crudest Version of the Mean-Field Theory

The most popular (and simple-minded) approximation is (34.1) or more concretely for the present example:

$$\langle \tanh[\beta h + \beta J (s_1 + \cdots + s_{2\mathfrak{d}})] \rangle \simeq \tanh[\beta h + \beta J \langle s_1 + \cdots + s_{2\mathfrak{d}} \rangle].$$
(34.7)

Therefore, writing $m = \langle s_0 \rangle$ and assuming the translational symmetry of the system, we obtain a closed equation

$$m = \tanh[\beta(2\mathfrak{d}Jm + h)].$$
(34.8)

$2\mathfrak{d}Jm + h$ may be understood as an effective magnetic field acting on s_0, so this is called the *mean field*. The equation such as (34.8) determining the expectation value of the order parameter self-consistently is called a *mean-field equation* or *consistency equation*.

---

[3] Recognize that this procedure is allowed thanks to the principle of equal probability; the reader might think this is using.a probability estimate for a particular microstate, but actually, the set of microstates specified by $\{s_0, s_1, \ldots, s_{2\mathfrak{d}}\}$ contains macroscopic number of microstates.

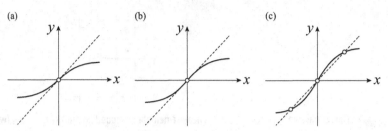

Fig. 34.3 The solution to (34.9) may be obtained graphically. The dashed line is $y = x$ and the black curve denotes $y = 2\partial\beta J \tanh x$. Solutions are given by the crossing points (white disks) of the dashed line and the black curve. (a): for $2\partial\beta J < 1$ (above the mean-field critical point T_c); (b): for $2\partial\beta J = 1$ (at the mean-field T_c); (c): for $2\partial\beta J > 1$ (below the mean-field T_c).

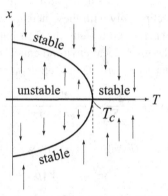

Fig. 34.4 The stability of the solution to (34.9) may be understood graphically. In this figure, for each T, the solutions x of (34.9) are given as the vertical positions. Connecting these points for all T, we can make the curves and line in black. The vertical arrows indicate how the perturbed solution evolves in time at various temperatures. For large T the directions of the arrows are dictated by the stability of the $m = 0$ state. The continuity of the stability property determines the arrows outside the curves. Then, since the arrows flip their directions upon crossing the solution curves (notice that arrows vanish only at the solutions), the arrow directions in the remaining regions are determined.

Let $2\partial\beta Jm = x$. For simplicity, assuming $h = 0$, we can rewrite (34.8) as

$$x = 2\partial\beta J \tanh x. \qquad (34.9)$$

This may be graphically solved (Fig. 34.3).

The bifurcation from the case with a single solution (Fig. 34.3(a)) to that with three solutions (c) occurs at $2\partial\beta J = 1$ (b).[4] Since $m \neq 0$ solutions appear only if $k_B T < 2\partial J$ we expect (b) gives the phase transition (or critical) temperature T_c. Near this point m increases as $|T - T_c|^{1/2}$ (i.e., the critical exponent $\beta = 1/2$ (Chapter 33)).

---

[4] **Bifurcation** A phenomenon in which the solution (or the solution set) changes its character qualitatively is called *bifurcation*. There are many types, and the one we encounter here is a *pitchfork bifurcation*. The so-called exchange of the stability of the branches (i.e., upon crossing the bifurcation parameter value, in our case T_c, the stability property of the solution branches switch) immediately tells us the stabilities of the solution branches as illustrated in the text and Fig. 34.4.

To confirm that the bifurcation actually signifies the phase transition (within the mean-field approximation), we must check that the nonzero solutions are the equilibrium solutions. That is, we must demonstrate that the $m \neq 0$ solutions have a lower free energy than the $m = 0$ case. The practically best way may be to study the bifurcation diagram (such as Fig. 34.4) and check the stability of various solutions under perturbations; if the state is a stable equilibrium, small deviation from the state will decay. We know for high temperatures the $m = 0$ phase is the true equilibrium state, so it is stable against perturbations. From this the stability of the $m \neq 0$ state is obvious from Fig. 34.4.

Our crude mean-field approach tells us that in any dimensional space, the nearest-neighbor interacting Ising models have a phase transition. This contradicts what we surveyed in Chapter 32 (and also at variance with the renormalization group study **Q33.1** and with the exact answer we will see soon). A lesson is that the phase transition predicted by a mean-field approach can be illusory. The approach ignores fluctuations that disfavor the order in the system, so the mean-field approach tends to encourage ordering in the system. In any case, as our general strategy says, the mean-field theory is not designed to understand phase transitions.

34.4* Improving the Mean-Field Approach

There is room to improve even the crude mean-field theory above. We know $s^2 = 1$. Using this, we can handle fluctuations in a better way. See, for example, **Q34.4**. Here, a particularly "lucky example" is illustrated that gives an exact result.

For the 1-Ising model without the magnetic field (34.7) reads as

$$\langle s_0 \rangle = \langle \tanh \left[\beta J (s_1 + s_{-1}) \right] \rangle, \tag{34.10}$$

where the nearest-neighbor spins of s_0 are denoted as s_1 and s_{-1}. Notice that $\tanh x$ can be expanded into an odd power series around $x = 0$:

$$\tanh x = x - \frac{1}{3}x^3 + \frac{2}{15}x^5 + \cdots . \tag{34.11}$$

We need odd powers of $s_{-1} + s_1$. For example,

$$(s_{-1}+s_1)^3 = s_{-1}^3 + 3s_{-1}^2 s_1 + 3s_{-1}s_1^2 + s_1^3 = s_{-1} + 3s_1 + 3s_{-1} + s_1 = 4(s_{-1}+s_1). \tag{34.12}$$

Analogously, any odd power of $s_{-1} + s_1$ is proportional to $s_{-1} + s_1$. Therefore, we must have the identity of the following form, if s_i takes only ± 1:

$$\tanh[\beta J (s_{-1} + s_1)] = A(s_{-1} + s_1). \tag{34.13}$$

A is determined by substituting ± 1 for the spins:

$$\tanh(2\beta J) = 2A. \tag{34.14}$$

Therefore, (34.10) reads

$$\langle s_0 \rangle = \frac{\tanh 2\beta J}{2} \langle s_{-1} + s_1 \rangle. \tag{34.15}$$

That is, we have obtained an exact relation (no approximation has been used):

$$m = m \tanh(2\beta J). \tag{34.16}$$

We know $|\tanh(2\beta J)| < 1$, so unless $T = 0$, $m = 0$. Therefore, there is no phase transition for $T > 0$.

34.5 When is the Mean-Field Approach Reliable?

Generally speaking, mean-field approaches may be relied upon, if the fluctuation effect is not decisive:

(1) If the spatial dimensionality is sufficiently high, then "spins" gang up against thermal fluctuations;
(2) If the first-order phase transition is with a big "jump," then fluctuations may not easily be able to fill the gap that must be jumped.

Thus, these cases are (often) amenable to mean-field approaches (at least qualitatively).

Perhaps, practically, we may summarize the use of the mean-field method as follows: we should not swallow the results of the method uncritically (especially as to the phase transitions), but since the method is easy to use in many cases, it is worth trying first.

34.6 Transfer Matrix Method

Let us consider a 1-Ising model with the Hamiltonian generally given by

$$H = -J \sum_i s_i s_{i+1} - h \sum_i s_i, \tag{34.17}$$

where the summations are over all the lattice points. Let us write down the partition function for this chain of length N containing N spins $s_1, \ldots, s_N$:

$$Z_N = \sum_{s_1,\ldots,s_{N-1},s_N \in \{-1,+1\}} e^{\beta[Js_N s_{N-1}+hs_N]} e^{\beta[Js_{N-1}s_{N-2}+hs_{N-1}]} \cdots e^{\beta[Js_2 s_1+hs_2]} e^{\beta hs_1}. \tag{34.18}$$

We can classify the spin configurations of this length N chain into two sets, the configuration sets with the Nth spin up and that down. We can define the end-spin fixed partition functions as

$$Z_N(\pm) = \sum_{s_1,\ldots,s_{N-1} \in \{-1,+1\}} e^{\pm\beta[Js_{N-1}+h]} e^{\beta[Js_{N-1}s_{N-2}+hs_{N-1}]} \cdots e^{\beta[Js_2 s_1+hs_2]} e^{\beta hs_1}. \tag{34.19}$$

Thus, we have

$$Z_N = Z_N(+) + Z_N(-). \tag{34.20}$$

We can analogously introduce the end-spin fixed partition function $Z_{N+1}(\pm)$ for the length $N + 1$ chain such as

$$Z_{N+1}(\pm) = \sum_{\{s_n\}_{n=1}^N} e^{\pm\beta[Js_N+h]} e^{\beta[Js_N s_{N-1}+hs_N]} \cdots e^{\beta[Js_2 s_1+hs_2]} e^{\beta hs_1} \quad (34.21)$$

$$= \sum_{s_N \in \{-1,+1\}} e^{\pm\beta[Js_N+h]} Z_N(s_N). \quad (34.22)$$

Therefore, if we introduce the vector

$$\mathbf{Z}_N = \begin{pmatrix} Z_N(+) \\ Z_N(-) \end{pmatrix}, \quad (34.23)$$

we may rewrite (34.22) as

$$\mathbf{Z}_{N+1} = \mathbf{T}\mathbf{Z}_N, \quad (34.24)$$

where $\mathbf{T}$, called the *transfer matrix*, is defined as

$$\mathbf{T} = \mathrm{Matr}(e^{\beta[Jss'+hs]}) = \begin{pmatrix} T_{++} & T_{+-} \\ T_{-+} & T_{--} \end{pmatrix} = \begin{pmatrix} e^{\beta J+\beta h} & e^{-\beta J+\beta h} \\ e^{-\beta J-\beta h} & e^{\beta J-\beta h} \end{pmatrix}. \quad (34.25)$$

Notice that

$$Z_N = (1,1)\mathbf{Z}_N. \quad (34.26)$$

Repeated use of the recursion (34.24) results in

$$\mathbf{Z}_{N+1} = \mathbf{T}^N \begin{pmatrix} e^{\beta h} \\ e^{-\beta h} \end{pmatrix}. \quad (34.27)$$

In this case the first spin is free to point up or down. For a ring of N spins ($s_1 = s_{N+1}$), as we see immediately, $Z_N = \mathrm{Tr}\,\mathbf{T}^N$. Since N is huge, we need not worry about the difference between N and $N + 1$.

34.7 How to Compute the Product of Matrices

The easiest method to compute (34.27) is to use an orthogonal transformation (or more generally unitary transformation) to convert $\mathbf{T}$ into a diagonal form:[5]

$$\mathbf{T} = U^{-1} \begin{pmatrix} \lambda_1 & 0 \\ 0 & \lambda_2 \end{pmatrix} U, \quad (34.28)$$

where λ_1 and λ_2 are eigenvalues of $\mathbf{T}$, and U is the orthogonal transformation needed to diagonalize $\mathbf{T}$. Introducing (34.28) into (34.27), we obtain

[5] If impossible, into a Jordan normal form.

Matrix diagonalization A necessary and sufficient condition for a matrix $\mathbf{T}$ to be diagonalizable is that it is *normal*: $\mathbf{T}^*\mathbf{T} = \mathbf{T}\mathbf{T}^*$, where the * denote Hermitian conjugation (see Appendix 17A). If all the eigenvalues are distinct, the matrix is normal.

$$Z_N = U^{-1} \begin{pmatrix} \lambda_1^N & 0 \\ 0 & \lambda_2^N \end{pmatrix} U \begin{pmatrix} e^{\beta h} \\ e^{-\beta h} \end{pmatrix}. \tag{34.29}$$

Therefore, we finally have the following structure:

$$Z_N = a\lambda_1^N + b\lambda_2^N, \tag{34.30}$$

where a and b are nonzero real numbers. If $\lambda_1 > |\lambda_2|$ (this is actually generally true; see below), a is positive, and, since $N \gg 1$, the first term dominates Z_N.

The free energy per spin is completely determined by the largest eigenvalue of the transfer matrix as

$$f = -k_B T \log \lambda_1. \tag{34.31}$$

Since λ_1 is not dependent on the boundary condition, the free energy per spin (this is the only quantity meaningful in the thermodynamic limit $N \to \infty$) is independent of the boundary conditions,[6] although the partition function (34.30) itself (i.e., a and b in the formula) depends on them.

34.8 Why There is no Phase Transition in 1-Space

Let us discuss why there is no phase transition for $T > 0$ in 1D finite-range interaction systems from the transfer matrix point of view. The free energy could exhibit a singularity if $Z \leq 0$, but this does not happen, because Z is a sum of positive terms. As long as $T > 0$, βJ is finite, so all the elements of the transfer matrix are without any singularity as a function of T (and h). The eigenvalues are *algebraic functions* of the matrix elements.[7] Therefore, as long as eigenvalues are finite, their singularities are branch points.[8] The branch points of the eigenvalues occur when they change their multiplicities (digeneracies), so the multiplicity of the largest eigenvalue is of vital importance. The key theorem we need is the following important theorem:

Theorem (Perron and Frobenius)[9]

Let A be a square matrix whose elements are all non-negative, and there is a positive integer n such that all the elements of A^n are positive. Then, there is a non-degenerate real positive eigenvalue λ such that

---

[6] The independence of the free energy per spin from the boundary condition in the thermodynamic limit is a general property of normal thermodynamic system.

[7] **Algebraic function** An algebraic function of $\{x_i\}$ is a function that can satisfy a polynomial equation whose coefficients are polynomials of $\{x_i\}$.

[8] **Branch point** For example, consider $\sqrt{z}$. This is real if $z > 0$ and has two values, but for $z = 0$ there is only one value. In this case, $z = 0$ is an example of the branch point. Take $x^2 - 2zx + 1 = 0$. The roots are algebraic functions of z: $x = z \pm \sqrt{z^2 - 1}$. Therefore, $z = 1$ is a branch point for x. Notice that $dx/dz = 1 \pm z/\sqrt{z^2 - 1}$, so at the branch point the derivative ceases to exist (for this example).

[9] For a proof, see, e.g., Seneta, E. (2013). *Nonnegative Matrices and Markov Chains*, 2nd edition. Berlin: Springer.

(i) $|\lambda_i| < \lambda$, where λ_i are eigenvalues of A other than λ (i.e., λ is the spectral radius[10] of A),

(ii) the elements of the eigenvector belonging to λ may be chosen all positive.

This special real positive eigenvalue giving the spectral radius is called the *Perron–Frobenius eigenvalue*.

Since the transfer matrix of the 1D nearest-neighbor Ising model has positive elements, the logarithm of its Perron–Frobenius eigenvalue gives the free energy per spin and there is no phase transition. More generally, if the number of states for each 1D element is finite and the interaction range is finite, then no phase transition occurs for $T > 0$, because the transfer matrix is finite dimensional.

34.9 Onsager Obtained the Exact Free Energy of 2-Ising Model

Onsager used the transfer matrix method to evaluate the partition function of the 2-Ising model on the square lattice exactly.[11]

Problems

Q34.1 Another Derivation of Mean-Field Theory

A mean-field approach may be obtained with the aid of a *variational principle* for the free energy. If the (density) distribution function of microstates is f (we consider the classical case), the Helmholtz free energy may be written as (recall **21.1**):

$$A = \langle H \rangle + k_B T \int d\Gamma f(\Gamma) \log f(\Gamma). \tag{34.32}$$

Here, the integration is over the whole phase space, and $\langle H \rangle$ is the expectation value of the system Hamiltonian with respect to f. Let us apply this to the Ising model on a $N \times N$ square lattice (there are N^2 spins). Its Hamiltonian is as usual ($s_i = \pm 1$ Ising spins):

---

[10] **Spectral radius** The radius of the smallest closed disk in the complex plane centered at the origin containing all the eigenvalues of an operator A is called the *spectral radius* of A.

[11] **Onsager's biography** See Longuet-Higgins, C. and Fisher, M. E. (1995). Lars Onsager: November 27, 1903–October 5, 1976, *J. Stat. Phys.*, **78**, 605–640. This is an Onsager biography everyone can enjoy. According to this article, Onsager applied the transfer matrix method to the strip of width 2, 3, and 4 lattice points, and constructed a conjecture from these results, then confirmed it for the width 5 strip and closed in on the general formula.

"His statistical mechanics were popularly known as 'Advanced Norwegian I' and 'Advanced Norwegian II'." He was fired more than once for his poor teaching, and his Nobel-prize winning dissertation intended for his PhD was rejected as insufficient from his alma mater. See Appendix 26A: Onsager's Theory of Irreversible Processes.

$$H = -J \sum_{\langle i,j \rangle} s_i s_j. \tag{34.33}$$

If we could vary f unconditionally and minimize A, then the minimum must be the correct free energy, but this is in many cases extremely hard or plainly impossible. Therefore, we assume an approximate form for f and the range of variation is restricted. For example, we could introduce a "single-body" approximation:

$$f = \phi(s_1)\phi(s_2) \cdots \phi(s_{N^2}), \tag{34.34}$$

where ϕ is a single-spin distribution function.

(1) Under this approximation write down A in terms of ϕ.
(2) Minimize A with respect to ϕ. The function ϕ must be normalized. What is the equation determining ϕ?
(3) Using the obtained formula, write down the magnetization m per spin to obtain the mean-field equation.

Solution

(1) We have only to compute each term honestly ($2N^2$ is the total number of pairs):

$$\langle H \rangle = -J \sum_{s_1,\dots,s_{N^2} \in \{-1,1\}} \left(\sum_{\langle i,j \rangle} s_i s_j \right) \prod_k \phi(s_k) = -J \sum_{\langle i,j \rangle} \langle s_i \rangle \langle s_j \rangle = -2JN^2 \langle s \rangle^2. \tag{34.35}$$

The entropy part reads

$$k_B T \sum_{s_1,\dots,s_{N^2} \in \{-1,1\}} \prod_k \phi(s_k) \sum_k \log \phi(s_k) = N^2 k_B T \sum_{s=\pm 1} \phi(s) \log \phi(s). \tag{34.36}$$

Combining these, we get

$$A = -2JN^2 \left(\sum_{s=\pm 1} \phi(s)s \right)^2 + N^2 k_B T \sum_{s=\pm 1} \phi(s) \log \phi(s), \tag{34.37}$$

where N and $N \pm 1$ need not be distinguished.

(2) Introducing a Lagrange's multiplier λ to impose the normalization condition for ϕ, we must minimize

$$A + \lambda \sum_{s=\pm 1} \phi(s) = -2N^2 J \left(\sum_{s=\pm 1} \phi(s)s \right)^2 + N^2 k_B T \sum_{s=\pm 1} \phi(s) \log \phi(s) + \lambda \sum_{s=\pm 1} \phi(s). \tag{34.38}$$

The minimization condition (the variational equation with respect to ϕ) reads

$$-4N^2 Js \langle s \rangle + N^2 k_B T(1 + \log \phi(s)) + \lambda = 0, \tag{34.39}$$

so we see

$$\phi(s) \propto \exp(4\beta Jms), \tag{34.40}$$

where $\langle s \rangle = m$.

(3) We get

$$m = \sum_{s=\pm 1} s\phi(s) = \frac{\sum_{s=\pm 1} s \exp(4\beta Jms)}{\sum_{s=\pm 1} \exp(4\beta Jms)} = \tanh 4J\beta m. \tag{34.41}$$

This is just the consistency equation.

Q34.2 The Gibbs–Bogoliubov Inequality and Mean Field

We have already encountered with the Gibbs–Bogoliubov inequality (see **Q19.3**):

$$A \leq A_0 + \langle H - H_0 \rangle_0. \tag{34.42}$$

Here, A is the free energy of the system with the Hamiltonian H, A_0 is the free energy of the system with the Hamiltonian H_0, and $\langle\ \rangle_0$ is the average over the canonical distribution with respect to H_0.

Let H be the $N \times N$ square lattice Ising model Hamiltonian (34.33) (without a magnetic field; even with it there is almost no change), and

$$H_0 = -\sum_i hs_i, \tag{34.43}$$

where h is the variational parameter in the present approach. Derive the equation for h that minimizes the right-hand side of (34.42).

Solution

Let us first compute A_0 and $m = \langle s_i \rangle_0$:

$$A_0 = -k_B T \log[2\cosh\beta h]^{N^2}, \tag{34.44}$$

$$m = \tanh\beta h. \tag{34.45}$$

Consequently, since the number of nearest-neighbor pairs is $2N^2$, (34.33) gives

$$\langle H \rangle_0 = -J(2N^2)\tanh^2\beta h. \tag{34.46}$$

Also we have

$$\langle H_0 \rangle_0 = -N^2 h \tanh\beta h. \tag{34.47}$$

Combining all the results, we can write the Gibbs–Bogoliubov inequality as

$$A \leq -N^2 k_B T \log[2\cosh\beta h] - J(2N^2)\tanh^2\beta h + N^2 h \tanh\beta h. \tag{34.48}$$

Differentiating the right-hand side with respect to h, we have

$$-N^2 \frac{\sinh\beta h}{\cosh\beta h} - 4\beta JN^2 \tanh\beta h \frac{1}{\cosh^2\beta h} + N^2 \tanh\beta h + N^2\beta h \frac{1}{\cosh^2\beta h} = 0, \tag{34.49}$$

so we obtain

$$4\beta J \tanh\beta h = \beta h. \tag{34.50}$$

From this we get

$$\tanh(4\beta J \tanh\beta h) = \tanh\beta h, \tag{34.51}$$

but if we use (34.45), this turns out to be our familiar formula:

$$m = \tanh 4\beta Jm.$$ (34.52)

Q34.3 Mean Field on Diamond Lattice

Consider Ising spins on the diamond lattice (without an external magnetic field h). The interactions of the spins are restricted to the nearest-neighbor pairs.

(1) Write down the fundamental equation for this system corresponding to (34.6).
(2) What is the critical point, if the simplest approximation is used: $\langle \tan(\cdots) \rangle = \tan(\langle \cdots \rangle)$?
(3) This is a three-dimensional system, so there is definitely a positive critical temperature T_c. What can we say about this true T_c from our result?

Solution

(1) This is quite the same as is explained in the text. On the diamond lattice one spin has only four nearest-neighbor spins, so

$$\langle s_0 \rangle = \langle \tanh[\beta J(s_1 + \cdots + s_4)] \rangle,$$ (34.53)

where $s_1, \ldots, s_4$ are the spins connected to s_0 with the lattice bonds.
(2) The naivest approach gives

$$\langle s_0 \rangle = \tanh[\beta J \langle s_1 + \cdots + s_4 \rangle].$$ (34.54)

Expecting the translationally symmetric magnetization,

$$m = \tanh 4\beta Jm$$ (34.55)

is the crudest mean-field equation. This means $x = 4\beta J \tanh x$, so $T_c = 4J/k_B$.
(3) Thermal fluctuation tends to be against ordering, so theories ignoring fluctuations overestimate the ordering effect, pushing the critical point up. Thus, we may conclude that the critical temperature of the Ising model on a diamond lattice must not be higher than $4J/k_B$. Actually, we may guess the true T_c is far less than this (the estimated value is $T_c = 2.7J/k_B$)[12].

Q34.4 Improving Q34.3

We have derived in **Q34.3** the following exact relation for the Ising model on a diamond lattice:

$$\langle s_0 \rangle = \langle \tanh[\beta J(s_1 + \cdots + s_4)] \rangle.$$ (34.56)

We wish to exploit the fact that $s^2 = 1$.

---

[12] Essam, J. W. and Sykes, M. F. (1963). The crystal statistics of the diamond lattice, *Physica*, **29**, 378–388. A more recent simulation result is $T_c = 2.704...J/k_B$ (Deng, Y. and Blöte, H. W. J. (2003). Simultaneous analysis of several models in the three-dimensional Ising universality class, *Phys. Rev. E*, **68**, 036125.

(1) Expanding tanh in a power series, show that

$$\tanh[\beta J(s_1 + \cdots + s_k)] = A(s_1 + s_2 + s_3 + s_4) + B(s_1 s_2 s_3 + s_1 s_3 s_4 + s_2 s_3 s_4 + s_1 s_2 s_4).$$
(34.57)

That is, any odd power of $(s_1 + s_2 + s_3 + s_4)$ is written as a sum of $(s_1 + s_2 + s_3 + s_4)$ and $(s_1 s_2 s_3 + s_1 s_3 s_4 + s_2 s_3 s_4 + s_1 s_2 s_4)$.

(2) Determine A and B by setting $s = \pm 1$ so that (34.57) holds.

(3) We assume the translational symmetry of the phases: $\langle s_0 \rangle = \langle s_1 \rangle = \cdots = m$. (34.56) with (34.57) reads

$$m = 4Am + 4B\langle s_1 s_2 s_3 \rangle.$$
(34.58)

There is no approximation up to this point, but, unfortunately, we cannot solve (34.58). Now, let us introduce the approximation

$$\langle s_1 s_2 s_3 \rangle = m^3.$$
(34.59)

Then, our "approximate" mean–field equation is

$$m = 4Am + 4Bm^3.$$
(34.60)

What is the condition that determines the phase transition? Is there any improvement?

Solution

(1) Checking first 2 or three terms in the expansion of tanh is enough, practically.

However, if we wish to "prove" (34.57), we can proceed as follows. Generally, we have an odd power $(s_1 + s_2 + s_3 + s_4)^m$, where m is an odd positive integer. If we expand this, we obtain the terms of the following form:

$$s_1^a s_2^b s_3^c s_4^d$$
(34.61)

for $a + b + c + d = m$ ($a, \ldots, d$ are nonnegative integers). There is a perfect permutation symmetry among $s_1, \ldots, s_4$ and among a, b, c and d. $a + b + c + d$ must be odd and so either one or three of the four terms must be odd. Since each even power gives one, we have only two surviving types: s_1 and $s_1 s_2 s_3$. Therefore, summing all these terms, we must have (34.57) thanks to the permutation symmetry of the spins.

(2) For the case with all s being $+1$:

$$\tanh(4\beta J) = 4A + 4B.$$
(34.62)

For one -1

$$\tanh(2\beta J) = 2A - 2B.$$
(34.63)

Other possibilities do not give any new relation. From these, we get

$$A = (1/8)(\tanh 4\beta J + 2\tanh 2\beta J), \quad B = (1/8)(\tanh 4\beta J - 2\tanh 2\beta J).$$
(34.64)

(3) We must solve $m = 4Am + 4Bm^3$. One way is to use the graphical method (recall **34.3**). This tells us that when the slope of $4Am + 4Bm^3$ at $m = 0$ is 1, bifurcation occurs. Hence, T_c is given by $4A = 1$ or

$$\tanh 4\beta J + 2\tanh 2\beta J = 2.$$
(34.65)

The reader need not solve this, but notice that the T_c obtained from this must be smaller than that obtained in **Q34.3**. Actually, $T_c \simeq 3.08 \, J/k_B$ is obtained, which is a considerable improvement over $4J/k_B$.

Q34.5 Transfer Matrix Exercise

At each lattice site of 1D lattice is a particle which has one ground state and one excited state. Only nearest-neighbor excited states can interact and the excitation energy required is ε (> 0). The Hamiltonian may be written as

$$\mathcal{H} = -J \sum_i s_i s_{i+1} + \varepsilon \sum_i s_i, \qquad (34.66)$$

where $s_i = 0$ denotes the ground state, and $s_i = 1$ the excited state. Find the free energy per particle (i.e., write down the transfer matrix and compute its eigenvalues).

Solution

The transfer matrix can be written as $T_{ij} = e^{\beta J s_i s_j - \beta \varepsilon s_i}$:

$$
T = \begin{array}{c|cc}
 & 0 & 1 \\
\hline
0 & 1 & 1 \\
1 & e^{-\beta \varepsilon} & e^{\beta J - \beta \varepsilon}
\end{array}
\qquad (34.67)
$$

Therefore, the characteristic equation reads

$$(1 - \lambda)(e^{\beta J - \beta \varepsilon} - \lambda) - e^{-\beta \varepsilon} = \lambda^2 - \lambda(1 + e^{\beta(J-\varepsilon)}) + e^{\beta(J-\varepsilon)} - e^{-\beta \varepsilon} = 0. \qquad (34.68)$$

Hence,

$$\lambda = \frac{1}{2}\left(1 + e^{\beta(J-\varepsilon)} \pm \sqrt{(1 - e^{\beta(J-\varepsilon)})^2 + 4e^{-\beta \varepsilon}}\right). \qquad (34.69)$$

$+$ gives the Perron–Frobenius eigenvalue, so we can read off the free energy.

Symmetry Breaking

Summary

* Symmetry may be described in terms of groups.
* Phase ordering is spontaneous symmetry breaking.
* Spontaneous symmetry breaking causes rigidity and a gapless excitation spectrum.

Key words

symmetry operations, group, subgroup, (spontaneous) symmetry breaking, rigidity, Nambu–Goldstone boson

The reader should be able to:

* Understand how to describe the symmetry of a system and its lowering unambiguously in terms of groups and subgroups.
* Understand spontaneous symmetry breaking and its consequences.

35.1 Ordering Means Lowering the System Symmetry

A phase transition is often between ordered and not-so-ordered phases. Ordering means that the system becomes less symmetric than the less ordered phase. For example, in a fluid phase the system has the full 3D translational and rotational symmetry, so we can move a portion of the fluid freely with negligible work, but if a crystal is formed, translational and rotational symmetries are lost, so moving its portion with no work cost is prohibitive.[1] Crystal (solid) phases (more ordered phases) have less symmetry than the corresponding fluid phases (less ordered phases).

---

[1] More microscopically, in a fluid phase particles can sit anywhere even if we fix its container, so the phase has a full translational symmetry. In contrast, in a crystal (solid) if its location is fixed, particles can sit only positions restricted by the crystal lattice, so the full spatial continuous translational symmetry is lost.

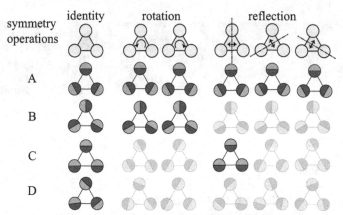

Fig. 35.1 Symmetry illustrated. There are two kinds of rotational operations and three kinds of reflection operations that keep A intact (symmetry group C_{3v}) other than "doing nothing" (identity operation). With lowering the symmetry the allowed symmetry operations become restricted. For B only rotations are allowed (symmetry group C_3, which is a genuine subgroup of C_{3v}), and for C only one reflection is allowed (C_σ). Without any symmetry (case D) only identity I keeps the figure intact. The symmetry of a system may be described by the totality (called the symmetry group) of symmetry operations that keep the system intact.

35.2 How to Describe the Symmetry

The symmetry of a system may be understood through *symmetry operations* (Fig. 35.1). It is clear that more symmetric objects allow more symmetry operations that keep the objects intact (invariant). The totality G of the symmetry operations that keep an object intact is called the *symmetry group* of the object.

35.3 Group and Subgroup

If $a, b \in G$, and if we write operating b first and then a next as the product ab, then $ab \in G$, so we can have an algebraic structure on G. We know (i) the identity $e \in G$ and (ii) the inverse operation a^{-1} of any operation $a \in G$ is again in G. Furthermore, (iii) $(ab)c = a(bc)$. If G satisfies these three conditions, G is called a *group*. If a subset $H \subset G$ is again a group (with the same multiplication rule), it is called a *subgroup* of G. Lowering of the symmetry of a system corresponds to restricting the original symmetry group to a genuine subgroup (a subgroup not equal to G).

35.4 Spontaneous Breaking of Symmetry

If an equilibrium state has a symmetry group which is a genuine subgroup of the symmetry group of the system Hamiltonian, we say the symmetry is *spontaneously broken*. Certainly, the symmetry is spontaneously broken below T_c for the 2-Ising model. In this case the symmetry that is broken is described by a discrete group (up–down symmetry).

Crystallization mentioned previously is another example. The Hamiltonian of the system is something like

$$H = \sum_i \frac{\boldsymbol{p}_i^2}{2m} + \sum_{i<j} \phi(\boldsymbol{r}_i - \boldsymbol{r}_j), \tag{35.1}$$

where ϕ is usually a binary interaction potential. Thus, the Hamiltonian has a full translational symmetry: nothing happens even if translation $\boldsymbol{r}_i \rightarrow \boldsymbol{r}_i + \boldsymbol{a}$ is applied to all the particles for any 3-vector $\boldsymbol{a}$ (in the thermodynamic limit).[2] However, we believe this system can crystallize, losing its translational symmetry, if the interaction potential is similar to the Lennard-Jones potential (see Fig. 11.1).[3] Thus, crystallization is a typical spontaneous symmetry breaking. In contrast to the 2-Ising model, the symmetry group in this case is continuous. That is, translations with any $\boldsymbol{a}$ or any spatial rotations keep the Hamiltonian intact.

35.5 Symmetry Breaking in a Heisenberg Magnet

Consider a (classical) Heisenberg magnet (the spin dimension three magnet model) as an example (cf. Fig. 35.2). Let $\boldsymbol{s}_i$ be the ith spin that is a unit 3-vector. In this case the system Hamiltonian

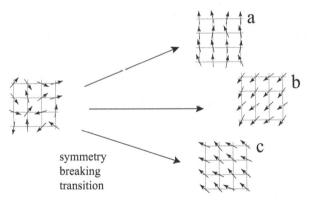

symmetry
breaking
transition

Fig. 35.2 Symmetry breaking results in an ensemble of symmetry broken phases a, b, c, . . . collectively representing the whole symmetry of the system. This illustration corresponds to a transition from a paramagnetic phase to a ferromagnetic phase (in 3-space). Spins sit on the lattice, but at each lattice site the spin can rotate (in the spin space). At higher temperatures spins can rotate freely, but below T_c spins predominantly point in a particular direction, which corresponds to the magnetization vector. There are continuously many distinct phases below T_c with distinct magnetization directions.

---

[2] The reader might ask how the boundary of the system is taken care of. We take a huge system (eventually the thermodynamic limit), or we may impose a periodic boundary condition.

[3] As already noted, we have not been able to prove this within statistical mechanics.

$$H = -J \sum_{\langle i,j \rangle} s_i \cdot s_j \tag{35.2}$$

has a full 3D rotational (O_3) symmetry of spins (in the spin space). (Fig. 35.2).[4] The disordered phase (paramagnetic phase) has no magnetization $m = 0$, so indeed the system is fully rotationally symmetric. However, below T_c, when ferromagnetic order emerges, m is a definite nonzero vector. Thus, the system symmetry is no longer 3D rotational but only the 2D rotation around the axis parallel to m (in the spin space). That is, the symmetry is lowered upon ordering, a typical example of spontaneous symmetry breaking (Fig. 35.2)[5].

35.6 Consequences of Symmetry Breaking: Rigidity

Take a Heisenberg magnet below its T_c with magnetization m being in the $+z$-direction. Let us choose one spin in front of us and rotate it by 90° to point in the $+x$-direction, and hold it. What happens? The spins around the rotated spin do not like this, because the interaction is energetically unfavorable. Thermal fluctuations rotate them and align them to the spin we hold. Needless to say, then, these reoriented spins will have uncomfortable relations with further outside spins. This outside relation is, however, "better"; the central spin never moves, so the discomfort is steady, but in the outer layers "discomfort" can be "relieved" by thermal reorientation of outer spins. Thus, the x-reoriented domain gradually widens, and eventually the macroscopic magnet changes its direction of magnetization. Since this state has the same energy as the original state (because H, the system Hamiltonian, is symmetric with respect to the global spin rotation), the final state will last forever, even if we stop holding the central spin. If we do not pay attention to what was actually happening between the two equilibrium states, what happens is just the rotation of the magnetization vector. This property – the whole system following the modification of its part – is called (generalized) *rigidity*.[6]

    The rigidity the most familiar to us is the rigidity of a solid. If we push one end of a solid, the other end also moves accordingly. We cannot do this for fluids. Only after translational symmetry is spontaneously broken can we have this ordinary rigidity of solid. If one end

---

[4] O_3-symmetry, needless to say, is a continuous symmetry. The spins live on a lattice, so there is no spatial rotational symmetry. Do not confuse the rotations in the spin space and in the actual space. We rotate every spin around its sitting lattice point.

[5] In this case, the macroscopic states with different m are understood as distinct phases just as gas and liquid phases in fluids. If m changes to m' ($\neq m$), this is a first-order phase transition between two distinct ordered phases, because m changes discontinuously.

[6] The change in a "small part" is in this case kept by an external means (e.g., by us). Then, the change eventually propagates to the whole system (i.e., any indefinitely large finite domain follows).
    Equilibrium states are stable against localized perturbations Notice that an equilibrium state is stable under any perturbation applied to any finite domain, if the perturbation effect is not maintained (e.g., by an external agent). Thus, even if we flip all the spins in a big but finite domain into the x-direction, if we leave the system alone, eventually this x-oriented domain would be swallowed by the surrounding z-oriented phase. Do not mix up these different situations.

is twisted, the other end follows as well. This is due to the spontaneous breaking of the rotational symmetry by crystallization.

Rigidity also occurs in the 2-Ising model: if we flip the central spin and hold it, eventually the magnetization would change its sign. Thus, whenever symmetry is spontaneously broken, rigidity emerges. However, symmetry breaking of continuous symmetries is much more dramatic, because any local small change (which is impossible for discrete symmetry cases) propagates to the other end.

35.7 Nambu–Goldstone Bosons: A Consequence of Breaking of Continuous Symmetry

If the spontaneously broken symmetry is continuous, and if the system interactions are short-ranged, we have another universal feature: the *Nambu–Goldstone bosons* (NG bosons). The NG bosons refer to long wavelength collective excitations in the ordered phase (like acoustic phonons equal to sound waves in solids; see Fig. 23.2) whose excitation energy tends to zero in the long-wavelength limit (gapless excitations).

All possible symmetry broken phases (see Fig. 35.2) have the same energy, because they can be transformed into each other with an element of the symmetry group of the system Hamiltonian that keeps the Hamiltonian intact. Consider a long 3D rectangular parallelepiped in a Heisenberg model, and assume that the magnetization changes continuously along its one axis (the x-axis in Fig. 35.3). Let us estimate the energy needed for such deformation of the magnetization per cross section perpendicular to the x-axis. Suppose over the distance L, the spin direction changes by 2π. If the lattice spacing is a, the angle between the adjacent spins is $2\pi a/L$. The spin interaction energy is given by the scalar product of the neighboring spins, so $1 - \cos(2\pi a/L) \propto (2\pi a/L)^2$ is the required energy for twisting between the neighboring spins. Therefore, the total energy change due to this twisting of spins from one end 0 to the other end L is given by

$$\left(\frac{2\pi a}{L}\right)^2 \times \frac{L}{a} = \frac{4\pi^2 a}{L}. \tag{35.3}$$

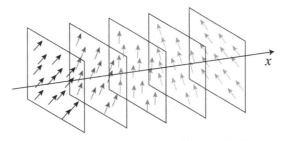

Fig. 35.3 The situation in which the spin directions (magnetizations) in the planes perpendicular to an axis (x-axis) change gradually.

Table 35.1	A summary of representative examples		
	solid	Heisenberg ferro	superfluid
Broken symmetry	3D translational	rotational (in spin space)	phase
Order	3D periodicity	ferromagnetism	superfluidity
NG boson	acoustic phonons	spin wave	second sound
Rigidity	rigidity	ferromagnetism	superfluidity

Fig. 35.4 However far away the + and − charges are apart, in this case (recall a parallel plate capacitor) the Coulomb interaction between the separated charges does not decay. [The figure only illustrates a chunk with a small cross section; imagine a big block containing the cylinder.]

That is, if we can deform the system continuously, longer wave deformations (fluctuations) require less energy to be excited. Thus, the Nambu–Goldstone bosons become possible.

In the case of the Heisenberg ferromagnet, precession of spins can propagate as a wave (*spin waves*) and its quantum is called *magnon*. In a crystal the vibration due to the mutual displacement of lattice cells can propagate as a wave and its quantum is our familiar (acoustic) phonons (recall **23.10**). They are the NG bosons due to crystallization.

35.8* The NG Bosons do not Exist for Long-Range Interaction Systems

However, as can be guessed from the previous explanation, if there is a long-range interaction, then the energy required by a long-wavelength excitation may not vanish. This indeed happens in plasmas. Suppose we displace plus charges relative to minus charges as in Fig. 35.4. The Coulomb interaction energy between the charge density fluctuations does not decrease with distance (remember the parallel plate capacitor). Therefore, the excitation energy has a lower cut off and the long-wave frequency does not converge to zero.[7]

35.9 Summary of Symmetry Breaking

We can summarize representative examples in Table 35.1. Although not discussed, superfluidity of ^{4}He is also added.[8]

[7] These excitations are called plasma oscillations.
[8] About this section, a strongly recommended reference is: Anderson, P. W. (1984). *Basic Notions of Condensed Matter Physics*, Boulder: Westview Press, chapter 2.

35.10 Symmetry Breaking Requires Big Systems

If the system is finite, there is no symmetry breaking.[9] Figure 35.2 implies the following difficulty: if we compute the partition function of a system as usual

$$Z = \sum e^{-\beta H}, \tag{35.4}$$

because the sum is over all the possible microstates, the resultant Z or the free energy of the system is completely symmetric; that is, its symmetry group is identical to that of the microscopic Hamiltonian. This statement is true if the system is finite, because the sum is a finite sum. Thus, taking the thermodynamic limit is absolutely needed to make a rational and simple framework to understand spontaneous symmetry breaking.

35.11 What Actually Selects a Particular Symmetry Broken State?

When the intrinsic symmetry is broken, how is a particular phase selected in the real world? This is selected by extremely small fortuitous external effects or even without such effects by intrinsic thermal fluctuations. If there is a weak external field (stray field), the system would react very sensitively to it. Therefore, if one wishes to study a particular phase with the aid of statistical mechanics, an appropriate weak field conjugate to the order parameter is introduced to the system Hamiltonian to select the phase. After computing its thermodynamic limit, the field is set to zero. This limit must be performed after the thermodynamic limit; if performed before the thermodynamic limit, the symmetry breaking field effect disappears. Symmetry breaking means that the thermodynamic limit and the conjugate-field zero limit are not commutative.

---

[9] **Symmetry breaking and size** To state more practically, the state with a broken symmetry has a lifetime. For example, for a very small crystal, thermal fluctuation could spontaneously rearrange the crystal axes. Needless to say, if a crystal is not very small such fluctuations occur only very rarely. The agreement of its behavior to the behavior in the thermodynamic limit is practically perfect, because the lifetime of a given orientation is very long. However, mathematically, or theoretically, the situation is idealized and simplified by taking the thermodynamic limit.

Summary

* First-order phase transitions occur if reduction of order enhances further reduction.
* This observation allows a simple model illustrating features of the first-order phase transition.
* Even if phase transition occurs, the ensemble equivalence of statistical mechanics holds; we may use any convenient statistical ensemble.

Key words

metastable state, unstable state, nucleation, ensemble equivalence

The reader should be able to:

* Know basic features of the first-order phase transition.
* Recognize clearly that metastable states are generally beyond equilibrium statistical thermodynamics.
* Remember that E is once continuously differentiable with respect to S, V, and other work coordinates.
* Illustrate why ensemble equivalence holds.

36.1 First-Order Phase Transition Example: Nematic Isotropic Transition in Liquid Crystals

In the case of liquid crystals, the ordering and volume change are coupled, and there is an isotropic liquid–nematic liquid crystal phase transition which is (weakly) first order. A liquid crystal consists of slender molecules which orient in random directions at higher temperatures but tend to align at lower temperatures, although the center of mass coordinates of the molecules are spatially distributed as randomly as in ordinary liquids. This partially (only directionally) ordered phase is called the nematic liquid crystal phase. If we increase its temperature, due to thermal expansion, the distances between molecules increase slightly. This enhances disorganization of the molecular orientation, weakening molecular interactions further. This in turn enhances volume expansion and enhances

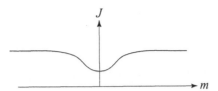

Fig. 36.1 Order-dependent coupling constant that induces a first-order phase transition. If the order is reduced (i.e., $|m|$ becomes small), the spin–spin interaction becomes weak. In such a model the order can decrease precipitously. $J \propto (1 - 2/3 \cosh 5m)$ is an example.

disorder. In this way increasingly order is lost, and a first-order phase transition, nematic-isotropic phase transition, ensues.[1]

36.2 Caricature Model of First-Order Phase Transition

The previous observation suggests a caricature model of first-order phase transition within the mean field approximation. For an Ising magnet model, suppose that if the magnetic order is reduced (i.e., $|m|$ becomes small), J in (34.9) decreases as illustrated in Fig. 36.1. We expect a first-order phase transition.

Let us review the mean-field approach for a square lattice (Chapter 34). Our starting point is the following equation

$$\langle s_0 \rangle = \left\langle \tanh \left[\beta J \left(s_1 + s_2 + s_3 + s_4 \right) \right] \right\rangle. \tag{36.1}$$

The naivest mean-field approach is

$$m = \tanh 4\beta Jm, \tag{36.2}$$

or setting $x = 4\beta Jm$, we must solve

$$x = 4\beta J \tanh x. \tag{36.3}$$

We replace J with the $J(x)$ in Fig. 36.1:

$$x = 4\beta J(x) \tanh x. \tag{36.4}$$

This modification is illustrated in Fig. 36.2.

---

[1] In contrast to the magnetic spin that has the head and tail (or north–south) distinction, the slender molecules making nematic liquid crystals are essentially symmetric around their centers of mass. Thus the order parameter is not a vector (actually it is a sort of tensor). The orientation of the molecules are maintained by repulsive interactions due to the molecular cores, so without reducing the number density there is no increase in disorder. This actually happens in some liquid crystals (kind of lyotropic liquid crystals), and no isotropic liquid phase is observed (that is, the nematic–isotropic phase transition is abolished).

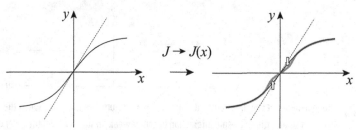

Fig. 36.2   Introduction of the m-dependent coupling constant: Replacing J with $J(x)$ illustrated in Fig. 36.1 corresponds to the modification from left to right (thick gray curve). The arrows highlight the changes. The vertical direction is slightly exaggerated.

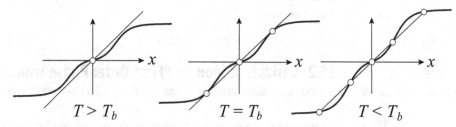

Fig. 36.3   A first-order phase transition occurs slightly below T_b. To determine the exact phase transition temperature, we need an analogue of Maxwell's rule. See Fig. 36.4 for the corresponding bifurcation diagram.

36.3 Bifurcations Exhibited by the Caricature Model

Let us study what happens if we lower the temperature, following the graphical method to solve (36.4). As β increases, the curve in Fig. 36.2 becomes steeper and shares three points with the diagonal line, and then five points as shown in Fig. 36.3; at T_b new nonzero fixed points appear.

The local stability of solutions may be read off from the bifurcation diagram Fig. 36.4 (recall Fig. 34.4). Below T_b there is a branch where m is not zero that is locally stable.

36.4 Metastable and Unstable States

In Fig. 36.4 the dark gray thick curves denote stable solutions (roots) and pale gray curves unstable solutions (in the sense that small perturbations added to the solution grow). Above T_b without question the $m = 0$ disordered phase is the equilibrium phase, and below T_X, again without doubt, the ordered (i.e., $m \neq 0$) phases are equilibrium phases. Between T_b and T_X the situation looks complicated. Thermodynamically, we expect there must be a phase transition from the $m = 0$ branch to the $m \neq 0$ branch somewhere between these two temperatures. The situation is analogous to the van der Waals gas; there should be a counterpart of Maxwell's rule (cf. Fig. 32.4) that determines the equilibrium phase transition point. Thus, if we come from the high temperature side slowly toward T_X a first-order phase transition occurs at the vertical thick line position in Fig. 36.4 (and *not* at T_b).

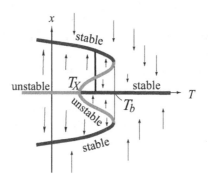

Fig. 36.4 The bifurcation diagram for the model that allows a first-order phase transition. The vertical arrows denote the evolving direction of perturbations to the fixed point values of x at various temperatures. We can see the local stability of the fixed points from the exchange of stability occurring at every bifurcation. To determine the exact phase transition temperature (denoted by the thick vertical line in the figure; within the mean-field theory) we need a rule parallel to Maxwell's rule.

If we rapidly cool the system, it is possible that we can stay on the dark gray line (i.e., the $m = 0$ line) below the phase transition point, which is the supercooled disordered phase and is *metastable*: it is stable against small perturbations (i.e., stable locally) but it is not really globally stable (does not correspond to the global free energy minimum). If we heat the system from an ordered phase (say, $m > 0$ phase) gradually, at the phase transition point order is lost and the $m = 0$ phase appears. However, if we heat the system rapidly, we could continue to stay on the dark gray curve, which is the superheated ordered state and is metastable. Thus we could have a *hysteresis*.

If we cool the disordered state really rapidly (temperature quench), then we could move the state along the pale gray line left to T_X. This state is unstable, so it rapidly organizes into an ordered phase.

Warning. This explanation pretends that the metastable and unstable branches are well defined as suggested by the illustration in Fig. 36.4. The picture seems generally true if the mean-field theory is exact (as in the Kac model with infinite-range interactions; recall **32.8**). For realistic short-range interaction systems, metastable states are indeed realizable, but generally they are not unique (history or preparatory procedure-dependent). There is no way to obtain the metastable states from the equilibrium statistical mechanics framework unambiguously. Extending the equilibrium free energy beyond the phase transition is generally illegitimate (the holomorphy of the thermodynamic potential is lost there). It goes without saying that unstable branches are beyond any reasonable grasp.

36.5 Phase Ordering Kinetics: Nucleation and Spinodal Decomposition

How phases change into each other is an interesting question, both pure and materials scientifically, because we could make various textures by arresting the transition process at an appropriate stage (by quenching). When a metastable disordered state orders, we expect seeds of ordered phases to appear in the ocean of a disordered phase as nuclei. The formation of nuclei is the rate-determining step. Once nuclei are formed, they grow rapidly and the phase transition is completed.

If a disordered phase is quenched into its unstable state, then immediately ordered domains appear everywhere in the space. However, ordered phases are usually not unique (due to symmetry breaking), so initially a fine mosaic state is formed. Then, each domain of a particular ordered phase tries to increase its size.[2]

36.6 First-Order Phase Transition due to External Field Change

Phase transitions can occur even if T is constant due to changes of other variables (say, P in the case of fluid; look at the P–T diagram). Again, this phase transition can be understood intuitively with the aid of magnets (but do not forget that the warning stated just above applies, so the picture outlined is crude).

Below T_c the 2-Ising model is in the up phase or down phase. If a small magnetic field is applied, then the direction of the spins of one phase is stabilized relative to the other phase. This means one phase is no longer a true equilibrium state but only a metastable state. Let us discuss the phase transition induced by this change with the aid of the mean-field theory. Let us assume J is constant, and consider (34.8), i.e.,[3]

$$m = \tanh\left(2d\beta Jm + \beta h\right). \tag{36.5}$$

To solve this equation we again use a graphic method (see Fig. 36.5; the corresponding bifurcation diagram is in Fig. 36.6). A large positive h shifts the tanh curve to the left.

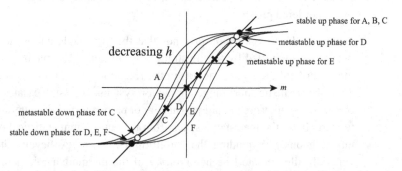

Fig. 36.5 Reducing h corresponds to A → F. If the magnetic field (z-component) is positively large (A), the up phase is stable. Between A and B even if h is reduced virtually nothing happens (the black dot in the first quadrant). If h is reduced further a metastable down spin state (white disk on the third quadrant) becomes possible (C). Also there is an unstable state (cross mark). If h is reduced further, the metastable down phase becomes stable, and the stable up phase becomes metastable (D; a white disk in the first quadrant). For E and F, h is so negative that the down phase is stable (black dots in the third quadrant). Look at the bifurcation diagram in Fig. 36.6(a).

---

[2] If the order parameter is conserved (i.e., the transport equation for the order parameter density obeys the conservation equation discussed in Chapter 9), it is called the *spinodal decomposition*.

[3] Recall that our convention is that $-hs$ is the potential energy of the spin, so "up" ($s = +1$) is favored, if $h > 0$.

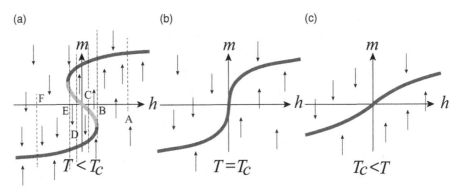

Fig. 36.6 Bifurcation diagram for (36.5). Dark gray curves denote locally stable solutions (i.e., thermodynamically (i.e., globally) stable or metastable states). The bifurcation diagram (a) corresponds to Fig. 36.5 with the corresponding letters. Pale gray portion denotes unstable solutions corresponding to the black crosses in Fig. 36.5.

In Fig. 36.5 A → F describes the effect of reducing the magnetic field favoring the up phase while keeping the temperature T ($< T_c$) constant. The corresponding bifurcation diagram (Fig. 36.6(a)) may be convenient for grasping the overall picture. Below B once the down phase domain is formed, it is *metastable* (i.e., if it is large enough, it lasts for a very long time). The stability switches between the up and down phases at $h = 0$ can be understood intuitively.

Suppose the system is initially in the down phase. If an upward magnetic field ($h > 0$) is applied, it becomes metastable, because the up phase is thermodynamically more stable (its free energy is less than that of the down phase). However, almost until B is realized in Fig. 36.5, big enough down spin domains persist. If the magnetic field is suddenly increased to A, the down phase becomes unstable and goes into the up phase in avalanches.

The picture just explained applies to many first-order phase transitions when we change the intensive variable (i.e., the field) that is conjugate to a density that jumps at the first-order phase transition. As can be guessed from the illustration in Fig. 32.7 for a fluid system (or a binary mixture system), pressure (or chemical potential) may be regarded as the intensive variable to induce first-order phase transitions.

We know there is no phase transition above T_c for fluids. This corresponds to the bifurcation diagram (c) on the right of Fig. 36.6. Just as the up phase turns into the down phase and vice versa smoothly, in the case of the fluid, very high density states may be converted into very low density states continuously through changing the pressure.

36.7 Statistical Thermodynamic Study of First-Order Phase Transitions

As the reader realizes, many statistical physics books describe the second-order phase transitions in detail, but the first-order phase transitions are rarely discussed except perhaps for the Clapeyron–Clausius formula. There is a reason for this; we must study two phases separately and find when the free energies of the two phases coincide. As mentioned already (recall the warning in **36.4**), since the free energy is very likely to have a singularity at the first-order phase transition point, it might be in principle possible to study the first-order phase transition through investigating a single phase only, e.g., the melting point from the liquid phase free energy only. However, the singularity is very subtle (losing holomorphy but still infinite-times differentiable, for example), not as conspicuous as the singularity associated with the second-order phase transition.

Still there have been attempts to find a criterion for particular kinds of first-order phase transitions. Perhaps the most famous is the *Lindemann rule*, stating that a solid melts when the ratio of the root-mean-square vibrational amplitude of the constituent of the crystal and the lattice spacing exceeds a certain critical value, say, 0.12. For a class of similar compounds or similar structured crystals the rule seems reasonable empirically.

These days, huge amounts of, e.g., melting point data allow one to make ad hoc statistical models. Fortunately, or unfortunately, understanding of a phenomenon is no longer required to be practical. Then, what is science for? We no longer need to try to understand materials-scientific fetish facts. Ultimately, the role of science may be to make us honest and to revise our Weltanschauung.

To conclude this book let us review an important message of statistical thermodynamics: the ensemble equivalence.

36.8 Phase Transition and Internal Energy Singularity

Let us review the meaning of "ensemble equivalence": the reader may use any convenient ensemble to compute any thermodynamic potential (especially E and S) they wish.

It has been stressed that the most fundamental macroscopic description of a macrosystem in equilibrium is in terms of thermodynamic coordinates (Chapter 12). The entropy as a function of the thermodynamic coordinates gives the most complete thermodynamic description of the system (the fundamental equation as Gibbs called it). In other words, if we know the internal energy as a function of entropy and work coordinates as $E = E(S, V, X, \ldots)$, we have a complete thermodynamic description of the system. Therefore, it might be a natural expectation that even if we compute the Helmholtz free energy $A = A(T, V, X, \ldots)$, we may not be able to obtain $E = E(S, V, X, \ldots)$ in its entirety. But, actually, from A we can fully reproduce E. If A is differentiable, of course we know the Gibbs–Helmholtz equation (18.35), but no differentiability is needed.[4]

However, since the thermodynamic coordinates are privileged variables (Chapter 12), we should lose something. Indeed, we lose some detailed information. Let us see what we can preserve and what we lose when we move from the thermodynamic coordinate system (in terms of $E = E(S, V, \ldots)$) to something else (in the illustration below, to $A = A(T, V, \ldots)$).

---

[4] It is solely due to the convexity of $-A$ as a function of T as we discussed in Chapter 16.

In terms of internal energy, a phase transition occurs where the convex function $E = E(S, V, \ldots)$ loses its smoothness. Here "smoothness" implies the analyticity as a multivariable function. Since a convex function is continuous, E cannot have any jump. Furthermore, as we see from the Gibbs relation,

$$dE = TdS - PdV + xdX + \cdots, \tag{36.6}$$

its continuous differentiability must be satisfied in the region of thermodynamic space meaningful to the system.[5] Thus, internal energy must be a C^1 (continuously differentiable) convex function of entropy and work coordinates. Consequently, the worst singularity is the loss of twice differentiability. For example, the constant volume specific heat can become not definable. We know at the critical point this indeed happens.

36.9 Internal Energetic Description of First-Order Phase Transition

Let us look at one variable S of E. Let us assume that work coordinates (such as the volume) are kept constant. Here we pay attention to a typical case in which the second-order derivatives have jumps.

Figure 36.7 illustrates E as a function of S. The slope of this curve is temperature T. Phase I and phase II coexist at one temperature T (given by the slope of the straight portion between "a" and "b"). These two phases are distinct and have different thermodynamic densities. Therefore, there is a first-order phase transition between phases I and II.

36.10 Legendre Transformation and Phase Transition

To understand the coexistence of two phases under constant temperature discussed here previously, it is convenient to use the thermodynamic potential one of whose independent

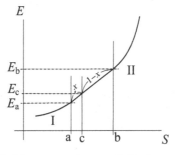

Fig. 36.7   In this illustration twice differentiability is lost at two points, a and b. The flat portion describes the phase coexistence and the internal energy at c may be written as $E_c = E(x) = (1 - x)E_a + xE_b$, where E_a (E_b) is the internal energy of state a (b) (the lever rule; recall Fig. 31.2). x is the fraction of phase II. This is the reason why the coexistence implies a flat portion.

[5] P and T never jump when we change thermodynamic variables. If $T = 0$, some pathological things may happen, but we know we never reach $T = 0$.

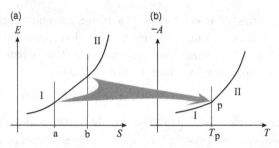

Fig. 36.8 Legendre transformation E to A (or $-A$). (a) is just the same as Fig. 36.7 and depicts E as a continuously differentiable function of S. E is linear between a and b, and the slopes at a and at b agree with the slope of the linear portion. Phase I occupies left of a, and phase II right of b, and the linear portion describes the coexistence of these two phases. The slope of the linear portion is the coexistence temperature T_p, corresponding to the break point p of the free energy graph in (b). All the coexisting phases between a and b are mapped to a point p by the Legendre transformation.

variables is temperature, that is, the Helmholtz free energy. It is obtained by the Legendre transformation with respect to entropy: $-A = \max_S[ST - E]$ (i.e., $E^* = -A$) (see Chapter 16). Thus, the free energy is convex *upward* as a function of temperature (in Fig. 36.8 the convex function $-A$ is illustrated).

If a first-order phase transition happens and if two phases can coexist, there is a "linear" portion in the graph of internal energy. The reason is explained in the caption of Fig. 36.7 (or recall Fig. 31.2). This is mapped to a single point by the Legendre transformation (Fig. 36.8). As can be seen from this, when two phases coexist, thermodynamic states that can be distinguished by thermodynamic coordinates (intuitively, the states distinguishable by different ratios of two phases; recall Fig. 12.1) are identified and mapped to a single point by the Legendre transformation. We lose the information about x in Fig. 36.7, the relative amount of coexisting phases, by the Legendre transformation. However, from Fig. 36.8(b), we can *still* completely reconstruct internal energy as a function of thermodynamic coordinates by the inverse Legendre transformation $E = \max_T[ST - (-A)]$ (i.e., $E^{**} = E$). This is the implication of the ensemble equivalence: we may use any statistical ensemble or any thermodynamic potential to study an equilibrium system even if there are phase transitions.

Problems

Q36.1 The Slope of Coexisting Curves: An Elementary Question

Discuss which slope is larger near the triple point on the TP-diagram of an ordinary fluid, the gas–liquid or the gas–solid coexistence curve?

Solution

The Clapeyron–Clausius equation tells us that

$$\left.\frac{dP}{dT}\right|_{SG} = \frac{S_G - S_S}{V_G - V_S} > \frac{S_G - S_L}{V_G - V_L} = \left.\frac{dP}{dT}\right|_{LG}, \qquad (36.7)$$

because the molar volumes of the solid and the liquid phases are almost the same, but because the entropy of the liquid is larger than that of the solid.

Q36.2 Lattice Gas

At each lattice site of a square lattice at most one particle can sit. If nearest-neighbor lattice points are occupied there is a stabilizing interaction, and the pair energy -4ε is assigned. The system is connected to a chemostat with the chemical potential μ of the particles. Study the phase transition of this system with the aid of the mean-field theory.

(1) Let n_i be the occupation number (0 or 1) of site i. Then, the Hamiltonian reads

$$H = -4\varepsilon \sum_{\langle i,j \rangle} n_i n_j. \tag{36.8}$$

In terms of the Ising spin $s_i = \pm 1$ at site i rewrite the Hamiltonian.

(2) Write down the grand canonical partition function for the system in terms of the Ising spins introduced in (1). Interpret the resultant system as a magnetic system, and discuss its phase transition.

Solution

(1) We have only to replace as $n_i = (1 + s_i)/2$. Therefore,

$$H = -\varepsilon \sum_{\langle i,j \rangle} \left[s_i s_j + s_i + s_j + 1 \right] = -\varepsilon \sum_{\langle i,j \rangle} s_i s_j - 4\varepsilon \sum_i s_i + \text{constant.} \tag{36.9}$$

Here, we have used the fact that the number of bonds is 2 times the number of the lattice sites.

(2) The grand partition function can be written in terms of $\beta(H - \mu N)$, so in terms of the spin variables, we have

$$H - \mu N = -\varepsilon \sum_{\langle i,j \rangle} s_i s_j - \left(4\varepsilon + \frac{1}{2}\mu \right) \sum_i s_i, \tag{36.10}$$

ignoring the constant terms. Comparing this with the Ising model, we see that the magnetic field is $4\varepsilon + \mu/2$. Thus, if we increase the chemical potential the magnetization m, which is related to the average site occupation in the original model as $\langle n \rangle = (1 + m)/2$, increases. Thus, the model can be understood in terms of the bifurcation diagrams in Fig. 36.6.

Q36.3 Legendre Transformation in Convex Analysis

When a phase transition occurs, the curve of $E(S)$ can have a linear part as a function of S (that is, E can change under constant $T = T_e$). Then, as discussed in the text, A as a function of T has a cusp at $T = T_e$ (that is, all the states corresponding to the flat part are collapsed to a point; the pointwise one-to-one correspondence can be lost by Legendre

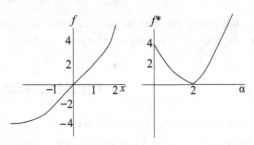

Fig. 36.9 The sketch of f and f^* for **Q36.3**.

transformation, if there is a phase transition). To illustrate this point, let us consider the following toy function

$$f(x) = \begin{cases} 2\tanh(x+1) - 2 & \text{for } x < -1, \\ 2x & \text{for } -1 \leq x \leq 1, \\ (x-1)^2 + 2x & \text{for } x > 1. \end{cases}$$

Sketch its Legendre transform $f^*(\alpha) = \max_x[\alpha x - f(x)]$. (Do not try to compute the explicit formula.)

Solution

We consider the Legendre transformation

$$f^*(\alpha) = \max_x[\alpha x - f(x)]. \tag{36.11}$$

However, it may be more convenient to refer to Fig. 16.3(b) for an intuitive way to guess the answer.

For $\alpha > 2$ this is easy, and we obtain $f^*(\alpha) = \alpha^2/4 - 1$. Between -2 and 2 of x the slope of f does not change and is 2, so $\alpha = 2$, which means $f^*(2) = 0$ is a cusp. For $\alpha < 2$, analytic calculation is not wise. We know α cannot be negative, and in the $\alpha \to 0$ limit, $f^* \to 4$, because $\lim_{x \to -\infty} f(x) = -4$. Since f^* is convex, we can easily sketch its overall shape as Fig. 36.9: Remember that convex functions are continuous.

Appendix **Introduction to Mechanics**

This book requires a rudimentary knowledge of quantum mechanics. The purpose of this Appendix is to make the reader somewhat familiar with the basic structure/principles of quantum mechanics. How we proceed is quite similar to our approach to statistical thermo-dynamics; mathematics and empirical facts are synthesized. For statistical thermodynamics the reader has only to understand the time-independent Schrödinger equation and energy eigenvalue problem **A.24** (the harmonic oscillator is discussed in **A.29**). Those who are not familiar with the bra-ket notation should read Appendix 17A first. The following exposition includes an outline of Fourier analysis in the Dirac notation.

A.1 How We Proceed

Everyone reading Dirac's *The Principles of Quantum Mechanics* should realize that respecting both the salient empirical facts and the mathematical aesthetics leads to the most natural exposition of an empirical theory. Therefore, the following exposition of mechanics is constructed around the empirical facts that suggest/justify the mathematical principles.

A.2 Remarks for Those Who are Familiar with Introductory Mechanics

For the reader who knows an outline of (analytical and quantum) mechanics, an outline of this Appendix is given here.

The first part **A.3–A.8** is a preparation of our language – Fourier analysis. This part explains the Fourier transformation in Dirac's notation.

Based on experimental facts and the thought experiments they inspire, Schrödinger's equation of motion is derived (**A.14**). At the same time the fundamental postulates of quantum mechanics are summarized (**Postulate 1–6**). The perturbation method (**A.31**) and the minimax principle for eigenvalues (**A.32**) as well as the uncertainty relation (**A.28**) are explained. The harmonic oscillator eigenenergies are obtained (**A.29**). For convenience 1/2 spins are also outlined (**A.30**).

We then proceed to the classical approximation. The reader should be familiar with the rudiments of analytical mechanics.[1] First, the Feynman path integral expression of

---

[1] For example, Landau, L. D and Lifshitz, E. M. (1982). *Classical Mechanics*, 3rd edition, Oxford: Butterworth-Heinemann.

the system evolution is derived (**A.33**), from which classical mechanics is deduced as a macroscopic approximation (**A.34**). The correspondence between Heisenberg's and Hamilton's equations of motion introduces the Poisson bracket and the canonical quantization condition (**A.37**).

Remark on the notation. In the first half of this Appendix, although the notation x is used for coordinates, this can be a vector of any dimension; xy should be interpreted as the scalar product, dx is a volume element.

A.3 Introduction to the Fourier Series

Here, we do analysis always in the very formal and intuitive fashion, so any change in the order of mathematical operations is legitimate as long as the results (formally) make sense.

If f is a periodic function with period 2ℓ, it can be expressed in the following (complex) *Fourier series* form:

$$f(x) = \sum_{n=-\infty}^{+\infty} c_n e^{in\pi x/\ell}, \tag{A.1}$$

where the coefficients can be computed as

$$c_n = \frac{1}{2\ell} \int_{-\ell}^{+\ell} dy f(y) e^{-in\pi y/\ell}. \tag{A.2}$$

The reader can check this easily from (A.1) by termwise integration. Combining these two formulas, we get

$$f(x) = \int_{-\ell}^{+\ell} dy f(y) \left[\frac{1}{2\ell} \sum_{n=-\infty}^{+\infty} e^{in\pi(x-y)/\ell} \right]. \tag{A.3}$$

That is, we get an expression of the δ-function (see Chapter 6) on $(-\ell, \ell]$:[2]

$$\delta(x - y) = \frac{1}{2\ell} \sum_{n=-\infty}^{+\infty} e^{in\pi(x-y)/\ell}. \tag{A.4}$$

A.4 Fourier Integral Representation of δ-Function

Take $\ell \to \infty$ limit. To this end let us introduce $k = n\pi/\ell$. Then,

$$\frac{1}{2\ell} \sum_{n=-\infty}^{+\infty} \sim \frac{1}{2\ell} \int_{-\infty}^{+\infty} dn \to \frac{1}{2\pi} \int_{-\infty}^{+\infty} dk. \tag{A.5}$$

---

[2] Notice that the outcome is a periodic function with period 2ℓ, so we need this restriction.

Therefore, (A.4), which is in 1-space, becomes generally in $\eth$-space as follows. That is, the $\eth$-dimensional δ-function may be expressed as

$$\delta(x - y) = \frac{1}{(2\pi)^\eth} \int_{\mathbb{R}^\eth} dk\, e^{ik(x-y)}. \tag{A.6}$$

A.5 Fourier Transformation

Using (A.6) and consulting (A.3), we get

$$f(x) = \int_{\mathbb{R}^\eth} dy\, \delta(x - y) f(y) = \frac{1}{(2\pi)^{\eth/2}} \int_{\mathbb{R}^\eth} dk\, e^{ikx} \frac{1}{(2\pi)^{\eth/2}} \int_{\mathbb{R}^\eth} dy\, e^{-iky} f(y) \tag{A.7}$$

in $\eth$-space. Thus, we have obtained the (symmetric form of)[3] Fourier transformation formula in $\eth$-space:[4]

$$f(x) = \frac{1}{(2\pi)^{\eth/2}} \int_{\mathbb{R}^\eth} dk\, \tilde{f}(k) e^{ikx} \iff \tilde{f}(k) = \frac{1}{(2\pi)^{\eth/2}} \int_{\mathbb{R}^\eth} dx\, f(x) e^{-ikx}. \tag{A.8}$$

$\tilde{f}(k)$ is called the *Fourier transform* of $f(x)$. Note that $f(x)$ is recovered from $\tilde{f}(k)$ by almost the same transformation (only with the sign change) making $\tilde{f}$ from f.

A.6 The Function $f(x)$ as Position Representation of $|f\rangle$

The *i*th component of a vector v (or ket $|v\rangle$) may be written as $v_i = \langle i|v\rangle$ ($\overline{v_i} = \langle v|i\rangle$) using a basis vector $|i\rangle$ (review Appendix 17A, if the reader is not familiar with the Dirac notation). Analogously, we wish to write

$$f(x) = \langle x|f\rangle, \quad \overline{f(x)} = \langle f|x\rangle. \tag{A.9}$$

That is, we regard a function (formally) as a vector in a vector space spanned by *position kets* $\{|x\rangle : x \in D\}$, where D is the domain of f. The function $f(x)$ may be called the *position representation* of $|f\rangle$.

---

[3] **Asymmetric form** For practical use of Fourier transformation, the following "asymmetric form" is often more convenient:

$$f(x) = \int_{\mathbb{R}^\eth} dk\, \hat{f}(k) e^{ikx} \iff \hat{f}(k) = \frac{1}{(2\pi)^\eth} \int_{\mathbb{R}^\eth} dx\, f(x) e^{-ikx}.$$

[4] As remarked at the beginning, x in this Appendix is generally a vector. To clarify the remark further, (A.8) in $\eth$-space is given here explicitly in terms of $\eth$-vectors $\boldsymbol{k}$ and $\boldsymbol{x}$:

$$f(\boldsymbol{x}) = \frac{1}{(2\pi)^{\eth/2}} \int_{\mathbb{R}^\eth} d^\eth \boldsymbol{k}\, \tilde{f}(\boldsymbol{k}) e^{i\boldsymbol{k}\cdot\boldsymbol{x}} \iff \tilde{f}(\boldsymbol{k}) = \frac{1}{(2\pi)^{\eth/2}} \int_{\mathbb{R}^\eth} d^\eth \boldsymbol{x}\, f(\boldsymbol{x}) e^{-i\boldsymbol{k}\cdot\boldsymbol{x}}.$$

The position kets are normalized as

$$\langle x|y\rangle = \delta(x - y). \tag{A.10}$$

Then, we may write

$$f(x) = \int dy\, \delta(x - y)f(y) \tag{A.11}$$

as

$$\langle x|f\rangle = \int dy\, \langle x|y\rangle \langle y|f\rangle. \tag{A.12}$$

Thus, we get the following *resolution of unity* (if we explicitly write the domain D for the function space of our interest):

$$1 = \int_D dx\, |x\rangle\langle x|. \tag{A.13}$$

We see that the scalar product of two functions has the following natural expression:

$$\langle f|g\rangle = \int_D dx\, \langle f|x\rangle\langle x|g\rangle = \int_D dx\, \overline{f(x)}g(x). \tag{A.14}$$

A.7 Fourier Analysis in Dirac Notation

Henceforth, we set $D = \mathbb{R}^\partial$. We introduce the wavenumber kets $\{|k\rangle\}_{k \in \mathbb{R}^\partial}$

$$\tilde{f}(k) = \langle k|f\rangle, \quad \overline{\tilde{f}(k)} = \langle f|k\rangle. \tag{A.15}$$

The wavenumber kets are normalized as

$$\langle k|l\rangle = \delta(k - l), \tag{A.16}$$

which implies, just as can be shown as in **A.6**, the following resolution of unity:

$$1 = \int_{\mathbb{R}^\partial} dk\, |k\rangle\langle k|. \tag{A.17}$$

Then, we must have

$$f(x) = \int dk \langle x|k\rangle \langle k|f\rangle, \quad \tilde{f}(k) = \int dx \langle k|x\rangle \langle x|f\rangle. \tag{A.18}$$

Comparing this with (A.8), we must conclude

$$\langle k|x\rangle = \frac{1}{(2\pi)^{\partial/2}}e^{-ikx}, \quad \langle x|k\rangle = \frac{1}{(2\pi)^{\partial/2}}e^{ikx}. \tag{A.19}$$

A.8 Momentum Ket

To do mechanics it is more convenient to introduce the momentum ket $|p\rangle$ with the normalization: $\langle p|p'\rangle = \delta(p - p')$. Since $p = \hbar k$ (see de Broglie wave **A.9**):

$$\langle x|p\rangle = \frac{1}{h^{3/2}} e^{ip \cdot x/\hbar}. \tag{A.20}$$

This can be checked easily as (introduce $k/2\pi = p/h$; here we are in 3-space):

$$\langle x|x'\rangle = \int d^3p \langle x|p\rangle \langle p|x'\rangle = \frac{1}{h^3} \int_{\mathbb{R}^3} d^3p\, e^{ip \cdot (x-x')/\hbar} = \frac{1}{(2\pi)^3} \int_{\mathbb{R}^3} d^3k\, e^{ik \cdot (x-x')} = \delta(x-x').$$
$$\tag{A.21}$$

With these preparations, let us look at key empirical facts and the mathematical structure suggested by them.

A.9 Matter Waves[5]

A free particle (i.e., a particle in a vacuum without interacting with anything) with momentum p (equal to mass m times the velocity v) can propagate as a plane wave of wavelength $\lambda = h/|p|$, where $h = 6.626 \times 10^{-34}$ Js is *Planck's constant*, in the same direction of p. In formulas, the plane wave has the form

$$\psi(r, t) = Ae^{i(k \cdot r - \omega t)} = Ae^{i(p \cdot r - Et)/\hbar} \tag{A.22}$$

with A being a complex amplitude and $E = p^2/2m$ the kinetic energy.

A.10 How Potential Energy is Involved[6]

Consider a box containing a charged particle with charge q. If the box is large enough, then the de Broglie wave of the particle is almost planar. If the voltage of the box is raised

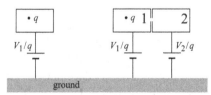

Fig. A.1 How to shift the potential energy; q is the charge of the particle.

[5] The **de Broglie wave** The idea that with every particle is associated a "material" wave was initially proposed by de Broglie. A photon of angular frequency ω (equals 2π times the frequency ν) has an energy $\hbar\omega$ and momentum h/λ, where λ is the wavelength and described as a plane electromagnetic wave $Ae^{i(k \cdot r - \omega t)}$. de Broglie proposed in 1924 (this is about 10 years after Bohr's atom model) that this holds not only for photons but also for any material particle. This was soon experimentally proved by Davisson and Germer (1927).

[6] The exposition here follows *Feynman Lectures* III.

by V/q, the total energy of the particle is shifted as $E \rightarrow E + V$ (Fig. A.1 Left), but the momentum does not change. Thus, the de Broglie wave associated with the particle should read as ψ_1 in (A.23).

A.11 Conservation of Energy Equals Constancy of Frequency of Matter Wave

If we connect two boxes with different potential energies (Fig. A.1 Right), the plane waves in box 1 and box 2 are respectively given by

$$\psi_1 = e^{i\{p_1 x - (p_1^2/2m + V_1)t\}/\hbar}, \quad \psi_2 = e^{i\{p_2 x - (p_2^2/2m + V_2)t\}/\hbar}. \tag{A.23}$$

For these waves to describe a wave propagating from 1 to 2 the frequencies should be common, so $p_1^2/2m + V_1 = p_2^2/2m + V_2$. That is, the invariance of the de Broglie wave frequency implies the conservation of energy.

A.12 Double-Slit Experiment and Superposition Principle

Suppose particles with a given momentum are sent to a double slit (such as used in Young's experiment) and are detected on a screen placed behind the slits. The locations where the particles are detected collectively construct the double-slit interference pattern consistent with the de Broglie wavelength of the particles.[7]

The double-slit experiment tells us that plane wave states associated with a single particle may be linearly superposed; the states going through one slit and those through the other may be superposable linearly. Thus, generally, we assume that:

Postulate 1 *The totality of states of a mechanical system is an inner product space (i.e., a vector space with a scalar product defined). Each vector corresponds to a particular state of the system.*

This vector space is most conveniently described as a ket space, as outlined in Appendix 17A. A ket $|a\rangle$ is a normalized ket, if $\langle a|a \rangle = 1$.

Postulate 1 is usually augmented with a postulate that a superposition of a state and the same state itself does not produce any new state. This means $|a\rangle + c|a\rangle$ is physically indistinguishable from $|a\rangle$ unless $1 + c = 0$, where $c \in \mathbb{C}$. Thus, a complex constant multiple of a ket expresses the same state as the original ket, so physically distinguishable states must correspond to "directions" (rays) or normalized kets.

---

[7] **Double-slit interference pattern is due to a single-particle effect** This is not an ensemble effect; if so, at dark fringes two particles must annihilate each other or at bright fringes two particles must create two more particles, fragrantly violating the conservation of energy or charge.

Postulate 1'   *For any nonzero $c \in \mathbb{C}$, $|a\rangle$ and $c|a\rangle$ represent physically indistinguishable states. Thus, a physical state corresponds to a 1-subspace spanned by a ket.*

The plane wave (A.22) may be written in terms of a position representation of a momentum ket $|p\rangle$ given by (A.20):

$$\psi(\boldsymbol{x}, t) \propto e^{-iEt/\hbar} \langle \boldsymbol{x} | \boldsymbol{p} \rangle \propto e^{(i/\hbar)(\boldsymbol{p} \cdot \boldsymbol{x} - Et)}. \tag{A.24}$$

The plane de Broglie wave corresponds to the momentum ket.

A.13 Evolution of Matter Wave for an Infinitesimal Time

A general state $|t\rangle$ at time t need not be a plane wave, but can be a superposition of plane waves as

$$\langle \boldsymbol{x} | t \rangle = \int d^3 \boldsymbol{p} \, e^{-iE(\boldsymbol{p})t/\hbar} \langle \boldsymbol{x} | \boldsymbol{p} \rangle \langle \boldsymbol{p} | 0 \rangle, \tag{A.25}$$

where $\langle \boldsymbol{p} | 0 \rangle$ is the amplitude (corresponding to the Fourier coefficient) of the plane wave $|p\rangle$ component in the initial state $|0\rangle$ at time $t = 0$ and $E(\boldsymbol{p})$ its energy. This expression is admissible in a vacuum, but if there is a potential energy, E depends not only on p but also on position x. Therefore, formula (A.25) cannot be used for a long time. If t is small (written as δt), however, the particle position should not change very much, so

$$\langle \boldsymbol{x} | t \rangle = \int d^3 \boldsymbol{p} \, e^{-iE(\boldsymbol{p}, \boldsymbol{x})\delta t/\hbar} \langle \boldsymbol{x} | \boldsymbol{p} \rangle \langle \boldsymbol{p} | t - \delta t \rangle, \tag{A.26}$$

should be accurate to $O[\delta t]$, where $E(\boldsymbol{p}, \boldsymbol{x}) = \boldsymbol{p}^2 / 2m + V(\boldsymbol{x})$. We can rewrite (A.26) with the aid of a resolution of unity (A.13) as

$$\langle \boldsymbol{x} | t \rangle = \int d^3 \boldsymbol{y} \int d^3 \boldsymbol{p} \, \langle \boldsymbol{x} | \boldsymbol{p} \rangle e^{-iE(\boldsymbol{p}, \boldsymbol{x})\delta t/\hbar} \langle \boldsymbol{p} | \boldsymbol{y} \rangle \langle \boldsymbol{y} | t - \delta t \rangle. \tag{A.27}$$

A.14 Derivation of Schrödinger's Equation

From (A.27) we obtain

$$\langle \boldsymbol{x} | t \rangle - \langle \boldsymbol{x} | t - \delta t \rangle = \int d^3 \boldsymbol{y} \int d^3 \boldsymbol{p} \, \langle \boldsymbol{x} | \boldsymbol{p} \rangle \left[e^{-iE(\boldsymbol{p}, \boldsymbol{x})\delta t/\hbar} - 1 \right] \langle \boldsymbol{p} | \boldsymbol{y} \rangle \langle \boldsymbol{y} | t - \delta t \rangle \tag{A.28}$$

$$= -\frac{i}{\hbar} \int d^3 \boldsymbol{y} \int d^3 \boldsymbol{p} \, \langle \boldsymbol{x} | \boldsymbol{p} \rangle E(\boldsymbol{p}, \boldsymbol{x}) \langle \boldsymbol{p} | \boldsymbol{y} \rangle \langle \boldsymbol{y} | t \rangle \delta t + o[\delta t], \tag{A.29}$$

or using the explicit form of E, we get *Schrödinger's equation*:

$$i\hbar \frac{\partial}{\partial t} \langle \boldsymbol{x} | t \rangle = \int d^3 \boldsymbol{y} \int d^3 \boldsymbol{p} \, \langle \boldsymbol{x} | \boldsymbol{p} \rangle \left[\frac{\boldsymbol{p}^2}{2m} + V(\boldsymbol{x}) \right] \langle \boldsymbol{p} | \boldsymbol{y} \rangle \langle \boldsymbol{y} | t \rangle. \tag{A.30}$$

Using (A.20) for the expression of the momentum bra and ket, the p integration can be computed as:[8]

$$\frac{1}{\hbar^3} \int d^3p \, e^{i\boldsymbol{p}\cdot(\boldsymbol{x}-\boldsymbol{y})/\hbar} \left[\frac{\boldsymbol{p}^2}{2m}\right] = \frac{1}{\hbar^3} \int d^3p \, \frac{1}{2m}\left(-i\hbar\frac{d}{d\boldsymbol{x}}\right)^2 e^{i\boldsymbol{p}\cdot(\boldsymbol{x}-\boldsymbol{y})/\hbar} = -\frac{\hbar^2}{2m}\Delta_{\boldsymbol{x}}\delta(\boldsymbol{x}-\boldsymbol{y}).$$
(A.31)

Therefore, the right-hand side of (A.30) reads

$$\int d^3y \left[-\frac{\hbar^2}{2m}\Delta\delta(\boldsymbol{x}-\boldsymbol{y}) + V(\boldsymbol{x})\delta(\boldsymbol{x}-\boldsymbol{y})\right]\langle\boldsymbol{y}|t\rangle = \left[-\frac{\hbar^2}{2m}\Delta + V(\boldsymbol{x})\right]\langle\boldsymbol{x}|t\rangle. \qquad \text{(A.32)}$$

If we write $\psi(\boldsymbol{x}, t) = \langle\boldsymbol{x}|t\rangle$ (called the *wavefunction*), we reach

$$i\hbar\frac{\partial}{\partial t}\psi(\boldsymbol{x}, t) = \left[-\frac{\hbar^2}{2m}\Delta + V(\boldsymbol{x})\right]\psi(\boldsymbol{x}, t), \qquad \text{(A.33)}$$

which is the standard form of *Schrödinger's equation*.[9] Its linearity is compatible with superposition (**A.12**). Notice that the operator (actually the position representation $\langle\boldsymbol{x}|H|\boldsymbol{x}\rangle$ of operator H called the *Hamiltonian*):

$$H = -\frac{\hbar^2}{2m}\Delta + V(\boldsymbol{x}) \qquad \text{(A.34)}$$

comes from the energy term. Thus, this must be related to the energy of the system.

A.15 Schrödinger's Equation in Kets

Since $\psi(\boldsymbol{x}, t) = \langle\boldsymbol{x}|t\rangle$, Schrödinger's equation reads:[10]

$$i\hbar\frac{d}{dt}|t\rangle = H|t\rangle, \quad -i\hbar\frac{d}{dt}\langle t| = \langle t|H. \qquad \text{(A.35)}$$

Therefore,

$$i\hbar\frac{d}{dt}\langle t|t\rangle = \left(i\hbar\frac{d}{dt}\langle t|\right)|t\rangle + \langle t|\left(i\hbar\frac{d}{dt}|t\rangle\right) = -\langle t|H|t\rangle + \langle t|H|t\rangle = 0. \qquad \text{(A.36)}$$

That is, $\langle t|t\rangle$ is invariant.

---

[8] In the following $\partial/\partial\boldsymbol{y}$ is ∇ with respect to $\boldsymbol{y}$, so $(\partial/\partial\boldsymbol{y})^2$ reads $\nabla\cdot\nabla = \Delta$.

[9] Note that $i\hbar$ instead of $-i\hbar$ in front of the equation is inherited from our convention adopted when we wrote the de Broglie wave as $e^{i(px-Et)/\hbar}$ instead of $e^{-i(px-Et)/\hbar}$. Thus, the choice is a historical accident.

[10] In the following formulas, the operator H is written in terms of the matrix element (A.34) as

$$H = \int d^3x \, |\boldsymbol{x}\rangle\langle\boldsymbol{x}|H|\boldsymbol{x}\rangle\langle\boldsymbol{x}| = \int d\boldsymbol{x} \, |\boldsymbol{x}\rangle\left[-\frac{\hbar^2}{2m}\Delta + V(\boldsymbol{x})\right]\langle\boldsymbol{x}|.$$

Multiplying $\langle\boldsymbol{x}|$ to (A.35), we immediately get (A.33).

A.16 Time-Evolution Operator is Unitary

Let us introduce the *time-evolution operator* $\mathcal{T}_t$ as $|t\rangle = \mathcal{T}_t|0\rangle$. If H is time-independent, we can solve (A.35) as

$$\mathcal{T}_t = e^{-itH/\hbar}. \tag{A.37}$$

We have $\langle 0|\mathcal{T}_t^*\mathcal{T}_t|0\rangle = \langle 0|0\rangle$, so $\mathcal{T}_t^*\mathcal{T}_t = 1$. We expect any state must obey the same evolution law, so the domain of $\mathcal{T}_t$ must be the whole state space. Therefore, $\mathcal{T}_t$ must be a unitary operator (cf. Appendix 17A for a proof). Therefore, generally we postulate:

Postulate 2   *(Unitarity of time evolution) The time evolution of the state ket of a system is a unitary transformation.*

A.17 Time Evolution of Many-Body Systems

For a many-body system with N particles, formally the Schrödinger equation is given by (A.35) with H replaced by its many-body version:

$$H = -\sum_j \frac{\hbar^2}{2m_j}\Delta_j + V\left(\boldsymbol{q}_1, \boldsymbol{q}_2, \ldots, \boldsymbol{q}_N\right), \tag{A.38}$$

where m_j is the mass of the jth particle, and Δ_j is the Laplacian for the position coordinates $\boldsymbol{q}_j$ of the jth particle.

A.18 Heisenberg's Equation of Motion

The expectation value of an observable A for a normalized state $|a\rangle$ is given by $\langle a|A|a\rangle$ as we will see in **A.27**. Therefore, the expectation value of A at time t may be written as (see (A.37)):

$$\langle A\rangle_t = \langle t|A|t\rangle = \langle 0|e^{iHt/\hbar}Ae^{-iHt/\hbar}|0\rangle. \tag{A.39}$$

Let us introduce

$$A(t) = \mathcal{T}_t^*A\mathcal{T}_t = e^{iHt/\hbar}Ae^{-iHt/\hbar}. \tag{A.40}$$

Then, we have

$$i\hbar\frac{d}{dt}A(t) = [A(t), H]. \tag{A.41}$$

This is called *Heisenberg's equation of motion.*
  We have

$$\langle t|A|t\rangle = \langle 0|A(t)|0\rangle. \tag{A.42}$$

Thus, we can describe the time evolution of a system in two ways: with time-dependent vectors (the left-hand side of (A.42); the *Schrödinger picture*) and with time-dependent observables (the right-hand side of (A.42); the *Heisenberg picture*). The equation of motion in the Schrödinger picture is Schrödinger's equation, and that in the Heisenberg picture is Heisenberg's equation of motion.

A.19 Mechanics Preserves Information/Entropy

According to von Neumann the information in a density operator ρ may be expressed as (see Chapter 21):[11],[12]

$$S(\rho) = -\mathrm{Tr}\, \rho \log \rho. \tag{A.43}$$

Mechanics is supposedly the most detailed description of the world, so it should preserve information. We expect (A.43) is time-invariant.

Since ρ is self-adjoint, we can take an orthonormal basis to diagonalize it as

$$\rho = \sum_i |i\rangle p_i \langle i|, \tag{A.44}$$

and $\log \rho$ is defined as

$$\log \rho = \sum_i |i\rangle \log p_i \langle i|. \tag{A.45}$$

The time evolution may be written as $\rho(t) = \mathcal{T}_t \rho \mathcal{T}_t^*$. Therefore, the time evolution of $\rho(t)$ is governed by the following *von Neumann equation*

$$i\hbar \frac{d}{dt}\rho(t) = H\rho - \rho H = [H, \rho]. \tag{A.46}$$

We immediately see an analogous formula for $\log \rho(t)$. Therefore, we get

$$i\hbar \frac{d}{dt} S(\rho(t)) \;=\; \mathrm{Tr}\,[H, \rho(t)] \log \rho(t) + \mathrm{Tr}\, \rho(t)[H, \log \rho(t)] \tag{A.47}$$

$$=\; -\mathrm{Tr}\,[\rho(t), H] \log \rho(t) + \mathrm{Tr}\, \rho(t)[H, \log \rho(t)] = 0. \tag{A.48}$$

Here, we have used $\mathrm{Tr}\,[A, B]C = \mathrm{Tr}\, A[B, C]$.[13*]

A.20 Mechanical Observables are Linear Operators[14]

We have seen that energy is an observable, and the Hamiltonian H is the corresponding operator by construction. It must be a linear operator to be compatible with (i) the state space is a vector space, and (ii) time evolution is consistent with superposition. Therefore, it

---

[11] The density operator ρ is defined in Chapter 19: A positive definite (i.e., $\mathrm{Tr}(\rho A) \geq 0$ if $\langle *|A|* \rangle \geq 0$ for any vector $|*\rangle$) and normalized (i.e., $\mathrm{Tr}\, \rho = 1$) linear self-adjoint operator is called a density operator.

[12] $S(\rho)$ is often written as $H(\rho)$ as well: H is the capital letter for η suggesting entropy.

[13*] This is because cyclic permutations are allowed inside a trace.

[14] Why are all observables linear? The author has never seen any explanation that is without any historical elements (i.e., various considerations in the so-called "old quantum theory"); here at least a reason is given for some of the observables.

is natural to expect the position and the momentum are also linear operators. Functions of linear operators may be interpreted as linear operators, so most classically meaningful mechanics-related observables may be interpreted as linear operators on the ket space.

Thus we postulate that observables are linear operators. This postulate will be completed as Postulate 5 in **A.23**.

A.21 Pure State and Mixed State

A state written as a single ket (i.e., linear combination of kets; any element in the state ket space) is called a *pure state*. In this case the density operator ρ can always be written as $\rho = |\cdot\rangle\langle\cdot|$ with the aid of an appropriate normalized ket $|\cdot\rangle$ and becomes a projection operator so $\rho^2 = \rho$. A *mixed state* is a probabilistic superposition of pure states, so the density matrix may be written as $\rho = \sum_k |k\rangle p_k \langle k|$,[15] with at least two $p_k > 0$. Here ρ is not a projection; $\rho^2 \neq \rho$. Any state of a quantum mechanical system actually observable by one-time observation is a pure state. Mixed states appear only when we wish to describe the system as an ensemble (i.e., to summarize the results of many experiments).

A.22 The Measurement Process and Decoherence[16]

The measurement process consists of interaction between the system and the measurement apparatus (which is, in reality, embedded in an external environment). They interact and the measurement apparatus settles down to a state, which is read by the observer. The ket space for the two interacting systems must be set up. To this end we assume

Postulate 3 *(Composition postulate) The state of the composite system is expressed as a ket of the tensor product space of the component system ket spaces.*[17]

To guarantee the objectivity of the measured results we demand

Postulate 4 *(Repeatability) An immediately repeated measurement yields the same outcome.*

We wish to observe an observable, $Q = \sum_i |s_i\rangle s_i \langle s_i|$,[18] with an apparatus A. Suppose that a measurement starts at t_0 and that, initially, the state of the system S is a pure state $\sum_i \alpha_i |s_i\rangle$. The measurement process reads (all the kets and bras are normalized):

---

[15] Where $p_k \geq 0$ and $\sum p_k = 1$ due to the definition of the density operator.

[16] The exposition around here relies on Zurek, W. H. (2003). Decoherence, einselection, and the quantum origins of the classical *Rev. Mod. Phys.*, **75**, 715–775

[17] That is, if $\{|a\rangle\}$ are the kets for system I and $\{|\alpha\rangle\}$ that for system II, the kets for the compound system I + II may be expressed as a linear combination $\sum c|a\rangle|\alpha\rangle$.

[18] That is, the normalized eigenvectors of Q are $|s_i\rangle$ that belongs to eigenvalue s_i.

$$\left(\sum_i \alpha_i |s_i\rangle\right) |A\rangle \rightarrow \sum_i \alpha_i |s_i\rangle |A_i\rangle, \tag{A.49}$$

where $|A\rangle$ is the state of the measurement apparatus before measurement and the arrow implies the time evolution due to the system–apparatus interaction. Notice that the result is still a pure state, but it is a superposition of $|s_i\rangle |A_i\rangle$ (we say these states are *entangled*). For a real apparatus, however, interaction with the environment is inevitable and decoherence ensues.

Decoherence is a loss of relative phase angles among the linear combination of pure states as in $\sum_i \alpha_i |s_i\rangle |A_i\rangle$. Here, the apparatus states A_i are assumed to be inevitably associated with its environmental factors to decohere:[19] in terms of the density matrix this decoherence reads

$$\left(\sum_i \alpha_i |s_i\rangle |A_i\rangle\right)\left(\sum_i \alpha_i^* \langle A_i| \langle s_i|\right) \xrightarrow{\text{decohere}} \sum_i p_i |s_i\rangle \langle s_i| |A_i\rangle \langle A_i|, \tag{A.50}$$

where $\sum p_i = 1$ ($p_i \geq 0$) (see **A.26**). That is, if we measure an observable using an apparatus (plus its environment), the coherent relation may be created initially but almost simultaneously the environmental effects choose a single apparatus state, say, A_i with an appropriate probability.[20] The resultant statistical ensemble (mixed state) means that one of the ensemble states is actually realized before we consult the apparatus.[21]

A.23   Observables Must be Self-Adjoint

The apparatus state A_i corresponding to s_i is observed for a state $\sum_i \alpha_i |s_i\rangle$ only with a certain probability. If A_i is observed, after the measurement, Postulate 4 (Repeatability) implies that the system must be in $|s_i\rangle$ and the newly repeated observation process must be

$$|s_i\rangle |A\rangle \rightarrow |s_i\rangle |A_i\rangle, \tag{A.51}$$

where the arrow denotes the unitary evolution due to the system–apparatus interaction. Since this evolution is unitary (Postulate 2):

$$\langle s_i|s_j\rangle \langle A|A\rangle = \langle s_i|s_j\rangle \langle A_i|A_j\rangle. \tag{A.52}$$

---

[19] That is, what is observed by A is dictated by the interaction with its environment; only the quantities the environment allows may be observed. The role of an apparatus is to prepare its environment to measure a particular observable (see the next footnote).

[20] **Respectable observables** Actually, an observable is a respectable observable only if its observed results A_i are einselected = environment-induced superselected. A good example is chiral molecules. Their ground states should not be chiral, but because of the spatial structural monitoring by the environment, we observe only environmentally stable chiral eigenstates. Also superposition of states with different constants of motion is hardly observable.

[21] The decoherence time scale is very short compared to the time after which memory states are typically consulted (i.e., copied or used in information processing).

Since the measurement apparatus must be able to distinguish states $|s_i\rangle$ and $|s_j\rangle$, $\langle A_i|A_j\rangle \neq 1$, so $\langle s_i|s_j\rangle = 0$. That is, the observable must be real and normal (i.e., Hermitian). Thus, generally, we demand:

Postulate 5 *Every observable is associated with a Hermitian (self-adjoint) operator.*[22]

Notice this means the totality of the eigenstates of any observable must be able to span the whole ket space V.

A.24 Energy Eigenstates and Time-Independent Schrödinger Equation

For statistical thermodynamics energy is of superb importance, and the Hamiltonian H must be the energy observable. Thus, the ket $|E\rangle$ satisfying

$$H|E\rangle = E|E\rangle, \tag{A.53}$$

that is, the eigenket of the system Hamiltonian, is interpreted as a system state with a definite energy E. Recalling **A.15**, we see this equation is the *time-independent Schrödinger equation*. Since H is self-adjoint,[23] the totality of its eigenkets can span the whole ket space V. Counting the number of energy eigenstates in a certain energy interval $(E - \Delta E, E]$ is the key issue of equilibrium statistical mechanics as we saw in Chapter 17.

A.25 Commutativity and Simultaneous Observation

If two observables A and B are commutative, then they are simultaneously diagonalizable (see Appendix 17A for a demonstration), so the immediate repeatability is compatible with the simultaneous observability of these two observables.

A.26 Probability of Observation: Quantum Mechanics and Probability

Usually, we postulate:

Postulate 6 *(Born's rule)* $p_a = |\langle a|\psi\rangle|^2$ *is the probability to observe state* $|a\rangle$ *when the system state is* $|\psi\rangle$.

---

[22] **Quantum jump** When only a discrete set of states in the Hilbert space is stable, the evolution of the system will look like a jump, but this is only apparent.

[23] Most physicists may accept this as "physically natural," but to show that a given H is self-adjoint mathematically can be very nontrivial.

Note that $p_i = |\alpha_i|^2$ for (A.50) according to this postulate. This is the key link between mathematics and physics. When probability is introduced in mechanics, an ad hoc rule (i.e., extra-mechanical convention) is needed.[24] Born's rule may be deduced, if the following is accepted: the probabilities for observing $|a\rangle$ and $|b\rangle$ for

$$|\psi\rangle = \alpha|a\rangle + \beta|b\rangle \tag{A.54}$$

are equal, if $|\alpha| = |\beta|$.

A.27 Expectation Value of an Observable

The expectation value of an observable $A = \sum_i |a_i\rangle a_i \langle a_i|$ when it is repeatedly observed for a state $|\psi\rangle$ (i.e., we prepare the same state $|\psi\rangle$ again and again and repeat the observation of A), the expectation value is

$$\langle A \rangle = \langle \psi|A|\psi \rangle = \sum_i \langle \psi|a_i\rangle a_i \langle a_i|\psi \rangle = \sum_i |\langle a_i|\psi \rangle|^2 a_i, \tag{A.55}$$

since $|\langle a_i|\psi \rangle|^2$ is the probability to observe the eigenvalue a_i for the state $|\psi\rangle$ according to Born's rule.

A.28 Uncertainty Relation

Let A and B be observables and $[A, B] = iC$. Note that C is also an observable. For a state $|\psi\rangle$ the expectation value of A (resp., B) is $\langle A \rangle = \langle \psi|A|\psi \rangle$ (resp., $\langle B \rangle = \langle \psi|B|\psi \rangle$). Write $\delta A = A - \langle A \rangle$ and $\delta B = B - \langle B \rangle$ and define the following ket for some $\alpha \in \mathbb{R}$:

$$|\phi\rangle = (\delta A + i\alpha\delta B)|\psi\rangle. \tag{A.56}$$

We have (note $[\delta A, \delta B] = iC$ as well):

$$0 \leq \langle \phi|\phi \rangle = (\Delta A)^2 + \alpha^2 (\Delta B)^2 - \alpha \langle C \rangle, \tag{A.57}$$

where $\Delta A = \sqrt{\langle \psi|(\delta A)^2|\psi \rangle}$, $\Delta B = \sqrt{\langle \psi|(\delta B)^2|\psi \rangle}$, and $\langle C \rangle = \langle \psi|C|\psi \rangle$. We choose α to minimize the above formula: $\alpha = \langle C \rangle/2(\Delta B)^2$. Thus we have arrived at the uncertainty relation

$$\Delta A \Delta B \geq |\langle C \rangle|/2. \tag{A.58}$$

---

[24] "we cannot derive probabilities from a theory that does not already contain some probabilistic concept," Schlosshauser, M. and Fine, A. (2005). On Zurek's derivation of the Born rule, *Found. Phys.*, **35**, 197–213.

    In classical mechanics it is usually injected into the intial condition, which is outside the control of mechanics.

Notice that the position representation of the momentum operator is

$$\langle x|p|y\rangle = \int d^3p \, \langle x|p\rangle p \langle p|y\rangle = \frac{1}{h^3} \int d^3p \, p \, e^{ip\cdot(x-y)/\hbar} \tag{A.59}$$

$$= \frac{1}{h^3} \int d^3p \, (-i\hbar\nabla_x)e^{ip\cdot(x-y)/\hbar} = (-i\hbar\nabla_x)\delta(x-y). \tag{A.60}$$

This implies (do not forget that x, p_x, etc. are operators):

$$[x,p_x] = [y,p_y] = [z,p_z] = i\hbar. \tag{A.61}$$

Thus, (A.58) gives us

$$\Delta x \Delta p_x \geq \hbar/2. \tag{A.62}$$

That is, for a given state $|\psi\rangle$, the particle position uncertainty Δx and the momentum uncertainty Δp_x are related and we cannot reduce both beyond the limit dictated by the uncertainty relation by modifying the state.[25]

A.29 Harmonic Oscillator

The Hamiltonian of a harmonic oscillator reads

$$H = \frac{1}{2m}\left(p^2 + m^2\omega^2 x^2\right). \tag{A.63}$$

Define the following operator called the *annihilation operator*, a

$$a = \frac{1}{\sqrt{2m\hbar\omega}}(p - i\omega m x). \tag{A.64}$$

Its adjoint a^+ $(= a^*)$ is called the *creation operator*. They satisfy

$$[a, a^+] = 1 \tag{A.65}$$

as can be checked with the aid of (A.61). In terms of these operators

$$H = \hbar\omega\left(a^+ a + \frac{1}{2}\right). \tag{A.66}$$

Therefore, we have

$$[H, a] = -\hbar\omega a. \tag{A.67}$$

Let $|\varepsilon\rangle$ be the eigenket of H belonging to the energy ε: $H|\varepsilon\rangle = \varepsilon|\varepsilon\rangle$. Then,

$$\hbar\omega\langle\varepsilon|a^+ a|\varepsilon\rangle = \langle\varepsilon|(H - \hbar\omega/2)|\varepsilon\rangle = (\varepsilon - \hbar\omega/2)\langle\varepsilon|\varepsilon\rangle \geq 0. \tag{A.68}$$

This implies that $\varepsilon \geq \hbar\omega/2$, so there must be the lowest eigenvalue. We can easily obtain $Ha^n|\varepsilon\rangle = (\varepsilon - n\hbar\omega)a^n|\varepsilon\rangle$ thanks to (A.67). Since $\varepsilon - n\hbar\omega < \hbar\omega/2$ is not allowed, for

---

[25] The uncertainty relation explained here is about the ensemble average of the observation of A or B for a fixed pure state; in every measurement A or B is measured and the results are averaged. There is also an uncertainty principle about the simultaneous observation results as well.

some positive integer n the equality $\varepsilon - n\hbar\omega = \hbar\omega/2$ holds and $a^{n+1}|\varepsilon\rangle = 0$. For such $|\varepsilon\rangle$, let us make a normalized ket $|0\rangle \propto a^n|\varepsilon\rangle$. Then, $H|0\rangle = (\hbar\omega/2)|0\rangle$. This implies that $|0\rangle$ is the ground state. On the other hand, if we introduce the normalized ket $|n\rangle \propto (a^+)^n|0\rangle$, $H|n\rangle = (n + 1/2)\hbar\omega|n\rangle$. Therefore, we conclude that the eigenvalues of the harmonic oscillator are $(n + 1/2)\hbar\omega$ $(n \in \mathbb{N})$ without any degeneracy. Here $\hbar\omega/2$ is the *zero point energy*.

A.30 Spins[26]

Electrons have an internal degree of freedom called the spin. It behaves as a 3-vector, S, when it couples with a magnetic field: its energy is $\propto -S \cdot B$. The components satisfy the following commutator relation (and the versions with cyclic permutation of x, y, z):

$$[S_x, S_y] = i\hbar S_z. \tag{A.69}$$

The eigenkets are $|+\rangle$ and $|-\rangle$ satisfying

$$S^2|\pm\rangle = (3\hbar^2/4)|\pm\rangle, \quad S_z|\pm\rangle = \pm\frac{\hbar}{2}|\pm\rangle. \tag{A.70}$$

If we introduce 2×2 Pauli matrices defined as

$$\sigma_x = \begin{pmatrix} 0 & 1 \\ 1 & 0 \end{pmatrix} \quad \sigma_y = \begin{pmatrix} 0 & -i \\ i & 0 \end{pmatrix} \quad \sigma_z = \begin{pmatrix} 1 & 0 \\ 0 & -1 \end{pmatrix}, \tag{A.71}$$

all the algebraic properties of the spin operators are recovered by the following expression:

$$S_x = \frac{\hbar}{2}\sigma_x, \quad S_y = \frac{\hbar}{2}\sigma_y, \quad S_z = \frac{\hbar}{2}\sigma_z. \tag{A.72}$$

Note that $\sigma_i\sigma_j + \sigma_j\sigma_i = 2\delta_{ij}$ (in particular $\sigma_x^2 = \sigma_y^2 = \sigma_z^2 = 1$). Consistent with the 2×2 expression of S or σ the eigenkets read

$$|+\rangle = \begin{pmatrix} 1 \\ 0 \end{pmatrix}, \quad |-\rangle = \begin{pmatrix} 0 \\ 1 \end{pmatrix}. \tag{A.73}$$

The rotation of a spin in the spin space with an angle, α,[27] can be expressed as

$$e^{-i(\alpha/2)\cdot\sigma} = \cos\frac{\alpha}{2} - i\frac{\alpha}{\alpha}\sin\frac{\alpha}{2}. \tag{A.74}$$

A.31 Perturbation Theory

Suppose the system Hamiltonian is H_0. The system is perturbed by a perturbation Hamiltonian λh, where λ is a "size" parameter for the perturbation: $H = H_0 + \lambda h$. Let

---

[26] There must be introductory entries for angular momentum in general, but since hydrogen atoms are at least outlined in any elementary physics and chemistry course, we go directly to spins.

[27] This equals rotation by angle $|\alpha|$ around the axis specified the directional vector $\alpha/|\alpha|$.

us assume the unperturbed eigenspace $\mathcal{B}$ for eigenvalue E_0 is g-dimensional: $H_0|j\rangle = E_0|j\rangle$ ($j \in \{1, \ldots, g\}$; we may assume $\langle i|j\rangle = \delta_{ij}$). Let us write the perturbed results as $E_0 \to E_0 + \lambda e_1 + \cdots$, $|i\rangle \to |i\rangle + \lambda|i, 1\rangle + \cdots$:

$$(H_0 + \lambda h)(|j\rangle + \lambda|j, 1\rangle + \cdots) = (E_0 + \lambda e_1 + \cdots)(|j\rangle + \lambda|j, 1\rangle + \cdots). \qquad (A.75)$$

Collecting powers of λ, we get

$$H_0|j\rangle = E_0|j\rangle, \qquad (A.76)$$

$$H_0|j, 1\rangle + h|j\rangle = E_0|j, 1\rangle + e_1|j\rangle, \qquad (A.77)$$

and so on. From the second equation (the lowest nontrivial order result) (A.77), for already chosen unperturbed eigenkets $|i\rangle$, $|j\rangle$ we get:[28*]

$$\langle i|h|j\rangle = \langle i|e_1|j\rangle = e_1\delta_{ij}. \qquad (A.78)$$

Therefore, the first-order perturbation result e_1 must be the eigenvalues of the operator h restricted to the subspace $\mathcal{B}$. In particular, if E_0 is non-degenerate (i.e., $g = 1$), $e_1 = \langle 1|h|1\rangle$.

A.32 Minimax Principle for Eigenvalues[29]

Let A be a self-adjoint operator defined on a vector space V. Let us arrange the eigenvalues of A in increasing order and number them (with their multiplicity taken into account), writing the kth eigenvalue as $\mu(k)$. There is a variational principle for $\mu(k)$. In particular, there is a variational principle for the ground state $\mu(1)$.

Minimax Principle[30]

For an arbitrary finite dimensional subspace $\mathcal{M}$ of V, let us compute

$$\lambda(\mathcal{M}) = \max_{|\varphi\rangle \in \mathcal{M}, \langle \varphi|\varphi\rangle = 1} \langle \varphi|A|\varphi\rangle. \qquad (A.79)$$

The minimum value of $\lambda(\mathcal{M})$ with $\mathcal{M}$ restricted to k-subspace is the kth eigenvalue:

$$\mu(k) = \lambda(k) \equiv \min_{\dim \mathcal{M} = k} \lambda(\mathcal{M}). \qquad (A.80)$$

☐ Let us demonstrate this. Write the orthonormal basis corresponding to $\{\mu(k)\}$ as $\{|k\rangle\}$. First of all, $\lambda(k) \leq \mu(k)$ is obvious.[31*] Thus, we have only to show $\lambda(k) \geq \mu(k)$.

Since $\mathcal{M}$ is a finite dimensional vector space, it must be contained in a subspace spanned by $\{|1\rangle, \ldots, |N\rangle\}$ for a sufficiently large N (recall our numbering of the eigenvectors of

---

[28*]Note $\langle i|H_0 = \langle i|E_0$. Notice that the ON basis $\{|i\rangle\}$ of $\mathcal{B}$ is already chosen to diagonalize the perturbation Hamiltonian h on $\mathcal{B}$; this is always possible.

[29] Ruelle, D. (1999). *Statistical Mechanics*, Singapore: World Scientific (original 1969) Section 2.5.

[30] In the following, precisely speaking, instead of min and max we should use inf and sup.

[31*]If we adopt the k-dimensional subspace spanned by $\{|1\rangle, \ldots, |k\rangle\}$ as $\mathcal{M}$, $\lambda(\mathcal{M}) = \mu(k)$, so the smallest value we look for by changing $\mathcal{M}$ cannot be larger than this value.

$A: A|k\rangle = \mu(k)|k\rangle$). Take a subspace $\mathcal{V}$ spanned by $\{|k\rangle,\ldots,|N\rangle\}$ (that is, the orthogonal complement of the $(k-1)$-dimensional subspace spanned by $\{|1\rangle,\ldots,|k-1\rangle\}$). Since $\mathcal{M}$ is with dimension k, $\mathcal{M}$, and $\mathcal{V}$ must share a vector which is notzero. Let us normalize it and call it $|0\rangle$. Since $\lambda(\mathcal{M})$ is defined through maximization,

$$\lambda(\mathcal{M}) \geq \langle 0|A|0\rangle, \tag{A.81}$$

but $|0\rangle = \sum_{i=k}^{N} |i\rangle\langle i|0\rangle$, so

$$\langle 0|A|0\rangle = \sum_{i=k}^{N} |\langle i|0\rangle|^2 \mu_i \geq \mu(k). \tag{A.82}$$

That is, for any k dimensional $\mathcal{M}$ we have $\lambda(\mathcal{M}) \geq \mu(k)$. This implies $\lambda(k) \geq \mu(k)$. Thus, we have shown that $\lambda(k) = \mu(k)$.

A.33 Path Integral Representation

Equation (A.27) reads

$$|t\rangle = e^{-iH\delta t/\hbar}|t-\delta t\rangle. \tag{A.83}$$

Using this repeatedly, we get

$$|t\rangle = e^{-iH\delta t/\hbar}|t-\delta t\rangle = \cdots = e^{-iH\delta t/\hbar}e^{-iH\delta t/\hbar}e^{-iH\delta t/\hbar}\cdots e^{-iH\delta t/\hbar}|0\rangle. \tag{A.84}$$

The position representation reads:[32]

$$\langle x_n|t\rangle = \int\cdots\int d^3x_{n-1}\cdots d^3x_0 \langle x_n|e^{-iH\delta t/\hbar}|x_{n-1}\rangle\langle x_{n-1}|e^{-iH\delta t/\hbar}|x_{n-2}\rangle\cdots$$
$$\langle x_1|e^{-iH\delta t/\hbar}|x_0\rangle\langle x_0|0\rangle. \tag{A.85}$$

We can compute each factor as (with an error of $O[\delta t^2]$)

$$\langle x|e^{-iH\delta t/\hbar}|y\rangle = \langle x|e^{-i(p^2/2m)\delta t/\hbar}|y\rangle e^{-iV(x)\delta t/\hbar} \tag{A.86}$$

$$= \int d^3p\, \langle x|p\rangle e^{-i(p^2/2m)\delta t/\hbar}\langle p|y\rangle e^{-iV(x)\delta t/\hbar} \tag{A.87}$$

$$= \frac{1}{h^3}\int d^3p\, e^{-i(p^2/2m)\delta t/\hbar}e^{ip\cdot(x-y)/\hbar}e^{-iV(x)\delta t/\hbar} \tag{A.88}$$

$$\propto \exp\left[\frac{i\delta t}{\hbar}\left(\frac{m}{2}\left(\frac{x-y}{\delta t}\right)^2 - V(x)\right)\right]. \tag{A.89}$$

Therefore, (A.85) reads (with an error of $O[\delta t]$)

$$\psi(x_n,t) \propto \int\cdots\int \prod_{j=0}^{n-1} dx_j e^{i(\delta t/\hbar)\sum_{j=1}^n [m\{(x_j-x_{j-1})/\delta t\}^2/2 - V(x_j)]}\psi(x_0,0). \tag{A.90}$$

[32] In the following, a one-particle system in 3-space is described, but extending the result to many-body systems is obvious.

Now, pay attention to the exponent:

$$\frac{i}{\hbar} \sum_{j=1}^{n} \left[\frac{1}{2} m \left(\frac{\boldsymbol{x}_j - \boldsymbol{x}_{j-1}}{\delta t} \right)^2 - V(\boldsymbol{x}_j) \right] \delta t. \tag{A.91}$$

Take $\delta t \, (= t/n) \to 0$ limit. Then, $(\boldsymbol{x}_j - \boldsymbol{x}_{j-1})/\delta t \to \boldsymbol{v}$ is the velocity of the particle at time $(j/n)t$, and the sum is a Riemann integral. Therefore, (A.91) reads

$$\frac{i}{\hbar} S[\boldsymbol{x}(\cdot)] \equiv \frac{i}{\hbar} \int dt \left[\frac{1}{2} m v^2(t) - V(\boldsymbol{x}(t)) \right]. \tag{A.92}$$

Here S, which is a functional of the trajectory $\boldsymbol{x}(\cdot)$, is called the *action* in classical mechanics.

Thus, we conclude that the time evolution operator $\mathcal{T}_t$ has the following "path integral" expression (the *Feynman path integral* expression):

$$\langle \boldsymbol{x} | \mathcal{T} | \boldsymbol{y} \rangle = \int \!\! - \!\! \int \mathcal{D}[\boldsymbol{x}(s)] \exp \left[\frac{i}{\hbar} \int_0^t ds \left(\frac{1}{2} m v^2(s) - V(\boldsymbol{x}(s)) \right) \right], \tag{A.93}$$

where the "integral" $\int\!-\!\int$ is over all the paths $\boldsymbol{x}(s)$ ($0 \le s \le t$) connecting $\boldsymbol{y}$ and $\boldsymbol{x}$ ($\boldsymbol{x}(t) = \boldsymbol{x}$, $\boldsymbol{x}(0) = \boldsymbol{y}$). This path integral representation was first introduced by Feynman.[33] The quantity

$$L(\boldsymbol{x}, \dot{\boldsymbol{x}}) = \frac{1}{2} m \dot{\boldsymbol{x}}^2 - V(\boldsymbol{x}) \tag{A.94}$$

is called the *Lagrangian*.

The picture described by (A.93), that is, the time evolution is a superposition of many different histories along various paths, is in perfect harmony with the superposition principle with which we started our topic **A.12**.

A.34 Macroscopic Particle: Classical Approximation

For an ordinary macroscopic particle, $S[\boldsymbol{x}(\cdot)]/\hbar$ appearing in [ ] of (A.93) is generally very large. This means $\exp(iS[\boldsymbol{x}(\cdot)]/\hbar)$ is an extremely rapidly oscillating function of the path $\boldsymbol{x}(\cdot)$. Then, it is natural that the trajectory we can actually observe for a large particle is characterized by the paths where phase changes are minimized so the phases of nearby trajectories reinforce each other constructively. Consider the following variational problem; we vary the path $\boldsymbol{x}(\cdot)$ fixing both ends of the trajectory:

$$\delta S = \delta \int_0^t ds \, L(\boldsymbol{x}, \dot{\boldsymbol{x}}) = \int_0^t ds \left(\frac{\partial L}{\partial \dot{\boldsymbol{x}}} \delta \dot{\boldsymbol{x}} + \frac{\partial L}{\partial \boldsymbol{x}} \delta \boldsymbol{x} \right) \tag{A.95}$$

$$= \int_0^t ds \left(-\frac{d}{ds} \frac{\partial L}{\partial \dot{\boldsymbol{x}}} + \frac{\partial L}{\partial \boldsymbol{x}} \right) \delta \boldsymbol{x} = 0. \tag{A.96}$$

---

[33] Barry Simon calls this "Feynman's poetry" (Simon, B. (2015). *A Comprehensive Course in Analysis*, Part I Real Analysis. Providence: AMS, p. 384). Mathematically respectable path integrals inspired by Feynman's talk are due to Kac. His idea is basically to define an integral on the path space $\{\Omega, P\}$ such as constructed in the fine-lettered portion of **3.16**.

Here, integration by parts has been performed with the condition that $\delta x(\cdot)$ vanishes at the both ends of the path which are fixed. $\delta x(\cdot)$ may be arbitrary in between, so we must conclude

$$\frac{d}{dt}\frac{\partial L}{\partial \dot{x}} - \frac{\partial L}{\partial x} = 0. \tag{A.97}$$

This is called *Lagrange's equation of motion*. The variational principle $\delta S = 0$ (A.96) deriving this equation is called the *action principle*. If we use the explicit form of L for a single particle in **A.33**, we get Newton's equation of motion:

$$m\frac{d^2}{dt^2}x = -\nabla V(x). \tag{A.98}$$

A.35 Classical Energy

The energy we have encountered above is $H = p^2/2m + V$. If we classically interpret $p = mv$, then $H = p^2/2m + V$ is classically conserved:

$$\frac{dH}{dt} = \frac{p}{m}\frac{dp}{dt} + \nabla V \cdot v = v \cdot \left(m\frac{dv}{dt} + \nabla V\right) = 0. \tag{A.99}$$

Here H is called the *Hamiltonian* of the system classically as well.

A.36 Hamilton's Equation of Motion

The independent variables of the Lagrangian are x and $\dot{x}$. From the equation of motion, we have

$$dL = pd\dot{x} - \operatorname{grad} V dx. \tag{A.100}$$

We have seen, however, p and x are a set of natural variables. We should apply a Legendre transformation (**16.8**) $L \to p\dot{x} - L = H$:

$$dH = \dot{x}dp + \operatorname{grad} V dx. \tag{A.101}$$

Therefore, the equation of motion reads

$$\frac{dx}{dt} = \frac{\partial H}{\partial p} = \frac{1}{m}p, \quad \frac{dp}{dt} = -\frac{\partial H}{\partial x} = -\nabla V. \tag{A.102}$$

These equations are called the canonical equations of motion or *Hamilton's equations of motion*.

A.37 Commutator and Poisson Bracket[34]

For a differentiable function f we note, quantum mechanically,

$$[x, f(p)] = i\hbar f'(p), \quad [f(x), p] = i\hbar f'(x). \tag{A.103}$$

Therefore, Heisenberg's equation of motion (A.41) gives

$$i\hbar\frac{dx}{dt} = [x, H] = \left[x, \frac{p^2}{2m} + V(x)\right] = i\hbar\frac{p}{m}, \tag{A.104}$$

$$i\hbar\frac{dp}{dt} = [p, H] = [p, V(x)] = -i\hbar V'(x). \tag{A.105}$$

Thus, we have the same form as Hamilton's equation of motion (A.102). This suggests that if we define the *Poisson bracket* $[\ ,\ ]_{PB}$ for classical observables $= [\ ,\ ]/i\hbar$ for the corresponding quantum observables, then Hamilton's equation of motion may be rewritten as

$$\frac{dx}{dt} = [x, H]_{PB}, \quad \frac{dp}{dt} = [p, H]_{PB}. \tag{A.106}$$

Equation (A.103) requires

$$[x, f(p)]_{PB} = f'(p), \quad [f(x), p]_{PB} = f'(x). \tag{A.107}$$

Thus, the following expression for the Poisson bracket works:

$$[f, g]_{PB} = \frac{\partial f}{\partial x}\frac{\partial g}{\partial p} - \frac{\partial f}{\partial p}\frac{\partial g}{\partial x}. \tag{A.108}$$

Multidimensional multiparticle versions are obtained straightforwardly from this. If the phase space is $\{q_1, \ldots, q_{3N}, p_1, \ldots, p_{3N}\}$, then

$$[f, g]_{PB} = \sum_{i=1}^{3N} \frac{\partial}{\partial q_i}\frac{\partial}{\partial p_i} - \frac{\partial}{\partial p_i}\frac{\partial}{\partial q_i}. \tag{A.109}$$

A.38 Canonical Coordinates

For a classical system consisting of N point masses, we need the spatial coordinates $q_1, \ldots, q_N$ and momenta $p_1, \ldots, p_N$. The set is called the *canonical coordinate system* of the classical mechanical system.

---

[34] Here, to simplify the exposition, the 1D system is discussed.

A.39 Canonical Quantization

Historically, classical mechanics was already perfected far before quantum mechanics. Replacing the Poisson bracket for classical observables Q with the commutator for the corresponding linear operators $\hat{Q}$ as (i.e., the exact reverse of our route to the Poisson bracket in **A.37**):

$$[Q_1, Q_2]_{PB} = Q_3 \rightarrow [\hat{Q}_1, \hat{Q}_2] = i\hbar\hat{Q}_3, \qquad (A.110)$$

we can translate classical mechanics to quantum mechanics. This is called the *canonical quantization*.

Warning: Classically, any canonical transformation preserves Poisson bracket relations, but this invariance does not propagate to quantum mechanics. We must stick to the Cartesian coordinate system for quantization.

Bibliography

References Consulted while Writing this Book

Gallavotti, G. (1999). *Statistical Mechanics: A Short Treatise*. Berlin: Springer.

Kubo, R., Ichimura, H., and Hashitsume, N. (1961). *Thermodynamics and Statistical Mechanics*. Tokyo: Shokabo.

Landau, L. D. and Lifshitz, E. M. (2013). *Statistical Physics*. Part 1. 3rd edition. Oxford: Butterworth-Heinemann.

Nakano, H. and Kimura, H. (1988). *Statistical Thermodynamics of Phase Transitions*. Tokyo: Asakura Shoten.

Peliti, L. (2011). *Statistical Mechanics in a Nutshell*. Princeton: Princeton University Press.

Tasaki, H. (2000). *Thermodynamics*. Tokyo: Baifukan.

Tasaki, H. (2008). *Statistical Mechanics I*. Tokyo: Baifukan.

Tasaki, H. (2008). *Statistical Mechanics II*. Tokyo: Baifukan.

Tasaki, H. and Hara, T. (2015). *Mathematics of Phase Transitions and Critical Phenomena*. Tokyo: Kyoritsu Publ.

Shimizu, A. (2007). *Principles of Thermodynamics*. Tokyo: University of Tokyo Press.

References Cited: Sources of Facts, Evidence, and Details (usually in footnotes)

Akhiezer, N. I. and Glazman, I. M. (2013). *Theory of Linear Operators in Hilbert Space*. Dover Books on Mathematics. Mineola, NY: Dover Publications.

Anderson, P. W. (1984). *Basic Notions of Condensed Matter Physics*. Boulder: Westview Press.

Andreas, B., Azuma, Y., Bartl, G. et al. (2011). Determination of the Avogadro constant by counting the atoms in a ^{28}Si crystal, *Physics Review Letters*, **106** (3): 030801.

Berryman, S. (2011). Ancient Atomism. In *The Stanford Encyclopedia of Philosophy* (Winter 2011 Edition). ed. E. N. Zalta, available online at http://plato.stanford.edu/archives/win2011/entries/atomism-ancient/

Bionumbers (the database of useful biological numbers) available online at http://bionumbers.hms.harvard.edu/default.aspx.

Boltzmann, L. (1877). Über die Beziehung zwischen dem zweiten Hauptsatze der mechanischen Wärmetheorie und der Wahrscheinlichkeitsrechinung respective den Sätzen über Wärmegleichgewicht, *Wiener Berichte*, **76**, 373–435.

Borovkov, A. A., Golovanov, P. P., and Ya Kozlov, V. et al. (1969). Ivan Nikolaevich Sanov (Obituary), *Russian Mathematical Surveys*, **24**, 4, 159.

Braun, K. F. (1888). Über einen allgemeinen qualitativen Satz für Zustandsänderumgen nebst einigen sick anschliessenden Bemerkungen. insbesondere über nicht eindeutige Systeme, *Annalen der Physik*, **269**, 337–353.

Broda, E. (1955). *Ludwig Boltzmann. Mensch, Physiker, Philosoph*. Wien: F. Deuticke.

Brown, P. C., Roediger III, H. L. and McDaniel, M. A. (2014). *Make it Stick: The Science of Successful Learning*, Cambridge (MA): The Belknap Press.

Browne, J. (1995). *Charles Darwin: Voyaging*. New York: Knopf.

Browne, J. (2002). *Charles Darwin: The Power of Place*. New York: Knopf.

Brush, S. G. (1968a). On Mach's atomism, *Synthese*. **18**, 192–215.

Brush, S. G. (1968b). A history of random processes: I Brownian movement from Brown to Perrin, *Archive of History of Exact Science*, **5**, 1–36.

Brush, S. G. (1983). *Statistical Physics and the Atomic Theory of Matter: From Boyle and Newton to Landau and Onsager*. Princeton: Princeton University Press.

Cercignani, C. (2001). The rise of statistical mechanics, in *Chance in Physics*. Lecture Notes in Physics, **574**, eds, J. Bricmont, D. Dürr, M. C. Gallavotti, G. C. Ghirardi, F. Petruccione, and N. Zanghi. Berlin: Springer, pp. 25–38.

Chaikin, P. M. and Lubensky, T. C. (1995). *Principles of Condensed Matter Physics*. Cambridge: Cambridge University Press.

Cover, T. M. and King, R. C. (1978). A convergent gambling estimate of the entropy of English, *IEEE Transactions Information Theory*, **24**, 413–421.

Cover, T. M. and Thomas, J. A. (1991). *Elements of Information Theory*. New York: Wiley.

Daussy, C. et al. (2007). Direct determination of the Boltzmann constant by an optical method, *Physical Review Letters*, **98**, 250801.

de Heer, J. (1957). The principle of le Chatelier and Braun, *Journal of Chemical Education*, **34**, 375–380.

Deng, Y. and Blöte, H. W. J. (2003). Simultaneous analysis of several models in the three-dimensional Ising universality class, *Physical Review E*, **68**, 036125 (9 pages).

Dirac, P. A. M. (1982). *The Principles of Quantum Mechanics*. Oxford: Clarendon Press.

Durrett, R. (1991). *Probability: Theory and Examples*. Pacific Grove: Wadsworth & Brooks/Cole.

Ebbinghaus, H.-D. (2007). *Ernst Zermelo: An Approach to his Life and Work*. Berlin: Springer.

Einstein, A. (1903). Eine Theorie der Grundlagen der Thermodynamik, *Annalen der Physik*, **316**, 170–187.

Einstein, A. (1905). Über die von der molekularkinetischen Theorie der Wärme geforderten Bewegung von in ruhenden Flüssigkeiten suspendierten Teilchen, *Annalen der Physik*, **17**, 549–560.

Einstein, A. (1910). Theorie der Opaleszenz von homogenen Flüssigkeitsgemischen in der Nähe des kritischen Zustandes, *Annalen der Physik*, **33**, 1275–1298.

Einstein, A. (1917). Zur Quantentheorie der Strahlung, *Physikalisches Zeitschrift*, **18**, 121–128.

Essam, J. W. and Sykes, M. F. (1963). The crystal statistics of the diamond lattice, *Physica*, **29**, 378–388.

Feller, W. (1957). *An Introduction to Probability Theory and its Applications*. Volume 1. New York: Wiley.

Feller, W. (1971). *An Introduction to Probability Theory and its Applications*. Volume 2. New York: Wiley.

Gibbs, J. W. (1902). *Elementary Principles in Statistical Mechanics, Developed with Especial Reference to the Rational Foundation of Thermodynamics*, New Haven: Yale University Press.

Girardeau, M. D. and Mazo, R. M. (1973). Variational methods in statistical mechanics, in *Advances in Chemical Physics*, vol. 24, eds, I. Prigogine and S. A. Rice, New York: Academic Press, pp. 187–255.

Golomb, S. W., Berlekamp, E. R., Cover, T. M. et al. (2002). Claude Elwood Shannon (1916–2002). *Notices of American Mathematical Society*, **49**, 8–16.

Graham, J. B. Aguilar, N., Dudley, R., and Gans, C. (1995). Implication of the late Paleozoic oxygen pulse for physiology and evolution. *Nature*, **375**, 117–120.

Guggenheim, E. A. (1945). The principle of corresponding states, *Journal of Chemical Physics*, **13**, 254–261.

Haag, R. (1996). *Local Quantum Physics: Fields, Particles, Algebras*, Second revised and enlarged edition, Berlin: Springer.

Havil, J. (2003). *Gamma: Exploring Euler's Constant*. Princeton: Princeton University Press.

Israel, R. B. (1979). *Convexity in the Theory of Lattice Gases*. Princeton: Princeton University Press.

Jarzynski, C. (1997). Nonequilibrium equality for free energy differences, *Physical Review Letters*, **78**, 2690–2693.

Jarzynski, C. (2004). Nonequilibrium work theorem for a system strongly coupled to a thermal environment, *Journal of Statistical Mechanics*, **2004**. P09005.

Jeans, J. (1952). *An Introduction to the Kinetic Theory of Gases*. Cambridge: Cambridge University Press.

Kadanoff, L. P. (1966). Scaling Laws for Ising Models near T_c. *Physics*, **2**, 263–272.

Klaers, J., Schmitt, J., Vewinger, F., and Weitz, M. (2011). Bose–Einstein condensation of photons in an optical microcavity, *Nature*, **468**, 545–548.

Kolmogorov, A. N. (2004; reprint of a book chapter in 1956). The theory of probability, *Theory of Probability and its Applications*, **48**, 191–200.

Konvalina., J. (2000). A unified interpretation of the binomial coefficients, the Stirling numbers, and the Gaussian coefficients. *American Mathematical Monthly*, **107**, 901–910.

Körner, T. W. (1988). *Fourier Analysis*. Cambridge: Cambridge University Press.

Landau, L. D. and Lifshitz, E. M. (1982). *Classical Mechanics*. 3rd edition. Oxford: Butterworth-Heinemann.

Landau, L. D. and Lifshitz, E. M. (1987). *Fluid Mechanics*. 2nd edition. Oxford: Butterworth-Heinemann.

Landsberg, P. T. (1972). The fourth law of thermodynamics, *Nature*, **238**, 229–231.

Langevin, P. (1908). Sur la théorie du mouvement brownien, *Comptes Rendus de l'Académie des Sciences*, **146**, 530–533.

Le Chatelier, H. L. (1884). Sur un énoncé général des lois des équilibres chimiques. *Comptes Rendus de l'Académie des Sciences*, **99**, 786–789.

Lemons, D. S. and Gythiel. A. (1997). Paul Langevin's 1908 paper "On the Theory of Brownian Motion" ["Sur la théorie du mouvement brownien." *Comptes Rendus de l'Académie des Sciences*. 146, 530–533 (1908)], *American Journal of Physics*, **65**, 1079–1081.

Lenard, A. (1978). Thermodynamical proof of the Gibbs formula for elementary quantum systems, *Journal of Statistical Physics*, **19**, 575–586.

Lenker, T. D. (1979). Caratheodory's concept of temperature, *Synthese*, **42**, 167–171 (1979).

Lieb, E. H. and Seiringer, R. (2010). *The Stability of Matter in Quantum Mechanics*. Cambridge: Cambridge University Press.

Lindley, D. (2001). *Boltzmann's Atom: The Great Debate that Launched a Revolution in Physics*. New York: The Free Press.

Longuet-Higgins, C. and Fisher, M. E. (1995). Lars Onsager: November 27. 1903-October 5. 1976, *Journal of Statistical Physics*, **78**, 605–640.

Maxwell, J. C. (1860). Illustrations of the dynamical theory of gases, *Philosophical Magazine*. **19** 19–32; **20**, 21–37.

McKean, H. (2014). *Probability: The Classical Limit Theorems*. Cambridge: Cambridge University Press.

Mendelssohn, K. (1973). *The World of Walther Nernst: The Rise and Fall of German Science 1864–1941* (ebook form from Plunket Lake Press, 2015).

Meulders, M. (2010). *Helmholtz from Enlightenment to Neuroscience*. Boston: MIT Press.

Mortici, C. (2011). On Gosper's formula for the gamma function, *Journal of Mathematical Inequalities*, **5**, 611–614.

Mukamel, S. (2003). Quantum extension of the Jarzynski relation: Analogy with stochastic dephasing, *Physical Review Letters*, **90**, 170604.

Niemeijer, Th. and van Leeuwen, J. M. J. (1973). Wilson theory for spin systems on a triangular lattice, *Physical Review Letters*, **31**, 1411–1414.

Oono, Y. (1989). Large deviation and statistical physics, *Progress of Theoretical Physics Supplement*, **99**, 165–205.

Oono, Y. (2013). *The Nonlinear World*. Tokyo: Springer.

Perrin, J. (1916). *Atoms*. London: Constable. translated by D. L. Hammick. Available online at: https://archive.org/details/atomsper00perruoft.

Priestley, H. A. (1990). *Introduction to Complex Analysis*. Oxford: Oxford University Press.

Rao, M. and Stetker, H. (1991). *Complex Analysis: An Invitation*. Singapore: World Scientific.

Rockafellar, R. R. (1970). *Convex Analysis*. Princeton: Princeton University Press (since 1997 in the series Princeton Landmarks in Mathematics).

Rosenhouse, J. (2009). *The Monty Hall Problem*. Oxford: Oxford University Press.

Ruelle, D. (1999). *Statistical Mechanics*. Singapore: World Scientific (original 1969).

Ruelle, D. (2004). *Thermodynamic Formalism: The Mathematical Structure of Equilibrium Statistical Mechanics* Cambridge: Cambridge University Press (Cambridge Mathematical Library).

Russell, B. (1945). *A History of Western Philosophy*. London: Simon and Schuster.

Sagawa, T. (2013). *Thermodynamics of Information Processing in Small Systems*. Tokyo: Springer.

Sagawa, T. and Ueda, M. (2008). Second law of thermodynamics with discrete quantum feedback control, *Physical Review Letters*, **100**, 080403.

Sagawa, T. and Ueda, M. (2009). Minimal energy cost for thermodynamic information processing: Measurement and information erasure, *Physical Review Letters*, **102**, 250602.

Saito, N. (1967). *Polymer Physics*. Tokyo: Shokabo.

Sanov, I. N. (1957). On the probability of large deviations of random variables. *Matematicheskii Sbornik*, **42**, 11–44.

Schlosshauser, M. and Fine, A. (2005). On Zurek's derivation of the Born rule, *Foundation of Physics*, **35**, 197–213.

Seneta, E. (2013). *Nonnegative Matrices and Markov Chains*. 2nd edition Berlin: Springer.

Sender, R., Fuchs, S., and Milo, R. (2016). Are we really vastly outnumbered? Revisiting the ratio of bacterial to host cells in humans, *Cell*, **164**, 337–340.

Simon, B. (2015). *A Comprehensive Course in Analysis, Part I Real Analysis*. Providence: AMS, p. 384.

Sinai, Ya. G. (1982). *Theory of Phase Transitions: Rigorous Results*. Oxford: Pergamon Press.

Slovej, J. P. (2011). The stability of matter in quantum mechanics, by Elliott H. Lieb and Robert Seiringer, Book review *Bulletin of American Mathematical Society*, **50**, 169–174.

Suksombat S., Khafizov, R., Kozlov, A. et al. (2015). Structural dynamics of *E. coli* single-stranded DNA binding protein reveal DNA wrapping and unwrapping pathways, *Elife*, **25**: 4, 1–53, doi: 10.7554/eLife.08193.

Tasaki, H. (1998). From quantum dynamics to the canonical distribution: general picture and a rigorous example, *Physical Review Letters*, **80**, 1373–1376.

Theobald, D. L. (2011). On universal common ancestry: Sequence similarity, and phylogenetic structure: the sins of P-values and the virtues of Bayesian evidence. *Biology Direct*, **6**, 60 (25 pages).

Tombari, E., Ferrari, C., Salvetti, G. B., and Johari, G. (2005). Endothermic freezing on heating and exothermic melting on cooling, *Journal of Chemical Physics*, **123**, 051104.

Tourchette, H. (2009). The large deviation approach to statistical mechanics, *Physics Report*, **478**, 1–69.

Widom, B. (1965). Equation of state in the neighborhood of the critical point. *Journal of Chemical Physics*, **43**, 3898.

Williams, R. (2009). September, 1911, the Sackur-Tetrode equation: how entropy met quantum mechanics, APS News: *This Month in Physics History*, September.

Yamamoto, Y. (2007–8). *Historical Development of Thoughts of Heat Theory*. Tokyo: Chikuma Shobo.

Zurek, W. H. (2003). Decoherence, einselection, and the quantum origins of the classical, *Review of Modern Physics*, **75**, 715–775.

Index

Printed in the United States
by Baker & Taylor Publisher Services